Lecture Notes in Computer Science   14048

Founding Editors

Gerhard Goos
Juris Hartmanis

The series Lecture Notes in Computer Science (LNCS), including its subseries Lecture Notes in Artificial Intelligence (LNAI) and Lecture Notes in Bioinformatics (LNBI), has established itself as a medium for the publication of new developments in computer science and information technology research, teaching, and education.

LNCS enjoys close cooperation with the computer science R & D community, the series counts many renowned academics among its volume editors and paper authors, and collaborates with prestigious societies. Its mission is to serve this international community by providing an invaluable service, mainly focused on the publication of conference and workshop proceedings and postproceedings. LNCS commenced publication in 1973.

Heidi Krömker

Editor

HCI in Mobility, Transport, and Automotive Systems

5th International Conference, MobiTAS 2023
Held as Part of the 25th HCI International Conference, HCII 2023
Copenhagen, Denmark, July 23–28, 2023
Proceedings, Part I

 Springer

Editor
Heidi Krömker
Technische Universität Ilmenau
Ilmenau, Germany

ISSN 0302-9743   ISSN 1611-3349 (electronic)
Lecture Notes in Computer Science
ISBN 978-3-031-35677-3   ISBN 978-3-031-35678-0 (eBook)
https://doi.org/10.1007/978-3-031-35678-0

This Springer imprint is published by the registered company Springer Nature Switzerland AG
The registered company address is: Gewerbestrasse 11, 6330 Cham, Switzerland

Foreword

Human-computer interaction (HCI) is acquiring an ever-increasing scientific and industrial importance, as well as having more impact on people's everyday lives, as an ever-growing number of human activities are progressively moving from the physical to the digital world. This process, which has been ongoing for some time now, was further accelerated during the acute period of the COVID-19 pandemic. The HCI International (HCII) conference series, held annually, aims to respond to the compelling need to advance the exchange of knowledge and research and development efforts on the human aspects of design and use of computing systems.

The 25th International Conference on Human-Computer Interaction, HCI International 2023 (HCII 2023), was held in the emerging post-pandemic era as a 'hybrid' event at the AC Bella Sky Hotel and Bella Center, Copenhagen, Denmark, during July 23–28, 2023. It incorporated the 21 thematic areas and affiliated conferences listed below.

A total of 7472 individuals from academia, research institutes, industry, and government agencies from 85 countries submitted contributions, and 1578 papers and 396 posters were included in the volumes of the proceedings that were published just before the start of the conference, these are listed below. The contributions thoroughly cover the entire field of human-computer interaction, addressing major advances in knowledge and effective use of computers in a variety of application areas. These papers provide academics, researchers, engineers, scientists, practitioners and students with state-of-the-art information on the most recent advances in HCI.

The HCI International (HCII) conference also offers the option of presenting 'Late Breaking Work', and this applies both for papers and posters, with corresponding volumes of proceedings that will be published after the conference. Full papers will be included in the 'HCII 2023 - Late Breaking Work - Papers' volumes of the proceedings to be published in the Springer LNCS series, while 'Poster Extended Abstracts' will be included as short research papers in the 'HCII 2023 - Late Breaking Work - Posters' volumes to be published in the Springer CCIS series.

I would like to thank the Program Board Chairs and the members of the Program Boards of all thematic areas and affiliated conferences for their contribution towards the high scientific quality and overall success of the HCI International 2023 conference. Their manifold support in terms of paper reviewing (single-blind review process, with a minimum of two reviews per submission), session organization and their willingness to act as goodwill ambassadors for the conference is most highly appreciated.

This conference would not have been possible without the continuous and unwavering support and advice of Gavriel Salvendy, founder, General Chair Emeritus, and Scientific Advisor. For his outstanding efforts, I would like to express my sincere appreciation to Abbas Moallem, Communications Chair and Editor of HCI International News.

July 2023   Constantine Stephanidis

HCI International 2023 Thematic Areas and Affiliated Conferences

Thematic Areas

- HCI: Human-Computer Interaction
- HIMI: Human Interface and the Management of Information

Affiliated Conferences

- EPCE: 20th International Conference on Engineering Psychology and Cognitive Ergonomics
- AC: 17th International Conference on Augmented Cognition
- UAHCI: 17th International Conference on Universal Access in Human-Computer Interaction
- CCD: 15th International Conference on Cross-Cultural Design
- SCSM: 15th International Conference on Social Computing and Social Media
- VAMR: 15th International Conference on Virtual, Augmented and Mixed Reality
- DHM: 14th International Conference on Digital Human Modeling and Applications in Health, Safety, Ergonomics and Risk Management
- DUXU: 12th International Conference on Design, User Experience and Usability
- C&C: 11th International Conference on Culture and Computing
- DAPI: 11th International Conference on Distributed, Ambient and Pervasive Interactions
- HCIBGO: 10th International Conference on HCI in Business, Government and Organizations
- LCT: 10th International Conference on Learning and Collaboration Technologies
- ITAP: 9th International Conference on Human Aspects of IT for the Aged Population
- AIS: 5th International Conference on Adaptive Instructional Systems
- HCI-CPT: 5th International Conference on HCI for Cybersecurity, Privacy and Trust
- HCI-Games: 5th International Conference on HCI in Games
- MobiTAS: 5th International Conference on HCI in Mobility, Transport and Automotive Systems
- AI-HCI: 4th International Conference on Artificial Intelligence in HCI
- MOBILE: 4th International Conference on Design, Operation and Evaluation of Mobile Communications

List of Conference Proceedings Volumes Appearing Before the Conference

1. LNCS 14011, Human-Computer Interaction: Part I, edited by Masaaki Kurosu and Ayako Hashizume
2. LNCS 14012, Human-Computer Interaction: Part II, edited by Masaaki Kurosu and Ayako Hashizume
3. LNCS 14013, Human-Computer Interaction: Part III, edited by Masaaki Kurosu and Ayako Hashizume
4. LNCS 14014, Human-Computer Interaction: Part IV, edited by Masaaki Kurosu and Ayako Hashizume
5. LNCS 14015, Human Interface and the Management of Information: Part I, edited by Hirohiko Mori and Yumi Asahi
6. LNCS 14016, Human Interface and the Management of Information: Part II, edited by Hirohiko Mori and Yumi Asahi
7. LNAI 14017, Engineering Psychology and Cognitive Ergonomics: Part I, edited by Don Harris and Wen-Chin Li
8. LNAI 14018, Engineering Psychology and Cognitive Ergonomics: Part II, edited by Don Harris and Wen-Chin Li
9. LNAI 14019, Augmented Cognition, edited by Dylan D. Schmorrow and Cali M. Fidopiastis
10. LNCS 14020, Universal Access in Human-Computer Interaction: Part I, edited by Margherita Antona and Constantine Stephanidis
11. LNCS 14021, Universal Access in Human-Computer Interaction: Part II, edited by Margherita Antona and Constantine Stephanidis
12. LNCS 14022, Cross-Cultural Design: Part I, edited by Pei-Luen Patrick Rau
13. LNCS 14023, Cross-Cultural Design: Part II, edited by Pei-Luen Patrick Rau
14. LNCS 14024, Cross-Cultural Design: Part III, edited by Pei-Luen Patrick Rau
15. LNCS 14025, Social Computing and Social Media: Part I, edited by Adela Coman and Simona Vasilache
16. LNCS 14026, Social Computing and Social Media: Part II, edited by Adela Coman and Simona Vasilache
17. LNCS 14027, Virtual, Augmented and Mixed Reality, edited by Jessie Y. C. Chen and Gino Fragomeni
18. LNCS 14028, Digital Human Modeling and Applications in Health, Safety, Ergonomics and Risk Management: Part I, edited by Vincent G. Duffy
19. LNCS 14029, Digital Human Modeling and Applications in Health, Safety, Ergonomics and Risk Management: Part II, edited by Vincent G. Duffy
20. LNCS 14030, Design, User Experience, and Usability: Part I, edited by Aaron Marcus, Elizabeth Rosenzweig and Marcelo Soares
21. LNCS 14031, Design, User Experience, and Usability: Part II, edited by Aaron Marcus, Elizabeth Rosenzweig and Marcelo Soares

47. CCIS 1836, HCI International 2023 Posters - Part V, edited by Constantine Stephanidis, Margherita Antona, Stavroula Ntoa and Gavriel Salvendy

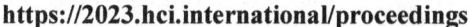

https://2023.hci.international/proceedings

Preface

Human-computer interaction in the highly complex field of mobility and intermodal transport leads to completely new challenges. A variety of different travelers move in different travel chains. The interplay of such different systems, such as car and bike sharing, local and long-distance public transport, and individual transport, must be adapted to the needs of travelers. Intelligent traveler information systems must be created to make it easier for travelers to plan, book, and execute an intermodal travel chain and to interact with the different systems. Innovative means of transport are developed, such as electric vehicles and autonomous vehicles. To achieve the acceptance of these systems, human-machine interaction must be completely redesigned.

The 5th International Conference on HCI in Mobility, Transport, and Automotive Systems (MobiTAS 2023), an affiliated conference of the HCI International (HCII) conference, encouraged papers from academics, researchers, industry, and professionals, on a broad range of theoretical and applied issues related to mobility, transport, and automotive systems and their applications.

For MobiTAS 2023, a key theme with which researchers were concerned was autonomous and assisted driving, as well as intelligent transportation systems. Topics covered in this area included designing driver training applications for adaptive cruise control, designing automation and intelligent car features, designing and evaluating intelligent assistants, exploring the cooperation between humans and agents, and understanding user behavior towards automation. Other papers focused on urban and sustainable mobility, exploring various aspects of urban transportation, such as autonomous shuttles, delivery robots, heavy vehicles in underground mines, urban air mobility, and electric vehicles and e-scooters, as well as sustainability perspectives, such as service ecosystems, vehicle sharing, route optimization, and traffic planning. In addition, a considerable number of contributions reflected the need for understanding driver behavior and performance, offering insights into the effect of technology on driving experience, exploring driver preferences for automated driving styles, and classifying driving styles based on driver behavior. The design of in-vehicle experiences for drivers and passengers was another key theme of research addressed in this year's proceedings, covering a range of topics, such as interface design and evaluation of alert systems and dashboards, data-driven design, game-based design, multimodal design featuring ultrasound skin stimulation and music, as well as design to alleviate motion sickness for passengers. Finally, a number of papers focused on the topic of accessibility and inclusive mobility, addressing accessibility models, independent mobility for people with intellectual disabilities, and understanding driver behavior in individuals with mild cognitive disabilities.

Two volumes of the HCII 2023 proceedings are dedicated to this year's edition of the MobiTAS conference. The first focuses on topics related to autonomous and intelligent transport systems for urban and sustainable mobility, while the second focuses on topics related to human-centered design of mobility and in-vehicle experiences.

Papers of these volumes are included for publication after a minimum of two single-blind reviews from the members of the MobiTAS Program Board or, in some cases, from members of the Program Boards of other affiliated conferences. I would like to thank all of them for their invaluable contribution, support, and efforts.

July 2023   Heidi Krömker

5th International Conference on HCI in Mobility, Transport and Automotive Systems (MobiTAS 2023)

Program Board Chair:

- Heidi Kroemker, *Technische Universität Ilmenau, Germany*
- Avinoam Borowksy, *Ben-Gurion University of the Negev, Israel*
- Angelika C. Bullinger, *Chemnitz University of Technology, Germany*
- Bertrand David, *Ecole Centrale de Lyon, France*
- Marco Diana, *Politecnico di Torino, Italy*
- Cyriel Diels, *Royal College of Art, UK*
- Chinh Ho, *University of Sydney, Australia*
- Christophe Kolski, *Université Polytechnique Hauts-de-France, France*
- Josef F. Krems, *Chemnitz University of Technology, Germany*
- Roberto Montanari, *RE:LAB srl, Italy*
- Matthias Rötting, *Technische Universität Berlin, Germany*
- Frank Ritter, *Penn State University, USA*
- Philipp Rode, *Volkswagen Group, Germany*
- Thomas Schlegel, *Hochschule Furtwangen University, Germany*
- Felix W. Siebert, *Technical University of Denmark, Denmark*
- Ulrike Stopka, *Technische Universität Dresden, Germany*
- Tobias Wienken, *CodeCamp:N GmbH, Germany*
- Xiaowei Yuan, *Beijing ISAR User Interface Design Limited, P.R. China*

The full list with the Program Board Chairs and the members of the Program Boards of all thematic areas and affiliated conferences of HCII2023 is available online at:

http://www.hci.international/board-members-2023.php

HCI International 2024 Conference

The 26th International Conference on Human-Computer Interaction, HCI International 2024, will be held jointly with the affiliated conferences at the Washington Hilton Hotel, Washington, DC, USA, June 29 – July 4, 2024. It will cover a broad spectrum of themes related to Human-Computer Interaction, including theoretical issues, methods, tools, processes, and case studies in HCI design, as well as novel interaction techniques, interfaces, and applications. The proceedings will be published by Springer. More information will be made available on the conference website: http://2024.hci.international/.

General Chair
Prof. Constantine Stephanidis
University of Crete and ICS-FORTH
Heraklion, Crete, Greece
Email: general_chair@hcii2024.org

https://2024.hci.international/

Contents – Part I

Urban Mobility

Sustainable Mobility

Contents – Part II

Accessibility and Inclusive Mobility

Autonomous and Assisted Driving

Human Factor Risks in Driving Automation Crashes

Yueying Chu and Peng Liu(✉)

Center for Psychological Sciences, Zhejiang University, Hangzhou, China
pengliu86@zju.edu.cn

Abstract. Driving automation systems are capable of continuously performing part or all of the dynamic driving task. Driving automation is intended to reduce the probability and severity of traffic crashes by minimizing manual operation. However, inadequate automation systems (including advanced driver assistance systems mounted on Level 2 vehicles and automated driving systems mounted on Level 3 or higher levels vehicles) and inappropriate human-automation interaction will threaten road safety. This study analyzed the factors of crashes related to driving automation from the perspective of human factor risks. We summarized the crashes and categorized the probable causes mentioned in six accident reports from National Transportation Safety Board. We extracted common causal factors related to human drivers, including inappropriate using ways of driving automation, human distraction or disengagement, and complacency (overreliance) on vehicle automation. Finally, we discussed the relationship between the extracted common causes and previous insights in the driving automation domain, such as the rationality of complacency as a causal factor, and provided potential countermeasures.

Keywords: Driving automation · Human errors · Human factor risks · Crash analysis · Complacency · Overtrust

1 Introduction

On March 18, 2018, in Arizona, an Uber developmental automated vehicle (AV) struck and fatally injured a pedestrian pushing a bicycle walking across N. Mill Avenue. According to the accident report from National Transportation Safety Board (NTSB) [1], 5.6 s prior to the crash, its automated driving system (ADS) first detected the pedestrian, however, recognized her as a vehicle, and then as an unknown object and a cyclist. When the system determined that the crash was inevitable and imminent, "the situation exceeded the response specifications of the ADS braking system" (p. v) [1]. It was the first AV crash involving the death of a pedestrian in the world. This crash shows that automated systems are not as safe as we think; in other words, automation may make mistakes and cause harms on roads.

It is reported that the majority of vehicle crashes (e.g., more than 90% in the USA [2] and China [3, 4]) are caused by human factors. The advent of driving automation may address this issue. The Society of Automotive Engineers (SAE) [5] identified driving automation as six levels: No Driving Automation (Level 0), Driver Assistance (Level 1), Partial Automation (Level 2), Conditional Automation (Level 3), High Automation (Level 4), and Full Automation (Level 5). Driving automation technology in Level 2 vehicles is called the advanced driver assistance system (ADAS), which requires the driver and the system to perform the dynamic driving task together. The driver must monitor the system's behavior and make appropriate responses to guarantee safety. Vehicles equipped with ADAS are commercially available. AVs refer to vehicles with Level 3 or higher levels equipped with the automated driving system (ADS). AVs have not yet been adopted on a large scale. In the future, ADS is expected to operate a vehicle by itself.

However, as indicated by the 2018 Uber AV crash, perfect automation is impossible to guarantee and difficult to achieve [6]. It is unavoidable that automation will go wrong. Due to limited even incorrect performance, current automation cannot fully ensure driving safety. On one hand, automation with limited capabilities cannot perform all tasks. It is difficult for designers and engineers to anticipate any possible dangerous situations and thus set up the corresponding functions of automation in advance. On the other hand, automation is unreliable because of its infrequently but fatally incorrect responses (e.g., misclassifying the pedestrian in the 2018 Uber AV crash). Thus, the driver and automation need to share vehicle control at present, also known as human-machine cooperative driving or human-machine co-driving [7].

Human-machine cooperative driving compensates for the possible problems of driving automation technology since the driver or operator plays a backup role. However, safety issues occur when the human and driving automation system operate the vehicle together. First, drivers' behavior may change compared to traditional driving (i.e., behavioral adaptation) influenced by the addition of driving automation systems; for example, they increase speed and reduce headway [8] unwittingly. Such behavioral changes, which are not conducive to traffic safety, make it difficult for automation to achieve its intended benefits. In addition, automation functions effectively in the conditions for which it was programmed, but it needs manual intervention in other circumstances [9]. The driver or operator needs to complete tasks they are not good at, and pays attention to the system operating status and the road environment for a long time; however, a human nature is that humans are not good at keeping vigilant for a long time [10, 11]. Human-machine cooperative driving brings other problems. For example, using automation may lead drivers to gradually lose their manual control abilities because they have fewer opportunities to drive manually [12]. Also, unjustified use of automation related to overtrust, overreliance, and complacency will occur [12].

Driving automation systems reduces the probability and severity of traffic crashes by minimizing manual operation [13], while road safety is still threatened by inadequate automation systems and inappropriate interaction between the system (including ADAS and ADS) and driver (or operator). This study analyzed and summarized the causal factors of crashes from the perspective of human errors. We discussed the probable causes combined with insights from the current literature, potential gaps of current focus in human-machine cooperative driving between research and practice, the controversial issue of complacency being considered as a causal factor, and finally offered potential countermeasures.

2 Driving Automation Crashes

We showed scenes and summarized probable causes of six crash cases, see Table 1. These cases had relatively complete analysis reports and were provided detailed information by NTSB, which is helpful for subsequent analysis. Brief crash event descriptions are as follows.

In Case A [14], on May 7, 2016, near Williston, Florida, a 2015 Tesla Model S (Level 2) was traveling east while a tractor-semitrailer truck was traveling west to make a left turn. The Tesla struck the right side of the semitrailer, causing extensive roof damage. After the crash with the semitrailer, the Tesla struck and broke a utility pole before coming to a halt. The Tesla driver and a passenger in the vehicle were killed in the crash. The driver of the semitrailer was not injured. It is the first reported traffic fatal crash involving Level 2 automation in the world.

In Case B [15], on January 22, 2018, in Culver City, California, a 2014 Tesla Model S (Level 2) drove on an interstate and a fire truck was parked in the high-occupancy vehicle lane with its emergency lights on. When another vehicle in front of the Tesla changed lanes to the right, the Tesla did not change lanes. Instead, the Tesla accelerated and struck the back of the fire truck. No one was inside the fire engine at the time of the crash, and no injuries were reported.

In Case C [16], on March 23, 2018, in Mountain View, Santa Clara County, California, a 2017 Tesla Model X (Level 2) struck a previously damaged and non-functional crash attenuator (a protecting equipment that slows the vehicle to reduce the impact of a crash) on the highway, causing the front and rear structures to separate. Then, the Tesla collided with two other vehicles, a Mazda 3 and an Audi A4. The Tesla driver died.

In Case D [17], on March 1, 2019, in Delray Beach, Palm Beach County, Florida, a truck traveling east attempted to cross the southbound lane, then turned left into the northbound lane. A 2018 Tesla Model 3 vehicle (Level 2) was traveling south at 69 mph and did not slow down or take any other action to avoid the truck before the crash. As a result, the Tesla driver died.

In Case E [1], on March 18, 2018, in Tempe, Arizona, a vehicle equipped with an Uber developmental automated driving system (Level 3) struck a pedestrian pushing a bicycle across N. Mill Avenue in the northbound lanes. The crash resulted in the death of the pedestrian and the Uber test operator was not injured. It was the first pedestrian fatality caused by an AV.

In Case F [18], on November 8, 2017, in downtown Las Vegas, Clark County, Nevada, an automated shuttle (2017 Navya Arma autonomous shuttle; Level 5) was driving on a designated test site in Las Vegas, Nevada. When the shuttle detected the truck, it began to decelerate. However, the shuttle came to an almost complete stop, and the truck in front of it continued to reverse, causing a minor crash. No one was injured in the crash.

Table 1. Scenes and probable cause of each crash case

Crash	Probable cause
Case A May 7, 2016 (2015 Tesla Model S) 	**Human (Driver):** Inattention due to overreliance on vehicle automation; prolonged disengagement from the driving task; using vehicle automation in ways inconsistent with guidance and warnings from the manufacturer **Other road user (truck driver):** Failure to yield the right of way to the car **Vehicle:** Operational design permitted the driver's inappropriate behavior
Case B January 22, 2018 (2014 Tesla Model S) 	**Human (Driver):** Inattention and overreliance on vehicle automation; disengagement from the driving task; using vehicle automation in ways inconsistent with guidance and warnings from the manufacturer **Vehicle:** Operational design permitted the driver's inappropriate behavior
Case C March 23, 2018 (2017 Tesla Model X) 	**Human (Driver):** Overreliance and complacency on vehicle automation; distraction likely from a cell phone game application **Vehicle:** Mistakenly steering the vehicle into a highway gore area; ineffective monitoring of driver engagement **Environment:** A damaged crash attenuator **Regulator:** The California Highway Patrol's failure to report the damage of the crash attenuator following a previous crash; systemic problems with the California Department of Transportation's maintenance division in repairing traffic safety hardware in a timely manner
Case D March 1, 2019 (2018 Tesla Model 3) 	**Human (Driver):** Inattention due to overreliance on vehicle automation; failure to react to the presence of the truck **Other road user (truck driver):** Failure to yield the right of way to the car **Vehicle:** Operational design permitted the driver's inappropriate behavior **Manufacturer:** Failure to limit the use of the system to the designed conditions **Regulator:** The National Highway Traffic Safety Administration's failure to develop a method of verifying manufacturers' incorporation of acceptable system safeguards for vehicles with Level 2 automation capabilities that limit the use of automated vehicle control systems to the designed conditions

(continued)

Table 1. (*continued*)

Crash	Probable cause
Case E March 18, 2018 (Uber developmental automated vehicle) 	**Human (Operator):** Automation complacency; failure to monitor the driving environment and the operation of the automated driving system because of the visual distraction throughout the trip by personal cell phone **Other road user (impaired pedestrian):** Crossing N. Mill Avenue outside a crosswalk **Vehicle:** Forward collision warning system and automatic emergency braking system in the Volvo ADASs were not active **Manufacturer:** The Uber Advanced Technologies Group's inadequate safety risk assessment procedures, ineffective oversight of vehicle operators, and lack of adequate mechanisms for addressing operators' automation complacency **Regulator:** The Arizona Department of Transportation's insufficient oversight of automated vehicle testing
Case F November 8, 2017 (2017 Navya Arma autonomous shuttle) 	**Human (attendant):** Not being in a position to take manual control of the vehicle in an emergency **Other road user (truck driver):** His action of backing into an alley, and his expectation that the shuttle would stop at a sufficient distance from his vehicle to allow him to complete his backup maneuver

3   Results

According to accident reports from the NTSB, inappropriate using ways of vehicle automation, human distraction or disengagement, and complacency (overreliance) are believed to be probable causes in most cases.

First, in Cases A, B, C, and D, the driver used vehicle automation in ways inconsistent with guidance and warnings from the manufacturer. In these four crashes, the driver's hands were always off the steering wheel when Autopilot was active, departing from system usage specifications. For example, in Case A, the "Autopilot hands on state" parameter remained at "Hands required not detected" for the great portion of the trip. During the trip, the system made seven visual warnings, and six of these visual warnings transitioned further to auditory warnings (i.e., chime), followed by a brief detection of manual operation that lasted one to three seconds. During the 37 mins that the ADAS was in operation, the system detected manual operation for only 25 s. Also, the driver used the system on State Road 24 (a non-preferred roadway for the use of Autopilot) even though the system was not designed for this type of road. However, although the accident reports of Cases C and D did not explicitly state that "the driver used vehicle automation in ways inconsistent with guidance and warnings from the manufacturer" in the "Probable Cause" part, a lack of steering wheel torque (force applied to the steering wheel to make it rotate about the steering column) was still detected. "No driver-applied steering wheel torque was detected by Autosteer" (p. 6) [16] in Case C and "no driver-applied steering wheel torque was detected for 7.7 s before impact, indicating driver disengagement" (p. 14) [17] in Case D. It is conceivable for hands to be just resting on the steering wheel without any torque. However, "a lack of steering wheel torque indicates to the vehicle system that the driver's hands are not on the steering wheel" (p. 6) [16].

Next, in Cases A, B, and D, the driver was (prolonged) disengaged from the driving task; in cases C and E, the driver or operator was reported to be distracted by the cell phone, leading to a crash. Inappropriate operational design (e.g., "Despite the system's known limitations, Tesla does not restrict where Autopilot can be used." (p. x) [16]; the system did not prevent drivers from using it improperly) permitted driver disengagement, causing them not to realize the approaching danger (e.g., the tractor-semitrailer truck in Case A, the stationary fire truck in Case B, and the truck trying to turn left in Case D) in time. Another safety issue is human distraction. In Case C, the driver "was likely distracted by a gaming application on his cell phone before the crash" (p. x) [16], making him not realize the system had steered the vehicle into a gore area of the highway not used for vehicle. Also, the human operator was "glancing away from the roadway for extended periods throughout the trip" (p. 1) and might watch a television show according to the phone records [1] in Case E.

Accident reports stated that irrational use or inattention resulted from complacency or overreliance. The NTSB concluded that the behavior of drivers who did not follow the owner's manual strongly indicated their overreliance on vehicle automation (e.g., [14]). For example, the driver in Case A used the system in roadways not satisfying the condition of "highways and limited-access roads with a fully attentive driver" (p. 74 in Tesla Model S Owner's Manual [19]; cited in [14]). Driver disengagement did not meet the requirement that they must "keep hands on the steering wheel at all times" (p. 74 in Tesla Model S Owner's Manual [19]; cited in [14]). As for Case E, the NTSB asserted that the human operator's prolonged visual distraction was a typical result of automation complacency and prevented her from spotting pedestrians in time to prevent a crash.

In addition, other road users' improper operation caused crashes. The truck drivers failed to yield the right of way to the vehicle (in Cases A and D) and incorrect evaluation of the shuttle stopping distance (in Case F); the pedestrian crossed N. Mill Avenue outside a crosswalk (in Case E). Similarly, vehicle system operational design, manufacturers, environments, and regulators also had some responsibility for the crash (refer to Table 1 for details).

4   Discussion

In this study, we explored and summarized causal factors of road crashes related to driving automation. Through a series of analyses of accident reports from NTSB, we found that the common causes focus on human errors, including inappropriate using ways of driving automation systems, human distraction or disengagement, and complacency (overreliance) on vehicle automation. The involved crashes indicate that human factor risks are present in human-automation interaction on current roads. Improper interaction results in an ineffective combination of human and machine strengths.

The current operational design of the driving automation system allows the person out of the loop of the dynamic driving task, which causes problems when the human has to regain control [12]. In our analysis, although the system uses the detection of steering wheel torque to measure whether the driver is in control of the steering wheel, it is difficult to ensure that the driver has maintained effective attention and is ready to perform the dynamic driving task. Additionally, in several cases (e.g., Case A), the system did not intervene (such as forced deactivation of ADAS) when the driver's hands left the steering wheel repeatedly. This "human out of the loop" operational design is not helpful for the driver to regain control of the vehicle in time, leading to a crash (e.g., the driver is too late to brake or steer to avoid the crash in Case C). Thus, driving automation systems should put more emphasis on the "human in the loop" design, enabling drivers to notice automation problems when systems (suddenly) reach the boundaries of their capabilities [20, 21].

Another causal factor is that drivers misuse the driving automation system and such behavior is detrimental to driving safety. Misuse refers to the user's incorrect use of automation [12]. For example, drivers used systems inconsistent with the requirement in the vehicle owner's manual, including activating systems on some types of roads which are not designed for and hands off the steering wheel during the system is enabled.

Misuse is frequently used in conjunction with complacency or overreliance [12]. In our study, the NTSB concluded that complacency or overreliance, which caused inappropriate use of driving automation systems by drivers or operators, was a probable cause of crashes. The concept of complacency was first introduced in aviation crash investigations. It is described as "self-satisfaction, which may result in non-vigilance based on an unjustified assumption of satisfactory system state" by the NASA Aviation Safety Reporting System [22]. However, this construct has not been given enough attention in theoretical research, as compared to other constructs that are important for human-automation interaction and traffic safety in automated driving. For instance, Heikoop et al. [8] measured the frequency with which important psychological constructs or pairs of constructs were discussed in the research on automated driving in order to develop a psychological model of automated driving. A total of 15 concepts were extracted from 43 articles, including mental workload, attention, feedback, stress, situation awareness, task demands, fatigue, trust, mental model, arousal, complacency, vigilance, locus of control, acceptance, and satisfaction. They were ranked according to the number of times that a link between the construct and another construct in the model is proposed in the extracted articles. Complacency was left out of the model because it was after the cut-off point set at 10 counts [8]. It might suggest that the importance of complacency is not fully recognized in current academic research. In practice, it has been identified in all of the current accident reports involving human drivers/operator (see Table 1). Thus, there may be a gap between research and practice regarding the role of complacency in the domain of driving automation.

In addition, there are strong debates on the concept of complacency (overreliance) and its explanatory power in the literature of human factors and ergonomics. First, there is no clear definition of complacency. Some researchers considered it as a "mental state"

[22] or a "psychological state" [23]. Others regarded it as insufficient monitoring and verification behaviors or their negative consequences [24, 25]. The failure to reach a consensus on the definition of complacency makes it potentially confusing to people when it is used in the crash analysis [26]. Second, it is hard to collect evidence of complacency and to measure it objectively [27]. For instance, research suggests using unreasonable usage behaviors of automation (or being distracted and missing automation failures) as an indicator of complacency while using automation. The NTSB [14] also came to the conclusion that the driver's behavior of disregarding the owner's handbook is clearly indicative of complacency or overreliance on driving automation. In all of the five crashes, complacency or overreliance was considered the probable driver cause. It is not a coincidence. Human drivers, if they are believed to cause crashes while using automation, will be finally blamed for their complacency with automation. However, researchers [28, 29] have argued that current evidence of so-called "complacency" (e.g., being distracted or improper automation usage behaviors) might be due to other factors. When drivers or operators are accused of being complacent or over-reliant, there are essentially system design flaws behind them [30].

In order to prevent crashes related to driving automation and find a more appropriate way to assign responsibility, future research could be directed at theoretical research as well as practical technology enhancement in the specific area of driving automation. First, the results of the crash analysis need to guide theoretical analysis of traffic crash factors in research. For example, further theoretical exploration is needed on the concept and the effect on traffic safety of automation complacency. Also, the driving automation domain has its own specificities. Drivers are ordinary consumers who may have problems when interacting with driving automation systems, including understanding the terminology in the owner's manual and staying focused on the system's operating conditions for long periods of time. It is important to develop driver training programs. Merriman et al. [31] stated that the occurrence of crashes related to driving automation technology is associated with driver attitudes, mental models, and trust in automation, and that automation may impair the driver's ability to recognize and avoid hazards. Therefore, training programs for drivers can be developed from the perspectives of workload, trust, and situational awareness, which can help drivers develop a reasonable perception and use of driving automation systems. From the perspective of technology, it is possible to equip with a multimodal driver monitoring system (e.g., based on steering wheel torque, eye movements, and facial expressions information) to improve monitoring accuracy while ensuring privacy and security. Also, engineers could upgrade the technology and improve system reliability to avoid drivers from bypassing the usage restrictions of driving automation systems through deceptive behaviors, such as placing heavy objects on the steering wheel to simulate manual operation. Importantly, manufacturers need to take into account the potential for emergency hazardous situations on the road and then establish adequate safety risk assessment procedures. Relevant government departments should develop appropriate safeguards to ensure that the use of driving automation systems or the testing of AVs is strictly conformed to system design requirements in order to reduce the abuse and misuse of the systems.

Acknowledgments. This research was supported by the National Natural Science Foundation of China (Grant No. 72071143 and T2192933).

References

1. National Transportation Safety Board: Collision between vehicle controlled by developmental automated driving system and pedestrian, Tempe, Arizona, 18 March 2018. Highway Accident Report NTSB/HAR-19/03. Washington, DC: NTSB (2019)

2. Dingus, T.A., et al.: Driver crash risk factors and prevalence evaluation using naturalistic driving data. Proc. Nat. Acad. Sci. 113(10), 2636–2641 (2016). https://doi.org/10.1073/pnas.1513271113

3. Huang, H., Chang, F., Schwebel, D.C., Ning, P., Cheng, P., Hu, G.: Improve traffic death statistics in China. Science 362(6415), 650 (2018). https://doi.org/10.1126/science.aav5117

4. Wang, X., Liu, Q., Guo, F., Fang, S., Xu, X., Chen, X.: Causation analysis of crashes and near crashes using naturalistic driving data. Accid. Anal. Prev. 177(10), 106821 (2022). https://doi.org/10.1016/j.aap.2022.106821

5. SAE International: Taxonomy and definitions for terms related to driving automation systems for on-road motor vehicles. Society of Automotive Engineering, USA (2021)

6. Parasuraman, R., Riley, V.: Humans and automation: use, misuse, disuse, abuse. Hum Factors 39(2), 230–253 (1997). https://doi.org/10.1518/001872097778543886

7. Marcano, M., Díaz, S., Pérez, J., Irigoyen, E.: A review of shared control for automated vehicles: theory and applications. IEEE Trans. Hum. Mach. Syst. 50(6), 475–491 (2020). https://doi.org/10.1109/THMS.2020.3017748

8. Heikoop, D.D., de Winter, J.C.F., van Arem, B., Stanton, N.A.: Psychological constructs in driving automation: A consensus model and critical comment on construct proliferation. Theor. Issues Ergon. Sci. 17(3), 284–303 (2016). https://doi.org/10.1080/1463922X.2015.1101507

9. Cummings, M.L., Clare, A., Hart, C.: The role of human-automation consensus in multiple unmanned vehicle scheduling. Hum. Factors 52(1), 17–27 (2010). https://doi.org/10.1177/0018720810368674

10. Mackworth, N.H.: The breakdown of vigilance during prolonged visual search. Q. J. Exp. Psychol. 1(1), 6–21 (1948). https://doi.org/10.1080/17470214808416738

11. Molloy, R., Parasuraman, R.: Monitoring an automated system for a single failure: vigilance and task complexity effects. Hum. Factors 38(2), 311–322 (1996). https://doi.org/10.1177/001872089606380211

12. de Winter, J.C.F., Petermeijer, S.M., Abbink, D.A.: Shared control versus traded control in driving: a debate around automation pitfalls. Ergonomics. in press (2022). https://doi.org/10.1080/00140139.2022.2153175

13. Wang, J., Zhang, L., Huang, Y., Zhao, J.: Safety of autonomous vehicles. J. Adv. Transp. 2020, 8867757 (2020). https://doi.org/10.1155/2020/8867757

14. National Transportation Safety Board: Collision between a car operating with automated vehicle control systems and a tractor-semitrailer truck near Williston, Florida, 7 May 2016. Highway Accident Report NTSB/HAR-17/02. Washington, DC: NTSB (2017)

15. National Transportation Safety Board: Rear-end collision between a car operating with advanced driver assistance systems and a stationary fire truck, Culver City, California, 22 January 2018. Highway Accident Report NTSB/HAB-19/07. Washington, DC: NTSB (2019)

16. National Transportation Safety Board: Collision between a sport utility vehicle operating with partial driving automation and a crash attenuator, Mountain View, California, 23 March 2018. Highway Accident Report NTSB/HAR-20/01. Washington, DC: NTSB (2020)

17. National Transportation Safety Board: Collision between car operating with partial driving automation and truck-tractor semitrailer Delray Beach, Florida, 1 March 2019. Highway Accident Report NTSB/HAR-20/01. Washington, DC: NTSB (2020)

18. National Transportation Safety Board: Low-speed collision between truck-tractor and autonomous shuttle, Las Vegas, Nevada, 8 November 2017. Highway Accident Report NTSB/HAR-19/06. Washington, DC: NTSB (2019)

19. Tesla Inc.: Tesla Model S Owner's Manual (2016)

20. Abbink, D. A., et al.: A topology of shared control systems—finding common ground in diversity. IEEE Trans. Hum. Mach. Syst. **48**(5), 509–525 (2018). https://doi.org/10.1109/THMS.2018.2791570

21. Abbink, D.A., Mulder, M.: Exploring the dimensions of haptic feedback support in manual control. J. Comput. Inf. Sci. Eng. **9**(1), 011006 (2009). https://doi.org/10.1115/1.3072902

22. Billings, C. E., Lauber, J. K., Funkhouser, H., Lyman, G., Huff, E. W.: NASA aviation safety reporting system. NASA-TM-X-3445 (1976). https://ntrs.nasa.gov/citations/19760026757

23. Weiner, B.: Social motivation, justice, and the moral emotions: an attributional approach. Lawrence Erlbaum Associates, Mahwah, NJ (2006). https://doi.org/10.4324/9781410615749

24. Mouloua, M., Ferraro, J. C., Kaplan, A. D., Mangos, P., Hancock, P. A.: Human factors issues regarding automation trust in UAS operation, selection, and training. In Mouloua, M., Hancock, P. A. (eds.), Human Performance in Automated and Autonomous Systems: Current Theory and Methods, pp. 169–190. CRC Press, London (2019). https://doi.org/10.1201/9780429458330-9

25. Parasuraman, R., Molloy, R., Singh, I.L.: Performance consequences of automation-induced "complacency." Int. J. Aerosp. Psychol. **3**, 1–23 (1993). https://doi.org/10.1207/s15327108ijap0301_1

26. Liu, P.: Automation complacency as causal to traffic crashes: Fact or fallacy? Accident Analysis and Prevention

27. Drnec, K., Marathe, A.R., Lukos, J.R., Metcalfe, J.S.: From trust in automation to decision neuroscience: Applying cognitive neuroscience methods to understand and improve interaction decisions involved in human automation interaction. Front. Hum. Neurosci. **10**, 290 (2016). https://doi.org/10.3389/fnhum.2016.00290

28. Boos, A., Feldhütter, A., Schwiebacher, J., Bengler, K.: Mode errors and intentional violations in visual monitoring of Level 2 driving automation. In: 2020 IEEE 23rd International Conference on Intelligent Transportation Systems (ITSC), pp. 1–7 (2020). https://doi.org/10.1109/ITSC45102.2020.9294690

29. Feldhütter, A., Härtwig, N., Kurpiers, C., Hernandez, J.M., Bengler, K.: Effect on mode awareness when changing from conditionally to partially automated driving. In: Bagnara, S., Tartaglia, R., Albolino, S., Alexander, T., Fujita, Y. (eds.) Proceedings of the 20th Congress of the International Ergonomics Association (IEA 2018). Advances in Intelligent Systems and Computing, vol. 823, pp. 314–324. Springer, Cham (2019). https://doi.org/10.1007/978-3-319-96074-6_34

30. Miranda, A.T.: Misconceptions of human factors concepts. Theor. Issues Ergon. Sci. **20**(1), 73–83 (2019). https://doi.org/10.1080/1463922X.2018.1497727

31. Merriman, S.E., Plant, K.L., Revell, K.M.A., Stanton, N.A.: What can we learn from automated vehicle collisions? A deductive thematic analysis of five automated vehicle collisions. Saf. Sci. **141**, 105320 (2021)

A Longitudinal Driving Experience Study with a Novel and Retrofit Intelligent Speed Assistant System

Yaliang Chuang[(⊠)], Tim Muyrers, Wanyan Zhang, and Marieke Martens

Eindhoven University of Technology, Eindhoven 5612 AZ, The Netherlands
{y.chuang,m.h.martens}@tue.nl, {t.h.c.muyrers,
w.zhang1}@student.tue.nl

Abstract. Speeding is the primary cause of traffic accidents. To improve road safety, the European Union started implementing a new regulation mandating that all new vehicles coming to the EU market must be equipped with an Intelligent Speed Assistance (ISA) system from 2022 onwards. However, the rule did not include existing vehicles on the roads. Our research aims to fulfill this gap by investigating user experiences and acceptance with a retrofit system developed by V-tron, a Dutch company. Seven participants signed up for our study and our technicians installed the ISA system on their cars. They then used the car for more than one month and reported their experiences weekly. We also recruited a driving school and conducted a focus group with five instructors. Using interview and questionnaire methods to collect their first-person experiences, we saw that all participants acknowledged the vision and potential of ISA systems in reducing speeding and improving traffic safety. While the retrofit system is easy to use, the technology needs to be improved in accuracy and robustness. The overruling mechanism also needs to minimize the latency and consider secondary users unfamiliar with the speed control. Three design concepts were proposed to improve user experiences and eventually promote the adaptation of ISA systems.

Keywords: Speeding · Driving Assistance System · Technology Acceptance Model · User Experience

1 Introduction

Driver assistance systems have been proven beneficial in several situations, both from a driver's and road environment perspectives. One is the intelligent speed assistants (ISA) system, which has been developed for over two decades. It has constantly been evolving, testing, and improving. A system must communicate with a driver properly to inform without causing unnecessary distraction or interference (Bakker & Niemantsverdriet 2016). For this reason, we see more and more systems being tested and introduced into legal road automobiles (Nidamanuri et al. 2021). Besides the need for safety checking (Dikmen & Burns 2016; Endsley 2017; Ingle & Phute 2016), achieving a high level of user acceptance is essential. This is essential for social support and the actual use of these systems.

When looking at the user response to advanced driver systems like adaptive cruise control (ACC), we see that acceptance and positive attitudes are significantly above 90% (Strand et al. 2016; Xiao & Gao 2010). While this shows potential for systems that aim at improving road safety by assisting in maintaining and regulating speed, there is still quite a significant difference between ACC and ISA systems that provide and control a vehicle's speed according to the situational speed limits.

While research on roads with dynamic speeds is limited, there is a growing interest in improving driver and other road users' safety through adjustable speeds. One great example is the study performed by Li et al. (2019). Their study used a car simulator to put drivers in a situation where poor visibility hindered their ability to drive safely at a normal road speed. Using a connected vehicle attached to a variable speed limit system, they explored the driver's acceptance and its influencing factors. Within their study, they confirmed that external factors could provide enough support to convince drivers of the importance of dynamic speed limits. They found support for on-board communication and on-road sign displays, with a slight preference for the latter.

Research on these signs generally does involve user acceptance but is limited to the willingness of drivers to adhere to the different speeds. For instance, Janssen et al. (2021) claim that asking for user feedback on dynamic speed sections would not provide actual results, as experiencing and imagining are sometimes significantly different. Furthermore, they address that technology is often only partially used as the inventor/designer intended (Bainbridge 1983). However, their conclusion is promising, indicating that while the willingness to adhere to lowered speed limits is low, it rises significantly when drivers feel the need for the reduced speed is meaningful.

Our study aims to investigate the user experience of the retrofit ISA system (see Fig. 1) developed by V-tron, a Dutch company. The system currently equips two technical components to detect the in-context speed limit, including a front camera for recognizing the road signage and the GPS location. In the near future, V-tron's ISA system can also retrieve dynamic speed limits from the city's traffic management system and adjust the driver's speed accordingly.

We want to understand (1) the effectiveness and usefulness users can perceive, (2) the usability of the system, (3) user experiences and social acceptability, and (4) the possible changes in their driving behaviors. Those insights can help improve the ISA system's designs and bring the products to the market to improve safety and reduce traffic accidents.

We used interviews and questionnaires to understand users' expectations, experiences, and behaviors. We conducted a pre-interview before they started using the system to collect their basic information and understand their driving preferences and experiences, commuting routines, expected expectations related to the system, etc. During the trial period, we sent periodical questionnaires to participants and asked them to report their experiences every week. After the testing phase, we conducted a post-interview to probe their overall experiences and discuss some issues we observed from analyzing their driving data and questionnaire responses.

In the following sections, we will first briefly review theories and measurement methods in related topics. Based on that, we will explain how we developed the interview scripts and questionnaires for collecting participants' subjective experiences. In

Sect. 3, we will present our study setup. The findings and design recommendations are summarized in Sects. 4 and 5, respectively.

Fig. 1. The V-tron ISA system was used in the study presented in this paper. It consists of a camera (right) that can recognize the speed sign on the road, a round-shape display (middle) that can show the current speed limit, and other safety-related information.

2 Background and Literature Review

2.1 The Definition of Acceptance

Acceptance, acceptability, and social acceptance are all terms used to describe "how potential users will react and act if a certain measure or device is implemented" (Vlassenroot et al. 2010, p. 167). If there is public/social support, the effectiveness and success of an initiative will increase. For the implementation of ISA, it is essential to know whether drivers and other public will accept the system and what factors influence the acceptance of this technology. Although the importance of investigating acceptance and acceptability is well recognized, there needs to be more consistency between studies as to what acceptance is and how it is measured [cited in Regan et al. (2014); Mitsopoulos et al. (2002)]. The definition of acceptance is the basis for the assessment structure and acceptance model, "without a definition, it is not possible to examine the validity and reliability of any assessment methods and/or models" (p. 12). Acceptance has been defined differently but partially overlapping in numerous studies.

Adell (2009) divides acceptance into five categories. The first category is "accept"; the second one considers the need and requirements, which are relevant to the usefulness of the system; the third type views acceptance as the aggregation of attitudes; the fourth category is the intention to use, and the fifth one emphasizes the actual use. They can be seen as an evaluation process from usefulness to actual use. It shows that acceptance is a multifaceted concept and that researchers select the propensity to conform and limit its scope (Regan et al. 2014).

Some researchers have further subdivided attitudinal acceptance and behavioral acceptance, conditional acceptance, and context-based acceptance, as well as distinguished between acceptance and support [see Ch. 2 in Regan et al. (2014) for a detailed review]. Although there are different definitions, what they have in common is that "acceptance and acceptability are ... based on the individual's judgment" (p. 12). Any assistance system only produces the desired effect when used by the user, which means that using the system is essential (Regan et al. 2014).

Since the users recruited in this project use the ISA system realistically, based on Schade and Schlag (2003), we ask users about their forward-looking judgments about the system and their expectations of using it during the baseline phase as acceptability and ask them about their attitudes and behaviors toward using this system during the testing phase as acceptance. The acceptance is closely related to the usage, and the acceptance will depend on how the users' requirements are integrated into the development of the system. Therefore, in our study, we take the definition of acceptance proposed by Adell (2009), "the degree to which an individual intends to use a system and, when available, to incorporate the system in his/her driving" (p. 31).

Another related concept, social acceptability, considers a broader range of factors affecting acceptance related to security and economics, but also some with cultural, social, and psychological significance (Otway & Von Winterfeldt 1982). Researchers have studied social acceptance and social acceptability to understand the impact of various potential social contexts and their specific factors on human interaction with technology (Uhde & Hassenzahl 2021). Wüstenhagen et al. (2007) propose a triangle of social acceptance of renewable energies, including socio-political, community, and market acceptance. According to Vlassenroot et al. (2010), "social acceptance is a more indirect evaluation of consequences of the system" (p. 168); it involves a broader range of factors beyond the direct operating system. Vlassenroot et al. proposed 14 indicators most likely to influence acceptance and acceptability for the ISA scenario, adding factors such as personal and social aims, social norms, responsiveness awareness, and affordability. Since their subjects were people with no experience using the ISA system, their findings were used by us more in the pre-interview stage of our study.

2.2 Assessment of Acceptance

Since the driver's judgment of the system is based on personal knowledge, understanding, and experience, this may differ from the influence of the system as measured by an external observer (Adell 2009). Several widely applied models exist to measure the impact on acceptance: the *technology acceptance model* (TAM) proposed by Davis et al. (1989); TAM2 by Venkatesh and Davis (2000), and TAM3 by Venkatesh and Bala (2008), which is the extension of TAM. There are also some other models that extended TAM with other methods, such as the *unified theory of acceptance and use of technology* (UTAUT, Venkatesh et al. 2003) with the integration of several widely used models and some comprehensive research. Lastly, Osswald et al. (2012) incorporated the influence of contextual information in TAM and proposed the *car technology acceptance model* (CTAM) based on the abovementioned UTAUT. Two crucial elements were introduced into the CTAM model: anxiety and perceived safety. After conducting a literature review, we selected the following influencing factors from several models that fit our research.

Individual Factors. From Rahman et al.'s (2018) survey, we know that age and gender are the most frequently cited demographic factors in the mobility domain, although some studies have concluded that gender has no significant effect on acceptance (Rahman 2016). Besides, Adell (2009) found the effect of age was more in the perception of usefulness, satisfaction, and keeping the system, with younger drivers rating lower in these categories than older drivers. Some studies also point out that a driver's experience with similar devices can affect the acceptance of new technologies (Höltl & Trommer 2013; Rödel et al. 2014).

Driving habit is a broad phenomenon that covers the choice of driving speed, distance to a preceding vehicle, overtaking other vehicles, and the tendency to commit traffic violations constitute behavioral tendencies of drivers. These habits are often described as 'driving style.' Collecting the driver's driving style helps to interpret which habits have more significant effects on their current driving behaviors and to compare them with the use of the ISA system afterward.

Expectations of the System. According to Compeau et al. (1999), expectations are the user's prediction of the system's purpose and function before actually using a system or product. It is also called outcome expectations, such as increased efficiency and improved quality. This item allows us to understand the user's vision of the new technology and validate it after use.

Perceived ease of use means the degree to which users expect the target system to be easily operated (Davis et al. 1989). It is called Effort Expectancy in the UTAUT model (Venkatesh et al. 2003) and the CTAM model (Osswald et al. 2012). According to Venkatesh et al., the effect of effort expectancy on intention is more pronounced among females and more senior users. In Davis' (1989) study, this factor was expected to be more prominent in the early stages of using a new system.

Perceived Usefulness presented in TAM (Davis et al. 1989) means the extent to which the user subjectively believes that using the target system will improve their performance. TAM theorized that perceived usefulness strongly influences users' *behavior intention* (BI), and perceived ease of use also significantly affects them but diminishes over time (Davis et al. 1989). Meanwhile, in their conclusion, perceived usefulness was also influenced by perceived ease of use because the easier the system is to use, the more useful it will be. In the UTAUT model (Venkatesh et al. 2003), it is called *performance expectancy*; the authors conclude that performance expectancy is a determinant of intention in most cases, and this relation is influenced by gender and age, which is more pronounced among males and younger users (Venkatesh et al. 2003).

Effectiveness is defined as "the extent to which a system performs its intended tasks" (Rahman et al. 2018, p. 136). Many studies recognized that system reliability is an essential factor that affects the effectiveness and act as a barrier to acceptance. Some common examples of poor reliability are false/nuisance alarms and accuracy when using the system.

Attitude is defined as "an individual's overall affective reaction upon using a system" (Osswald et al. 2012, p. 54). This factor is intended to reflect the user's perception of the use of the system and its impact, going beyond an assessment based on pure functionality. The initial attitude toward the system mainly affects workload, emotional state, and usage (Adell 2009). Venkatesh et al. (2003) argue that attitudes toward the use of technology are not theoretically a direct determinant of behavior intent. But in CTAM, this element is considered because we cannot assess the effect in a car context in advance (Osswald et al. 2012).

Anxiety and Perceived Safety are the two crucial elements in CTAM (Osswald et al. 2012). *Anxiety* was defined in the car context as "the degree to which a person responds to a situation with apprehension, uneasiness or feelings of arousal" (p. 55). *Perceived* *safety* was defined as a judgment of individual driving skills and a sense of safety in relation to other drivers (p. 55).

Social influence is also called the *subjective norm* in TAM2 (Venkatesh & Davis 2000). It refers to the influence of people who are important to the user's view of whether he/she should perform the behavior or not. Vlassenroot et al. (2010) observed that "peers, e.g., co-workers or specific other road users, will influence the attitudes and behavior of individuals" (p. 176). Venkatesh et al. (2003) also found that "the role of social influence in technology acceptance decisions is complex and subject to a wide range of contingent influences" (p. 452). Based on the TAM2 model (Venkatesh & Davis 2000), people incorporate social influence into their perception of usefulness, i.e., gaining status and influence through the use of systems and thereby improving their job performance. When people gain direct experience over time, they rely less on social information in forming their intentions and instead make judgments based on the potential benefits that come with use.

2.3 Summary

Based on the context and purpose of this study, we integrated the theoretical model described above. As this study focuses on the driver's actual use of the system, we divided all relevant factors into before exposure to the ISA system and while using the system. Individual factors and expectations of the system are investigated before the system is activated. Perceived *ease of use, usefulness, effectiveness, attitude, anxiety, safety,* and *social influence*, which are direct determinants in the driving environment, are investigated when drivers use the system. The figure below shows the specific meaning of these factors in the driving context.

3 Study Setup

In order to investigate (1) the effectiveness and usefulness users can perceive, (2) the usability of the system, (3) user experiences and social acceptability, and (4) the possible changes in their driving behaviors. We interviewed two user groups in this study: A. voluntary people with their vehicles; B. five tutors of a driving school at Helmond. For type A users, we used interviews and questionnaires to understand their experiences. We organized focus group discussions for the B group to know their observations and suggestions for the tested ISA system.

In total, we recruited 12 participants, including seven people with their vehicles and five instructors from the driving school where the ISA system was installed on four cars used for driving lessons.

3.1 Three Phases for Individual Participants

(1) Collecting Participants' Expectations and Prior Experiences. We recruited our participants through the project website and social media channels of the local communities. After signing up for our study, we arranged a meeting with them to explain our study of collecting their subjective experiences. When they agreed with our plan and signed the consent form, we asked them for some basic information related to this study, including demographic info, gender, and diving experiences. We asked them the average annual kilometers they had driven and their driving style preferences (van Huysduynen et al., 2015), which may help to interpret the events that occurred during the test. We also collected contextual information, like how often the user drove and how familiar they were with the road conditions, which could affect their driving experiences with a new assistant system in place. At the end of the pre-interview, we asked participants about their expectations of the ISA system.

(2) A Weekly Questionnaire Survey for Three Weeks. During the testing period of three to four weeks, we collect their experiences by asking them to fill in a short questionnaire weekly. The interactive questionnaire was implemented LimeSurvey. It covers six dimensions of TAMs, including Perceived ease of use, Perceived Usefulness, Effectiveness, Attitude, Anxiety and Perceived Safety, and Social Influence. There are multiple-choice questions that a participant can easily express their attitudes and experiences. We also used Likert scale to design questions to investigate how much they agreed with the specific criteria. Taking "information accuracy" as an example, we asked, "The information from the system was accurate," and they can drag a slider from 1 = " strongly disagree" to 7 = "strongly agree" to indicate their experience. In case of negative feedback, additional open-ended questions would appear to ask participants for more details and explanations.

To protect participants' personal information while filling in a digital questionnaire, we customize a link for every individual participant and send it to their mailbox on the day of their convenience. In this way, they don't need to fill in their personal information every time, and we can proceed with their inputs anonymously.

(3) Post-Interview. After the testing phase, we conduct a post-interview to collect participants' total experiences, improvement feedback, and possible changes in their driving behaviors. To host the post-interview smoothly and efficiently, we ask participants to fill in a summary questionnaire. It covers the Car Technology Acceptance Model and some specific questions based on the results participants reported in the weekly questionnaire. We synthesize their responses and produce a UX (user experience) curve for each participant to investigate what makes the changes in their experiences and opinions throughout the entire study. We also learned improvement ideas from the participants and discussed possible marketing proposals to understand which subscription models and pricing are affordable for the target customers. The main questions include the following:

- **Effectiveness:** Do you think driving a car with this system will make you drive differently? Followed-up question: In what ways? Under what conditions?
- **Usefulness:** How useful would you find the technology? Would it serve a purpose for you?
- **Usability:** Can you think of any potential problems or concerns that you might have in using the system? For instance: Source of distraction? Potential for over-reliance? Reliability issues? Issues with the look and feel of the warnings?
- **Social acceptability:** How would you feel if it were compulsory for you to fit this technology into your vehicle?
- **Total experiences:** From 1 to 10, how satisfied are you with the system? Can you also explain why?

3.2 Focus Group Study with Instructor Participants

In this study, we organized a focus group discussion with five instructors at Rijschool VOC van Oijen in Helmond to learn about their experiences from the second-person view. They are the secondary target users of the ISA system, and we want to understand their observations on their students' usage of the system. This can help us to understand their expert opinions and examine whether the driving school could be a proper market for the system. We start the discussion with the TAM questions covering five dimensions: Effectiveness, Usefulness, Usability, Social acceptability, and Affordability, and probe their opinions collectively.

4 Findings

Due to technical issues and the delay of needed equipment, our driving experiment started in the second week of December. Due to the Christmas holiday, many participants couldn't respond to our questionnaires. As a result, we received 18 responses from participants' questionnaires. In this section, we first present their feedback with diagrams and explanations, followed by a more detailed analysis of the interview data discussed later in this section.

To keep track of the feedback gathered from different participants, we used the following acronyms to mark the source of the specific quote:

- PreIn (pre-interview): the data we collected from the first interview conducted before the ISA system installed in the participant's vehicle was activated.
- PxQy (the yth weekly questionnaire): the responses we collected with the yth weekly questionnaire participant x filled out.
- PosIn (post-interview): the data we collected from the second interview we did when the participant completed the four-week driving experience experiment.

4.1 Subjective Responses

First, all users perceived the ISA System as easy to use throughout the experiment. Although three participants (5 [28%] responses) were neutral about this option, all other users found the system very easy to use.

Secondly, when we asked participants their perceived usefulness (including, *the system is useful* and *using this system will improve my driving performance*), their responses were a bit disputed. Half of them felt positive. They indicated that when the system works well, it helps them to understand the current speed limit and to be more focused when driving. From a more macro perspective, the system can contribute to road safety. And half of the responses were neutral or opposed because of problems with the system that influenced their judgment on perceived usefulness. The two main issues reported by the participants are the information misalignment, and the system does not react quickly. Fortunately, users widely agreed that if the system worked well, it would be beneficial for assisting their driving.

Thirdly, participants expressed some complaints. "Information Accuracy" has the most popular comments: except for one user who rated it neutral, all other users reported

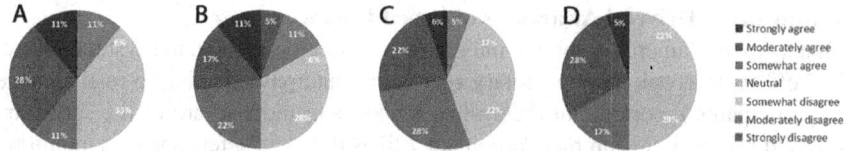

Fig. 2. The percentage of participants' perceived values on (A) usefulness, (B) improving driving performance, (C) providing helpful information, and (D) feedback clarity. The amount of total responses is eighteen.

inaccurate information (one user thought the information was accurate in the first week and changed to disagree after two weeks of use). Incorrect speed limits may lead to danger, so this point directly affects the user's overall attitude and perceived usefulness of the system. Since the system provides the wrong speed limit, and users need to override it, half of the responses consider the system annoying. We discussed all those issues with the participants in the post-interviews and focus group discussions, and the results were reported in the following sections.

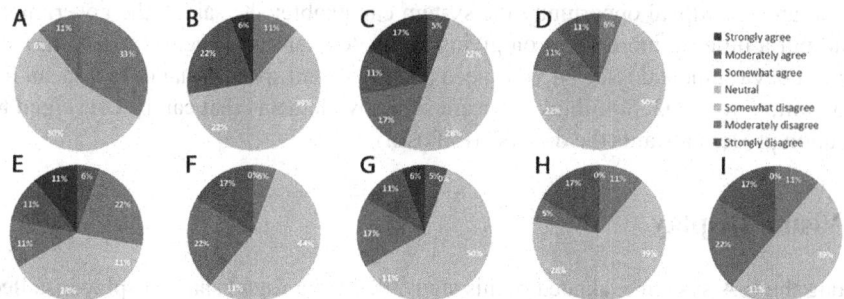

Fig. 3. The percentage of participants' perceived values on (A) the accuracy of information provided, (B) capability for avoiding traffic accidents, (C) improvement of safety, (D) believability, (E) assistance, (F) annoyingness, (G) Desirability, (H) pleasantness, and (I) overall satisfaction. The amount of total responses is eighteen.

4.2 Positive User Experiences

In our interviews, most participants agreed that the ISA system could be an effective assistance system for improving road safety. For instance, Participant 2 (P2) mentioned that the situational speed limit could also educate the drivers on why the authority limits the speed to a certain degree in specific areas. This information can persuade drivers to care about other road users' safety and adjust their driving speed accordingly. The user interface is straightforward to understand without distraction. When used in an old car without an embedded information system, the system can be an easy and affordable upgrading solution for providing situational information and speed control. We synthesize their positive feedback into the following themes.

1. Improving Drivers' Awareness of Speed Limits

Thanks to the camera sensor and integrated information system, the system reduced drivers' efforts to check the momentary speed limit and pay attention to road signage. Some participants reported that the system made them more aware of the speed and the safety it brings. For example, Participant 7 likes that "the safety speed is monitored automatically" (P7Q2). By using this system, drivers can be more aware of the speed limit and calmer, "The system monitors speed so I do not commit a violation and can concentrate more on traffic" (P7Q2).

2. Reducing or Changing Speeding Behaviors

Before the testing starts, all participants express positive expectations of the system's assistance in avoiding speeding and preventing potential accidents or traffic tickets. During the experiment period, the system acted well in restricting their speeding. Furthermore, some participants also found additional benefits in informing them of the correct speed limits and triggering them to reflect on their existing behaviors. For instance, when P7 drove the regular commuting route in the second week of testing, he noticed that the speed limit shown on display was lower than he thought (i.e., 50 km/h). He said, "I now realize that I've been driving too fast in the past" (P7Q2). Most participants acknowledged that the system raised drivers' mindsets back to the limit. P2 further pinpointed an educational opportunity the system can enable; she said, " the government would put a little bit more focus on giving the pedestrians and cyclists room, may be aware of environmental issues if you speed and break and speed, what does it mean for the pollution, for CO_2. So I think there are so many elements that can be converged to education process towards the drivers" (P2PosIn).

5 Visual Display

Among the ISA system evaluated in this study, there is a round shape display installed on the dashboard within the participants' vehicles. All participants indicated seeing speed restrictions but no other driving assistance alerts. The placement, size, visibility, and clarity of the display were well received by all the participants. P1 said, "The visual display was placed correctly and easy to see" (P1PosIn). Based on the combined feedback from the participants, the display works correctly and appropriately. The current information suits the way the screen communicates with the participants. However, slight improvements can be made by making the display adapt better to dark conditions (see driving school feedback), and if additional, more urgent information is displayed, sound warnings could be a beneficial option as well. If the future system wants to provide contextual information for educating the driver on the reactionaries of the speed limit set by the authorities (P2PreIn), a larger display or auditory feedback will be needed.

5.1 Issues that Need to Be Improved

Through this real-world evaluation, our participants reported several issues that affected their driving experiences. We synthesized their repones and the problems they encountered into the following five themes:

1. Comparability Issue: Some participants reported that the start-up process was sometimes slow. For example, P2 mentioned, "It could take up to half an hour for the system to turn on after starting a drive. Sometimes it didn't turn on at all" (P6PosIn). "System frequently does not turn on or drops out for a few minutes while driving" (P1Q2). Another point of interest can be found in the different cars available. P5 indicated how electric vehicles with regenerative capabilities brake significantly when power to the pedal is cut "Electric cars break quite hard when releasing the pedal and regenerating energy" (P5PreIn). Two other participants also mentioned this question (P3Q2, P8Q2). P9 thought, "The over-intervention of the system, not functioning and not thought through for an electric vehicle" (P9Q2). This could also lead to dangerous situations, especially when the system behaves unexpectedly.

2. Misalignment of the Speed Limit Information: In this trial, all participants reported that in some areas, the system's speed limit setting was different from that shown on road signage or in mobile apps, such as Google Maps or Fitsmeister (a Dutch mobile app for assisting driving). For instance, P5 noticed several inaccurate speed limits when he drove around the city, especially when there were multiple roads close to each other and every road had different speed limits. P5 said,"where I live here, a lot of situations were 30, 50, and 80 (speed limit), the parallel roads just very close... On the right side, that's the different road, that is 30, and you are on the 50 roads. The GPS is not precise enough to recognize 30 or 50; it shows 30 instead of 50..." (P5PosIn). P3 reports that "on the highway, it often shows 50 or 70 of the speed limit; fortunately it was on cruise control. Otherwise, it would have reduced the speed suddenly" (P3Q2). Another technical problem is the map accuracy. All participants indicate a need for updated maps, paired with better speed recognition "my advice would be to make sure that the speeds of the system are correct so that the user experience becomes much more pleasant. There will be fewer irritations" (P2Q1). P5 further suggested assessing the speed limit dynamic because "different cars have different dashboards, it's always different in bandwidth - if the 50 was 50, or if it's 47 or 53... so which should not be limited to exactly 50 but allows for a 10% variation" (P5PosIn).

3. The Latency and Unexpected Acceleration while Overruling the System: Some participants reported that the reaction time was not as quick as expected when they pushed the paddle to temporarily deactivate the system. P2 said, "unexpecting situation when you need to overtake other cars, the system's reaction is slow" (P2Q1). During the interview, we also heard that "when I want to overtake a slower vehicle and change to the inner lane, my car can't go beyond the limit quickly. This latency could be dangerous because this delay might increase the chance of being hit by other vehicles" (P2PosIn). On the other hand, P9 reported extreme acceleration caused when overruling the system, "I must kick the system out, which results in an immediate torque of 100% in the vehicle. This, in turn, manifests itself in an extreme acceleration of at least 10km/h extra" (P9Q2).

4. Safety Concerns: In some situations, the restricted speed limit might increase the danger of driving. P9 reported that "Often the car limits itself which leads to severe speed reduction, and on days like today where the road surface is extremely wet, the car has to "pedal" through it leading to slipping tires." (P9Q2).

5. Trust in the System: Most participants did not fully trust the system due to the above four issues. P2 said, "sometimes the speed [limit] given by the system does not correspond with the reality, which is a major issue when trusting the system" (P2Q1). P1 also mentioned, "Trust in the system will only occur when it is always correct" (P1PosIn).

6. Impact on Other Road Users: A recurring theme within the interviews with all participants was that they felt annoyed about being rushed by other road users. Using the system in an isolated environment (meaning only you have the system, not the drivers around you) indicated that participants frequently experienced speeding cars, and they resorted to tailgating and other dangerous driving behavior. For instance, P4 said, "I have experienced people driving extremely close to me, and at times overtaking me on stretches of road where this wasn't allowed" (P4PosIn). This situation sometimes impacts other road users. P6 noted that "it is irritating because you still want to go with the traffic if the speed is lower, then it irritates other road users" (P6Q2). When a truck or aggressive driver pushed a user's car in the back, they felt very uncomfortable (P3PosIn). Sometimes, "it could be dangerous" (P2PreIn) if the other cars behind us do not always notice that we are reducing speed quite rapidly" (P8Q2). To improve communication with other drivers, many participants wished to have a signal on their vehicles to tell others their speed was restricted within the safety range. P2 shared a workaround approach, "when I was driving slower than other cars, and I could not immediately pick up the speed, I needed to press the flash to inform others 'sorry, I cannot speed over 70'" (P2PosIn).

5.2 Findings from the Focus Group Discussion with Five Instructors

The focus group discussion was held on Dec. 8th in the office at Rijschool VOC van Oijen in Helmond. Five instructors participated in this two-hour section, including four full-time driving instructors and one supervisor/team leader who is also a part-time driving school instructor. V-tron installed the ISA-System on four school vehicles in September, and all participants had the experience that students used them during the driving lessons. Throughout the focus-group discussion, the researchers observed and noted strong coherence and support between participants. The main findings and points of interest will be discussed below.

The Benefits and Disadvantages of Having a Mandatory ISA System. We first discuss the European Vision of having the ISA system on all vehicles. One said, "If everyone has a system and the speeds are correct, then yes, I think the biggest irritation is all gone" (FCP5, participant #5 of the focus group session). While their collective knowledge also included driving a truck, which would benefit from a similar system. "The only difference between a truck and a passenger car is (…) the braking deceleration of a truck is not as great as the braking deceleration of a passenger car" (FCP1). According to the two participants who owned a motorcycle license, motorcycles might be the only road-legal vehicle that an ISA-system could not safely fit. One said, "A motorcycle is a balanced vehicle, and suppose I'm in a corner and the gas is reduced in one go, then you have a chance that the vehicle will crash due to an imbalance" (FCP2). When reflecting on their driving and the collective road safety enforced by the ISA system, one participant worried that their driving experiences might be hindered. He said, "from my own point of view, I agree that making ISA systems mandatory would reduce driving pleasure, but from the safety point of view, it would make sense. Those speeds don't count for nothing, do they?" (FCP1).

System Inaccuracies Lead to Decreased Acceptance and Commitment. Similar to the inaccurate problems reported and discussed in the previous section, all instructors shared several observations on the wrong speed limit shown in the display. For instance, FCP3 said, "I wrote down a couple of observations that the system was wrong when there was an overpass or close parallel road." Two participants continued to share their ideas of how it could be improved. They indicated that intelligent systems could be installed in several cars to collect data, and based on group behavior, correct/appropriate speeds on the road could be gathered. "Collecting data, yes, that does not matter a lot to me" (FCP2). "If we can contribute to road safety in Helmond and a better road network, yes, then I would like to contribute to that" (FCP3).

Other unmet requirements hinder their acceptance and commitment to the study. The first one is the moment of activation of the speed limit. FCP1 said, "Our students also need to look ahead. Yes, you see that there is a 50 sign, so you have to release your gas pedal. [...] When you see an increased speed sign, you can already start increasing speed, so you should actually react before the board. Now [the ISA system] only responds two or three seconds after the sign" (FCP1). The other one is the latency while overruling the system. One instructor said, "[A student needs to] be able to overrule the system when it makes mistakes, more directly, so not those two seconds, but immediately" (FCP5). Considering those two unmet needs and the inaccuracy of the speed limit, the instructors only briefly turned on the system for their pupils to experience it for a short time. Then, they turn it off completely during the actual driving lessons. "The students also indicated almost unanimously that they would have preferred systems off" (FCP4). All instructors acknowledged how incorrect speed reductions caused by the system could lead to dangerous situations (FCP3).

They are driving Autonomy and Possible Challenges of Adapting New Assistant System. New driving assistant systems are constantly developed and improved. Most aim to improve driving capabilities and safety. This is something the instructors all confirmed. One example is the navigation systems that display both speed limits as well as current vehicle speed. "It shows ... how fast you are allowed to drive; then it says how fast I drive and the faster I drive it changes color, say from green to orange to red. (...) I always find it [quite] useful" (FCP1). However, they further discussed the effect these systems can have and how they regularly observe their students to fully trust and follow a system instead of basing their decision-making on their observations. "[When using the ISA system] they will no longer pay attention to the signs, then it will decrease road safety because there will be mistakes" (FCP3). This is confirmed by FCP2, who said "Because the students are still too much focussed on watching that screen than reading the speed signs" (FCP5). While observed in many students, this effect is not precisely the same in all drivers. The participants explain how self-confidence changed their reliance on the systems and how easy it was for them to stay in control while driving. "Some students have a little more self-confidence than others and one student can handle this better than the others" (FCP2).

6   Discussions and Design Recommendations

In this research, all participants acknowledged the vision and potential of ISA System in reducing speeding and improving traffic safety. One driving instructor said, "I think almost everyone would like to have the ISA system if it works well" (FCP4). They helped us identify several technical and social issues that need improvement and consideration in further developing the ISA-System into a robust product. Based on their feedback and comments, we further specify some concrete recommendations for technology improvement, infrastructure, and citizen education that the city hall and relevant organizations can collaborate on to improve overall traffic safety by utilizing and gradually introducing the ISA System to prospective drivers.

6.1   Technology Improvement

Improving the Accuracy of the Speed Limit. The product development team needs to thoroughly validate the system's output with the actual speed limit on the road. One approach is to work with the city and follow a well-established test procedure: Link to the document. According to the protocol, manufacturers need to ensure that their system can correctly identify 90% of the speed limits along a 400–500 km test route consisting of a mix of motorways, highways, and urban roads. The instructors from the driving school are pleased to help. Providing them with a method to indicate inaccuracies through, for example, an alert button could help the developers to focus on map areas that could benefit from revision precisely.

Solving the Reaction Time Issues. It includes the slow starting up and the latency while overruling. A typical case worth investigating is the lane-change situations on different roads, such as on low-speed roads or high-speed highways. On the other hand, from the driving school instructors, we learned that there was some delay in showing the information on the display when a speed sign was already visible in the front. This is a crucial requirement for their driving teaching practice.

Re-examining the Speed Control in Different Weather Conditions (such as heavy rain and snow). In our study, we found one case that severe speed reduction might lead to slipping tires in highly wet situations. The technical teams need to test the possible effects of variant weather to ensure the system can function all the time. Whenever there is any uncertainty, the system could automatically deactivate.

Re-evaluating the Existing Design of Overruling by Pushing the Pedal. According to the focus group and the feedback of many interviewees, whether automatic speed limitation by pedal is, the best solution still needs further investigation. Because the system automatically slows down, and the user needs to step on the pedal again to override the speed limit, the user feels that it is not them but the system that controls the car. We propose a new design for installing two steering wheel buttons (see Fig. 4 middle). When a driver wants to overrule the system, he can press both buttons simultaneously and use the pedal to speed seamlessly without latency.

Providing Deactivate Mechanisms for Other Users. Providing an easy function for temporarily disabling the ISA system when the vehicle is used by other users unfamiliar with the system, including other family members or students in the driving school. However, this flexibility should encourage the primary user to refrain from overrule the

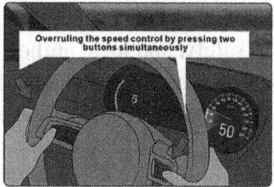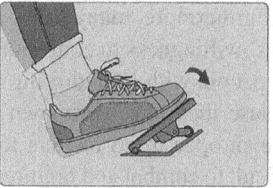

Fig. 4. A scenario of using the new design buttons for overruling the speed control by pressing buttons on the steering wheel.

system frequently. To do so, we propose a concept (see Fig. 5) of providing a monthly quota (e.g., ten times) for users to temporarily deactivate the ISA function for others to use the car without the need to get used to it. We expect that this mechanism could help to increase users' acceptance of the ISA system and eventually to improve road safety.

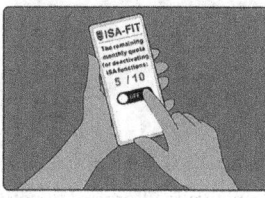

Fig. 5. A scenario of using the mobile application and service to temporarily deactivate the ISA function when the other family member or friend uses the car.

Having a Reasonable Margin of Speed Limit (e.g., 10%). Most participants indicated it is fine and normal if their speed is controlled below 110% of the speed limit. This means that the threshold could be considered to align with the users' mental models, especially when the system is not sure on the correct setting. This will also give some flexibility when the road conditions are empty (e.g., no other visible cars in the front).

Checking and Improving the Comparability with Variant Vehicles. The system is still in development, and unexpected new situations will regularly occur. We recommend exploring the system on different cars, including Electric, Gas, and other less occurring vehicles. Consider these and adapt possible system features to suit these vehicles better.

6.2   Redesigning the Interaction Design to Increase and Promote Acceptance

Drivers with a tendency to speed might often do so consciously and will regularly use an overrule option whenever presented to them. The system could be redesigned in a way that the system can let drivers feel in control and gradually negotiate to take over the control rather than reducing the speed automatically. This strategy could increase users' awareness of speeding and promote their acceptance. We got one valuable idea from participant P1: "I would continue to feel more in control if the system would first notify me visually, followed by an auditory alert before finally regulating my pedal" (P1PostIn). This, however, could work counterproductive, as a certain level of speeding would become easier and more accessible by users.

Furthermore, many participants believe that accepting and using the system requires users' willingness to contribute to safety, and local authorities might need to play a role. For people with long driving experience, it might be harder to fully follow the system because they have their own habits, "You need to re-educate yourself… you need to have an open mind for that to drive according to the system" (P2PosIn). This issue is difficult to combat and requires a governmental solution/approach. The city can consider organizing campaigns or educational programs to let citizens understand and experience the ISA system and gradually increase its adoption.

Based on participants' feedback and suggestions, we propose a design concept (see Fig. 6) of providing visual and auditory feedforward (Chuang et al. 2018; Chuang 2020) to notify the driver of the change in speed limit and explain why the authority set it that way. We hope the additional explanation can reinforce drivers' understanding and action without speeding.

Fig. 6. A scenario of providing informative feedforward to notify the upcoming change of speed limit ahead and explain why the authority set the limit for the area. When the driver reaches the exit of the area, the system will speak out to notify the new change of limit.

7   Conclusion

While ISA systems show potential to reduce speeding and improve driving safety, our findings suggest improving the technology and developing the adaptation campaign from broader perspectives of the community and city. Regarding inaccurate or misalignment issues, overruling should be an easily accessible feature that drivers should use without hesitation. This also applies to over-reliance on being able to drive the speed limit while a slower speed might be needed for safety reasons. Our design proposals utilized multimodal feedback and feedforward for communicating essential information to the driver, especially in newer and, thus, less experienced drivers. If the system was to become mandatory, we recommend regulators first introduce the system as obligatory in driving schools before the system becomes mandatory in all privately owned cars.

Acknowledgments. This paper is part of an activity that has received funding from the European Institute of Innovation and Technology (EIT). We want to thank all participants and the project partners, including Tractebel, the City of Helmond, POLIS, V-tron, CTAG, and TN-ITS.

References

Adell, E.: Driver experience and acceptance of driver support systems: a case of speed adaptation [Unpublished doctoral dissertation]. Lund University (2009)

Bainbridge, L.: Ironies of automation. Automatica **19**(6), 775–779 (1983)

Bakker, S., Niemantsverdriet, K.: The interaction-attention continuum: considering various levels of human attention in interaction design. Int. J. Des. **10**(2), 1–14 (2016)

Chuang, Y.: Designing the expressivity of multiple smart things for intuitive and unobtrusive interactions. In: Proceedings of the conference on designing interactive systems, pp. 2007–2019. ACM (2020). https://doi.org/10.1145/3357236.3395450

Chuang, Y., Chen, L.-L., Liu, Y.: Design vocabulary for human–IoT systems communication. In Proceedings of the SIGCHI conference on human factors in computing systems, no. 274. ACM (2018). https://doi.org/10.1145/3173574.3173848

Compeau, D., Higgins, C.A., Huff, S.: Social cognitive theory and individual reactions to computing technology: a longitudinal study. MIS Q. **23**(2), 145–158 (1999)

Davis, F.D.: Perceived usefulness, perceived ease of use, and user acceptance of information technology. MIS Q. **13**(3), 319–340 (1989)

Davis, F.D., Bagozzi, R.P., Warshaw, P.R.: User acceptance of computer technology: a comparison of two theoretical models. Manage. Sci. **35**(8), 982–1003 (1989)

de Craen, S., de Niet, M.: Extra informatie op matrixborden: Mogelijkheden en effecten. Additional information on matrix boards: Possibilities and effects. SWOV (2002). https://swov.nl/nl/pub licatie/extra-informatie-op-matrixborden-mogelijkheden-en-effecten

Dikmen, M., Burns, C.M.: Autonomous driving in the real world: experiences with tesla autopilot and summon. In: Proceedings of the 8th International Conference on Automotive User Interfaces and Interactive Vehicular Applications, pp. 225–228. ACM (2016). https://doi.org/10.1145/3003715.3005465

Endsley, M.R.: Autonomous driving systems: a preliminary naturalistic study of the Tesla Model S. J. Cogn. Eng. Decis. Making **11**(3), 225–238 (2017)

Höltl, A., Trommer, S.: Driver assistance systems for transport system efficiency: influencing factors on user acceptance. J. Intell. Transport. Syst. **17**(3), 245–254 (2013)

Ingle, S., Phute, M.: Tesla autopilot: semi-autonomous driving, an uptick for future autonomy. Int. Res. J. Eng. Technol. **3**(9), 369–372 (2016)

Janssen, C., Donker, S., Hessels, R., Hooge, I., Kenemans, L., van der Stigchel, S.: De invloed van frequente aanduiding van maximumsnelheid boven de weg op het gedrag van de weggebruiker: Literatuuronderzoek [The influence of frequent indication of speed limits above the road on road user behaviour: Literature study]. Utrecht University (2021). https://dspace.library.uu.nl/handle/1874/415442

Li, J., Xu, W., Zhao, X.: Technology acceptance comparison between on-road dynamic message sign and on-board human machine interface for connected vehicle-based variable speed limit in fog area. J. Intell. Connected Veh. **2**(2), 33–40 (2019)

Md Isa, M.H., Md Deros, B., Abu Kassim, K.A.: A review of empirical studies on user acceptance of driver assistance systems. Global J. Business Soc. Sci. Rev. **3**(3), 39–46 (2015). https://ssrn.com/abstract=3002220

Mitsopoulos, E., Regan, M. A., Haworth, N.: Acceptability of in-vehicle intelligent transport systems to Victorian car drivers (Report No. 02/02). Royal Automobile Club of Victoria (RACV) (2002). https://www.monash.edu/__data/assets/pdf_file/0008/217961/racv-0202-its.pdf

Nidamanuri, J., Nibhanupudi, C., Assfalg, R., Venkataraman, H.: A progressive review-emerging technologies for ADAS driven solutions. IEEE Trans. Intell. Veh. **7**(2), 3122898 (2021). https://doi.org/10.1109/TIV.2021.3122898

Osswald, S., Wurhofer, D., Trösterer, S., Beck, E., Tscheligi, M.: Predicting information technology usage in the car: Towards a car technology acceptance model. In: Proceedings of the 4th international conference on automotive user interfaces and interactive vehicular applications, pp. 51–58. ACM (2012). https://doi.org/10.1145/2390256.2390264

Otway, H.J., Von Winterfeldt, D.: Beyond acceptable risk: on the social acceptability of technologies. Policy Sci. **14**(3), 247–256 (1982)

Pianelli, C., Saad, F.: Acceptabilité du LAVIA. In Erlich, J. (ed.), Carnet de route du LAVIA: Limiteur s'adaptant à la vitesse autorisée, pp. 47–52. PREDIT (2006). https://hal.science/hal-00542631

Rahman, M.M.: Driver acceptance of advanced driver assistance systems and semi-autonomous driving systems [Doctoral dissertation]. Mississippi State University (2016). https://scholarsjunction.msstate.edu/td/1575

Rahman, M.M., Strawderman, L., Lesch, M.F., Horrey, W.J., Babski-Reeves, K., Garrison, T.: Modeling driver acceptance of driver support systems. Accid. Anal. Prev. **121**, 134–147 (2018)

Regan, M., Horberry, T., Stevens, A. (eds.) Driver acceptance of new technology: theory, measurement and optimisation. CRC Press (2014). https://doi.org/10.1201/9781315578132

Rödel, C., Stadler, S., Meschtscherjakov, A., Tscheligi, M.: Towards autonomous cars: the effect of autonomy levels on acceptance and user experience. In: Proceedings of the 6th international conference on automotive user interfaces and interactive vehicular applications, pp, 1–8. ACM (2014). https://doi.org/10.1145/2667317.2667330

Schade, J., Schlag, B.: Acceptability of urban transport pricing strategies. Transport. Res. Part F: Traff. Psychol. Behav. **6**(1), 45–61 (2003). https://doi.org/10.1016/S1369-8478(02)00046-3

Strand, N., Karlsson, I. M., Nilsson, L.: End users' acceptance and use of adaptive cruise control systems. In: Di Bucchianico, G., Vallicelli, A., Stanton, N.A., Landry, S.H. (eds.) Human factors in transportation (pp. 263–274). CRC Press (2016). https://doi.org/10.1201/9781315370460

Uhde, A., Hassenzahl, M.: Towards a better understanding of social acceptability. In Extended abstracts of the SIGCHI conference on human factors in computing systems (Article No. 323). ACM (2021). https://doi.org/10.1145/3411763.3451649

Van Der Laan, J.D., Heino, A., De Waard, D.: A simple procedure for the assessment of acceptance of advanced transport telematics. Transport. Res.: Emerg. Techno. **5**(1), 1 (1997). https://doi.org/10.1016/S0968-090X(96)00025-3

van Huysduynen, H.H., Terken, J., Martens, J.-B., Eggen, B.: Measuring driving styles: a validation of the multidimensional driving style inventory. In: Proceedings of the 7th international conference on automotive user interfaces and interactive vehicular applications, pp. 257–264. ACM (2015). https://doi.org/10.1145/2799250.2799266

Venkatesh, V., Bala, H.: Technology acceptance model 3 and a research agenda on interventions. Decis. Sci. **39**(2), 273–315 (2008)

Venkatesh, V., Davis, F.D.: A theoretical extension of the technology acceptance model: four longitudinal field studies. Manage. Sci. **46**(2), 186–204 (2000)

Venkatesh, V., Morris, M.G., Davis, G.B., Davis, F.D.: User acceptance of information technology: toward a unified view. MIS Q. **27**(3), 425–478 (2003). https://doi.org/10.2307/30036540

Vlassenroot, S., Brookhuis, K., Marchau, V., Witlox, F.: Towards defining a unified concept for the acceptability of intelligent transport systems (ITS): a conceptual analysis based on the case of intelligent speed adaptation (ISA). Transport. Res. Part F: Traff sychol. Behavior **13**(3), 164–178 (2010). https://doi.org/10.1016/j.trf.2010.02.001

Wüstenhagen, R., Wolsink, M., Bürer, M.J.: Social acceptance of renewable energy innovation: an introduction to the concept. Energy Policy **35**(5), 2683–2691 (2007)

Xiao, L., Gao, F.: A comprehensive review of the development of adaptive cruise control systems. Veh. Syst. Dyn. **48**(10), 1167–1192 (2010)

Characterizing and Optimizing Differentially-Private Techniques for High-Utility, Privacy-Preserving Internet-of-Vehicles

Yicun Duan[✉], Junyu Liu, Xiaoxing Ming, Wangkai Jin, Zilin Song, and Xiangjun Peng

User-Centric Computing Group, Ningbo, China
yicunduan@gmail.com

Abstract. Recent developments of advanced Human-Vehicle Interactions rely on *Internet-of-Vehicles (IoV)*, to achieve large-scale communications and synchronizations of data in practice. *IoV* is highly similar to a distributed system, where each vehicle is considered as a node and all nodes are grouped with a centralized server. In this manner, concerns of data privacy are rising, since privacy leak possibly occurs when all vehicles collect, process and share personal statistics (e.g. driver's heart rate, skin conductance and etc.). Therefore, it's important to understand how to efficiently apply modern privacy-preserving techniques on *IoV*. In this work, we first present a comprehensive study to characterize modern privacy-preserving techniques for *IoV*, and then propose a *Differential Privacy(DP)* privacy-protection framework specialized for unique characteristics of *IoV*. Our characterization focuses on *DP*, a representative set of mathematically-guaranteed mechanisms for both privacy-preserving processing and sharing of sensitive data. It demystifies the tradeoffs of deploying *DP* techniques, in terms of service quality and privacy-preserving effects. The lessons learned from characterization reveal the importance of data utility in *DP*-protected *IoV* and motivate us to examine new opportunities. To better balance tradeoffs and improve service quality, we introduce HUT, for high-utility, batched queries under *DP*-protection on IoV. We quantitatively examine the benefits of HUT, and experimentally show that, in an *IoV* context, HUT can reduce information loss by 95.69% while enabling strong mathematically-guaranteed protection over sensitive data. Based on our characterization and optimizations, we identify key challenges and opportunities for future studies, to enable privacy-preserving *IoV* with low service quality degradation.

Keywords: Differential Privacy · Internet of Vehicles · Human-Vehicle Interaction

1 Introduction

The technical evolution from *Vehicle Ad-hoc Networks (VANETs)* to *Internet-of-Vehicles (IoV)* makes possible the frequent, large-scale and efficient communi-

H. Krömker (Ed.): HCII 2023, LNCS 14048, pp. 31–50, 2023.
https://doi.org/10.1007/978-3-031-35678-0_3

cations and synchronizations of data in practice. Unlike *VANETs*, *IoV* views multiple vehicles as nodes within a distributed system, where groups of nodes are powered by a centralized server with abundant computation resources to enable diverse vehicle/server-side services, such as transportation scheduling [5,7], healthcare applications [15], lane-changing warnings [20,27], and driver statistics analysis [4,16,31,33,35,38,39]. However, *IoV* also significantly raises the concerns of privacy, due to centralized data sharing and processing. For instance, computation-intensive tasks (e.g., DNN-based applications [41]) are usually outsourced to centralized servers where computation resources are abundant, which results in sensitive data to be exposed outside vehicles during data sharing. The design pre-requisites of several key applications (e.g., vehicle accident identification [25]) demand centralized processing of sensitive statistics collected from all vehicles on *IoV*, and this may lead to privacy leaks when the centralized servers are attacked.

Therefore, it's important to explore and exploit privacy-preserving techniques in the era of *IoV*. However, since such topic lacks enough attention in previous research, it's still unclear how modern privacy-preserving techniques perform in the context of *IoV*. In this work, we focus on evaluating and optimizing *Differential Privacy* (*DP*), a set of mathematically-guaranteed mechanisms for privacy-preserving processing and sharing of sensitive data, on *IoV*. We achieve this goal through two coherent steps: First, we investigate and characterize the pros and cons of four representative *DP* mechanisms for processing sensitive data on *IoV*. Second, we leverage newly-revealed insights, from our characterization, to optimize *DP*-protected algorithms for high-utility *IoV* (i.e., *IoV* with high service quality).

To conduct characterization, we leverage a platform [18,26] to simulate *IoV* environment, where we test four state-of-the-art *DP* mechanisms. In our emulation, we reproduce the classic vehicle-server interaction pattern in *IoV*: simulated centralized servers perform queries to retrieve sensitive data from vehicle nodes, and then conduct data processing. For our testing on this emulation platform, we select four mechanisms for *DP*-enabled processing, and evaluate their performance on representative *IoV* workloads. We characterize the effects of these techniques, from the aspects of privacy-preserving effects and service quality. Based on our characterization results, we identify the most significant bottleneck of providing *DP*-enabled *IoV* services: poor data utility under strong protection. Therefore, our subsequent optimization for *DP* on *IoV* aims to address such issue, by improving data utility while remaining privacy-preserving protection.

To this end, we propose HUT, a novel approach to enable High-UTility, batched queries under *DP* protection for *IoV*. The key insight of HUT is to synergistically combine unique characteristics of *IoV*, where queries within *IoV* are usually unbalanced and batched, with a carefully-redesigned *DP*-enabled processing algorithm to improve data utility. Our HUT algorithm has two advantages, compared with prior approaches. First, HUT is the first algorithmic design for *DP*-protected *IoV*, which brings huge benefits on data utility in *IoV*. Second, HUT doesn't bring extra overheads in terms of mathematically-guaranteed foundation for DP protection, and therefore ensure the strong protection of the sensitive data. We further quantitatively examine the benefits of HUT, and the results show that: HUT can efficiently reduce information loss by 95.69%/71.71% for simple/counting query on *IoV*. To the best of our knowledge, HUT is the first work to balance tradeoffs between data utility and privacy protection in an *IoV* context.

We summarize our contributions as follows:

- We identify the privacy violation threats within *IoV*, and introduce the *DP* mechanism to realize privacy-preserving *IoV* in practice.
- We characterize *DP*-enabled data processing and sharing on sensitive data, in the context of *IoV*. We investigate, implement and evaluate state-of-the-art *DP* mechanisms over *IoV*.
- We propose HUT, a novel approach achieving High-UTility, batched queries under *DP*-enabled protection for *IoV*, by synergistically combine unique characteristics of *IoV* and *DP*.
- We design experiments to verify the outstanding service quality improvement brought about by HUT. We also point out possible opportunities for future studies, by enabling privacy-preserving *IoV* with low overheads for service quality.

The rest of this paper is organized as follows. Section 2 introduces the background and motivation of our work. Section 3 elaborates detailed characterization methodologies, and comprehensively analyzes the results. Section 4 demonstrates our proposed novel *DP*-enabled algorithm HUT and verifies the data utility enhancement through experiments. Section 5 discusses several takeaways and future work.

2 Background and Motivation

In this section, we introduce background and motivation of this work. We first outline the concept and application of *Internet-of-Vehicles* (*IoV*) (Sect. 2.1). Next, we introduce *Differential Privacy* (*DP*), a series of privacy-preserving mechanisms (Sect. 2.2). Finally, we identify our motivation and goal in this work (Sect. 2.3).

2.1 The Concept and Application of *Internet-of-Vehicles*

Internet-of-Vehicles (*IoV*) is a large-scale connected network for efficient cooperation among humans, vehicles, things, and environments. As shown in Fig. 1, all agents, including humans, vehicles, and things, connect with each other on *IoV*. The vehicle is the primary unit of a complicated agent. All interactions from other agents (e.g. humans, things and etc.) have to go through the interfaces from vehicles. The centralized servers (i.e., one part of the "Thing" in Fig. 1) assist the vehicles to provide computation/memory-intensive services, such as

AI jobs and vehicle accidents identification. For AI jobs like object detection [41], their high requirements of parallel computing resources exceed the capability of most in-vehicle computers, which makes it necessary to offload such tasks to centralized servers through data sharing. The other applications like vehicle accidents identification [25] usually perform data analysis on statistics collected from multiple vehicles, and, therefore, necessitate the centralized data gathering and processing. However, such behaviors accompany with the high risk of privacy disclosure, since the data transformation process and centralized servers are highly likely to become the targets of adversarial attacks.

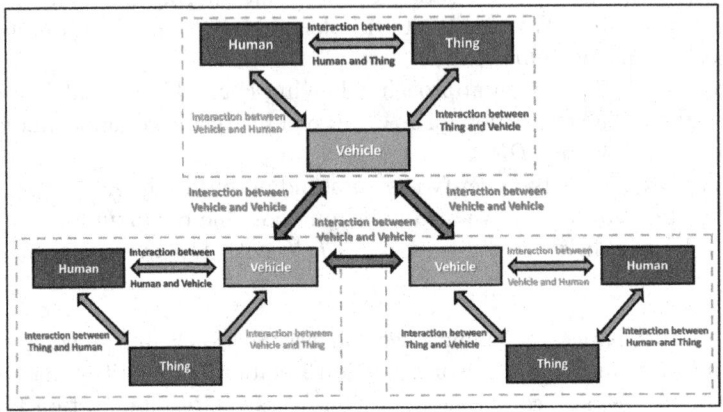

Fig. 1. The architecture of *Internet-of-Vehicles*: (1) Human: people in vehicles and other people in the context of *IoV*, such as pedestrians; (2) Vehicle: all vehicles that consume or provide services/applications of *IoV*; (3) Thing: anything except human and vehicles, such as roads and remote servers;

2.2 *Differential Privacy*

Differential Privacy (*DP*) is a series of general privacy-preserving techniques, which are mathematically guaranteed for privacy protection effects. *DP* can ensure the statistical analysis over gross dataset will not hint the truth of any individual person's privacy, which is formally described as the following definition:

Definition 1. *(ϵ-DP) A randomized function K_f satisfies ϵ-DP, if $\forall$ dataset D_1 and D_2 with at most one element difference, and $\forall S \subseteq Range(K_f)$, the K_f function obeys the below in-equation:*
$$Pr[K_f(D_1) \in S] \leq exp(\epsilon) \times Pr[K_f(D_2) \in S]$$

In practice, we process data by adding noise to realize ϵ-DP. The scaling factor of noise is equal to the fraction $\Delta f(D)/\epsilon$ (where D is the dataset and f

is a query function). We describe the detailed parameter settings in $\Delta f(D)/\epsilon$ as follows: ❶ The parameter ϵ should be manually selected according to the anticipated privacy protection degree. The smaller the ϵ is, the more distorted the post-processed data are; ❷ The value of parameter $\Delta f(D)$ (namely $L1 - sensitivity$) is determined by the query function f.

The reason why we select DP as the foundation of our privacy-preserving IoV is that: DP is task-agnostic, which means it can protect both data sharing and processing process on IoV. This feature makes DP stand out from other task-specific privacy-preserving mechanisms, such as Oblivious Inference [6,12] and Authentication [2,10,21]. Therefore, we believe DP is an ideal candidate for privacy protection on IoV, and it's worthy the efforts to characterize its tradeoffs in detail.

2.3 Motivation and Goal

Although DP possesses a wide application range (e.g., Online Learning [1], Streaming Algorithm Design [13] and Randomized Response Algorithm [11]), credited to its task-agnostic feature, it's still unclear how DP can introduce tradeoffs in the context of IoV. Therefore, our first goal is to investigate representative mechanisms of DP and experimentally characterize its advantages and disadvantages in the context of IoV. Then, as our second goal, we aim to design and develop new method to better balance the tradeoffs by synergistically combine IoV with DP.

3 Characterizing DP-Enabled Protection of Sensitive Data on IoV

For DP-enabled data sharing and processing on IoV, centralized servers perform DP-protected queries to retrieve sensitive data from vehicle nodes, and then execute computation without privacy violation concerns, due to DP's strong theoretical guarantee. In our characterization for this process, we focus on the influence brought about by the DP mechanisms, from the aspects of service quality (measured by relative/absolute error) and privacy-preserving effects (denoted as coefficient ϵ). To realize our characterization, we first establish a simulated IoV environment, where we emulate the basic vehicle-server interaction pattern in IoV. Then, we select four state-of-the-art DP-enabled processing mechanisms which are to be tested on our simulated IoV. Next, we elaborate the IoV query/workload types covered by our experiments while introducing how we setup the dataset, measure the query error and select ϵ/query structure. Finally, we characterize the effects of these techniques and analyze the results comprehensively.

3.1 Emulating IoV

We emulate IoV using the APIs provided by [18], a general-purpose and portable emulation platform specialized for Human-Vehicle Interaction applications on

IoV, to reproduce the vehicle-server interaction pattern in real-world *IoV*. In this simulated *IoV*, vehicles serve as data holders and centralized servers run Human-Vehicle Interaction applications. The virtual network connections between them enable centralized servers to query data from vehicles and feed the retrieved data into running applications. For a *DP*-enabled case, the queries on vehicles are orchestrated and interpolated by *DP* mechanisms, which provides strong and robust protection against adversarial attacks on data sharing and processing. The central aim of our characterization is to measure the trade-offs between privacy protection effects and service quality during *DP*-enabled protection.

3.2   Four Representative *DP* Mechanisms

In this section, we cover four state-of-the-art mechanisms for enabling *DP*-protected queries, which we would like to test in the context of *IoV*. We denote these mechanisms using words reflecting their key ideas, which are "Fourier", "Wavelet", "DataCube" and "Hierarchical".

Fourier. The design purpose of Fourier is specifically for workloads having all K-way marginals, for given K [3]. This mechanism transforms the cell counts, via the Fourier transformation, and obtains the values of the marginals based on the Fourier parameters. If the workload is not full rank, there are eliminations of the unnecessary queries from the Fourier basis, in order to reduce the sensitivity.

Wavelet. The design goal of Wavelet is to support multi-dimensional range workloads. This mechanism applies the Haar wavelet transformation to each dimension [37]. Note that Wavelet is expected to be applied in a hybrid manner, if using ϵ-differential privacy. Such a hybrid algorithm depends on the number of distinctive values in the selected dimension: when the number is small, the identity strategy is applied to improve Wavelet's performance.

DataCube. The target applicability of DabaCube is to adaptably support marginal workloads [8]. To provide deterministic comparison point, the BMAX algorithm is applied in this case. This algorithm chooses a subset of input marginals, to minimize the maximum errors of the answer to the input workloads. DataCube is naturally adaptable with $(\epsilon, 0)$-differential privacy.

Hierarchical. The aim of Hierarchical [14] is to answer workloads of range queries. The key idea is to use a binary tree structure of queries: (1) the first query is the sum of all cells; and (2) the rest of the queries recursively divide the first query into subsets. As the representative comparison point, we apply binary hierarchical strategies (although higher orders are possible). The strategy in [14] supports one-dimensional range workloads, but is capable to be adapted to multiple dimensions, in a similar manner to Wavelet [37].

3.3   Experimental Settings

In this section, the experimental settings of our characterization are demystified. We first discuss how to setup BROOK dataset [17,23,29] to fit the requirement of testing DP-protected queries. Then, we introduce the workload/query types of our testing (i.e., all-range query and one-way marginal query). Subsequently, we demonstrate how we utilize absolute error and relative error to measure the information loss in our testing queries. In the end, we disclose our ϵ and query structure configurations.

Dataset Setup. We leverage BROOK dataset [29], a multi-modal and facial video dataset for data-driven Human-Vehicle Interaction, as the source of driving statistics in our IoV simulation (supported by [30,36]. There are 11 dimensions[1] of data in the current BROOK dataset, among which we select two dimensions, Heart Rates and Skin Conductivity, of data to be deployed on our simulated vehicles. The vehicle-side data are reformulated as column vectors for the convenience of linear queries.

All-Range Query and One-Way Marginal Query. We choose all-range query and one-way marginal query as our testing workloads, since they are commonly-used linear queries in DP-enabled protection on IoV [40]. All-range query refers to the sum of all elements in input vector. For instance, the all-range query result for a vector $(x_0, x_1, \cdots, x_n)^T$ would be $\sum x_i$, $i \in \{0, 1, \cdots, n\}$. One-way marginal query is the counting sum along a certain dimension of input vector. We take the Heart-Rates-to-Skin-Conductivity vector as example. The one-way marginal query on it along "Heart Rates $= A$" would be $\sum x_i$, where A is a constant and index i refers to the element whose Heart Rates attribute is equal to A.

Absolute Error and Relative Error. We use absolute error and relative error to comprehensively measure the information loss in our queries. Absolute error can be formally expressed as: $\|Q(\mathbf{x}') - Q(\mathbf{x})\|$, where $Q(\cdot)$ is the query function, $\mathbf{x}$ is the original data vector and vector $\mathbf{x}'$ contains DP-processed data. In a similar way, we can define the relative error as: $\frac{\|Q(\mathbf{x}') - Q(\mathbf{x})\|}{\|Q(\mathbf{x})\|}$.

ϵ Value and Query Structure Selection. We follow the practice of representative DP-protected querying experiments [22] to select the ϵ values and query structures used in our characterization. ϵ values directly reflect the privacy protection extent of DP mechanisms. When the ϵ is close to 0.0, the DP-protected queries provide the robustest protection. However, smaller ϵ gives rise to additional noise and degrades the utility of data. For choosing ϵ values, we pick

---

[1] 11 dimensions include Facial Video, Vehicle Speed, Vehicle Acceleration, Vehicle Coordinate, Distance of Vehicle Ahead, Steering Wheel Coordinates, Throttle Status, Brake Status, Heart Rate, Skin Conductance, and Eye Tracking.

five frequently-used ϵ settings, which are 0.1, 0.2, 0.5. 1.0 and 2.5, with guaranteed privacy protection effects gradually degrading. Query structure refers to the shape of data vector on which we perform DP-protected/unprotected queries. The granularity of query structure is proportional to its shape's complexity. We say a vector of shape 2^5 is more fine-grained than a 4×8 vector, since the 2^5 vector possesses more internal clusters. These clusters are hard-coded configurations for some DP mechanisms like Wavelet [37]. For our experiments, we perform even division over our 32-dimension inputs to obtain 4×8, $4 \times 4 \times 2$, $4 \times 2 \times 2 \times 2$ and 2^5 query structures.

3.4   Characterization Results

We use four DP mechanisms (i.e., Fourier, Wavelet, DataCube and Hierarchical) to protect the data sharing and processing on our simulated IoV, and accumulate the testing results shown in Figs. 2 and 3. We summarize seven key observations, according to these results.

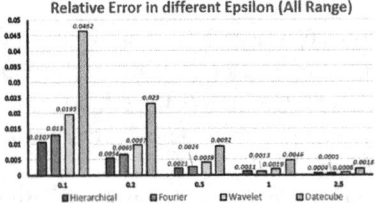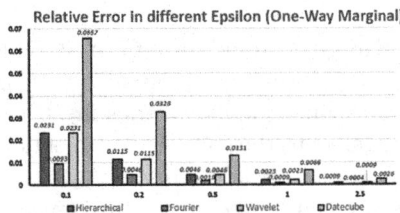

Fig. 2. Relative Error of Four DP-protected Query Methods: We evaluate the relative error of DP mechanisms over different ϵ values, with the query structure fixed to 32. The values lying on horizontal axis are the ϵ values controlling the privacy protection degree. A smaller ϵ will lead to a better protection effect.

Key Observation 1: The relative error has a negative correlation with ϵ values. The Fig. 2 demonstrates that all the four DP methods reach their highest relative error when the smallest ϵ value (i.e., 0.1) is set. The relative error degrades as the ϵ increases.

Key Observation 2: DataCube lags behind all the other three DP methods, regardless of ϵ settings. As shown in Fig. 2, Hierarchical, Fourier and Wavelet outperform DataCube by a large marginal in all cases. DataCube might be an unsatisfactory choice for protecting all-range query and one-way marginal query.

Key Observation 3: Under different ϵ settings, Fourier performs best for one-way marginal query, while Hierarchical overwhelms it on the testing of all-range query. Figure 2 reports that Fourier achieves $2 - 3\times$ less relative error for one-way marginal query, compared with Hierarchical. But this advantage vanishes when it comes to all-range query testing. On all-range query, Hierarchical outperforms Fourier by a marginal of about 20%.

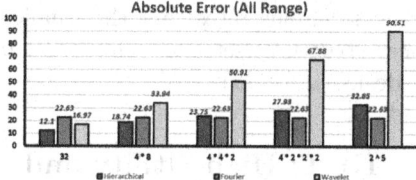 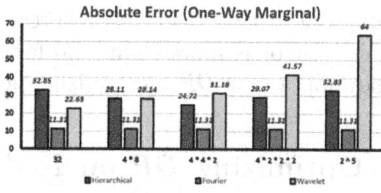

Fig. 3. Absolute Error of Three DP-protected Query Methods: We test the absolute error of DP mechanisms over various granularity of query structure, while keeping the ϵ to be 0.5. The horizontal axis displays the granularity of query structure. The DataCube [8] is not shown, because DataCube follows a set of specific cluster separation regulations which contradicts our cluster divisions.

Key Observation 4: Under various query structure configurations, the variation patterns of four DP methods' absolute errors differ with each other. As shown in Fig. 3, the absolute errors of Fourier remain the same no matter how we change the query structures. For all-range query, the absolute errors grow as the query structures become more fine-grained, except for Fourier. Under one-way marginal query testing, Hierarchical's absolute errors first decrease and then recover back to the original level, but, simultaneously, Wavelet suffers a consistent absolute error increase.

Key Observation 5: Fourier outperforms hierarchical and wavelet for one-way marginal query, under any query structure setting. Figure 3 depicts the scene that, for all query structure settings, Fourier is far ahead of other methods in terms of absolute errors.

Key Observation 6: There doesn't exist an optimal DP mechanism to protect all-range query on different query structures. Figure 3 illustrates that, for fine-grained query structures such as $4 \times 4 \times 2$, $4 \times 2 \times 2 \times 2$ and 2^5, Fourier achieves lower absolute errors for all-range query than Hierarchical does. However, Hierarchical beats Fourier by performing better on coarse-grained query structures like 32 and 4×8.

Key Observation 7: All *DP* protection mechanisms show low data utility when strong protection is applied, and there are urgent needs to deliver high-utility *DP* without affecting the extent of the overall protection. As shown in Fig. 2, all the four *DP* mechanisms we test can't ensure data utility (low relative error) when the protection degree is high (i.e., ϵ value is 0.1). Even though Fourier reports the smallest relative error (0.0093), it still can't be applied in practical *IoV* scenarios, since the batched query will accumulate the errors to reach an unacceptable level. Besides, this issue can't be addressed through adjusting query structures. The Fig. 3 demonstrates that the smallest absolute error is more than 10.0 for all query structures settings. This urges us to develop a new *DP* mechanism specially designed for *IoV*.

4   Optimizing *DP* on *IoV* with HUT: High Utility and Strong Protection

To balance the tradeoffs of data utility and privacy protection, we propose HUT, a new algorithm to ensure high utility within *IoV* while the strong protection on sensitive data is still guaranteed. Though there are several works on improving data utility under *DP* protection, they fail to incorporate the unique characteristics of *IoV* [14,32,34]. Based on our characterization and recent studies on *IoV* [19], we reveal that a unique and significant characteristic of *IoV* is that, query sequences are usually formed as unbalanced batches in practice. These batches are formed due to a practical issue: vehicles require frequent queries of relevant data for decision making in time. These unbalanced batches have different levels of sensitivities that can incur low data utility on small-value batches [14].

We first compare the state-of-the-art mechanisms aiming to balance the tradeoffs between data utility and privacy protection. We then give an overview of our algorithm design and discuss the novelty and privacy protection effects. Finally, we show the experiments designs and results of the decreased information loss compared to other state-of-art *DP* algorithms.

4.1   Prior Attempts for Data Utility Under *DP* Protection

Prior works aim to balance the tradeoffs between data utility and privacy protection, but fail to address the issues of batches within *IoV*. [14] introduces Constraint Inference, which applies a constraint on the order of a sorted data set to decrease unnecessary noise; [32] introduces Micro-Aggregation, which uses a fixed-size cluster to cluster data set before adding noise, to reduce the information loss; and [34] introduces K-aggregation, which aggregates small-value data to form new data when the summation of small-value data falls in a certain threshold. The ignorance prohibits *DP* to be applied effectively in the context of *IoV*, which demands new insights for bridging *DP* with *IoV*.

4.2   HUT Design Overview

We give an overview of our HUT design for ensuring *DP*-protection with high data utility. HUT consists of three steps: 1) we leverage Micro-aggregation and K-means algorithm to group small-value batches together, so that we can mitigate the negative effects of uniform DP protection; 2) we deploy *DP* to preserve privacy of the pre-processed data; and 3) we exploit Order Constraint (OC) to re-sort DP-protected data, so that we can achieve high data utility. Figure 4 compares HUT with two other state-of-the art *DP* protection methods, and we elaborate the differences in details as follows.

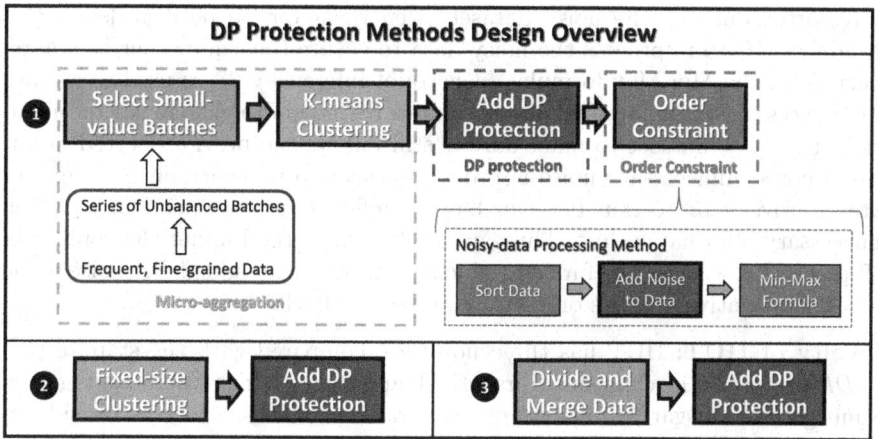

Fig. 4. An illustration of three *DP*-protection methods. ❶ refers to HUT (our proposal), in which the details of the noisy-data processing method(OC) [14] is illustrated; Two state-of-the-art methods are illustrated (❷ [32] and ❸ [34]).

1) **Micro-Aggregation:** HUT first apply Micro-Aggregation to cluster small-value raw data, so that we can mitigate the negative impacts from *DP* in terms of data utility. This is because *DP* adds uniform random noises to the dataset, and small-value data will have lower Signal-to-Noise Ratio (SNR), compared with large-value data. Thus, the data utility is impacted dependently according to its exact values. Therefore, a mechanism to mitigate such impacts is essential. We achieve so as follows. Before directly using *DP* techniques on raw data for protection, HUT sets a threshold to divide small-value data and the rest into different batches (as To-be-clustered batches and Not-clustered batches). Next, we apply K-Mean clustering on data within the threshold (the To-be-clustered batches), to automatically group such small-value data further into smaller intra-batches (i.e. Micro-Aggregation).

Compared with [32], HUT effectively reduces the sensitivity and variations of small-value data as they are upper bounded by the centroid of the respective cluster. In [32] (shown as ❷ in Fig. 4), a fixed-size clustering method is applied and this results in a consistent distance function on a total order relation, but incurs large bias on dataset which has highly uneven data distribution. Such bias on uneven distributed data could harm the integrity of dataset, leading to additional information loss. Thus, on the contrary, we leverage K-means clustering algorithm to perform the clustering to mitigate the bias.

2) **DP Protection:** we then add Laplace noise on the data to achieve ϵ-*DP*. The details of *DP* protection is introduced in Sect. 2.

3) **Order Constraint:** HUT finally utilizes a noisy-data processing method, Order Constraint(OC), to minimize unnecessary noise. A general feature of noisy-data is that after adding noise (especially large-scale noise) on a previously sorted dataset, the noisy dataset often turns out to be disordered. [14] introduces OC to re-process the noisy data to ensure the same order as before. Specifically, the Min-Max formulas (constraint inference) [28] is used to remove the unnecessary noise, applying least squares regression to change the data values in the noisy dataset to make sure it still follows the previous sorted order. The removed unnecessary noise is proved to have no influence on the effect of dataset privacy protection but can have significant improvement on reducing unnecessary information loss. Thus, we apply this method under the context of *DP* protection as a step to further reducing unnecessarily-added *DP* noise while having no negative impacts on privacy protection itself.

Novelty of HUT: HUT has three novelties, compared with the state-of-the-art *DP* protection mechanisms. First, HUT applies two-stage Micro-aggregation to mitigate the negative impacts on fine-grained batches, which is considered as a unique characteristic of *IoV*; Second, HUT leverages K-Means Clustering, rather than fixed-size clustering, to improve the utility of protected data; and third, HUT selectively enables lightweight constraints to be synergistic with other designs, for both DP protection and data utility.

No degradation on *DP* privacy protection: In short, there are no negative impacts on *DP* protection when applying HUT: since *DP* protection and K-means clustering are not conflict with the mathematical guarantee of *DP*, we focus on Micro-Aggregation algorithm. First, Micro-Aggregation in HUT does not impact the *DP* protection since K-Means Clustering constrains only small-valued data; and second, Micro-Aggregation in HUT does not change the data, and therefore no negative impacts are incurred on *DP* protection.

4.3 Evaluation Setup

- **Selected Dataset:** We use similar methodology as described in Sect. 3.3, with the following considerations in addition. In our experiments, we specifically handcraft a dataset that contains drivers' facial images and the corresponding vehicle speeds, as a proof-of-concept. Figure 5 demonstrates an example of our handcrafted dataset.

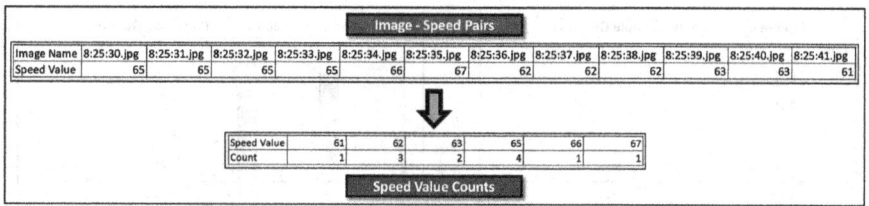

Fig. 5. A demo of the two different data set settings. The first situation is querying speed values from the Image-Speed Pair which simulates Simple Query, while the second special situation is querying the counting numbers from Speed Value Counts, which simulates Counting Query.

- **Configurations of Proposed Methods:** There are three key parameters for our method, the scale of DP ϵ, the threshold p and the number of clusters k for Micro-Aggregation. We testify our design in various parameter settings, including $\epsilon \in \{0.008, 0.01, 0.02, 0.05\}$, $k \in \{5, 8, 10, 15\}$ and $p \in \{30\%, 35\%, 40\%\}$, in which $k \in \{5, 8, 10\}$ are used for simple queries and $k \in \{5, 10, 15\}$ are used for counting queries to satisfy different data set configurations. We conduct 20 experiments for each parameter setting and take the mean as the final value of information loss.

- **Query Settings:** We examine the data utility of DP-protected dataset mainly on one-way marginal queries, since the majority of queries incurred in IoV are formed as unbalanced batches. For more detailed study, we rigorously divide one-way marginal queries into two representative types. The first query type is a simple query with facial images as input, and query for the speed data in our handcrafted dataset. These queries simulate the unit-length querying process for intelligent in-vehicle systems to continuously retrieve data for decision-making in a fine granularity.
  The second type of query is counting query, which simulates the statistical analysis function of in-vehicle systems. The counting query is similar to query data from a contingency table (the bottom image in Fig. 5). Different from the first general situation, this one already has a fixed sensitivity of 1, making it more challenging for HUT because sensitivity reduction is limited.

- **Evaluation Metric:** We use the Mean Square Error (MSE), extended from the relative error in Sect. 3.3, as the evaluation metric of data utility and information loss. We denote the query response on DP-protected data as d', and the query response on raw data as d. MSE is calculated by equation $\text{MSE} = \frac{1}{n} \sum_{i=1}^{n} (d'_i - d_i)^2$.

4.4 Evaluation Results

We compare HUT with the state-of-the-art DP algorithms to examine its effectiveness of data utility. Since the average information loss of method ❷ and ❸

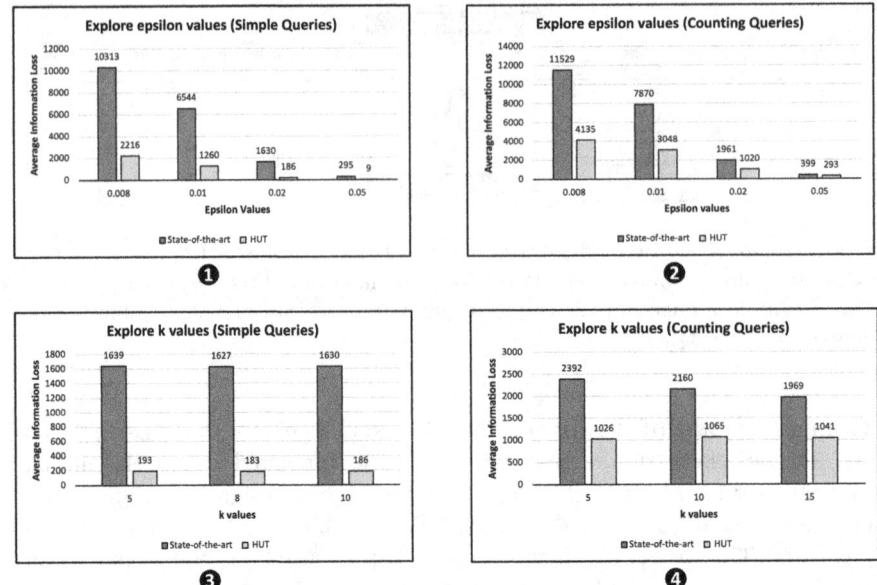

Fig. 6. Experiment results of HUT comparing with other state-of-the-art algorithms. ❶ shows the average information loss of different epsilon values for simple queries (when k = 10, threshold = 35%). ❷ shows the average information loss of different epsilon values for counting queries (when k = 10, threshold = 35%). ❸ shows the average information loss of different k values for simple queries (when epsilon = 0.02, threshold = 35%). ❹ shows the average information loss of different k values for counting queries (when epsilon = 0.02, threshold = 35%).

in Fig. 4 are highly similar under all experiment conditions, we select the overall best performer ❸ to compare the information loss with HUT.

The results of simple query and counting query in different ϵ values with fixed k clusters and threshold portions are presented as ❶ in Fig. 6 and ❷ in Fig. 6. Both figures prove the effectiveness of Micro-aggregation in increasing SNR for small-value data and mitigating the negative effects of DP protections, in a common setting (simple query) and challenging setting (counting query). For simple query, our method achieves maximum 98% (when ϵ=0.008) and minimum 79% (when ϵ=0.05) reduction information loss reduction compared to the state-of-the-art, with k=10 and 30% threshold. For counting query, the max information loss reduction percentage is 65% (when ϵ=0.008) and the minimum percentage is 27% (when ϵ=0.05). We note that when the noise magnitude grows (i.e. ϵ values decreases), the performance of our algorithm degrades in simple query, but improves in counting query. The opposite trend in two query settings proves two things: 1) Applying DP on raw data could significantly harm normal data utility, as simple query, a primary indexing operation could cause relatively severe information loss; 2) Aggregation, as one kind of counting operations, could effectively avoid over-protection in extreme conditions, but is less effective when data is not over-protected.

❸ in Fig. 6 and ❹ in Fig. 6 presents the information loss of simple query and counting query in different K settings, when ϵ and threshold are fixed at 0.02,35%. We obtain the key observations that an effective choice of number of clusters for K-mean clustering fall in the range between 5 and 15, which have the optimal information loss reduction. In addition, our empirical study shows that any reasonable threshold values would incur negligible perturbations on the information loss, as different threshold values cause maximum 1% variance for simple query and 6% for counting query.

Table 1 reports the best percentage of average information loss compared to the state-of-the-art mechanism in both simple query and counting query for each ϵ value, when the number of clusters k and threshold for aggregation varies for each selected ϵ. We believe exploiting the mitigation of DP's negative effects in different ϵ settings is essential, as a small scale of ϵ changes could cause relative large fluctuations in the information loss.

Table 1. Best information loss reduction on the state-of-the-art mechanism in different ϵ settings

	Epsilon	K	Threshold	% of reduction
Simple Query	0.008	10	30%	78.51%
	0.01	8	30%	80.74%
	0.02	8	30%	88.75%
	0.05	5	30%	**95.69%**
Counting Query	0.008	10	35%	64.13%
	0.01	5	30%	**71.71%**
	0.02	5	35%	45.22%
	0.05	10	40%	50.46%

5 Discussions

In this section, we summarize our key observations from characterizations of *DP*-enabled mechanisms and key takeaways from our proposed HUT approach. We first describe our most significant observations and conclusions in detail (Sect. 5.1). Then we elaborate our takeaways and summaries of our proposed HUT approach (Sect. 5.2). In addition, we suggest both challenges and opportunities for privacy-preserving *IoV* (Sect. 5.3).

5.1 Major Observations from *DP* Mechanisms Characterization

We summarize the main conclusions made from our characterization of four *DP*-enabled mechanisms. This conclusion reveal one of most significant challenges of applying *DP*-enabled mechanism in *IoV*: **All *DP* protection mechanisms show possibly insufficient data utility when strong protection is applied, and there are no one-size-fits-all mechanisms of *DP* for all applications of *IoV*.**

5.2   Takeaways and Summaries from Our Proposed HUT Approach

We present HUT, a new algorithm to enable H̲igh U̲T̲ility for *DP*-enabled protection in the context of *IoV*. Our key insight is to take advantage of the fact that, the interactions between centralized servers and edge vehicles are usually frequent and fine-grained, which are combined into a series of unbalanced batches. We evaluate the effectiveness of HUT against the state-of-the-art *DP* protection mechanism, and the results show that HUT can provide much lower information loss by 95.69% and simultaneously enable strong mathematically-guaranteed protection of sensitive data.

Our future works aim to explore the scalability and robustness of HUT. For scalability, we expect to examine whether HUT can be applied for data with broad range or unevenly distributed dataset; as for robustness, HUT still suffers from the instability of K-mean clustering method, which can produce query results with relatively large variance. Therefore, we can expect a tradeoff between different clustering algorithms, in terms of the robustness of HUT.

5.3   Challenges and Opportunities for Privacy-Preserving *IoV*

First, since there are no one-size-fits-all solution for privacy-preserving *IoV*, by using *DP*. We believe there are a large amount of opportunities to explore privacy-preserving techniques for *IoV* in practice. Second, for data processing on sensitive data, we already narrow the large design scope into a restricted region, by examining the value of ϵ in diverse settings. We believe this is very useful for follow-up studies, to avoid redundant workloads of attempted settings. Third, for data sharing on sensitive data, we already demonstrate that application-specific *DP* protections can potentially destroy the service quality. Therefore, we believe there are a huge amount of possibilities to co-design the application and privacy-preserving mechanisms, to ensure high service quality with privacy protection at the same time.

6   Conclusions

In this work, we characterize four state-of-the-art *DP* mechanisms on *IoV*, identify their drawbacks, and propose a new *DP*-protection method, HUT, specialized for *IoV*. Our characterization ensures that there is not a one-size-fits-all *DP* mechanism in the context of *IoV*. All the previous methods don't consider the frequently-used batched queries in *IoV*, and impose overdose noise on the protected data, which considerably degrades the data utility. To address this issue, we introduce HUT, for high-utility, batched queries under *DP*-protection for

IoV. Credited to HUT's special fine-tune for *IoV*'s characteristics, our newly-proposed approach can not only preserve privacy protection degree but also improve data utility. Our results show that HUT can maximally reduce information loss by 95.69% for simple query on *IoV*.

Acknowledgements. We thank the anonymous reviewers from AutomotiveUI 2022 and HCI 2023 for their valuable feedback. We thank all members from User-Centric Computing Group for their valuable feedback and discussions. Earlier versions of this work was released at [9,24].

References

1. Abernethy, J.D., Jung, Y.H., Lee, C., McMillan, A., Tewari, A.: Online learning via the differential privacy lens. In: Wallach, H., Larochelle, H., Beygelzimer, A., d' Alché-Buc, F., Fox, E., Garnett, R. (eds.) Advances in Neural Information Processing Systems. vol. 32. Curran Associates, Inc. (2019). https://proceedings.neurips. cc//paper/2019/file/c36b1132ac829ece87dda55d77ac06a4-Paper.pdf
2. Aman, M.N., Javaid, U., Sikdar, B.: A privacy-preserving and scalable authentication protocol for the internet of vehicles. IEEE Internet Things J. 8(2), 1123–1139 (2021). https://doi.org/10.1109/JIOT.2020.3010893. https://doi.org/ 10.1109/JIOT.2020.3010893
3. Barak, B., Chaudhuri, K., Dwork, C., Kale, S., McSherry, F., Talwar, K.: Privacy, accuracy, and consistency too: A holistic solution to contingency table release. In: Proceedings of the Twenty-Sixth ACM SIGMOD-SIGACT-SIGART Symposium on Principles of Database Systems, pp. 273–282. PODS 2007, Association for Computing Machinery, New York, NY, USA (2007). https://doi.org/10.1145/1265530. 1265569
4. Bi, Z., Ming, X., Liu, J., Peng, X., Jin, W.: FIGCONs: exploiting FIne-Grained CONstructs of Facial Expressions for Efficient and Accurate Estimation of In-Vehicle Drivers' Statistics. In: International Conference on Human-Computer Interaction (2023)
5. Chen, L.W., Chang, C.C.: Cooperative traffic control with green wave coordination for multiple intersections based on the internet of vehicles. IEEE Trans. Syst. Man Cybern. Syst. 47(7), 1321–1335 (2017). https://doi.org/10.1109/TSMC.2016. 2586500
6. Chou, E., Beal, J., Levy, D., Yeung, S., Haque, A., Fei-Fei, L.: Faster cryptonets: leveraging sparsity for real-world encrypted inference. CoRR abs/1811.09953 (2018). https://arxiv.org/abs/1811.09953
7. Dandala, T.T., Krishnamurthy, V., Alwan, R.: Internet of vehicles (iov) for traffic management. In: 2017 International Conference on Computer, Communication and Signal Processing (ICCCSP), pp. 1–4 (2017). https://doi.org/10.1109/ICCCSP. 2017.7944096
8. Ding, B., Winslett, M., Han, J., Li, Z.: Differentially private data cubes: optimizing noise sources and consistency. In: Proceedings of the 2011 ACM SIGMOD International Conference on Management of Data, pp. 217–228 (2011)
9. Duan, Y., Liu, J., Jin, W., Peng, X.: Characterizing differentially-private techniques in the era of internet-of-vehicles. Technical Report-Feb-03 at User-Centric Computing Group, University of Nottingham Ningbo China (2022)

10. Gabay, D., Akkaya, K., Cebe, M.: Privacy-preserving authentication scheme for connected electric vehicles using blockchain and zero knowledge proofs. IEEE Trans. Veh. Technol. **69**(6), 5760–5772 (2020). https://doi.org/10.1109/TVT.2020.2977361

11. Ghazi, B., Golowich, N., Kumar, R., Manurangsi, P., Zhang, C.: Deep learning with label differential privacy. In: Ranzato, M., Beygelzimer, A., Dauphin, Y., Liang, P., Vaughan, J.W. (eds.) Advances in Neural Information Processing Systems. vol. 34, pp. 27131–27145. Curran Associates, Inc. (2021). https://proceedings.neurips.cc//paper/2021/file/e3a54649aeec04cf1c13907bc6c5c8aa-Paper.pdf

12. Gilad-Bachrach, R., Dowlin, N., Laine, K., Lauter, K.E., Naehrig, M., Wernsing, J.: Cryptonets: Applying neural networks to encrypted data with high throughput and accuracy. In: Balcan, M., Weinberger, K.Q. (eds.) Proceedings of the 33nd International Conference on Machine Learning, ICML 2016, New York City, NY, USA, 19–24 June 2016. JMLR Workshop and Conference Proceedings, vol. 48, pp. 201–210. JMLR.org (2016). http://proceedings.mlr.press/v48/gilad-bachrach16.html

13. Hasidim, A., Kaplan, H., Mansour, Y., Matias, Y., Stemmer, U.: Adversarially robust streaming algorithms via differential privacy. In: Larochelle, H., Ranzato, M., Hadsell, R., Balcan, M., Lin, H. (eds.) Advances in Neural Information Processing Systems. vol. 33, pp. 147–158. Curran Associates, Inc. (2020). https://proceedings.neurips.cc//paper/2020/file/0172d289da48c48de8c5ebf3de9f7ee1-Paper.pdf

14. Hay, M., Rastogi, V., Miklau, G., Suciu, D.: Boosting the accuracy of differentially private histograms through consistency. Proc. VLDB Endow. **3**(1–2), 1021–1032 (2010). https://doi.org/10.14778/1920841.1920970

15. Hua, L., Anisi, H., Por, Y., Alam, M.: Social networking-based cooperation mechanisms in vehicular ad-hoc network- a survey. Vehicular Commun. **10**, 57–73 (2017). https://doi.org/10.1016/j.vehcom.2017.11.001

16. Huang, Z., et al.: Face2multi-modal: In-vehicle multi-modal predictors via facial expressions. In: Adjunct Proceedings of the 12th International Conference on Automotive User Interfaces and Interactive Vehicular Applications, AutomotiveUI 2020, Virtual Event, Washington, DC, USA, 21–22 September 2020, pp. 30–33. ACM (2020)

17. Jin, W., Duan, Y., Liu, J., Huang, S., Xiong, Z., Peng, X.: BROOK Dataset: a playground for exploiting data-driven techniques in human-vehicle interactive designs. Technical Report-Feb-01 at User-Centric Computing Group, University of Nottingham Ningbo China (2022)

18. Jin, W., Ming, X., Song, Z., Xiong, Z., Peng, X.: Towards Emulating Internet-of-Vehicles on a Single Machine. In: AutomotiveUI '21: 13th International Conference on Automotive User Interfaces and Interactive Vehicular Applications, Leeds, United Kingdom, 9–14 September 2021 - Adjunct Proceedings, pp. 112–114. ACM (2021). https://doi.org/10.1145/3473682.3480275

19. Kazhamiaka, F., Zaharia, M., Bailis, P.: Challenges and opportunities for autonomous vehicle query systems. In: 11th Conference on Innovative Data Systems Research, CIDR 2021, Virtual Event, 11–15 January 2021, Online Proceedings. http://www.cidrdb.org/ (2021). http://cidrdb.org/cidr2021/papers/cidr2021_paper18.pdf

20. Khatoun, R., Gut, P., Doulami, R., Khoukhi, L., Serrhrouchni, A.: A reputation system for detection of black hole attack in vehicular networking. In: 2015 International Conference on Cyber Security of Smart Cities, Industrial Control System and Communications (SSIC), pp. 1–5 (2015). https://doi.org/10.1109/SSIC.2015.7245328
21. Kilari, V.T., Yu, R., Misra, S., Xue, G.: Robust revocable anonymous authentication for vehicle to grid communications. IEEE Trans. Intell. Transp. Syst. **21**(11), 4845–4857 (2020). https://doi.org/10.1109/TITS.2019.2948803
22. Li, C., Hay, M., Miklau, G., Wang, Y.: A data- and workload-aware algorithm for range queries under differential privacy. Proc. VLDB Endow. **7**(5), 341–352 (2014). https://doi.org/10.14778/2732269.2732271
23. Liu, J., Duan, Y., Bi, Z., Ming, X., Jin, W., Song, Z., Peng, X.: BROOK Dataset: a playground for exploiting data-driven techniques in human-vehicle interactive designs. In: International Conference on Human-Computer Interaction (2023)
24. Liu, J., Jin, W., He, Z., Ming, X., Duan, Y., Xiong, Z., Peng, X.: HUT: enabling high-utility, batched queries under differential privacy protection for internet-of-vehicles. Technical Report-Feb-02 at User-Centric Computing Group, University of Nottingham Ningbo China (2022)
25. Mehrish, A., Subramanyam, A., Kankanhalli, M.: Multimedia signatures for vehicle forensics (2017). https://doi.org/10.1109/ICME.2017.8019488
26. Ming, X., et al.: Enabling efficient emulation of internet-of-vehicles on a single machine: practices and lessons. In: International Conference on Human-Computer Interaction (2023)
27. Panichpapiboon, S., Pattara-atikom, W.: A review of information dissemination protocols for vehicular ad hoc networks. Commun. Surv. Tutor. IEEE **14**, 1–15 (2012). https://doi.org/10.1109/SURV.2011.070711.00131
28. Patro, S.G.K., Sahu, K.K.: Normalization: a preprocessing stage (2015)
29. Peng, X., Huang, Z., Sun, X.: Building BROOK: a multi-modal and facial video database for human-vehicle interaction research, pp. 1–9 (2020). https://arxiv.org/abs/2005.08637
30. Song, Z., Duan, Y., Jin, W., Huang, S., Wang, S., Peng, X.: Omniverse-OpenDS: enabling agile developments for complex driving scenarios via reconfigurable abstractions. In: International Conference on Human-Computer Interaction (2022)
31. Song, Z., Wang, S., Kong, W., Peng, X., Sun, X.: First attempt to build realistic driving scenes using video-to-video synthesis in OpenDS framework. In: Adjunct Proceedings of the 11th International Conference on Automotive User Interfaces and Interactive Vehicular Applications, AutomotiveUI 2019, Utrecht, The Netherlands, 21–25 September 2019, pp. 387–391. ACM (2019). https://doi.org/10.1145/3349263.3351497
32. Soria-Comas, J., Domingo-Ferrer, J., Sánchez, D., Martínez, S.: Enhancing data utility in differential privacy via microaggregation-based k k -anonymity. VLDB J. **23**(5), 771–794 (2014). https://doi.org/10.1007/s00778-014-0351-4
33. Sun, X., et al.: Exploring personalised autonomous vehicles to influence user trust. Cogn. Comput. **12**(6), 1170–1186 (2020). https://doi.org/10.1007/s12559-020-09757-x
34. Tai, B., Li, S., Huang, Y.: K-aggregation: Improving accuracy for differential privacy synthetic dataset by utilizing k-anonymity algorithm. In: Barolli, L., Takizawa, M., Enokido, T., Hsu, H., Lin, C. (eds.) 31st IEEE International Conference on Advanced Information Networking and Applications, AINA 2017, Taipei, Taiwan, 27–29 March 2017, pp. 772–779. IEEE Computer Society (2017). https://doi.org/10.1109/AINA.2017.97

35. Wang, J., Xiong, Z., Duan, Y., Liu, J., Song, Z., Peng, X.: The importance distribution of drivers' facial expressions varies over time! In: 13th International Conference on Automotive User Interfaces and Interactive Vehicular Applications, pp. 148–151 (2021)

36. Wang, S., et al.: Oneiros-OpenDS: an interactive and extensible toolkit for agile and automated developments of complicated driving scenes. In: International Conference on Human-Computer Interaction (2022)

37. Xiao, X., Wang, G., Gehrke, J.: Differential privacy via wavelet transforms. IEEE Trans. Knowl. Data Eng. **23**(8), 1200–1214 (2010)

38. Xiong, Z., et al.: Face2Statistics: user-friendly, low-cost and effective alternative to in-vehicle sensors/monitors for drivers. In: Kromker, H. (eds) HCI in Mobility, Transport, and Automotive Systems. HCII 2022. Lecture Notes in Computer Science, vol. 13335, pp. 289–308. Springer, Cham (2022). https://doi.org/10.1007/978-3-031-04987-3_20

39. Zhang, Yu., Jin, W., Xiong, Z., Li, Z., Liu, Y., Peng, X.: Demystifying interactions between driving behaviors and styles through self-clustering algorithms. In: Krömker, H. (ed.) HCII 2021. LNCS, vol. 12791, pp. 335–350. Springer, Cham (2021). https://doi.org/10.1007/978-3-030-78358-7_23

40. Zhao, P., Zhang, G., Wan, S., Liu, G., Umer, T.: A survey of local differential privacy for securing internet of vehicles. J. Supercomput. **76**, 8391–8412 (2020). https://doi.org/10.1007/s11227-019-03104-0

41. Zhao, Z.Q., Zheng, P., Xu, S.T., Wu, X.: Object detection with deep learning: a review. IEEE Trans. Neural Netw. Learn. Syst. **30**(11), 3212–3232 (2019). https://doi.org/10.1109/TNNLS.2018.2876865

How Will Automated Trucks Change the Processes and Roles in Hub-to-Hub Transport?

Svenja Escherle[1][✉][iD], Anna Sprung[2][iD], and Klaus Bengler[1][iD]

[1] Technical University of Munich, Boltzmannstr. 15, 85748 Garching b. München, Germany
`svenja.escherle@tum.de`
[2] MAN Truck & Bus SE, Dachauer Str. 667, 80995 München, Germany

Abstract. Hub-to-hub transport will be one of the first use cases for automated driving in the transportation industry. As current in-hub processes are based on a driver being present, it is important to investigate how automated trucks are to be integrated in future in-hub processes when they arrive driverless. Following on from previous research, the in-hub tasks that will still have to be carried out manually in and around the truck were identified. By conducting individual workshops with seven experts of the logistics industry, four different processes were developed for the integration of automated trucks in future in-hub processes. The results show that hub personnel will still be needed for all of the processes. The processes developed differ in that they either include a trailer exchange or truck and trailer are not separated. The respective future roles and responsibilities are described for each process, and the necessary prerequisites and possible problems are discussed.

Keywords: Automated Trucks · Hub-to-hub Transport · Process Development · Task Distribution · Future Roles

1 Theoretical Background

Automated driving in the transportation industry offers the potential to successfully overcome current challenges, such as increasing driver shortages [1]. In SAE level 4, the automated driving system can completely take over the driving task within the defined operational design domain (ODD) [2, 3]. If system limits are exceeded, the system is able to reach a risk-minimized state and therefore does not have to be monitored. This even eliminates the need for a driver inside the vehicle. Hub-to-hub freight transport is cited as one of the first use cases for automated trucks and it is expected to occur by 2030 [4, 5]. Hub-to-hub transport refers to the transportation of goods between two or more logistics centers that are built close to and are connected by the highway. Therefore, the most part of the transport occurs on standardized and normed highways. In the case of automated hub-to-hub transport, the ODD of the automated system covers the route between two logistics centers via highways. The transportation industry is particularly interested in the automation of hub-to-hub transport in order to overcome driver shortages, but also for its expected benefits on an economical, environmental and social level [6]. For example, the total cost of ownership for trucks is expected to decrease in the long term, due to increased fuel efficiency and reduced labor costs [6, 7] out weighing the increased costs for sensors and automation. Furthermore, automated trucks are expected to improve traffic flow and reduce congestion, leading to better panning of transport times [6]. To benefit from these advantages, the technical development of automated driving functions is currently being pushed forward [8]. Although the technical implementation of these automated driving functions forms the basis for automated transport, a successful automation of hub-to-hub transport requires more than developing automated driving functions for the ODD. A successful introduction of automated driving in this domain depends on synchronized measures in automated driving technology, infrastructure, logistic processes and human resources. Above all, it is important to investigate how future incoming and outgoing goods processes of the hubs will have to be designed for trucks arriving driverless at the hubs. Currently, the incoming and outgoing goods processes at hubs are designed for the presence of truck drivers. For example, the driver is included in registration at the hub, handling of freight documents, or loading and unloading of the truck [1, 9–11].

This paper addresses the fact that, in the future, these incoming and outgoing goods processes will have to be redefined when automated trucks arrive without a driver at the hubs. As an introduction to this topic, the results of two previous studies are presented in the following sections. Firstly, the design of current hub processes that include the help of a truck driver is addressed in more detail [11]. Secondly, the development of the German logistics environment up until the arrival of automated trucks on the road by 2030 is presented [12].

1.1   Current Tasks and Processes

As mentioned above, current incoming and outgoing goods processes are designed based on a driver being present. Escherle et al. [11] conducted a task- and process analysis of four different logistic hubs in Germany and identified an overlying process that could be observed in each of the hubs. The authors identified four consecutive phases: registration, unloading goods, loading goods, and exiting the hub (Fig. 1). The driver is responsible for driving the truck on the hub premises between these phases. Depending on the transported goods and the organization in the hub, there can be several unloading and loading stations [9, 11].

For registration the tasks to be done vary depending on the hubs' state of digitalization [11]. Where the state of digitalization is low, more manual tasks have to be performed by the driver: The driver has to park, get out of the truck, walk to a registration office, register and hand over documents to a person working in the office. The office worker interacts with a system and informs the driver about the unloading and loading points he or she has to drive to.

The results of the task and process analysis also show that drivers often have to cope with difficult traffic situations while driving on the hub premises. These are the result of multiple different traffic participants, missing lane markings or unclear traffic regulations.

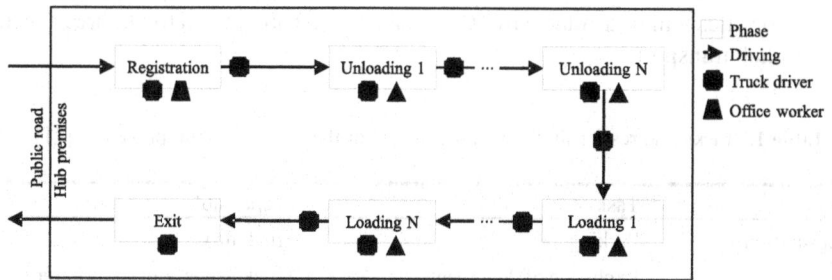

Fig. 1. Phases of incoming and outgoing goods processes of hubs in Germany based on the findings of Escherle et al. [11]. The symbols refer to the roles involved in each phase.

In the unloading phase, the driver brings the truck into the correct position. The tasks in this phase depend on the body that is used with the truck. If swap bodies are used, the truck driver opens the locks for detaching the body from the truck. Once the body is disconnected from the truck freight documents are exchanged between the driver and a hub worker and the driver can drive on to the loading point. The subsequent unloading process of the swap body is carried out by hub personnel [9, 11]. If fixed bodies are being used with the truck, truck and driver stay for the whole unloading process. In this case, the driver has to open the trailer and remove the load securing. Then the driver waits while the hub personnel unload the required part of the goods, e.g. using forklifts. The driver is then responsible for securing the remaining goods and closing the trailer. At the end of the unloading phase, freight documents are again exchanged between hub personnel and the driver. Unloading can take place at several unloading points until the trailer is empty.

After the unloading phase, the driver drives the truck to its loading position. The loading phase includes similar tasks to the unloading phase: either the driver has to attach a preloaded swap body to the truck, or the driver has to open the trailer and secure the load, while hub personnel are then responsible for loading the trailer, e.g. with forklifts, and then the driver closes the trailer again. In either case, the freight documents have to be exchanged between the driver and the hub personnel to complete the loading phase. Then the driver drives the truck off the hub premises. These tasks of the current process and the responsible roles are summarized in Table 1.

For automated trucks arriving driverless in the future, several challenges arise from the design of the current processes [11]. These include spontaneous changes in the unloading or loading position that are currently communicated to the driver in an ad hoc manner. Especially where fixed truck bodies are used, processes tend to be less standardized and the driver is spontaneously involved at a deeper level in the process and asked to help with other tasks. Furthermore, deficiencies in the hub infrastructure such as missing lane marks or the many traffic participants could be problematic for driverless trucks in the future. Digitalization plays an important role in the integration of automated trucks into the hub processes [11]. Especially in the registration phase and for document exchange at the loading and unloading points digital solutions might offer great potential to substitute tasks currently assigned to the driver. However, it remains open to question as to whether these digital solutions as well as the outcomes from other trends in logistics will be available by 2030 when the first automated trucks are expected in hub-to-hub transport.

Table 1. Tasks and responsibilities in the phases of the current in-hub processes [11].

Phase	Task	Responsible
Registration	Parking	Truck driver
	Exchange of documents	Truck driver + office worker
	System registration	Office worker
	Information on (un)loading points	Office worker to truck driver
Driving	Steering truck	Truck driver
	Navigation on hub	Truck driver
Unloading	Park truck in correct position	Truck driver
	Open trailer / detach swap body	Truck driver
	Unloading	Hub personnel
	Load securing	Truck driver + hub personnel
	Close trailer	Truck driver
	Document exchange	Truck driver + hub personnel
Loading	Park truck in correct position	Truck driver
	Open trailer / attach swap body	Truck driver
	Loading	Hub personnel
	Load securing	Truck driver + hub personnel
	Close trailer	Truck driver
	Document exchange	Truck driver + hub personnel

1.2 Future Logistics Environment of Automated Trucks

To identify which of these current tasks will remain in place for the future in-hub processes involving automated trucks, an understanding of the logistics environment by 2030 in Germany is crucial. Current trends in logistics include globalization [13–16], digitalization [17–19], environmental as well as social sustainability requirements [16, 20], and combined transport [21–23]. Further, the adaption of logistic processes and the development of infrastructure play an important role in logistics [4, 24, 25] as do changes in society and the availability of skilled personnel [20, 26]. To investigate the future logistics environment of automated trucks, Escherle et al. [12] conducted a Delphi-based scenario study for the year 2030.

The results indicate that by 2030, environmental and social sustainability will be an increasingly important factor that has to be considered when designing future in-hub processes for automated trucks. In this context, the proportion of trucks with electric drives will increase significantly by 2030. According to this study, the transportation of goods in Germany will remain focused on road-transport. A major shift to combined transport with rail is not expected [12]. In terms of digitalization, the results show that tracking and tracing as well as digital connection along the supply chain will be widespread, and freight and customs documents will be digital across large parts of Europe. Digital coordination of trucks on hub premises will be possible by 2030, however, smaller hubs might still not be able to provide the required infrastructure.

The first application area for automated trucks will be in hub-to-hub transport [12]. However, only a small percentage of trucks will be automated by 2030. Consequently, automated trucks as well as manually driven trucks will be used to deliver goods to the hubs. Regarding in-hub processes, unloading and loading of the truck will still be carried out manually and require personnel by 2030. Unfortunately, the lack of skilled personnel is expected to increase even more in the future [12]. Where public infrastructure is concerned, only limited changes will occur before 2030. The main changes in infrastructure are expected to be the provision of the required charging infrastructure for battery electric vehicles in hub-to-hub transport. As the charging network might not be sufficient to cover all areas of operation, charging might also have to take place at the hubs [12].

1.3 Considered Scenario

Even if it seems counterintuitive, automated driving within the controlled environment of hubs might be more complex than driving between hubs. Given the many different traffic participants and difficult traffic situations [11], driving inside a hub is comparable with driving in a city. Therefore, it is expected that by 2030 automated driving on hub premises will not be possible for many hubs without more substantial changes in hub infrastructure. However, hub operators are not willing or financially not able to make big investments in hub infrastructure in order to enable automated driving [9].

Based on the presented research, we address the following scenario for 2030: automated trucks are used in hub-to-hub transport. The automated truck arrives without a driver at the hub. However, automated driving on the hub premises is not possible. There is mixed transport, meaning automated as well as manually driven trucks deliver goods to the hubs. Given the progress in digitalization, the registration and the coordination of the trucks at the hub takes place digitally. Moreover, all of the freight and customs documents are digital and can be accessed through a digital system. The tasks related to unloading and loading the truck still have to be carried out manually.

As Escherle et al. [11] suggest, investigation is required to determine which of the current tasks will still have to be carried out manually in the future scenario (2030), how the in-hub processes with automated trucks should be configured, and who will be responsible for what task.

1.4 Research Questions

Automation not only replaces but also changes activities that are performed by humans or it introduces new tasks. These tasks should be designed to match human capabilities and requirements [27]. Thus, automated trucks may also change the future roles and tasks of the people involved in the delivery process. It is therefore necessary to identify which of the tasks in and around the automated truck will still have to be carried out manually in the scenario under consideration and who will be responsible for them. In this context, the organization of future incoming and outgoing goods processes with automated trucks needs to be addressed. Therefore, the research questions of this paper are:

1. How should the incoming and outgoing goods process for automated trucks in hub-to-hub transport be designed?
2. What impact does this process have on the roles and task distribution on the hubs?

These questions about the process that follows the arrival of an automated truck at a hub have received little attention in the literature to date. However, it is crucial that these questions are clarified before automated trucks enter the market to enable automated hub-to-hub transport.

2 Method

In a first step, the remaining manual tasks for the future scenario were identified. To do so, the tasks relating to the current processes at the hubs [11] were examined under the future conditions of the considered scenario (see 1.3). Based on the different aspects of the future state of development, it could be defined whether a task will be digitalized or still require manual operations. Subsequently, seven individual workshops with experts from the logistics industry were conducted. The workshops took place online using a virtual whiteboard and each one lasted approximately one hour.

2.1 Expert Panel

In total, seven experts participated in the workshops, six male and one female. The experts were employees of German car and truck manufacturers as well as logistics service providers. The mean age was $M = 43$ years ($SD = 9.65$) and their average work experience in the logistics field was $M = 18.5$ years ($SD = 8.16$). Table 2 shows their main areas of expertise.

Table 2. Main area of expertise (multiple answers possible).

Main area of expertise	Number of experts
Logistics management	2
Site planning	2
Inbound logistics	2
In-house logistics	4
Innovation	3
Process planning	2
Transport planning	2

2.2 Content of the Workshops

In the introduction, the participating expert was informed about the content and procedure of the workshop and subsequently signed the consent form which was then digitally sent to the workshop leader. After demographic data has been collected, the scenario described in Sect. 1.3 was presented to the experts. The current processes and tasks (Sect. 1.1) were explained followed by the remaining manual and unassigned tasks for the future in-hub processes (Table 3). After the experts had been provided with this information, they were asked how they would organize the in-hub process with automated trucks in the presented scenario based on the information provided and their own experience.

The experts each developed different processes for the integration of automated trucks in the future scenario. For each process, the related task allocation and responsibilities were discussed. Further, the prerequisites, possible problems and limiting factors were documented for each of the developed processes and the associated future job profiles were addressed.

2.3 Data Analysis

The processes developed were documented in detail during the course of each expert workshop. This documentation was further reviewed after the workshops were concluded: for each process, the underlying characteristics were identified and extracted. Following the inductive category development [28] four clusters based on the underlying characteristics could be identified for the processes. With the experts' description of the different processes in the workshops, the four overlying clusters could be further specified. The advantages, disadvantages, corresponding future tasks and roles named by the experts were compiled for each of the four clusters and the corresponding four processes were visualized.

3 Results

The results below present which tasks will be digitalized and which tasks will still require manual operation in the considered scenario. Furthermore, the processes developed for the integration of automated trucks and the related future roles and task allocations are presented.

3.1 Remaining Manual Tasks

With reference to the various phases of the current in-hub processes (Fig. 1), the remaining tasks and the responsibilities in the considered scenario are summarized in Table 3. As Table 3 shows, tasks previously assigned to the truck driver will partially be replaced by the digital system. The remaining manual tasks and responsibilities of the truck driver in terms of unloading and loading the truck as well as the task of driving the truck on the hub premises need to be reassigned, which was addressed in the workshops.

Table 3. Tasks and responsibilities in the phases of the future processes with automated trucks.

Phase	Task	Responsible
Registration	Recognition of vehicle	Digital system
	Exchange of documents	Digital system
	System registration	Digital system
	Information on (un)loading points	Digital system
Driving	Steering truck	**Unclear**
	Navigation on hub	**Unclear**
Unloading	Park truck in correct position	**Unclear**
	Open trailer / detach swap body	**Unclear**
	Unloading	Hub personnel
	Load securing	**Unclear**, poss. hub personnel
	Close trailer	**Unclear**
	Document exchange	Digital system
Loading	Park truck in correct position	**Unclear**
	Open trailer / attach swap body	**Unclear**
	Loading	Hub personnel
	Load securing	**Unclear**, poss. hub personnel
	Close trailer	**Unclear**
	Document exchange	Digital System

3.2 Future Processes with Automated Trucks

A general overview of the processes developed in the expert workshops for the integration of automated trucks into future in-hub processes is shown in Fig. 2. The results show the handling of the trailer as a central distinctive feature. Either automated truck and trailer remain connected or are separated from each other. If truck and trailer remain connected, the truck is used for moving the trailer on the hub. This process is further distinguished between manual steering and remote steering of the truck. If truck and trailer are disconnected, the trailer is moved by a shunting vehicle inside the hub, which is either steered manually or remotely. The resulting four possible processes are further described below. Overall, the processes not including trailer exchange were rated most suitable for the integration of automated trucks by four experts, the processes including trailer exchange by three of the seven experts.

In line with the described scenario, each of the processes presented below starts with the automated truck arriving driverless at the hub and its parking in a predefined position. The arrival of the truck is registered automatically by the system and the hub personnel are notified.

Process 1: Truck and Trailer Stay Connected, Truck is Steered Manually. This processes was described by 5 of the 7 experts and is visualized in Fig. 3.

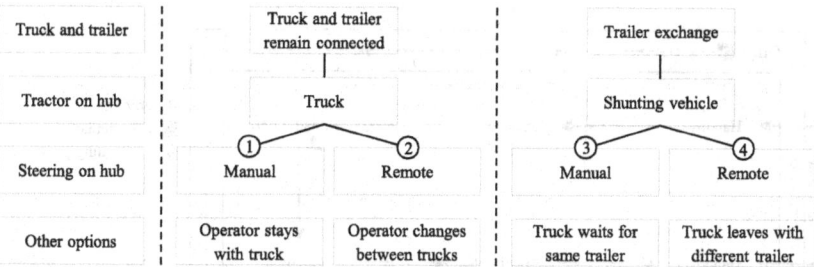

Fig. 2. Overview of the processes developed for the integration of automated trucks into future in-hub processes by the experts of this study.

In process 1, a hub driver gets into the truck and steers it manually on the hub premises. This hub driver remains with the same truck for the whole process and takes over all of the tasks that are currently performed by the conventional driver. The hub driver therefore is responsible for driving the truck to the correct loading and unloading points, and for opening and closing the trailer. Moreover, the driver helps the hub personnel with loading and unloading, and is responsible for correct load securing. When the unloading and loading phases are completed, the hub driver drives the truck back to the handover site, gets out and activates automated driving mode. The truck then leaves the hub in automated driving mode, and the hub driver repeats the process on the next arriving automated truck. The distribution of tasks in process 1 is also listed in Table 4.

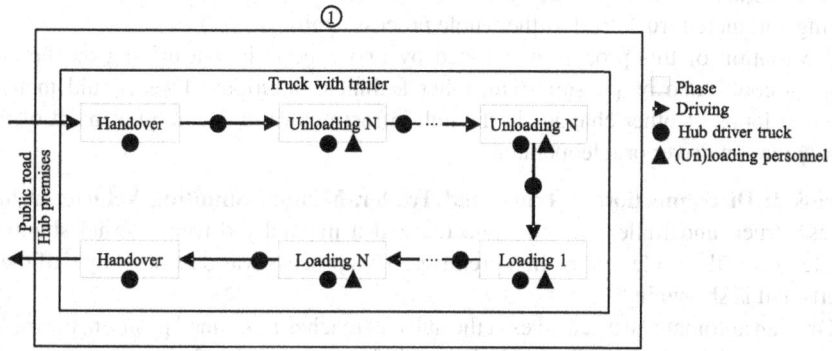

Fig. 3. Visualization of Process 1: Truck and trailer stay connected, truck is steered manually. The symbols refer to the roles involved in each phase.

A variation of this process was also discussed by two of the experts: The hub driver could also change between trucks when they are in the process of (un)loading. However, it could not be clarified how the driver would get from one truck to another. For big hubs in particular, additional vehicles or some sort of shuttle service would be necessary. The experts, therefore, considered this alternative not applicable.

Process 2: Truck and Trailer Stay Connected, Truck is Steered Remotely. This process was described by 6 of the 7 experts. The process is shown in Fig. 4.

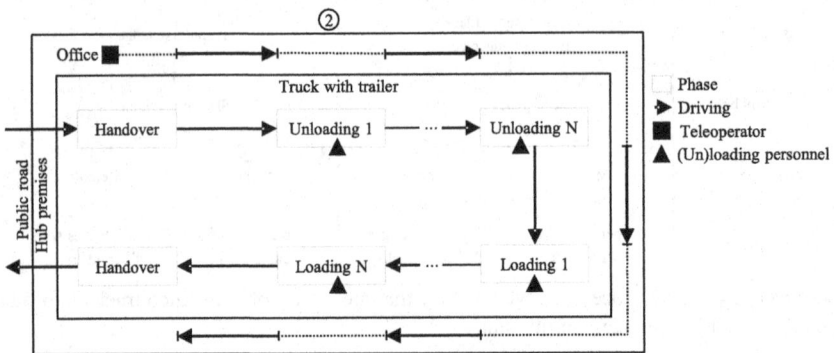

Fig. 4. Visualization of Process 2: Truck and trailer stay connected, truck is steered remotely. The symbols refer to the roles involved in each phase.

In this process, the truck is steered remotely on the hub premises by a teleoperator. The teleoperator connects with the arriving truck and steers it to the first unloading point. There the truck is unloaded by the unloading personnel of the hub. The hub personnel are responsible for opening and closing the truck, for (un)loading and load securing. Meanwhile, the teleoperator can switch to another truck that needs to be remotely steered on the hub premises. When the (un)loading is completed the teleoperator drives the truck to the next point where the (un)loading procedure is repeated. Once loading is completed, the teleoperator drives the truck back to the handover site, and activates automated driving mode. The truck drives off the hub premises and the teleoperator connects with the next arriving automated truck to start the whole process again.

A variation of this process mentioned by two experts is that driving on the hub premises could also be possible with other technical solutions. These could include induction loops or other changes in the hub infrastructure to enable movement of the truck without a driver or teleoperator.

Process 3: Disconnection of Truck and Trailer, Manual Shunting Vehicle. In this process, truck and trailer are disconnected and a manually driven internal shunting vehicle moves the trailer on the hub premises. This process was described by 6 of the 7 experts and is shown in Fig. 5.

Once an automated truck arrives at the hub and reaches its defined position, the trailer is disconnected from the truck with the help of hub personnel. A hub driver transports the trailer with an internal shunting vehicle and brings it to the correct (un)loading points. The (un)loading is done by hub personnel who are also responsible for opening and closing the trailer, as well as for load securing. While the trailer is being (un)loaded, the hub driver of the shunting vehicle can move other trailers on the hub. When loading of the trailer is completed, the hub driver takes the trailer back to the truck, where they are reconnected by personnel on site.

For this process the experts mentioned two alternatives. Either the truck stays at the hub until the same trailer is unloaded, loaded and brought back again. This waiting time could be used to recharge electric trucks. The other option is that the trailer is exchanged. In this case, after disconnection of the transported trailer, the truck is immediately connected to another preloaded trailer and can drive off directly. The trailer that was transported to the hub is unloaded afterwards.

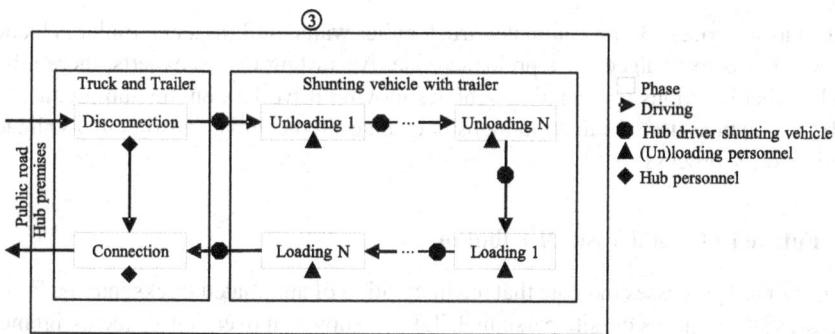

Fig. 5. Visualization of Process 3: Disconnection of truck and trailer, manual shunting vehicle. The symbols refer to the roles involved in each phase.

Process 4: Disconnection of Truck and Trailer, Teleoperated Shunting Vehicle. This process differs from process 3 in that the internal shunting vehicle is steered remotely by a teleoperator. 2 of the 7 experts described this process which is visualized in Fig. 5.

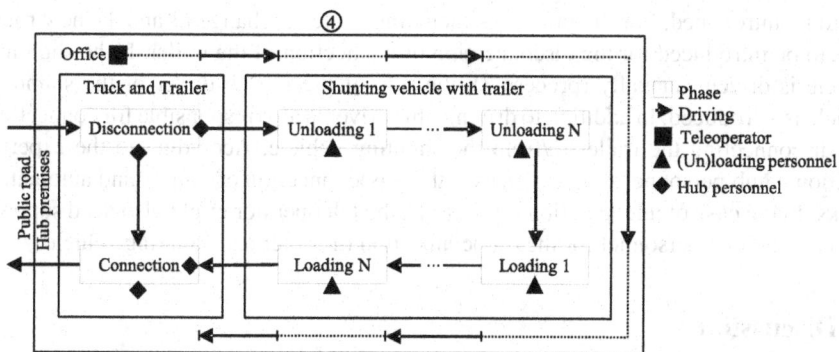

Fig. 6. Visualization of Process 4: Truck and trailer are disconnected, teleoperated internal shunting vehicle drives the trailer on the hub premises. The symbols refer to the roles involved in each phase.

The trailer is disconnected from the automated truck with the help of on-site hub personnel. The trailer is then connected to an internal shunting vehicle that is steered remotely by a teleoperator. According to the experts, hub personnel might be needed to connect the trailer and shunting vehicle. During the following in-hub process, trailer and shunting vehicle remain together. Therefore, the teleoperator drives the trailer to the (un)loading points and can switch steering between different shunting vehicles during waiting times. The opening, (un)loading, load securing and closing of the trailer is done by hub personnel at the (un)loading points. The teleoperator then takes the shunting vehicle and trailer back to the automated truck. Here, the trailer is disconnected from the shunting vehicle and connected to the truck. Hub personnel might be needed for this step.

Just as in process 3, the automated truck either waits until the same trailer is loaded or the trailer is exchanged for a preloaded one. According to two experts, there might also be other solutions for shunting vehicles moving driverless on the hub premises in the future, such as use of induction loops or changes in the hub infrastructure as already mentioned for process 2.

3.3 Future Roles and Task Distribution

The presented processes indicate that the integration of automated trucks into the in-hub processes still requires on-site personnel. Table 4 shows an overview of the assignment of tasks and responsibilities in the different processes. The task assignment of process 1 is most similar to the current process. In this case, the original driver is replaced by a new role of an in-hub driver. All of the original driver tasks are assigned to this in-hub driver. The role of the (un)loading personnel remains unaffected.

In processes 2 - 4, the person responsible for driving does not stay with the trailer for un(loading). In these processes, the (un)loading personnel are assigned additional tasks. Instead of just (un)loading the trailer, the personnel also have to open and close the trailer and are solely responsible for correct load securing.

With the processes including teleoperation (2 and 4), the new role of the teleoperator has to be introduced. For the processes including trailer exchange (3 and 4), new roles have to be introduced for the disconnection or connection of the trailer. If the shunting vehicle is driven manually (process 3), the role of the in-hub driver of the shunting vehicle is introduced. In addition to driving, this driver is also responsible for connecting and disconnecting the trailer to/from the shunting vehicle. According to the experts, additional hub personnel should take over the (dis)connection of trailers and automated trucks. In the case of teleoperation (process 4), the teleoperator might also need support from on-site hub personnel for the (dis)connection of trailer and shunting vehicle.

4   Discussion

The presented processes are in line with other scenarios in the literature [9], and are subject to different prerequisites and possible problems that were discussed with the experts in the workshops. Furthermore, the roles in future processes and the corresponding job descriptions were further elaborated. The different aspects are described below.

4.1   Future Processes: Prerequisites, Advantages and Open Questions

General. According to the experts, enough available open space on the hub premises is one of the main requirements for the implementation of implement any of the presented processes. In particular, enough space is required for the handover site or the trailer

Table 4. Task distribution, responsibilities and corresponding roles in the different processes.

Phase	Task	Processes				
		current	1	2	3	4
Arrival	Disconnect truck + trailer	-	-	-	◆	◆
	Connect trailer + shunting v.	-	-	-	●	◆■
Driving	Steer	■	●	■	●	■
	Navigate on hub	■	●	■	●	■
Unloading	Park in correct position	■	●	■	●	■
	Open trailer	■	●	▲	▲	▲
	Unloading	■▲	●▲	▲	▲	▲
	Load securing	■▲	●▲	▲	▲	▲
	Close trailer	■	●	▲	▲	▲
Loading	Park in correct position	■	●	■	●	■
	Open trailer	■	●	▲	▲	▲
	Loading	■▲	●▲	▲	▲	▲
	Load securing	■▲	●▲	▲	▲	▲
	Close trailer	■	●	▲	▲	▲
Departure	Disconnect trailer + shunting v.	-	-	-	●	◆■
	Connect trailer + truck	-	-	-	◆	◆
Number of different roles involved		2	2	2	3	3

■ Truck driver   ● Hub driver truck   ● Hub driver shunting vehicle   ■ Teleoperator
▲ (Un)loading personnel   ◆ Hub personnel

exchange. Furthermore, this transfer area also has to have the required infrastructure, including clear lane markings, predefined positions that the automated trucks park in, the digital system for registration and planning, as well as an interface between this system, truck, and involved personnel. Depending on hub conditions, the installation of such a transfer area might require high investment. The experts consider it possible that for some hubs the space currently used as a waiting area for manual trucks can be considered for this step. To reduce the risk of congestion and long waiting times, the arrival and processing of trucks should be well planned. There also has to be a solution to the problem of where an automated truck should wait if the transfer area is full. Therefore, an interface between the digital hub system, truck and hub personnel is also required.

Furthermore, the experts address the issue of liability. In the current process the conventional truck driver as well as the (un)loading personnel of the hubs are both responsible for correct loading and load securing. With no conventional truck driver involved in the future process, the responsibility for loading and load securing has to be transferred to the hub personnel in its entirety. This transfer of liability also requires change in German law. In this context, the traceability of the whole loading process needs to be ensured. In case of inspections, damages or accidents with the load the hub personnel responsible for loading and load securing need to be documented in the system.

Further, with the presented processes for automated trucks more in-hub personnel are required than is the case for the current situation. Therefore, enough personnel for driving the truck on the hub premises, for (dis)connecting trailers and (un)loading is a prerequisite. An open question in this context is who will employ the required personnel. The experts consider both an external service provider and the hub as possible solutions. With hub personnel being exclusively responsible for processing automated trucks in the hubs the common problem of drivers getting lost on the hub premises can be minimized.

Another requirement is that the automated truck still has to provide the option for manual control. In process 1, the hub driver has to get inside the truck to steer it on the hub premises. In processes 3 and 4, manual control might still be necessary for reconnecting truck and trailer. In this step, the truck has to drive backwards to the trailer, meeting it at the correct angle and height. The experts doubt that automation will be sufficiently advanced to cover this use case by 2030 and therefore expect manual driving will still be necessary. Moreover, the experts agree that there will be other use cases where automated trucks have to be driven manually. Examples include technical mistakes, accidents or simply driving the truck on roads where automation is not yet possible. In conclusion, a cabin with steering devices for manual driving is still needed for automated trucks in 2030.

According to the experts, one important factor regarding mixed transport is that the future incoming and outgoing goods process is the same for both manual and automated trucks. Two different processes would not be efficient are not conducive and would result in unnecessary high complexity.

Processes Without Trailer Exchange. Regarding processes 1 and 2 where truck and trailer remain connected, it has to be taken into account that hub personnel are driving the truck on the hub premises. Depending on the ownership of the trucks, this will involve driving of third-party property. In this case, regulations for the procedure in case of damage and respective tracing of responsibilities have to be defined.

Further, it should be noted that because the truck is integrated into the in-hub processes, the time share of automated driving between hubs is therefore less than in the processes with trailer exchange. However, the experts see the general advantage that the in-hub processes require less time and fewer personnel (Table 4) because the trailer exchange step is not a part of the process.

Processes With Trailer Exchange. Regarding the processes that include a trailer exchange (process 3 and 4), a sufficient number of shunting vehicles have to be available. Providing these shunting vehicles will represent a high investment for many hubs. This investment will be even higher for teleoperated shunting vehicles (process 4) for two reasons: firstly, each teleoperated shunting vehicle stays connected to one trailer during the in-hub process because the connection requires the help of an on-site person. Therefore, more shunting vehicles are necessary for the same amount of trailers compared to manually driven shunting vehicles that can switch between trailers while they are being (un)loaded. This is also the case when the shunting vehicle is not teleoperated but the driverless driving on the hub premises is enabled in some other way. Secondly, teleoperation in general requires the enabling infrastructure which entails even greater investment.

If the truck takes a trailer other than the one it delivered, it can leave the hub in less time when the outgoing trailer is already prepared and preloaded. In conclusion, the time share of automated driving is higher and in total more freight can be transported. Also, the experts note that idle times for trailers are not as expensive as for trucks. To profit from this advantage in processes 3 and 4, the trailer has to be preloaded by the hub personnel, which requires correct planning and time management. Moreover, two trailers have to be available per truck.

Processes With Teleoperation. As mentioned above, teleoperation requires the enabling infrastructure such as the underlying technology, a stable connection, available work space and corresponding training and education for the teleoperator. These prerequisites might require higher investments according to the experts and depend on the development of the technology. It still has to be assessed whether this option is applicable for the different kind of hubs. Also discussion is needed about how the process is handled when technical problems with the system occur. According to the experts, a back-up solution with on-site personnel might be necessary for these cases. In general, processes with teleoperation are desirable but might take longer to implement compared to processes with manually driven vehicles.

4.2 Future Roles and Characteristics

Depending on the process, new roles will have to be introduced. These include the in-hub drivers of the truck or shunting vehicle, hub-personnel for (dis)connecting trailers or the role of the teleoperator. The role of (un)loading personnel already exists in the current process. Depending on which process will be implemented to integrate automated trucks, the (un)loading personnel may be assigned additional tasks. These also include physically demanding tasks such as opening and closing the trailer.

In general, the roles in the future processes have the following advantages: the related future jobs will have the hub as a fixed work location and therefore allow proximity to home and family as well as regular working hours. Also, in the future roles nobody will have to deal with overcrowded highway parking lots as in the current role of truck driver.

Looking at a typical future workday in the presented processes, the workers repeat the same cycle of tasks for each arriving truck. For the hub driver in process 1, this means he or she will have to get in and out of the truck many times each working day, because the in-hub driver is involved in the (un)loading process. This is also the reason for a more diverse set of tasks and responsibilities compared to the in-hub drivers of process 3. It should also be noted that the latter drivers stay with their shunting vehicle the whole time and can settle into it, while the hub driver in process 1 has to change trucks very frequently. The teleoperator in processes 2 and 4 will remain in the same work place. As this will be a completely new job, this new profile is associated with many open questions.

4.3 Limitations

The processes were developed based on the opinions and experience of the experts involved. Even though the experts were selected based on their expertise in the relevant field, there is still the possibility that some factors of the future processes could have been missed due to the expert selection. However, the developed processes are in line with the current trends in logistics [12] and the scenarios found in the literature [9].

Furthermore, the developed processes are based on the future scenarios of Escherle et al. [12]. Therefore, the processes of this paper are based on the following prerequisites: trucks are able to drive automated between the hubs, freight and customs documents are digitalized and a digital system for the registration and coordination of trucks as well as an interface to the personnel on the hubs is available. As Escherle et al. [12] also show, this might not be the case for every hub by 2030 and the processes would have to be adjusted accordingly. Moreover, two of the developed processes include teleoperation. Whether or not this technology will be available and ready for use in hubs by 2030 was not part of the scenario study [12] and is therefore worthy of further investigation.

5 Summary and Outlook

In this research, seven expert workshops were conducted to address the future integration of automated trucks into in-hub processes. The results can be clustered into four main processes that are differentiated based on the following two factors. Either the process includes the exchange of the trailer or not and either the tractor vehicle on the hub is driven manually or remotely. Each of the processes was described and the associated roles were described with their assigned tasks and responsibilities in the process. Moreover, the advantages and disadvantages of the processes were elaborated.

The processes developed in the expert workshops clearly indicate that on-site personnel are still needed for the integration of automated trucks into the in-hub processes. Viewed from a distance, the future processes do not differ significantly from the current incoming and outgoing processes. The requirement that goods have to be transported to different points on the hub will remain unchanged. Even if in-hub driving can be implemented with teleoperation technology or some other solution that enables truck driving driverless on the hub-premises, the underlying process will remain the same compared to processes with manual control.

Viewed in more detail, the roles and responsibilities do change slightly for the different processes. In processes where the in-hub driver is with the trailer for (un)loading he or she can take over related tasks and responsibilities. If this is not the case, the tasks have to be assigned to hub personnel at the different (un)loading points.

As every hub has different prerequisites, an individual decision will have to be made about which process is most suitable and applicable for integrating automated trucks. The fact that hub operators want to avoid major changes and reconstruction measures for the integration of automated trucks [9] could serve as a criterion for deciding which process to adapt for a particular individual hub.

With processes including trailer exchange, waiting times can be reduced [29] and the automated truck can leave the hub premises after a short time. In this way, the truck can fulfil its purpose (automated driving) most efficiently and more goods can be transported in the same time. However, the trailer exchange concept requires more space and investment in a sufficient number of shunting vehicles. Also, in trailer exchange processes, more hub personnel are needed to handle the (dis)connection of trailer and truck. Technology that allows automated (dis)connection, including an automatic fold out of the trailer supports, would remove the necessity for additional personnel. Therefore, trailer design and functionality represents a further field of research that should be focused on in this context.

Processes that do not include trailer exchange have the advantage that they require less investment, are easy to adapt to current processes and therefore are especially suitable according to the experts. In these processes, there is one step less to be organized and one step less that might potentially be the source of further problems and errors.

Regardless of which of the processes is implemented, the experts recommend that the in-hub processes for manual and automated trucks should be the same.

In the presented scenario and the developed future processes, the role of driver changes from a role spending many hours inside one truck to one in which only short distances are driven inside a hub. Depending on the implemented process, the driver has to change between trucks very often (process 1), has to drive a shunting vehicle (process 3) or even steer remotely from a teleoperation work place (processes 2 and 4). In this context, it should be noted that the hub driver either has to get in and out of a truck many times (process 1) or stays inside a shunting vehicle for most of the time (process 3). As a result, the processes have different requirements for the cabin design of each vehicle. Future research should therefore investigate how the driver cabin or the shunting vehicle should be designed to support the workers in their future tasks and to support their wellbeing.

It is not only the role of the driver that will change, but also other roles that are included into the incoming and outgoing goods processes of the hubs. For example, in processes 2–4, the scope of tasks increases for the personnel at the (un)loading points. Future research should therefore address whether the respective personnel will actually have the capacity and time to carry out these additional tasks and whether these tasks are in line with human capabilities and requirements [27]. Also of interest is which additional qualifications might be required for the related job and the newly introduced roles in the future processes. In particular, the role of teleoperator does not yet exist in current processes and raises many new research questions. These include, for example, how the work place of the teleoperator should be designed to enable safe steering on the hub premises.

As already mentioned, personnel will still play a crucial part in future processes for the integration of automated trucks. However, by the time automated trucks are on the roads, the labor shortage will have become more acute [12]. Future research should therefore investigate whether the related job profiles will be attractive for future workers, which education will have to be provided and how the attractiveness can be enhanced.

Acknowledgements. This research was funded by the German Federal Ministry for Economic Affairs and Climate Action within the Project RUMBA (19A20007E). The authors are solely responsible for the content of this publication.

References

1. Müller S, Voigtländer F.: Automated trucks in road freight logistics: the user perspective. In: Clausen, U., Langkau, S., Kreuz, F. (eds.) Advances in Production, Logistics and Traffic. ICPLT 2019. Lecture Notes in Logistics. Springer, Cham (2019). https://doi.org/10.13140/RG.2.2.31144.83200
2. Czarnecki, K.: Operational Design Domain for Automated Driving Systems - Taxonomy of Basic Terms: Waterloo Intelligent Systems Engineering (WISE) Lab, University of Waterloo (2018)
3. SAE International. Taxonomy and Definitions for Terms Related to Driving Automation Systems for On-Road Motor Vehicles (J3016). USA and Switzerland (2021)
4. Gräter, A., Harrer, M., Rosenquist, M,, Steiger, E.: Connected, Cooperative and Automated Mobility Roadmap. 10th ed. Brüssel: ERTRAC (2022)
5. Nowak, G., Viereckl, R., Kauschke, P., Starke, F.: Charting your transformation to a new business model: the era of digitized trucking (2018). https://www.ttm.nl/wp-content/uploads/2018/10/The-era-of-digitized-trucking-charting-your-transformation.pdf. Accessed 08 Aug 2021
6. Fritschy, C., Spinler, S.: The impact of autonomous trucks on business models in the automotive and logistics industry–a Delphi-based scenario study. Technol. Forecast. Soc. Chang. **148**, 1–14 (2019). https://doi.org/10.1016/j.techfore.2019.119736
7. Merfeld, K., Wilhelms, M.-P., Henkel, S., Kreutzer, K.: Carsharing with shared autonomous vehicles: Uncovering drivers, barriers and future developments – A four-stage Delphi study. Technol. Forecast. Soc. Chang. **144**, 66–81 (2019). https://doi.org/10.1016/j.techfore.2019.03.012
8. Eichberger, A., Szalay, Z., Fellendorf, M., Liu, H.: Advances in automated driving systems. Energies. **15**, 3476 (2022). https://doi.org/10.3390/en15103476
9. Inninger, W., Schellert, M., Schulz, H.: Analyse der Randbedingungen und Voraussetzungen für einen automatisierten Betrieb von Nutzfahrzeugen im innerbetrieblichen Verkehr. FAT-Schriftenreihe, 312 (2018)
10. Flämig, H.: Autonomous vehicles and autonomos driving in freight transport. In: Maurer, M., Gerdes, J.C., Lenz, B., Winner, H. (eds.) Autonomous Driving, pp. 365–385. Springer, Berlin, Heidelberg (2016). https://doi.org/10.1007/978-3-662-48847-8_18
11. Escherle, S., Haentjes, J., Sprung, A., Bengler, K.: Automatisierbarkeit von Lkw im Hub-to-Hub-Verkehr: Eine Prozessanalyse. In: Gesellschaft für Arbeitswissenschaft, editor. Frühjahrskongress 2023 Nachhaltig Arbeiten und Lernen - Analyse und Gestaltung lernförderlicher und nachhaltiger Arbeitssysteme und Arbeits- und Lernprozesse; 1.-3.3.2023; Hannover: GfA Press (2023)
12. Escherle, S., Darlagiannis, E., Sprung, A.: Automated trucks and the future of logistics - a delphi-based scenario study. Logist. Res. **16**, 1–21 (2023)
13. Ritz, J.: Mobilitätswende – autonome Autos erobern unsere Straßen. Wiesbaden: Springer Fachmedien Wiesbaden (2018)
14. Zanker, C.: Branchenanalyse Logistik: Der Logistiksektor zwischen Globalisierung, Industrie 4.0 und Online-Handel. Düsseldorf: Hans Böckler Stiftung (2018)
15. Bonnes, M.: Der Weg zum umfassenden Dienstleistungspartner. In: Voß, P.H. (ed.) Logistik – die unterschätzte Zukunftsindustrie, pp. 169–178. Springer Fachmedien Wiesbaden, Wiesbaden (2020). https://doi.org/10.1007/978-3-658-27317-0_14
16. Pernestål, A., Engholm, A., Bemler, M., Gidofalvi, G.: How will digitalization change road freight transport? Scenarios tested in Sweden. Sustainability. **13**, 304 (2021). https://doi.org/10.3390/su13010304

17. Kersten, W., Seiter, M., von See, B., Hackius, N., Maurer, T.: Chancen der digitalen transformation: trends und strategien in logistik und supply chain management. Hamburg: DVV Media Group GmbH (2017)
18. Konrad, K., Wangler, L.U.: Tailor-made technology: the stretch of frugal innovation in the truck industry. Procedia Manufact. **19**, 10–17 (2018). https://doi.org/10.1016/j.promfg.2018.01.003
19. Wurst, C.: Chancen von Logistik 4.0 nutzen. Controll. Manage. Rev. **64**, 34–39 (2020). https://doi.org/10.1007/s12176-019-0084-8
20. Witten, P., Schmidt, C.: Globale Trends und die Konsequenzen für die Logistik der letzten Meile. In: Schröder, M., Wegner, K. (eds) Logistik im Wandel der Zeit – Von der Produktionssteuerung zu vernetzten Supply Chains. Springer Gabler, Wiesbaden (2019). https://doi.org/10.1007/978-3-658-25412-4_14
21. Kovacs, G., Kot, S.: New logistics and production trends as the effect of global economy changes. PJMS. **14**, 115–126 (2016). https://doi.org/10.17512/pjms.2016.14.2.11
22. Dong, C.: A supply chain perspective of synchromodality to increase the sustainability of freight transportation. 4OR **16**(3), 339–340 (2017). https://doi.org/10.1007/s10288-017-0367-x
23. Kourounioti, I., Kurapati, S., Lukosch, H., Tavasszy, L., Verbraeck, A.: Simulation games to study transportation issues and solutions: studies on synchromodality. Transp. Res. Rec. **2672**, 72–81 (2018). https://doi.org/10.1177/0361198118792334
24. Schiller, T., Maier, M., Büchle, M.: Global truck study 2016: the truck industry in transition (2017). https://www2.deloitte.com/content/dam/Deloitte/de/Documents/operations/Deloitte%20Global%20Truck%20Study%202016.pdf. Accessed 02 June 2021
25. Schuckmann, S.W., Gnatzy, T., Darkow, I.-L., von der Gracht, H.A.: Analysis of factors influencing the development of transport infrastructure until the year 2030 — A Delphi based scenario study. Technol. Forecast. Soc. Chang. **79**, 1373–1387 (2012). https://doi.org/10.1016/j.techfore.2012.05.008
26. Schönberg, T., Pisoke, M., Illi, A., Hollacher, J.: Advancing the future of logistics: FreightTech Whitepaper 2020. München (2020)
27. Burgess-Limerick, R.: Human-systems integration for the safe implementation of automation. Mining Metall. Explor. **37**(6), 1799–1806 (2020). https://doi.org/10.1007/s42461-020-00248-z
28. Mayring, P.: Qualitative content analysis. Forum: Qualit. Soc. Res. **1**, 1–10 (2000). https://doi.org/10.1093/acprof:oso/9780190215491.003.0004
29. Dashkovskiy, S., Suttner, R.: Reduction of waiting time in logistics centers by trailer yards. IFAC-PapersOnLine. **50**, 7959–7963 (2017). https://doi.org/10.1016/j.ifacol.2017.08.899

The Influence of Situational Variables Toward Initial Trust Formation on Autonomous System

Priscilla Ferronato, Liang Tang[✉], and Masooda Bashir

University of Illinois at Urbana-Champaign, Champaign, USA
{ltang29,mnb}@illinois.edu

Abstract. Autonomous vehicles (AV) are predicted to change our current transportation system, however, how and when they become fully adopted is still an uncertain matter. One essential aspect to consider is how people form trust towards AV. In the context of AV, trust in technology is critical for safety considerations. Although humans are capable of making instinctive assessments of the trustworthiness of other people, this ability does not directly translate to technological systems. The rising complexity of autonomous systems (AS) (e.g., cruise control) requires the operators to calibrate their trust in the system to achieve their safety and performance goals. As such, a detailed understanding of how trust develops, and especially the underlying mental processes, will facilitate the prediction of how trust levels influence behavior mode and decision-making strategies when interacting with AV. To investigate this in the context of AV, we conducted interviews and follow-up surveys to examine users' current behavior with an analogous system (cruise control) and explored its relationship with the perception of AV trustworthiness. Our findings suggest that external factors play a role in the adoption of cruise control (an analogous system), while internal factors determine non-adoption. Trustworthiness in AV is affected by external factors, users' trust in others, and their knowledge of advanced vehicle technology.

Keywords: Human-centered computing · User studies · Autonomous System

1 Introduction

Autonomous vehicles (AV) are anticipated to bring a revolution to our transportation system, representing a technological leap forward that reshapes the public's view of mobility [21]. AV systems (e.g. pilot assist and cruise control) are increasingly part of our lives. Yet, the interaction between human operators and AV systems are far from perfect and seamless. Mismatches between operators' trust and the objective trustworthiness of the system often occur in this process; these mismatches have been identified as a major factor of AV accidents, according to a recent report by the California Department of Motor Vehicles [39] and a study about the tragedies with Tesla Autopilot and Uber self-driving car [28]. A major barrier for mitigating this problem, however, is the incomplete understanding of the mechanisms underlying humans' trust in technology. While great attention has been devoted to the more dynamic aspects of trust, such as

H. Krömker (Ed.): HCII 2023, LNCS 14048, pp. 70–89, 2023.
https://doi.org/10.1007/978-3-031-35678-0_5

trust calibration facilitating design and methods, the knowledge of the operator's initial trust formation is limited. Bridging this gap, this study examines the influence of situational variables towards initial trust formation in the context of human-AV interaction by semi-systematic interviews, with a focus on the external (e.g. road conditions, weather, and task complexity) and internal variables (monitoring behavior, expertise, and mood).

In the context of autonomous vehicles (AV), trust in technology is critical for safety considerations. Although humans are capable of making instinctive assessments of the trustworthiness of other people, this ability does not directly translate to technological systems. Increasingly complex systems like AV require operators to calibrate their trust in automation in order to achieve performance and safety goals. The calibration of human trust is therefore essential to avoid situations of mismatched trust levels and to increase the security and accuracy of the system [30]. Trust predicts not only whether an automated system is used but also how it is used. Parasuraman and Riley [40] categorized the interaction with automation into four styles—use, disuse, abuse, and misuse—that can be linked to the operator's trust in the automation. The authors particularly highlight the negative effects of misuse, or inappropriate overtrust, when the operator's trust exceeds the automated system's capabilities. However, trust calibration is sometimes a spiral of overtrust and distrust [58] due to the complex mental mechanisms that underlie it. Frictions exist in drivers' coping with AV' automation limitations [13], as complex and safety-critical AV systems require operators to maintain full awareness of the situation and be prepared to intervene, which is challenging for human beings [3]. Monitoring behavior drains humans' attention resources. During such behavior, mistuned trust in the system can easily be established and lead to failures in error detection and intervention. AV' wide accessibility among users with different levels of experience has exposed the general public to these risks. The theme of trust in AV has been examined with a wide range of research focuses, especially in mental models of trust [2, 58], consequences of mistuned trust [28, 44], trust calibration methods [13], and technologies to enhance situation awareness [32]. Yet, critical knowledge regarding the underlying cognitive construct remains unknown, especially those aspects that cannot be directly observed due to the opaqueness of the human mind.

The initial trust is a reliable indicator of trust relationships in later phases as it reveals users' cognitive construct of uncertainty and risk when adopting a new technology [31]. Understanding how trust forms, especially the underlying mental process, will help predict how trust levels influence behavior and decision-making when interacting with autonomous vehicles (AV). Research has shown that people's initial trust in AV depends on their experiences with analogous technology, such as sophisticated vehicle technology and computers [29]. People tend to refer to their previous experience with technology when explaining the benefits and concerns of AV, confirming that they will integrate their previous interaction strategies into the adoption of new technology [46]. To achieve a holistic view of AV trust formation, several steps need to be taken that consider the three dimensions of human-automation trust: dispositional trust, situational trust, and learned trust [20]. Research has already identified the influence of individual differences and psychological factors on the formation of dispositional trust [12]. The next crucial step is to understand the role of situational factors in trust formation. Such situational factors may set the theoretical foundation for future research on dynamic learned trust and facilitate a human-centered approach to designing effective human-AV interaction with calibrated trust.

The situational trust dimension has two broad sources of variability: the external environment (e.g. workload, distractors, perception of risk) and the internal context-dependent characteristics of the operator (e.g. mood, attentional capacity, subject matter expertise). These variable factors are important "not only because they directly influence trust, but also because they determine the degree of influence that trust has on behavior towards automation" [20]. Empirical studies of the situational trust in automation often focus the investigation of current behavior, attitudes, and feelings towards the use of similar types of systems for more realistic and contextualized insights. Following this strategy, this study has conducted structured interviews with car owners regarding their interactions, opinions, and feelings with analogous systems (cruise control). Addressing the urgent need for further exploring trust in AV and the existing gaps in research, the central focus of this study was answering: What is the relation between the behavior with analogous systems and situational trust formation in AV? Facilitating and guiding this exploratory investigation, these two research questions are investigated in detail:

RQ1. What are the internal variables influencing the initial situational trust in the analogous system?

RQ2. What are the external variables influencing the initial situational trust in the analogous system?

2 Research Background

Trust is a complex and multifaceted concept that develops over time, as it has a long history and has been discussed in a variety of disciplines. The primary understanding of trust was developed on the basis of human-human trust relationships where trust is associated with the expectancy that another person, or an institution, will act in a certain manner [45]. Trust in autonomous systems, on the other hand, is a distinct process conditional on different variables that have increasingly been investigated by the Human-Computer Interaction (HCI) community. In this study, we adopt the conceptual framework of trust in automation proposed by Hoff and Bashir [20].

2.1 Trust in Automation

Human Trust in Automation. Trust in automation can be defined as the willingness of humans to rely on automated systems in situations characterized by uncertainty and vulnerability [21, 32]. Related research has great importance since trust was identified as a safety-critical factor in human-technology interactions decades ago. Ideally, human trust should be proportional to the capabilities of automation, but, in reality, trust mis-calibration is often observed. Humans' inherent cognitive mechanism does not provide the best support for tracking the performance, process, or purpose of automation over an extended time span. Vigilance decrements during supervision of systems with high perceived reliability often induce overtrust, leading to slow detection of failures, poor situational awareness, and delayed intervention. Distrust can similarly cause users to abandon necessary technical support, even when the situation exceeds human information processing capacity. According to Lee and Dee, "trust and its effect on reliance are part of a closed-loop process: interaction with the automation influences trust and trust influences the interaction with the automation" [30]. If users distrust the automation, they may gain limited experience with it and trust is unlikely to grow. Likewise, an overtrusted system tends to be monitored less frequently, so trust continues to grow despite occasional errors [30]. For example, according to Wintersberger and Riener, the general public's trust in autonomous vehicles (AV) dropped after news of a Tesla Autopilot accident, but it was regained as the concern faded away months later [58]. Thus, a more stable trait underlying human trust in automation, such as the initial formation mechanisms, needs to be understood.

Trust Formation. Trust formation between humans has been well-examined, while many gaps of knowledge remain regarding the understanding of this human-automation trust formation. Fortunately, parallels exist between the two mechanisms. In some sense, initial trust formation can be analyzed as a decision-making process based on the assessment of the perceived trustworthiness, not necessarily aligned with objective trustworthiness and experiences in similar situations. The formation of trust involves both thinking and feeling. Emotion is a reason for trust to have such a profound impact on human behavior [30]. As an enduring tendency of individuals, the affective aspect of trust formation was covered in a previous study of how dispositional factors influence trust [12]. The other aspect of thinking nevertheless demonstrates many features in analogic or analytic judgments. The former recruits' societal norms and the public's opinion in assessing trustworthiness, while the latter involves rational evaluation based on the salient characteristics of the technology [30]. Cognitive resources also play a role in this process, which lead people to deploy a two-way strategy just as they do in other decision-making scenarios. When there is sufficient time and mental resources, people tend to adopt the analytic approach to seek the optimal option. Limited mental resources alternatively lead people to affective and analog approaches, which are more efficient but less rational. The requirements in urgent decision-making and intervention in interaction with AV often forced the operators to take the analog approach, which is highly susceptible to context variables. Understanding how the situational factors impact trust formation in AV therefore will be a crucial step.

Trust in Automation Theoretical Model. This study adopts the Human-Automation trust model proposed by Hoff and Bashir in 2015. Its structure highly matches our focus on situational factors while providing a strong theoretical base of the initial stage of trust development. Addressing a wide range of automation systems and situations, this model defined three broad dimensions based on the categories of driven factors underlying the trust: the human operator, the environment, and the automated system itself. These variables respectively reflect the three different layers of trust, previously identified by Marsh and Dibben [33]: dispositional trust, situational trust, and learned trust.

Dispositional trust represents an individual's overall tendency to trust automation, independent of the context or the type of the automation system (AS). It refers to long-term tendencies arising from both biological and environmental influences, and it is generally stable within the course of a single interaction although these tendencies can change gradually over time (e.g., cultural values, age, and personality traits).

This dimension is related to the human operator and is highly dependent on tendencies related to individual differences. Situational trust, however, depends on the specific context of the human-AS interaction. The environment and the operator's mental state both exert a strong influence on situational trust. Hoff and Bashir suggest two sources of variability in situational trust: external variables, related to the system and the environment, and internal variables, which are context-dependent characteristics of the operator [20]. The external variables are perceived by the human operator and are influenced by the context.

Learned trust is closely related to situational trust since both are guided by past experience and system performance. Hoff and Bashir emphasize the difference between subject matter expertise (related to situational trust) and past experience with automation (related to initial learned trust). Therefore, for a better understanding of the factors that operate in this dimension, Hoff and Bashir categorize them as initial learned trust and dynamic learned trust [20]. Substantial research has shown that human operators adjust their trust in automation to reflect its real-time performance [30]. Within the context of a single interaction, most of these variables are stable whereas performance is not. Hoff and Bashir thus complete their theoretical model by explaining the influence of the system's performance and how operators adjust their trust in automation to correspond to its ongoing performance, which is affected by factors such as reliability, validity, predictability, dependability, system failure, and usefulness of the system. Moreover, reliability and validity are important antecedents of trust. Reliability refers to the consistency of an automated system's functions, and validity refers to the degree to which an AS performs its intended task. Predictability refers to the extent to which automation performs in a manner consistent with the operator's expectations, and dependability refers to the frequency of automation breakdowns or error messages [37].

2.2 Situational Trust in Autonomous Vehicles

Fully operated autonomous vehicles (AV), or level 5, are not expected to be available on the market for the next ten to twenty years. Until fully implemented, the current scenario of mixed traffic (co-existence of vehicles at different levels of automation) requires drivers to cope with automation limitations and act as monitoring (level 2 of automation) or fallback authority (level 3) [53], increasing the complexity of variables affecting situational trust. Moreover, this scenario also limits the development of empirical research; many studies have adopted simulations and other investigation methods, like analogous systems, to better understand situational trust in AV. The following section presents previous findings related to the external and internal variables of situational trust, either in AV or analogous systems. We conducted a literature review to examine the impact and significance of situational variables on trust. The reviewed literature covers studies on trust in AV as well as analogous systems such as aviation and other forms of automated vehicles.

Vehicle's Interface: The analysis of the design of semi-autonomous vehicles found that the interface of the AS makes participants believe that the system performs better than it actually does, which increases the level of trust and relates to situations of overtrust [13].

External Variables

System Capability: The capability of the automation is important and significantly related to trust formation, in which lower error rates lead to more trust in the system [50]. This is different from the perception that the system will behave perfectly, like the 'halo effect' induced by semi-autonomous cars' interfaces earlier mentioned.

Task Complexity: The analysis of trust shows that participants who were provided with the uncertainty of information trusted the automated system less than those who did not receive such information, which indicates a more proper trust calibration [16].

Weather: A human subject study with an AV simulation found that weather conditions do not have a significant impact on humans' trust during driving mode [51]. On the other hand, when analyzing an analogous system, previous findings show that snow decreases the level of trust in automated braking systems. This shows contradictory findings in a simulation and a real-world environment study.

Brand: Trust in an AV is associated with the vehicle brand. This is "a dual constitution", since, if participants perceive the brand of the vehicle as trustworthy, the level of trust in the AV increases [17, 45].

Automated Features: More technological features available in the car increases the level of trust. A previous study found that drivers present overall higher trust in the vehicle if it presented an automated safety braking system [25].

Internal Variables

Knowledge and Expertise: Overall, increased expertise with a system generally enhances trust development [39, 41, 42, 50]. Similar results for familiarity affecting trust have been found with AV analogous systems, like collision warning systems [26] and takeover performance [18].

Familiarity: The level of comfort with automation may be dependent on familiarity automation [50, 54]. Koustanai, Cavallo, Dalhomme, and Mas [26] suggest that individuals familiar with how the collision warning system functions tend to trust the system more than those who were unfamiliar with the system.

Situational Awareness: Situational awareness is defined as "the perception of the elements in the environment within a volume of time and space, the comprehension of their meaning, and the projection of their status in the near future" [11]. Moreover, system awareness is an operator's dynamic understanding of what is happening around them [48]. Therefore, previous studies stated that the knowledge and expertise of drivers increases their situational awareness [15, 34]. However, automated driving systems and automated driving tasks have shown to lower drivers' situational awareness [8]. Studies specifically about trust in automation have found that the lack of awareness leads to mistrust [43].

Mental Workload and Driving Performance: One of the most frequently expressed purposes of automation, and its initial main goal, is to reduce the operator's workload. Previous studies have found that automation is successful on reducing workload, both mental workload [57] and workload related to the implementation of action [23], which is linked to the situational external variable. Increases in mental workload associated with the automated systems have been shown to lead to degradations of trust in combat identification tasks [56]. The use of automated driving systems supports the reduction of drivers' mental workload and improved driving performance [4]. However, participants driving in semi-automated mode showed a reduction in the physiological activation than when compared to drivin in manual mode, resulting in slower drivers' response times [4].

Monitoring: Semi-autonomous mode reduces the level of driver monitoring [4]. The same study concluded that, by reducing drivers' level of monitoring, it increased their distraction, which they deemed automation-generated distraction. Research has shown that a higher experience level may impair an operator's ability to monitor for unanticipated states in highly reliable systems [2].

Mood: Attitudes toward automation influence overall awareness of system behavior. For example, positive attitudes and emotions (e.g., happiness) influence trust, liking, and reliance on a system; of course, they may also lead to overreliance [2, 36]. Conversely, errors or difficulty accessing information from the system often lead to negative attitudes and subsequent disuse [14]. Commitment is discussed within the literature in terms of feelings of satisfaction. As is the case in many other applications, satisfaction influences human perception of automation. Here, satisfaction with the automation is fostered by the quality of information and service that the automation provides [28]. Most importantly, satisfaction has been shown to be positively correlated with trust development with both vehicle automation [10] and combat identification [56]. Studies also suggest that there may be a direct relationship between mood and effect on trust development [35, 52]. Finally, Costa et al. [7] show that humans' trust is correlated with stress levels.

Level of Attention: Previous studies have indicated that operators with lower attentional control interact differently with and rely more heavily on automated systems compared to those with higher attentional control [5, 6]. If the workload on the driver is too low during periods of automation, they may experience passive fatigue due to low cognitive load and lack of direct control over the task [9]. This passive fatigue can lead to a degradation of overall driver performance. Increased vehicle automation has also been linked to reduced driver vigilance, as evidenced by increased braking and steering reaction times in response to sudden critical events [38, 49].

3   Method

This research utilizes a qualitative approach with semi-structured interviews to gain a comprehensive understanding of behavior, attitudes, opinions, and emotions towards AV-similar technology. The semi-structured interviews were designed to minimize research bias and differences in context by presenting all participants with the same questions in the same order. The aim of the interviews was to gather memorable information, both positive and negative, about past experiences of overtrust, distrust, and mistrust and the internal and external variables associated with these experiences.

To recruit participants, an electronic invitation was distributed to a midwestern US university staff mailing list and as a post on social media platform. The goal was to recruit a similar number of men and women and a diverse variety of participants from a professional background. Seventeen participants were recruited, and their occupations included schoolteacher, business owner, master's student, university professor, office administrator, librarian, student, lawyer, and retiree. For compensation, participants recruited were paid a \$20 Amazon gift card. Table 1 provides the demographic information about participants.

Table 1. Description of participants demographics.

Description		Percentage
Gender	Female	58.9% (10)
	Male	41.1% (7)
Age	18-23	0% (0)
	24-38	58.8% (10)
	39-53	29.4% (5)
	54-72	5.9% (1)
	>72	5.9% (1)
Education	High school or less	0
	College	2 (11.8%)
	Bachelor	6 (35.3%)
	Advanced (Master or Doctoral)	8 (47.1%)
Car Owner	Yes	16 (94.1%)
	No	1 (5.9%)

Due to the COVID-19 pandemic, the interviews were conducted virtually using Zoom conferencing technology and lasted an average of 45 min. The interviews were recorded and transcribed using Otti.ai, a machine learning transcription software. Each transcription was manually reviewed and corrected, if necessary, before the analysis process. The interview was divided into several parts, as follows: Part 1 was the introduction, in which the interviewer reiterated the consent form and used ice-breaking questions to initiate the conversation. Part 2 focused on the adoption of cruise control (CC) and individual preferences. Part 3 consisted of the closure of the interview.

After the interview, participants had 36 h to respond to a brief questionnaire that aimed to assess their perceived trustworthiness of AV and intention of monitoring AV. The perceived trustworthiness of AV scale is based on the work of Jessup et al. [22], which is a 7 point-Likert scale ranging from completely disagreeing to completely agreeing. The Complacency-Potential Rating Scale (CPRS) was also used to measure the perceived trustworthiness of automation.

Figure 1 illustrates the analysis process of this study. The first step consisted of the analysis of the questionnaire responses and the transcription of the interviews. The transcription was made using a machine learning software and reviewed manually by the researchers. Second step was classification of participants into "High AV Trust" or "Low AV Trust" based on the scores of the perceived AS trustworthiness questionnaire. Third step related to the categorization process of the transcriptions. Finally, the fourth step was the exploratory analysis of the categories and what were the particularities and trends when triangulating these categories with the participants' classification of trust in AV level.

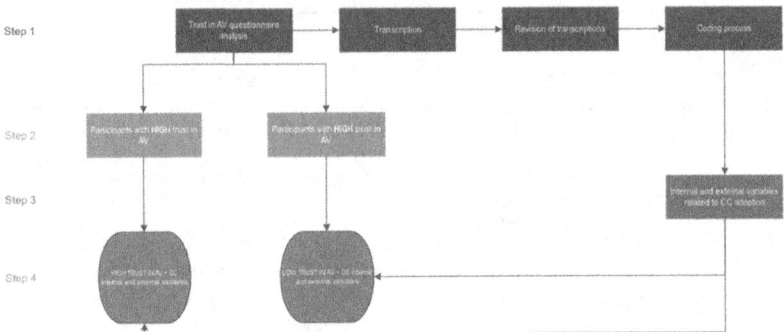

Fig. 1. Data categorization and analysis process

4   Result and Analysis

The analysis process initiated with the categorization of the participants into High Trust in AV and Low Trust in AV based on their scores on the perception of AV's trustworthiness. The scores ranged from 7 to 42, the mean score was 26.17, the mode was 30, the median was 27, and σ with 7.7. Participants with scores above 27 were classified as High Trust in AV and scores lower than 27 were classified as Low Trust in AV, as described in Table 2. Two out of 17 participants presented the median score, while 8 participants (47.1%) were categorized as High Trust and 7 participants (41.2%) as Low Trust in AV.

To identify and organize the main themes discussed by participants, the transcriptions of the interviews were thoroughly investigated through the development of a classification schema, according to the steps as follow: First, the researchers read the transcripts several times to find similarities and labeled them with preliminary codes. These codes included an initial set of themes related to the main research question of this study: adoption of CC, non-adoption of CC, expertise. These initial themes were analyzed to find similarities; they then were grouped into subcategories such as internal and external situational variables and situations that lead to monitoring behavior. In the next subsections, the subcategories developed for each research goal are explained in detail regarding their relationship with the participant's trust in AV classification.

Table 2. Categorization of participants based on the perception of AV trustworthiness.

High Trust	Low Trust
P2	P1
P3	P5
P4	P6
P9	P14
P10	P15
P11	P16
P12	P17
P13	

4.1 The Use of Cruise Control

Cruise Control Adoption. This section analyzes how participants adopt CC. The focus of the analysis is on the external and internal situational variables that influence situational trust in AS, based on the model proposed by Hoff and Bashir [20] and findings from our brief literature review, and followed the order below:

- *Categorization 1.1:* Identify the interests of participants and their knowledge of automobile technology.
- *Categorization 1.2:* The main reasons why participants decide to adopt or not adopt CC.
- *Categorization 1.3:* The reasons for adoption of CC organized into either external or internal situational variables.

The data analysis and categorization began by understanding the participants' interest and knowledge in automobile technologies. The hypothesis was that participants with a higher interest in car technology and technology in general would have a higher level of cruise control (CC) adoption. This aligns with previous findings that higher levels of perceived trust in technology result in more favorable perceptions and adoption of technological innovations [1, 24, 55]. Most participants had a similar level of knowledge about driving technologies and none showed a high level of interest in cars (such as following specialized automobile news or being a car collector). Although they did not follow any specialized media, six out of 17 participants reported conducting online research before buying their current vehicles, like searching for other owners' reviews and car safety ratings on the National Highway Traffic Safety Administration (NHTSA).

"Before buying this car I searched for information about safety on the NHTSA and on the internet to help me with my decision." (P12)

Results show that 5 out of 17 participants (29.4%), who are more engaged with the CC system – using it on more occasions and for longer periods of time – have cars with more technological features, such as lane keeping alerts and automated braking systems.

These participants also show that they are comfortable using CC and they would use it more if it was possible, as mentioned by participant 4: "*I use [it]every time that I know I can speed up a little and there isn't traffic. . . You know Illinois, everything is flat! And my car also breaks if I'm too close to another car, which makes it even more comfortable and I want to use it more*" (P4).

Figure 2 shows that, from the results of Categorization 1.1, knowledge and expertise on vehicles automated systems relates to the type of technology available on participants' cars. This also related to adoption and engagement of CC. As stated by P2, "*I went on a trip last week and I barely drove. The car was doing everything for me, controlling the speed, breaking and lanes. I try to use cruise control and all features of my car every time that I can or it is possible...,*" participants are more engaged with CC and more willing to use it if their cars have other types of automated systems.

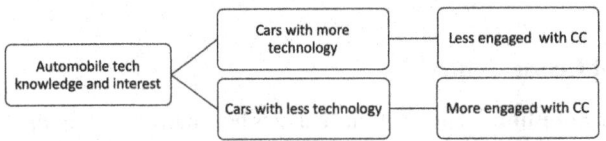

Fig. 2. Results of the categorization of participants' interest and knowledge about automobile technology

These categories were also analyzed regarding their relationship with the participants' trust classification. Table 3 gives the trust categorization and the information if participants did or did not own a car with advanced technological features. The observed discrepancy in trust level seems to well echo the closed-loop relationship between the trust level and the interaction with automation proposed by Lee and See [30]. Less experience with an automation system provides limited information regarding the system's capabilities, so the trust of users will not grow that consequently makes them less likely to use the unfamiliar system. The opposite situation is also observed. The highly trusted system tends to be less supervised, so the users will perceive fewer errors and keep growing their trust regardless of the occasionally occurring issues. In this study, participants' familiarity with advanced technologies, such as automated braking systems, was considered a factor in minimizing perceived risks and uncertainties related to CC, thus increasing trust in its performance and benefits.

Internal and External Variables. As a preliminary step in the analysis of the reasons why participants do or do not adopt CC, a word cloud image was developed to visualize the top 30 words related to the use of CC (Fig. 3) The process to create this image involved multiple steps on stop-words elimination (e.g. I, think, car, yes, ok, among others). It is possible to observe that "helps" is the most recurrent word, followed by "night," "speed," "weather," and "traffic." These words showed three main external variables that are situational to the context of use: the environment (weather, time of the day, and traffic) and process (speed control). Furthermore, this preliminary analysis shows that external variables seem to be highly related to the adoption of CC. External variables are situational factors that relate to the environment and specific characteristics of the context of use (e.g. environment: weather and road conditions) or characteristics of the system (e.g. process and performance: ease of control, complexity, and workload). Internal variables are situational attributes of the driver, such as their mood, attentional capacity, and stress level, among others. These are situational states because they might change over the course of a single or multiple interaction(s) with the system. For instance, a driver can become very sleepy during a long drive, reducing her/his attentional capacity and concentration. These variables were identified and analyzed regarding their relationship with the adoption of CC, as illustrated by Fig. 4, and their correlation with the perception of AV's trustworthiness.

Table 3. Identification of participants trust classification correlated with the presence of advanced autonomous system in their owned vehicles

#	Trust in AV	Car Automation
P1	Low	no tech
P2	High	tech
P3	High	tech
P4	High	tech
P5	Low	no tech
P6	Low	no tech
P7	Median	no tech
P8	Median	no tech
P9	High	no tech
P10	High	no tech
P11	High	tech
P12	High	no tech
P13	High	tech
P14	Low	no tech
P15	Low	no tech
P16	Low	tech
P17	Low	no tech

The preliminary conclusion that external variables have a greater impact on the adoption of CC can be confirmed through a detailed word cloud analysis of internal and external variables. The visibility of the road and the type of road were the top two external

Fig. 3. Top 30 more recurrent words related to the use of cruise control

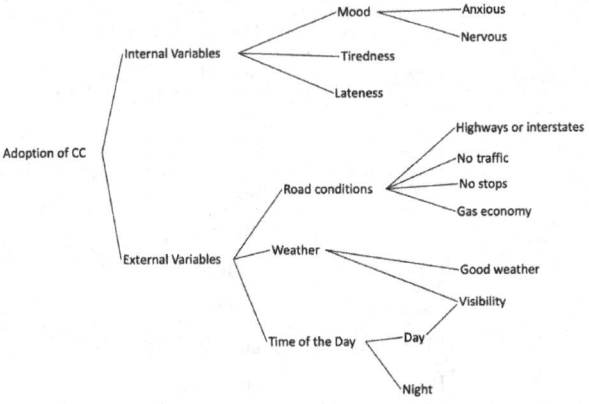

Fig. 4. Categories and subcategories of situational variables related to the adoption and non-adoption of CC

factors that participants considered when deciding to adopt CC. Additionally, 58.82% of participants emphasized the importance of light traffic along with road conditions.

> "Even if the weather is good and I'm driving in the interstate to Chicago for example and there is too much traffic I don't use cruise control." (P1)

> "I don't like to use CC when there is too much traffic because you need to always to break or speed up, and I need to keep turning on the cruise control." (P13)

The relationship between the traffic conditions and the adoption of CC can also relate to the process and performance of the system. This external variable (traffic condition) requires from the participant more or less monitoring behavior of the environment and the system itself (e.g. changing the speed more frequently). The increased monitoring behavior consequently relates to the purpose of the CC system, that is to support the driver on the speed control.

To clarify the impact of internal and external variables on the adoption of cruise control, these variables were used to categorize the reasons for non-adoption. When analyzing the influence of internal variables on the adoption or non-adoption of cruise control, it was noted that stress, tiredness, and nervousness were more frequently linked to avoiding cruise control. Although the majority of participants said they do not use cruise control when they are tired, a small minority (3 participants) reported using it to remain more focused on driving. Contrarily, external variables were more commonly linked to the use of cruise control. All participants stated that they would not activate cruise control during snow or icy road conditions. Most participants only considered heavy rain as a concern, while light rain would not affect their decision to use cruise control. Furthermore, heavy rain and snow were both considered as situational factors that would lead to non-adoption. Additionally, visibility limitations were also considered as a factor affecting the adoption of cruise control.

The next phase of the analysis involved examining the interplay between internal and external variables and their correlation with the participants' level of perceived trustworthiness in autonomous vehicles (AV). Participants were asked what they would do if they were driving on an interstate at night and feeling very tired. For the majority of participants (15 out of 17), including those who were classified as having high trust in AV, nighttime did not play a role in their decision to use or not use cruise control (CC). This was due to either good road visibility or the belief that nighttime was a better time to use CC because of less traffic. All 3 participants who would use CC at night were classified as having low trust in AV. However, the small sample size makes it difficult to determine the impact of low trust perception, and further research is needed.

The internal factor of tiredness presents a more considerable influence in the decision to adopt CC when compared to the external one (time of the day). Only 3 participants (2 high trust and 1 low trust) stated that the use of CC when tired helps them to concentrate more on the road instead of the speed. On the other hand, 6 out of 8 High Trust in AV participants claimed that they would not adopt CC in the given scenario because driving in manual mode would make them feel more engaged with the car and consequently more awake.

"I do think CC can be dangerous if you're tired because with cruise control you're doing less things by not taking control of the speed, then you have less physical contact with the car, and I think this makes you even more sleepy." (P5)

"I don't mind if it is night or day, as long as there is not too much traffic and the road is good, I will use CC. But if I am tired, I think cruise control will make me more sleepy, you know, . . . Because I will be more relaxed, I don't need to pay attention to the speed. So, if I am tired, I won't use it, doesn't matter if it is night or day." (P8)

Moreover, the decision for using CC also relates to the perception of timing reaction. For example, participants use the buttons on the steering wheel to control the speed when it is necessary to change only a few miles up or down. Most participants (14 of 17) claimed that, when they need to break, they immediately use the breaks instead of the control buttons:

> "... *usually when to slow down, I break and then reset if I'm going to...*". This action is also associated with the reaction timing of the car" (P6).

> "... because slowing down with the buttons takes a lot longer time than slamming on the brakes." (P3)

> "... if I could stop suddenly, like I'll press the brake, you know if I really have to slow down fast, but if someone in front of me is just going like, you know, few, few miles per hour like slower then I'll just slow the cruise I'll stay on cruise control, just slow down." (P17)

When participants control their speed using the pedals instead of the buttons, this was noted as an automatic reaction without conscious thought, even if it resulted in turning off the cruise control. Additionally, participants associated using cruise control (CC) with both quick and slow reactions while driving, indicating that its use affects their attention to the road and ability to adjust speed, as shown in Fig. 5, regardless of their level of trust in autonomous vehicles (AV).

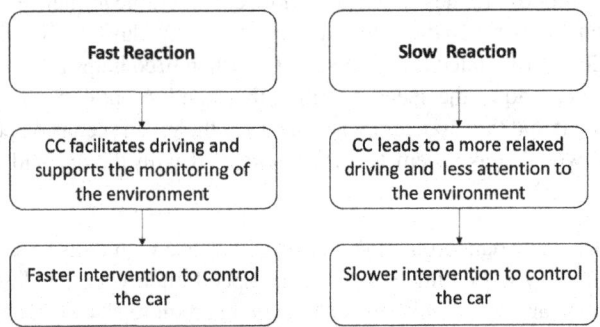

Fig. 5. Participants' association of cruise control adoption and their capacity of reaction

Overall Perception of Cruise Control. All participants had positive views of cruise control (CC), regardless of their level of trust in autonomous vehicles (AV). Figure 6 displays the top 30 words mentioned by participants when discussing their general perception of CC, such as "comfort," "ease of driving," and "improved concentration." For instance, participant P2 said "*It's really, really comfortable*" and also mentioned other common automated driving technologies, such as setting the car's cruise control to the speed limit and not having to drive themselves. Physical comfort was also noted, particularly during long drives, as stated by participant P6: "What I like about it is that it can give your leg a break." These attributes are related to the purpose of the system, while the perceived ease of use relates to its performance and process. The reliability of controlling the speed was another important element brought up by 6 participants,

Fig. 6. Word cloud developed to illustrate the top 30 words related to the perception of adopting cruise control

which also relates to the system performance, as illustrated by the comment of P7 *". . . that's probably the main reason why I use it. It is to just keep my speed steady."*

5 Discussion

The results of the semi-structured interviews suggest that familiarity with similar systems, as an internal variable, impacts trust in AV. Participants who own cars with more advanced autonomous systems (e.g. automated braking and lane-keeping systems) than the CC system under investigation showed higher trust in AV. Further research is needed to examine the concept of familiarity, not only in relation to more advanced automation but also the specific system under investigation. Familiarity plays a crucial role in situational trust processes by helping humans understand unknown situations before interacting with autonomous systems. Participants tend to act in ways that are familiar to them. For example, when deciding between controlling speed with buttons on the steering wheel or manually with the pedals, they usually choose the pedals because that is what they' are accustomed to. Familiarity also leads to increased comfort, which can result in a lack of situational awareness and overtrust. Previous research has shown that, when drivers feel comfortable, they are more likely to feel safe, engage in other activities, become distracted [47], and participate in hazardous driving behaviors [19].

Situations of overtrust have been identified as a major reason for AV incidents due to drivers' failures in monitoring the system and intervening in the control of AV [13, 27]. Our results show that the perception of time available for intervention is one of the main reasons why CC will or will not be adopted by participants, which relates both to the performance of the system and the performance of the user. For instance, participants stated that they will not use the CC system if they think they are in a situation where they will need to quickly take back control of the vehicle (under poor conditions such as bad weather or too much traffic). On the other hand, participants who were more engaged and comfortable with the use of CC (consequently higher perception of AV trustworthiness) were also less worried about their reaction time on taking the control back.

Regarding research question 1: "What are the internal variables influencing the initial situational trust in the analogous system?" and research question 2: "What are the external variables influencing the initial situational trust in the analogous system?", it is possible to conclude that participants classified as "High Trust in AV" are more influenced by internal variables than external ones when deciding to not adopt CC (i.e. tiredness). External variables (i.e. Time of day and traffic) did not impact the decision of adopting CC by "High Trust in AV" participants as it did with participants classified as "Low Trust in AV." This provides insight for future studies that must consider how different situational variables might influence levels of trust in different ways. A new model of trust calibration thus must be developed acknowledging the fact that trusting in AV is not only a simple YES or NO response. Rather, it must involve the dynamic differences of trust levels and how situational factors contribute differently. This acknowledgement is an important aspect for the design of features addressing certain types of behaviors with the main goal being the calibration of users' trust in AV.

Our work makes several contributions to the HCI community, including: (i) A review of situational variables that influence the initial trust in automation; (ii.) A user- centric study for the investigation of trust in automation. Upon elucidating the role of situational factors in trust formation, we will proceed to fill in the theoretical blanks in the role of the learned trust in initial trust (iii.) Providing implications and guidelines for future researchers to design better human-centered autonomous systems. Although this study presents exploratory results and has limitations in terms of sample size and diversity as well as methodology, the investigation of current behavior with analogous systems proved to be a valid approach, affording important qualitative insights for understanding human and AV interactions. Further research must focus on the investigation of how levels of trust are affected by different situational variables in different ways. Since trust in AV is a dynamic and multidimensional process, future studies must also investigate the relationship between the current behavior with these analogous systems and dispositional trust in AV by developing a data triangulation with the results of the situational trust dimension.

Finally, the results of this investigation reveal that situational factors involve complex cognitive processes that deeply influence users' perceptions and judgments. Understanding users' current behavior with analogous systems can therefore provide useful insights for the design of parameters calibrating different levels of trust in AV. In turn, this influences the decision of adopting (or not) AV technology and successful interaction between users and automation.

References

1. Mukherjee, A., Nath, P.: Role of electronic trust in online retailing. Eur. J. Mark. **41**, 1173–1202 (2007)
2. Bailey, N.R., Scerbo, M.W., Freeman, F.G., Mikulka, P.J., Scott, L.A.: Comparison of a brain-based adaptive system and a manual adaptable system for invoking automation. Hum. Fact.: J. Hum. Fact. Ergon. Soc. **48**, 693–709 (2006)
3. Bainbridge, L.: Ironies of automation, pp. 129–135. Analysis, Design and Evaluation of Man-Machine Systems (1983)
4. Biondi, F.N., Lohani, M., Hopman, R., Mills, S., Cooper, J.M., Strayer, D.L.: 80 mph and out-of-the-loop: effects of real-world semi-automated driving on driver workload and arousal. Proceed. Hum. Factors Ergon. Soc. Annual Meeting **62**, 1878–1882 (2018)

5. Chen, J.Y.C., Terrence, P.I.: Effects of imperfect automation and individual differences on concurrent performance of military and robotics tasks in a simulated multitasking environment. Ergonomics **52**, 907–920 (2009)
6. Chen, J.Y., Barnes, M.J.: Supervisory control of multiple robots. Hum. Factors: J. Human Factors Ergon. Soc. **54**, 157–174 (2012)
7. Costa, A.C., Roe, R.A., Taillieu, T.: Trust within teams: the relation with performance effectiveness. Eur. J. Work Organ. Psy. **10**, 225–244 (2001)
8. de Winter, J.C.F., Happee, R., Martens, M.H., Stanton, N.A.: Effects of adaptive cruise control and highly automated driving on workload and situation awareness: a review of the empirical evidence. Transport. Res. F: Traffic Psychol. Behav. **27**, 196–217 (2014)
9. Duffy, V.G.: Stress, workload, and fatigue. In: Hancock, P.A., Desmond, P.A. (eds.) Lawrence Erlbaum Associates, publishers, Mahwah, NJ, 2001, p. 682, hardcover: ISBN 0–8058–3178–9, $75. Hum. Factors Ergon. Manufact. **11**, 189–190 (2001)
10. Donmez, B., Boyle, L.N., Lee, J.D.: The impact of distraction mitigation strategies on driving performance. Hum. Factors: J. Hum. Factors Ergon. Soc. **48**, 785–804 (2006)
11. Endsley, M.R.: Toward a theory of situation awareness in dynamic systems. Hum. Factors: J. Hum. Factors Ergon. Soc. **37**, 32–64 (1995)
12. Ferronato, P., Bashir, M.: An examination of dispositional trust in human and autonomous system interactions. In: Kurosu, M. (eds) Human-Computer Interaction. Human Values and Quality of Life. HCII 2020. Lecture Notes in Computer Science(), vol 12183. Springer, Cham (2020). https://doi.org/10.1007/978-3-030-49065-2_30
13. Frison, A.-K., et al.: In UX we trust. In: Proceedings of the 2019 CHI Conference on Human Factors in Computing Systems (2019)
14. Gao, J., Lee, J.D., Zhang, Y.: A dynamic model of interaction between reliance on automation and cooperation in multi-operator multi-automation situations. Int. J. Ind. Ergon. **36**, 511–526 (2006)
15. Gugerty, L.J., Tirre, W.C.: Individual differences in situation awareness. Situation awareness analysis and measurement. Individual differences in situation awareness. Situation Awareness Analysis and Measurement, pp. 249–276 (2000)
16. Helldin, T., Falkman, G., Riveiro, M., Davidsson, S.: Presenting system uncertainty in Automotive UIS for Supporting Trust calibration in Autonomous Driving. In: Proceedings of the 5th International Conference on Automotive User Interfaces and Interactive Vehicular Applications (2013)
17. Hengstler, M., Enkel, E., Duelli, S.: Applied artificial intelligence and trust—the case of autonomous vehicles and medical assistance devices. Technol. Forecast. Soc. Chang. **105**, 105–120 (2016)
18. Hergeth, S., Lorenz, L., Krems, J.F.: Prior familiarization with takeover requests affects drivers' takeover performance and automation trust. Hum. Factors: J. Hum. Factors Ergon. Soc. **59**, 457–470 (2016)
19. Hoedemaeker, M., Brookhuis, K.A.: Behavioural adaptation to driving with an adaptive cruise control (acc). Transport. Res. F: Traffic Psychol. Behav. **1**, 95–106 (1998)
20. Hoff, K.A., Bashir, M.: Trust in automation. Hum. Factors: J. Hum. Factors Ergon. Soc. **57**, 407–434 (2014)
21. Peng, Y.: The ideological divide in public perceptions of self-driving cars. SSRN Electron. J. **29**, 436–451 (2020)
22. Patent, V., Searle, R.H.: Qualitative meta-analysis of propensity to trust measurement. J. Trust Res. **9**, 136–163 (2019)
23. Jou, Y.-T., Yenn, T.-C., Lin, C.J., Yang, C.-W., Chiang, C.-C.: Evaluation of operators' mental workload of human–system interface automation in the Advanced Nuclear Power Plants. Nucl. Eng. Des. **239**, 2537–2542 (2009)

24. Kim, K.J., Park, E., Shyam Sundar, S.: Caregiving role in Human-Robot Interaction: a Study of the mediating effects of perceived benefit and social presence. Comput. Hum. Behav. **29**, 1799–1806 (2013)

25. Koglbauer, I., Holzinger, J., Eichberger, A., Lex, C.: Drivers' interaction with adaptive cruise control on dry and snowy roads with various tire-road grip potentials. J. Adv. Transp. **2017**, 1–10 (2017)

26. Koustanaï, A., Cavallo, V., Delhomme, P., Mas, A.: Simulator training with a forward Collision Warning System. Hum. Factors: J. Hum. Factors Ergon. Soc. **54**, 709–721 (2012)

27. Kundinger, T., Wintersberger, P., Riener, A.: (over) Trust in Automated Driving. In: Extended Abstracts of the 2019 CHI Conference on Human Factors in Computing Systems (2019)

28. Lee, H., Kim, J., Kim, J.: Determinants of success for application service provider: an empirical test in small businesses. Int. J. Hum Comput Stud. **65**, 796–815 (2007)

29. Lee, J.D., Kolodge, K.: Exploring trust in self-driving vehicles through text analysis. Hum. Factors: J. Hum. Factors Ergon. Soc. **62**, 260–277 (2019)

30. Lee, J.D., See, K.A.: Trust in automation: designing for appropriate Reliance. Hum. Factors: J. Hum. Factors Ergon. Soc. **46**, 50–80 (2004)

31. Li, X., Hess, T.J., Valacich, J.S.: Why do we trust new technology? A study of initial trust formation with organizational information systems. J. Strateg. Inf. Syst. **17**, 39–71 (2008)

32. Mahadevan, K., Somanath, S., Sharlin, E.: Communicating Awareness and intent in autonomous vehicle-pedestrian interaction. In: Proceedings of the 2018 CHI Conference on Human Factors in Computing Systems (2018)

33. Mezgar, I., Kincses, Z.: The role of trust in Information Technology Management. In book: Knowledge and Information Technology Management, pp. 283–304 (2003)

34. Matthews, M.L., Bryant, D.J., Webb, R.D., Harbluk, J.L.: Model for situation awareness and driving: application to analysis and research for intelligent transportation systems. Transport. Res. Record: J. Transport. Res. Board **1779**, 26–32 (2001)

35. Merritt, S.M.: Affective processes in human–automation interactions. Hum. Factors: J. Hum. Factors Ergon. Soc. **53**, 356–370 (2011)

36. Merritt, S.M., Heimbaugh, H., LaChapell, J., Lee, D.: I trust it, but I don't know why. Hum. Factors: J. Hum. Factors Ergon. Soc. **55**, 520–534 (2012)

37. Merritt, S.M., Ilgen, D.R.: Not all trust is created equal: dispositional and history-based trust in human-automation interactions. Hum. Factors: J. Hum. Factors Ergon. Soc. **50**, 194–210 (2008)

38. Neubauer, C., Matthews, G., Langheim, L., Saxby, D.: Fatigue and voluntary utilization of automation in simulated driving. Hum. Factors: J. Hum. Factors Ergon. Soc. **54**, 734–746 (2011)

39. California Department of Vehicles: AUTONOMOUS VEHICLES: The Autonomous Vehicle branch of DMV oversees and regulates autonomous vehicle testing and deployment on California roads. https://www.dmv.ca.gov/portal/vehicle-industry-services/autonomous-vehicles/

40. Parasuraman, R., Riley, V.: Humans and automation: use, misuse, disuse, abuse. Hum. Factors: J. Hum. Factors Ergon. Soc. **39**, 230–253 (1997)

41. Rajaonah, B., Anceaux, F., Vienne, F.: Study of driver trust during cooperation with Adaptive Cruise Control. Le Travail Humain **69**, 99 (2006)

42. Rajaonah, B., Anceaux, F., Vienne, F.: Trust and the use of adaptive cruise control: a study of a cut-in situation. Cogn. Technol. Work **8**, 146–155 (2006)

43. Reig, S., Norman, S., Morales, C.G., Das, S., Steinfeld, A., Forlizzi, J.: A field study of pedestrians and autonomous vehicles. In: Proceedings of the 10th International Conference on Automotive User Interfaces and Interactive Vehicular Applications (2018)

44. Rogers, M., Zhang, Y., Kaber, D., Liang, Y., Gangakhedkar, S.: The effects of visual and cognitive distraction on driver situation awareness. In: Harris, D. (eds) Engineering Psychology and Cognitive Ergonomics. EPCE 2011. Lecture Notes in Computer Science(), vol 6781. Springer, Berlin, Heidelberg (2011). https://doi.org/10.1007/978-3-642-21741-8_21

45. Rotter, J.B.: Interpersonal Trust scale. PsycTESTS Dataset (1967)

46. Rousseau, D.M., Sitkin, S.B., Burt, R.S., Camerer, C.: Not so different after all: a cross-discipline view of trust. Acad. Manag. Rev. **23**, 393–404 (1998)

47. Rudin-Brown, C.M., Parker, H.A.: Behavioural adaptation to Adaptive Cruise Control (ACC): Implications for preventive strategies. Transport. Res. F: Traffic Psychol. Behav. **7**, 59–76 (2004)

48. Salmon, P.M., et al.: Measuring situation awareness in complex systems: comparison of Measures Study. Int. J. Ind. Ergon. **39**, 490–500 (2009)

49. Saxby, D.J., Matthews, G., Warm, J.S., Hitchcock, E.M., Neubauer, C.: Active and passive fatigue in simulated driving: discriminating styles of workload regulation and their safety impacts. J. Exp. Psychol. Appl. **19**, 287–300 (2013)

50. Schaefer, K.E., Chen, J.Y., Szalma, J.L., Hancock, P.A.: A meta-analysis of factors influencing the development of trust in automation. Hum. Factors: J. Hum. Factors Ergon. Soc. **58**, 377–400 (2016)

51. Sheng, S., et al.: A case study of trust on Autonomous Driving. In: 2019 IEEE Intelligent Transportation Systems Conference (ITSC) (2019)

52. Stokes, C.K., Lyons, J.B., Schneider, T.R.: The impact of mood on Interpersonal Trust: implications for Multicultural Teams. Trust in Military Teams, pp. 13–30 (2018)

53. Templeton, B.: A Critique of NHTSA and SAE "Levels" of self driving. https://www.templetons.com/brad/robocars/levels.html

54. van den Broek, E.L., Westerink, J.H.D.M.: Considerations for emotion-aware consumer products. Appl. Ergon. **40**, 1055–1064 (2009)

55. Venkatesh, T.: Xu: consumer acceptance and use of information technology: extending the unified theory of acceptance and use of technology. MIS Q. **36**, 157 (2012)

56. Wang, L., Jamieson, G.A., Hollands, J.G.: The effects of design features on users' trust in and reliance on a combat identification system. PsycEXTRA Dataset (2011)

57. Wiegmann, D.A., Eggman, A.A., ElBardissi, A.W., Parker, S.H., Sundt, T.M.: Improving cardiac surgical care: a work systems approach. Appl. Ergon. **41**, 701–712 (2010)

58. Wintersberger, P., Riener, A.: Trust in technology as a safety aspect in highly automated driving. i-com **15**, 297–310 (2016)

Learning Design Strategies for Optimizing User Behaviour Towards Automation: Architecting Quality Interactions from Concept to Prototype

Naomi Y. Mbelekani[✉] and Klaus Bengler

TUM School of Engineering and Design, Chair of Ergonomics, Boltzmannstr. 15, 85748
Garching b. München, Germany
ny.mbelekani@tum.de

Abstract. Automated vehicles are equipped with systems for task operation and furnished with complex sensor-based systems for detecting and avoiding crashes. However, because technology is unruly and prone to error due to its low maturity level, the human is required to be an active member and collaborator in the operation of the vehicle. As a result, users are required to be constantly vigilant for instances where automation may fail and request to intervene. Yet, notably, some users have a tendency to perceive automation as a holy grail and partake in undesirable user behaviours linked to misuse, over trust, and even high-level complacency, etc. Consequently, industry faces safety criticalities, resulting in either low to high-level risk taking behaviours. Thus, focus should be on taming users' knowledge to fit the level of automated driving, trucking, flying, farming systems' capabilities and limitations by investing in exceptionally ergonomic-inspired strategies that promote desirable user behaviours, such as the intended use of automation and interaction with its human-machine interfaces (HMIs) over the sequence of time. As a result, $N = 20$ air and ground vehicle industry experts views on training to use and learning strategies for optimizing safety and risk-free human-automation interaction and use (HAI/U) were considered. The paper devises ergonomically enthused learning design strategies in support of deprogramming-risky behaviours and reprogramming-safe taking behaviours towards automation, by bearing in mind long-term effects.

Keywords: Automated vehicles · Learning design strategies · Long-term effects · Safe and risk-free behaviour · Automation-induced effects · Experts views

1 Introduction

Automation is seen as a multidimensional technology that exhibits tremendous potential for the vehicle industry, as it enhances user experiences on transit. Thus, the industry is experiencing a surge in automation advances, which will rigorously mature for years to come. Due to these fast-paced advances (from manual, assisted, highly automated, to fully automated), automation use and interaction with human-machine interfaces

© The Author(s), under exclusive license to Springer Nature Switzerland AG 2023
H. Krömker (Ed.): HCII 2023, LNCS 14048, pp. 90–111, 2023.
https://doi.org/10.1007/978-3-031-35678-0_6

(HMIs) seems more complex and dynamic. We can argue that the predecessor and current automated vehicles (AVs) have no match to their future successor, as they ubiquitously evolve with differing levels of automation (LoA), HMIs, and mode complexity, etc. Even so, as industries move towards these progressive AVs, there are pressing issues that need to be addressed, for instance long-term effects as users learn, unlearn and relearn to use. Thus, we should consider risk-taking/undesirable behaviours associated with long-term effects, in an effort to keep pace with safety taking trends. Moreover, co-risk factors due to such technology, as it has the potential to be homogeneously problematic and unconstructive if not properly used (see [9; 10]). In essence, long-term studies are crucial to comprehend users learning effects over time (from 'short-term effects', 'mid-term effects' to 'long-term effects'), in hopes to design tractable strategies. Furthermore, some industries have established safety frameworks in the form of "Robot Safety Laws" [25], Responsibility Sensitive Safety (RSS) by Intel/Mobileye [20] to digitize reasonable boundaries on the behaviour [25], Instantaneous Safety Metric (ISM) by the National Highway Traffic Safety Administration and the Transportation Research Centre [26], to name a few. The normalisation of efficient strategies means engaging not just with risk factors, which given the complexity of AV, will be radically indeterminate, but also with the processes and practices, attention and intention, inscribed and implicit knowledge, overt and covert behaviour, automation-induced efficiencies, inefficiencies and deficiencies over time.

1.1 User Knowledge and Behaviour Towards Automation

The advent of automation assures prospects for users and those itinerant in the same space, however, this comes with its huddle of challenges. Over the past years, developers have mainly focused on developing new and better algorithms and hardware specifically intended to push innovation, but fall short on research that explore long-term effects, long-term learning, behaviour adaptations/changes (BAC). In other cases, studies employ measures that are explicitly chosen to highlight the strengths of new systems without illuminating the limits as they are used over an extended period. As such, despite the rapid progression of AVs in the research realm, advances in exploring extended learning effects and BAC lag behind. Thus, we need to consider automation-induced effects and user behaviour, the gap between the innovative technology and the state of practice or rather the context of use over time. In essence, there is a need for authenticated human-automation interaction and use (HAI/U) metrics driven by the need for consistent and informative evaluation of knowledge to corroborate safe functionality and interactive patterns. Such evaluations are critical for advancing the underlying safety models for HAI/U, and for guiding designers and users to meter wrongful expectations. We thus consider several literature lenses that support, on one hand, exploring the technical automated system behaviour, such as the system limitations and capabilities, high-level system architectures, performance, planning and modelling. While on the other hand, exploring knowledge mechanisms on user behaviour based on mental models, cognition and comprehension, as a way to facilitate learning, information processing, decision-making processes, prediction/problem-solving processes, trust, acceptance, and performance over time. The out-of-the loop performance issues are another sub-set of user behaviour challenges, as the lack of situational awareness (SA) can be detrimental to safety. SA relates to the conscience of the surroundings and the comprehension of what information is useful to perform a task, thus divided into three levels: perception of the environment, comprehension of the current situation and projection of the future status [22]. When exploring the context of operation takeover (initiated by the human) and handover or requested-to intervene (RtI) (initiated by the automation), users need to be vigilant and quickly take back control of the vehicle when necessary. Thus, users need to always be situationally aware and avoid being out-of-the loop [8; 4], as this leaves users "handicapped in their ability to takeover manual operations in the event of automation failure" [4]. Thus, developers are faced with devising effective strategies that minimise risk, foster safety, assure efficiency, and applied intended use. Thus, as automation becomes ubiquitous, we need to reduce the possibility of catastrophe due to poor human-automation partnerships [9]. A flawed partnership can be described with examples of misuses and disuses [13], with misuse described as "failures that occur when people inadvertently violate critical assumptions and rely on automation inappropriately", and disuse as "failures that occur when people reject the capabilities of automation" [9]. In effect, user behaviours aligned with misuse and disuse somewhat depend on user knowledge on capabilities and limitations, emotional state, attitudes, personality, awareness, and trust, importantly. Trust in automation focuses on humans' level of confidence/reliance and expectation for the system to perform a required action, thus the quality/fluidity of the interaction/use greatly depends on the degree of trust. Equally, over-trust results in misuse or an overuse of a system while distrust/mistrust results in disuse or a refusal to use [9]. In other words, trust should not exceed system capabilities [8]. Thus, if we can mould the right trust continuum, we might be able to counter and negate undesirable user behaviours. It is essential that users are fully aware of the conditions that they need to engage and disengage automation. This means devising strategies that mobilise awareness, assembles knowledge (e.g. limitations, capabilities), and illuminate potential automation pitfalls, and in turn negate undesirable user behaviours over time. Hence, developers should expend additional design efforts to support users with problems related to hardware and software boundaries, environmental constraints, HMI design, and dynamic task complexity. Correct knowledge processing is critical for safety-taking behaviours, thus we should consider ergonomically inspired strategies that enhance knowledge and competency. Author [24] argued a pressing need to develop learning approaches in order to enable users to appropriately use this emerging technology. Thus, learning approaches might be one of the inevitable tasks for successful introduction of automation [24].

1.2 Training and Learning as a Strategy

The term 'easy' when referencing automated systems brands them as intuitive and intelligent to the degree of not needing training or even supervision, which can somewhat be misleading. Thus, branding automated systems as easy can be considered misnomer, if not an oxymoron, as easy only happens when tasks, human psyche and real-world complexities are sufficiently constrained, and when the system does not need supervision [18]. Author [2] argued, "all bets are off if system interfaces should ever become so simple that learning is virtually instantaneous and training unnecessary", and thus designed what they called "the training wheels interface" to aid user learning. It is important to consider that users, AVs and the environment in which they operate are not self-contained as well as self-sufficient, and thus should not be left to chance. As human history with technology lament that, those ignorant of history are doomed to repeat the same mistakes, and additionally it "becomes easy for innovators to argue that 'this time it is different' and that past pathologies of technological development," such as novel risks and other unintended consequences will be prevented [18]. Reference [19] argues for experts' accountability by reasoning that "in this more ambiguous world, it may be far more difficult to distinguish safe and unsafe actions or conditions, and it becomes more complicated to define responsibility or negligence on the part of those 'in charge', meaning the developers." Author [19] noted that "one specific way in which expert assumptions about the conditions of implementation of a technology are concealed from view, even as hypotheses, is the regular labelling of major accidents as 'human error' by downstream actors, such as 'stupid' or complacent operators." Besides, the premise that humans can faultlessly use automation and not be prone to distraction, illogic or complacency without proper training or continuous learning could be considered a deeper form of 'developer error' or 'designer fallacy'. In essence, just because developers claim 'easy-to-use', does not mean learning should be neglected nor ignored, as this may paralyze safety. Essentially, learning as regards to automation and the human takes precedes and thus imperative to consider the situational reality of 'risk-learning' in conjunction with 'safety-learning', as effects based on extended learning, as a give and take process over long-term use.

To some extend we can argue that machine learning and human learning can be viewed as two sides of the same coin of quality interactions, which promotes efficient vehicle operation, thus as the automation is taught efficient ways to interact with humans, the human is also taught efficient ways to interact with automation. Author [30] discusses the importance of machine learning systems in effective user interactions with machines, thus, offering examples of agents as 'rule-making' and 'rule-following'. When considering a nascent HAI/U, it is important that the automation as a machine learns the world around it and then adjust its behaviour to suit the human's reality, unpredictability, intents and expectations. Thus, considering machine learning and artificial intelligence (AI) using deep neural networks to somewhat mitigate undesirable behaviours during HAI/U. Thus, it is essential to work out the rules, for example, in a situation where a user tries to misbehave, a risk mitigation system could learn this behaviour and mitigate it. For instance, a rule extraction, 'rule learning', or 'rule induction' process can be considered, where a system can be trained by extracting patterns from vast datasets [18]. This also considers deep-learning software and AI that 'learns' the anthropological meaning of what it means to be human with automation, nature of user behaviour and its changeability over the sequence of time. Although, it is important to consider that human intractability is a nuanced phenomenon within the AI domain as human are complex beings. Moreover, as training can be direct (see [11] for recommended training guidelines for AV operation), it can also be indirect through system design. For instance, some of the skills gained in life were unconsciously learned by osmosis. For instance, considering psychological conditioning theories (e.g. Ivan Pavlov, B.F. Skinner, etc.) inspired to mould behaviour and mental maps. The ability to intuitively model patterns of behaviour that support a quality-output could be developers' greatest assets, thus learning by role modelling. Moreover, training protocols should address the need for users to "focus their attention on the roadway at all times but particularly during automation use", and the need for training "that incorporate individual differences in learning and with multiple stakeholder involvement" [11]. Training can be helpful in developing correct mental models on how to behave over time (e.g. when to activate, deactivate, takeover, etc.).

2 Method

As part of an industry-centred study with a user-centred design in mind, a sample of twenty ($N = 20$) ground and aircraft vehicle experts were consulted. The experts (Smart farming/Agriculture 46%, Aircrafts 17%, Trucks 8%, and Cars 29%) work for international organisations. In order to relate views, we assigned each expert a pseudonym based on their industry domain, for example, *Car* Experts (CE), *Truck* Experts (TE), *Smart Farming* Experts (SFE), and *Aircraft* Experts (AE). The interviews were based on different information themes and topics segments (with a focus on expected risks and benefits of automation, human and ergonomic factors, long-term effects, training and learning, etc.), for which this paper was derived. Experts where asked among others questions: how and/or what approaches could be used to nudge users to use automation as intended, short of misuse and undesirable behaviour? As the aim was to suggest applicable strategic concepts, the data was analysed using content analysis, resulting in anecdotal evidence on learning design strategies. The steps taken to analyse the data after transcription was applied, were: (1) data familiarisation, (2) theme selection and coding, (3) reviewing themes, (4) categorising the themes for reporting (5) overall data comparison and integration, (6) reporting the findings. In writing this report, *phase 1* entailed compiling interview extracts from the experts based on the specific information topic, then *phase 2* we consulted available literature on the topic, and then in *phase 3* we compared and integrated literature finding and experts' views to formulate the learning design strategies. The contributed expert views convey foci related to learning by considering direct (conventional) and indirect (unconventional) means as many developers' prognosis of higher LoA nears.

3 Results

Learning design strategies are based on the premise that knowledge acquisition is a critical element for safe and risk-free behaviour for long-term HAI/U, and describes a process for augmenting performance (a quality function) which includes ergonomic-inspired strategies that consider human factors, automation system factors, long-term effects, safety, risk, environment, training, context of use and behaviour changeability, etc. The aim is to support the human and increase knowledge on situational concerns. SFE4 stated, "maybe the combination of the two extremes, either completely automate it on one side and driver orientated training on the other side, to kind of combine them both, which could probably be learned by other industries, because they are just focusing on one perspective maybe." CE1 stated that, in the automotive industry "the system is designed, put it in the market and people are expected to know how to use it," and further noted, "a lot of stuff is almost like self-learning over time." Users eventually acquaint themselves with AVs as a process of domestication. A blend of human assisted-learning juxtaposed with machine-assisted learning is vital.

3.1 Human Assisted Learning

Experts mentioned that people's learning patterns, habits and processes vary as different people learn to use automated system in different ways. Training was recognised as essential, especially as automation matures and becomes complex for the less tech savvy or tech ignorant user. It was mentioned that researchers/designers need to consider why users behave the way they do and from there formulate strategies that can assist them to efficiently use automation. Most experts were concerned about designing self-explaining systems that prevent undesirable user behaviours, yet some noted that training could be used to bridge the knowledge gap between knowing and doing. For automated driving and trucking, CE2 described that "training can have a lot of value, if you are able to train people through an advertisement, video or whatever, it can have a lot of value; can make them understand the correct mental model." Lately the need for training has received a lot of consideration based on whether "to train or not to train" as well as "how to train". For instances, TE2 reasoned that they "had a lot of discussions on the need or no need for training", and further elaborated that "there are studies showing that training is important and there are studies showing that training is not important". As a result, due to these peculiar different findings, TE2 speculated that the different result may be biased based on who does the research, and that they "may be some confirmation bias or just more studies are needed". In assessing the significance of training, CE2 conducted a study that focused on users with and without training, and as a result, those that had received training exhibited a 'higher-level knowledge ability' and those without exhibited a 'lower-level knowledge ability'. CE2 noted that "people with no training, many of them did not realize that when that alert occurred, that they were not supposed to grab the wheel, most of them thought that alert was actually a collision alert, even though it was not." For those that got training, CE2 explained that the mere fact that they communicated to them that the system has to be supervised had useful results and "all of them instantly figured out what was going on." Seeing that they do not have highly automated trucks other than in concept vehicles, TE1 noted they "do not have an established approach for training" however, "we already have training for drivers that drive our normal trucks, and that is more focused on fuel efficient driving." The expert emphasised that they have the infrastructure in place and that "the markets are very involved in this with designing workshops, and so it would not be strange to include automation training" (TE1). TE1 described the need for long-haul logistics training that will send highly automated trucks to another hub. The lack of automation training highlighted points of contention for some experts, as according to CE5, "that is why we are doing these learning project now, to try to give them information and help them to use their ADAS, but it is quite a challenge to deal with people that are not skilled for that." Essentially, training was seen as multidimensional, thus including users and all in the field, for instance, "not just the OEMs' market campaigns, but also the people that write or discuss about them in different social media, in different papers, online papers, whatever" (TE2). TE2 detailed that "to learn, you probably need more and the right information for not only the user but also anyone talking about the systems, so that they talk about them in the right way." For automated tractors, experts noted the importance of providing training to users and salespersons, as those who sell the systems should be fully knowledgeable about capabilities and limitations. SFE6 stated, "we are doing trainings in our training centre for salespersons, it is necessary to explain the system and its benefits." Moreover, "to get an overview of which buttons are where and to take a look at the how to videos so that they have foundational knowledge" of functions (SFE6). SFE6 stated that "we have virtual learning, which is an ongoing process with VR glasses," so "you see and feel with your eyes", and "the customer or salesperson see all while at home on the screen, it is a long process because you have to test what is possible." Training is seen to consist of how to use the systems efficiently in different conditions. Apart from handbook and physical trainings, it is also via virtual webinars, for example, "there are many online tutorials if you want to go deeper" (SFE7), "we have manual and online courses, and manual" and "also have the demonstrations on site and the see with the hands approach" (SFE1). SFE1 noted, "the demonstrations and the learning by doing is the most efficient way." SFE7 mentioned that "we have a complete virtual world", and "we have a terminal simulation, so that you can use all the functionalities of the automated systems in the carb." Additionally, SFE4 noted they "have the screen/cockpit of the tractor sitting somewhere in the room, the user interface so that you can train on there." In addition, SFE3 highlighted that their training is divided into two parts, "so one is the *theoretical part*, which is approximately 50% of giving some background information and describing some theoretical material, and then there is a *practical part* on the vehicle." Furthermore, SFE3 noted that most users do not find the theory part interesting but rather prefer the practical experience on the machine. Usually, when users buy a new tractor, they get one day of training at their farm to set it up, thus "you can drive around three hours with the trainer as he helps you to understand everything" (SFE7). The expert mentioned that they do a virtual field tests with the machine and an analysis thinking (SFE4). SFE4 noted, "sometimes we give a tractor to a farmer to try out, and if he comes back and says it was too complicated or I did not understand this and that, then we might change it, the user interface should be obvious." SFE6 noted that it is important to know an overall function of the tractor, know how the tractor works, engine or transmission, and know where all material is located. The expert highlighted that they are planning a completely new strategy with life sequences, e.g. transmission rooms for 2/3 h in the morning and in the afternoon for three days (SFE6). For automated flying, training is considered vital as the industry employs rigorous and systematic learning approaches. AE1 stated, "we have a very detailed training concept." The experts emphasised that training is a prerequisite and part of a legal agreement with OEMs and regulatory bodies (e.g. European Union Aviation Safety Agency: EASA, etc.). For example, "a manufacturer tells the training organizations precisely in a detailed way how to train the pilots and what content to cover" (AE1), and in addition, "EASA says a pilot needs to demonstrate the handling of an engine failure every six months, so in each simulator we will train at least engine failure recovery." AE2 stated, "if you change from Airbus to a Boeing aircraft, you typically have to go to an upgrade training, or even do a different type rating course that involves hours in the simulator." Moreover, the expert noted that they train for abnormal operation, complexities, malfunction, workload management, SA, automation use, interaction with interface, limitations and capabilities, etc. Also, the "pilot needs to recover the airplane manually" (AE1). AE1 illustrated that, "the Strasburg accident with the Airbus A320, the pilots were not sure about what the autopilot was doing and there was a lack of interference between the interface to the autopilot, and the human pilot was not sure about what he was telling the autopilot to do, he was assuming something completely different." The expert noted the mismatch between the automation intention and the information presented by the IU, in that "the autopilot was functioning properly according to its design, but there was a problem with the UI, because the information to the pilot was completely different." Experts felt that training helps with situational complexities, "if you have a problem with your car, you pull off to the side, shut down the engine and inspect. For us, if we have a problem, we need to keep flying." This complexity in handling failure procedure, handling system error, and understanding requires a repetition after a certain timely interval, if the pump fails for example, handling difficulties. AE1 felt that "relearning or recurrent training of this kind of malfunction and to handle the procedure in a real time environment, not only theoretically is important." AE2 showed uncertainty over the lack of training in some domains and noted whether developers "are not far too optimistic in what they assume about human performance, vigilance and diversion in these kind of things." The expert claimed, "if you then automate more, then you will have a bigger issue with mind wandering and also secondary tasks" (AE2). AE1 stated, "young pilots starting with education get used to this kind of automation in early stages of their education, and once they are done with flight school, they start making a 'type rating', the initial training on the A320." They go through several stages of learning (AE1): *the first step (theory and computer-based training, CBT)* "is theoretical education, mostly by manuals and interactive computer-based training. These interactive CBTs enable the student pilots to kind of operate the system and see what happens. For some aircrafts, we even have an autopilot or flight-management system trainer, it is like a dummy, which you can load on your computer and can push some virtual buttons to see what happens". *The second step (simulated cockpit)* "is called a flat panel trainer, which is a classroom filled with touch screens in the position where you usually have your switches and displays in the cockpit. The haptic is completely different because you do not have the push button and the switches, you have soft keys on the TFTs, but the function is identical. So, the pilots come into the classroom with a flat panel trainer and operate the airplane. The system operation is demonstrated as well as the use of automation, the autopilot and the flight director can be operated. For example, the pilot can play the hydraulic system on the overhead panel and see the effect, what is happening, which valve closes, which pump is not producing pressure in detail. The automation in the flat panel trainer simulates the use of all the systems correctly". *The third (aircraft simulator) and fourth step (real aircraft)* "is a full flight simulator and actual flight operation. Now the pilots have their theoretical knowledge, for example, on how to deal with failures, learned about normal and abnormal operation of the system, and now they fly doing the real take-off in the simulator with motion. They use the skills learned, we train to proficiency." AE1 noted, "the target is to train to proficiency," thus "if I see uncertainties, some kind of misbehaviour or disoperation, I repeat it, so that the pilots operate without mistakes."

3.2 Machine Assisted Learning

Consequently, experts inferred that if humans want to misbehave, they will misbehave, that is uncontrollable even when you have provided physical training. TE2 noted that "if a system affords a different behaviour that would triumph on the other one, I mean, if you say you can only drive 50 kms an hour here, but the vehicle can drive in 70, and they are in a rush, they might do it because it affords it to some extent." The expert argued, "in the end, the systems needs to be designed for the intended use and designed to avoid unintended use, because people will misuse the product" (TE2). For example, "if you have something that is not a chair but you can sit on it, and you have an urge to sit then you will do it" (TE2). The expert further emphasised, "that is why we designed the microwaves in the truck far-off, so they cannot reach them from their steering wheel, because we do not want drivers to microwave their food while driving." Thus, "we tried to avoid unintended use of microwaves while driving, because we do not think it is enough to say you are not supposed to microwave your food while driving, because if the driver can do it then he might" (TE2). SFE2 noted that people never have time, but these systems are often complicated, so the intention is they "have to be as self-explaining as possible, and more of like learning by doing." SFE4 noted, "it can be virtual…, it would be nice to see what the consequences are and to experience the consequences of you doing so." CE2 described that users "learned that the light bar had different colours, and they associated those colours with the steering, they learned these things just by experience, by seeing it happen, through the driver monitoring system that interacted with them." The expert mentioned, "within minutes, they became very compliant with the alerts." However, explained that after everything goes back to normal, users have a tendency to misbehave, and as a learned response, "they are again like alright, let me find my phone, but this time I am going to be faster, I am not going to let that happen" (CE2). CE2 argued that "they learn the speed, they learn at speed – how fast do I have to, how often does it flash, how fast do I have to look up, and they actually get a little faster as they learn, and then that makes them feel more comfortable, because they do not want it to escalate." CE1 noted, "it is more like if they see a use for it, they will and if they do not, they will not use it." CE2 argued that, "the only learning that I can count on is the learning that occurs when people use the feature" and SFE5 felt "practical learning is the best way to achieve successful learning". Further described, "if we see a lot of misuse and failures, assuming that whoever designed the system is at least reasonably competent, then it's going to be more on the part of, I don't really see a use for this tech thing" (CE1). CE6 felt that "you might give the car to other people or sell it, or get a rental car" and TE2 noted, "truck drivers might drive a truck that their owners bought from another owner, and they did not get that particular training." Additionally, CE3 noted, "many car brands think you go in and you buy a car, then you own a car, and then you get used to it and can use it" however "we know if you are going somewhere new and then you get a rental car, then it's like how does this work, it's tricky." Further, "there are things like rental cars and people borrow cars" (CE2). CE2 stated "car dealers are not owned by the manufacturer, in fact, by law car manufacturers cannot force car dealers to do anything" thus, "we cannot make them train buyers and cannot make them train them right, we cannot even stop them from lying, which they do about these systems …they will say anything to sell the car". For instance, "we can't rely on training only; we can barely rely on people even knowing how to drive a car", and "we are judged based on if an ordinary person got in this car would they be safe" (CE2). Thus, CE6 noted, "when you get a car, someone really explains the systems to you and covers what the system does, how you are supposed to use it."

For in-vehicle persuasive AI interfaces, the concept of persuasion was considered, as experts felt it would help shape correct mental models. Furthermore, noted that it should be intuitive and versatile to different user types, thus offering extended learning over time. CE1 noted "in-vehicle guides where it says to the user, are you interested in the lane keeping aid system, I will tell you how it works; I will talk you through activation, so, a robot voice out of the centre stack." CE1 explained that people interested in lane keeping would ask it, "Okay, now what do I do, and the system says, well put the car in the middle and push this button, and say done when you're done, so now you should experience this." CE3 stated that "people are different, they need to negotiate what they say, how they act, and what they experience" thus, "if we get more competent systems that gets to know you a bit more or understands how you behave, like you can hopefully adapt it a bit more." However, experts stressed that there are no good examples for this kind of system yet. CE4 considered emotions but was "not sure of products on the market which already have that kind of learning algorithms". Accordingly, "digital virtual assistants sounds like an interesting idea, but we have not seen really good examples" (CE6). TE2 stated, "we do not have that adaptive interface, like the first time you turn on you get it, which is sophisticated and straightforward." SFE1 proposed intelligent interfaces that go beyond current systems, and TE2 suggested ideas similar to gamification, "the better you understand it." CE6 considered "something like virtual assistants, like someone explaining the system to you." CE2 suggested, "to design the UI so that a person who knew nothing, if that feature got engaged, that they would be safe, then as they use the feature that is where any learning has to happen." Additionally, CE1 distinguished "interfaces where it's not like the existing interfaces we see in current vehicles, but it's quite unique, using the wind screen as a form of cockpit and different other elements or objects in the vehicle as some sort of communication channels." CE6 considered a computer-generated assistant system, however, thought this could be tricky to design, as different users may have different expectations, "for example, middle aged white men sometimes have a problem with a female voice out of the navigation system." CE3 noted "if you are new to this kind of functionality, you may need more information", therefore, "for people to even dare to use that kind of mode, they need information on what is the car capable of and what is it not, because if people get scared they will not use it." CE1 explained a "very simple and straightforward dialogue with the computer assistant." However, the expert noted that in their study, people who were not so interested, had issues with following its guides, and further noted that "when the interest is not there, people are extremely slow to learn or not interested in learning, so you can have all the helpers in the world in place, but if there is no engagement, then that's that" (CE1). AE1 stated, "the system needs to talk to me and make itself transparent, show me what it is doing right now." AE2 seemed a bit sceptical, "whether these are things that I actually need or whether these are just costly add-ons that I have to pay." Experts expressed the need for a system designed to assist pilots in areas where air traffic/flying is challenging, for example, the system would help redirect them during flying. AE3 mentioned that if the Autopilot is behaving abnormal, the system has to know this. Moreover, "the automated system gives the pilot hints, so autopilot flight system has something like a supervisor, which tells the pilots, right now there was a calculation error for example; maybe you must not trust the flight system for the next minute" (AE3). The expert stressed that pilots "always propose that manufacturers should implement something like a risk block in the cockpit, where pilots can see what the risk for dying is at the moment" (AE3). For example, it says "you did not sleep that well, and it says today it's 10 times risk to die because of your sleeping experience" and also says "the automation system is performing fine, and everything's fine" then the system says, "pilot Steve as you can see today it's pretty safe to fly." AE3 explained, "if you switch off the Autopilot, then you are notified that it's around 100 times higher the risk of dying in the aircraft, or having a huge accident, instead of flying or keeping the Autopilot system on." Further, "so the chance that you are going to die in the aircraft is 100 times higher if you fly manually instead of automatically." AE1 mentioned threat and error management system, "threat and error management is a major subject, and a major part of training, so, how to recognise threats, to avoid and deal with crew errors, and how to avoid undesired aircraft states." AE1 mentioned, "two main components of the threat error management, which means unexperienced pilots or first time automation users will make errors, the experienced pilot will see them and he will intervene, he will tell the unexperienced co-pilot, look at this situation you should use this mode." According to AE1, "first time use or unexperienced pilots will make some errors when using automation and the use of automated systems will not always make sense for them." AE3 proposed a system that "would be a coordination of time slots on the ground with an automated system", for example, "says, I have to adjust the hold light in order to get into the parking stand at the correct time."

For in-vehicle monitoring technology, using a user's estimated engagement state for safe transition of control, this includes systems (cameras, haptics, sensors, etc.) that monitors user behaviour for crash avoidance, engagement, fatigue, drowsiness, distraction, and other impairments. Experts conveyed the importance of designing in-vehicle monitoring systems that will keep humans safe in perilous situations. Purposefully, monitoring enables real time adjustments to warning parameters when users are identified as being distracted, disengaged, or drowsy, etc. CE2 emphasised that "even if there is no training, using behaviourism, we created a system that kept them safe" and furthermore, "the feature kept them safe, even though we instructed the participants to do secondary tasks…, and they still kept their eyes on the road." However, "even if there's no training, we were able to create a system using behaviourism that kept them safe" (CE2), with inconveniences (punishments) and conveniences (rewards). CE2 emphasised that they "were able to identify things that the drivers found rewarding or would pursue, and in fact, spoiler alert, it is the hands off." In addition, having the vehicle steer while your hands are off for extended periods is seen as a motivating factor, find value in. Further, "on the punishment side, we have a wide variety of punishments, like haptic alerts, but we also figured out that if every time it escalates too far, when the system disengaged, and if the driver wants the hand-off again they have to search for the button and re-engage the feature, that's a punishment they want to avoid" (CE2). Moreover, "the vehicle will not maintain headway to vehicles ahead that also is perceived as punishment." As a result, it was evident that "there is this whole set of punishments that we are able to determine and then we have this reward that we were able to determine, and so by sequencing the interaction, rewards work better than punishment" (CE2). The systems systematise the interaction in such a way that it incentivizes the user to act accordingly. Further, "the thing that makes it work is the providing of rewards before we do the punishments" (CE2). An illustration of "the rat with the food pellet" was provided. CE2 emphasised that the way that they "organise the interaction, it incentivizes the customer to keep their eyes on the road and allows them to keep their hands off the wheel." CE2 stressed that "we figured out exactly how to manage the users behaviour, how to change their behaviour in real time." TE2 argued that "the systems need to be designed for the intended use and designed to avoid unintended us." Thus, crash avoidance technologies were seen as "quite useful if you're in a traffic jam" and further noted that "anything that can offer assistance is extremely helpful and extremely pleasant" (AE2). AE2 noted "anything that helps me to not run over pedestrian, that's extremely helpful", and "these are extremely helpful if they are well designed and intuitive enough in the way they operate and the feedback that they give to me."

4 General Discussion

As a general discussion based on experts' views, we gathered that there are inconsistent mental models between developers, users, and situational contexts. Author [23] argues that "most computer programmes are still highly deterministic (finite state machines) that reflect the views and values of the programmers." These systems are designed based on expert's level (and mental model), yet with the intention to be used by non-experts. Resulting in non-equivalent/unbalanced mental models, e.g. the designer's mental model (expert level) and user's mental model (none expert level). To add, [3] mentioned the lack of information relating between the user, the driving task, and the environment. There is also a lack of knowledge on how weather conditions (e.g. snowy, rainy, foggy, etc.) and the environment (e.g. infrastructure, mountainous, road markings, traffic signs, etc.) affects the context of use in real life over time (Fig. 1).

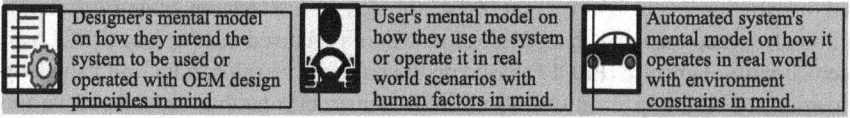

Fig. 1. A mismatch of mental models (there is a necessity for a mediation)

For the automotive industry, training is far more unruly, because OEMs downplay the aim for training and rather emphases designing for easy to use. However, what is neglected is that system complexity and design limitations can be circumvented by augmenting training, the differences in HMIs may certainly necessitate specific training, supporting with manual skills, as well as training appropriate trust for specific conditions. Even though the aircraft industry has maintained user training on correct use of automation [15], they are still not short of challenges, as also noted by experts. Both ground and air vehicles experts mentioned challenges they have experienced in their domains. In essence, the root/trunk of the problems they encounter have two branches, which are system design and human factors issues; moreover, each branch of issues has its own twigs of concerns. For example, on the part of human factors, noted knowledge paradigm for AV operations, complacency, skills issues, trust, and disuse, etc. For both ground and air vehicles, experts noted that based on the high level of variability in system design (especially as automation matures), it may appear that each automated system may require a unique set of knowledge and skills training or learning guided by the mode or task of operation, LoA, HMI design, etc. Reference [12] advocated for employing smart automation assistance (e.g., highlighting of relevant roadside information, manoeuvre interventions and corrections), and to enable automation assisted "manual" driving to assist in the driving condition.

Experts argued that automated systems need to be robust and easy to use, if not easier, but how that 'easy' is designed for, was met with suppositions, as some experts noted that 'easy' could mean different things to different people (e.g. designers and users). As a result, what may be considered easy by one person may not be considered easy by another, thus, 'what is easy' and to 'whom is it easy for'? In essence, tackling the concept 'easy' is a challenge, as one user's easy may be another user's difficult. In hindsight, to assume 'inclusively easy' is to deny putting the effort towards training protocols. For example, Tesla claimed that their Nvidia-powered Tesla Vision deep neural network "deconstructs the car's environment at greater levels of reliability than those achievable with classical vision processing techniques" and their hope is that this increase in brainpower will compensate for a lack of formal training [18]. However, the OEM has experienced major crashes in the past years due to human factors. This kind of thinking can tarnish the importance of providing training or supporting the user with resources for learning, as users may interpret this as saying the AV is without errs and intuitive. The differences in nuance inferred by such perspectives should not distract us from a larger concern, which is with the evolution of automation, the complexity of human nature, and the intricate task environment in which they operate. The AV related adversities/crashes that have occurred in the past years represents unsystematic real world experiments on user behaviours towards automation. When exploring how users learn to use (or misuse) automated systems, it is vital to consider the context of user archetypes (e.g., based on experience, personality, emotions, attitudes, social framing, cognition, etc.). In addition, the processual and changeable character of perceived-ease-of-use (PEOU) and perceived-ease-of-misuse (PEOM) based on the human's learning maps and knowledge disposition. As learning is an overall unstable construct, one that depends on various factors, such as long-term effects, user type, designer type, system type, and environment characteristics over time. Thus, long-term learning on user behaviour towards automation is based on the variation in human factors and knowledge, and thus important to consider (Fig. 2).

So, paying attention to information processing across time (cross-sectional comparison of short-term to long-term) and the correlation between knowledge and behaviour. Essentially, these knowledge dispositions inform us on how users' cognizance the intended use of a system and their mental modelling of the interaction strategy, thus

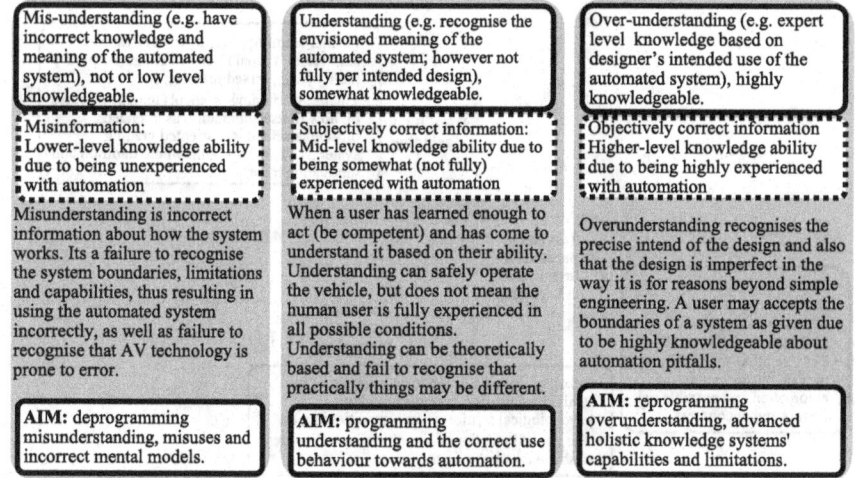

Fig. 2. Ergonomically thinking *Mis-Over-Under—stand* knowledge compositions.

their distinction is a matter of seeing, thinking and behaving. Conceivably, it is important that users are fully knowledgeable about the system on why it is the way it is, and of its essence, place or role in the grand scheme of things, effects, and situations over long-term use, as knowledge contributes to inferences on how it is used.

Further, automation like other technologies afore it, is never the exact same in each AV and the claim that it is generally easy to use (especially during its introduction stage) typically disguise a dogmatic ideology that is user predictability, user forbearance or self-control, deregulatory, de-standardisation, and no training and learning strategies needed. Which we believe to be negligent as automation is in its development (or infancy) stage and is not yet intuitive nor robust to all human factors. In today's technology infused society, learning is a vital aspect of knowledge acquisition and formation. In essence, learning can be seen in a broad range of techniques (information related processes) that can be considered to support long-term learning by users and assessment of the process, regularly in a collaborative manner between the trainer and trainee. In this context, the aim is to improve the learning process, to be a transformative force that can generate radically new ways of knowledge, knowing, and behaving, through different learning design strategies or learning experience (LX) design, in formal, non-formal and informal learning contexts targeted towards comprehending how humans learn with automation. This knowledge yields implementable strategies, thus we propose 'covert learning' (in-vehicle intelligent systems, etc.) and 'overt learning' (training programs that follow a grading system) strategies (Fig. 3).

We subscribe to the principle that practice makes perfect, through learning and rehearsal, the law of practice and performing, as the user learns the correct mental model to avoid confusion and unwanted errors. For instance, repetition of actions has the possibility of reinforcing learning and facilitating the development of appropriate behaviours [5]. Thus, learning should be considered valuable, and this understanding on how human learn can serve as initial benchmark when developing learning strategies

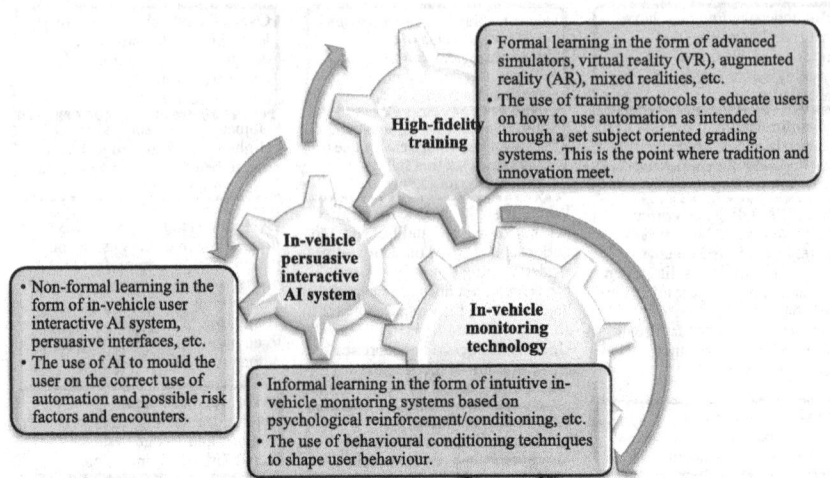

Fig. 3. Praxis of LX strategies for optimizing user behaviour towards automation

in order to optimize overall safety. The described strategies by experts make use of multiple forms of simplified system designs that are robust and intuitive in nature, with as little learning as possible, but the approaches varies and depends on matters such as ergonomic and economic logistics as well as engineering perspectives. We argue that training could lead to better uses, thereby, improving performance and safety, which is an important currency for HAI/U with upcoming HAVs. Thus, "if human factors would only use the mind-set and methods of yesterday to solve the problems of today, it would inadvertently contribute to the complexity of tomorrow and would be in a strong dilemma" [6]. Author [21] argue that the "technology that will fully enable autonomous driving is still in the future, which requires developers to envision this future in order to be able to design for it." Moreover, if the plan is to counter co-risks, foster calibrated trust and acceptance of, and pleasurable experiences with AVs, then these issues may need to be profusely addressed. As stated by [27], interaction design always deals with an imagined future and ergonomic design thinking affords the opportunity to understand the current state, "researchers generate new knowledge by understanding the current state and then suggesting an improved future state in the form of a design." Thus, "it involves deep reflection in iteratively understanding the people, problem, and context around a situation that researchers feel they can improve" [27]. Therefore, imbedding different user requirements in the minds of the designers, by what [21] accentuate as "reframing the problem and identifying the ultimate solution from an infinite set of possibilities."

4.1 High Fidelity Training Techniques

As automation matures and its HMIs become more sophisticated, we believe that current training protocols provided at driving/flying schools will not be sufficient in training users for imminent AVs. When people hear the word training, they automatically image something direct as a human teaching on how to execute a task or how to behave, like a driving school. As experts stated, training should become a prominent approach in making sure that automation is used correctly, as believed that the provision of advanced automated systems are intended to vary aspects of operations including attention allocation, performance, and mental load. Experts felt that even though automation is becoming ubiquitous, there are still no established protocols to train users on correct use/interaction. Aircraft experts were more positive about training compared to those in other industries, as they indicated that training helps with pilots' performance and skills when using automated flight system/Autopilot, which is intertwined with safe operation. Some experts felt that training plays a significant role on vigilance and monitoring/supervising the system, experience (familiarity with the system or a similar system), mode confusion, and knowing the difference. The experts' views are consistent with previous research indicating that user training can be useful for generating a better understanding of system designs. Author [11] suggested training the purpose of using automated systems (risks, benefits), understanding LoA (capabilities, limitations), transition between automated and manual mode and handling critical situations (malfunctioning), familiarity with system components and placement (sensor, radar, camera, etc.), and understanding limitations of systems.

Experts reported that users' familiarity peaks after they have gained experience with automation, thus, it is critical for trust to be calibrated according to experiences. It is argued that AV training protocols need to cover the requisite material and address reasons why users need to be vigilant, particularly during automation use. Further, expectations and user predilections to secondary tasks need to be calibrated through clear and engaging retraining/reskilling that is distributed over time, not as a one-time thing. Author [11] explained that users might "become more aware of system limitations with prolonged exposure; however, previous studies have shown that safety critical misunderstandings of system limitations persist over time", nonetheless, demonstrated that "brief experience with the system after training does not sufficiently alter misconceptions about the boundaries of system operational design domain." Thus, the retraining should be highly considered. This could be organised in a phased approach similar to the aircraft training process, through providing the user with periodic performance feedback if necessary, to determine changes in skills. For example, an-in-vehicle 'RoboCop' (virtual driving cyborg) that suggests a timed AV operating training based on a number of incorrect task compliance and misuses, among others. In developing training materials, different OEMs and regulatory bodies should be consulted for information on design inclusive protocols, which include predominant use errors and misconceptions. The training protocols should aim to educate users on human factors and automation-based errors or failures using advanced simulators, VR/AR/mixed realities, so that users can practically be saturated in the experience. This will lead to holistic training program that encompasses the strengths and occasional deficiencies of both human and automation. Thus, smart infrastructures for training users should be considered. Simulators, VR/AR/mixed realities are tools already used by researchers and industry experts for research purposes to allow participants to experience the processes associated with automation, and this process can be influential at smart driving schools. These approaches are similar to gaming settings (e.g. Grand Theft Auto), as users navigate through a computer-generated world (or a metaverse) that is expressed using mixed realities, and includes reinforcement learning as one way of accelerating knowledge. Reference [3] employed heuristic knowledge of human factors, ergonomics, and psychological theory to propose solution areas to human supervisory control problems of sustained attention, as well as a reference point towards the utilisation of conditioning learning principles, for example, gamification and/or selection or training techniques. These advanced approaches in training users can be seen as an attractive tool for learning, allowing users to exhibit normal and abnormal behaviours in a safe environment. Additionally, new licencing regulations might need to be democratised.

4.2   In-Vehicle Monitoring Technology

Some experts argued in favour of employing psychological conditioning techniques to improve user behaviour towards automation. Author [16] noted that some people avoid learning as they lack motivation, and some "people may interpret that material in ways we may not expect (and may not like)". As a result, reinforced learning should be considered, as behaviours are learned through a process of conditioning. Author [29] suggest that "direct manipulation aids initial learning and that previous experience is a moderate aid in learning". Additionally, humans are "believed to learn (and forget) to different degrees and in different ways or put that learning to unexpected uses which thwart our object as teachers or designers" [16]. Moreover, feedback is believed to be "an all-pervasive and fundamental aspect of behaviour" [14]. Thus, behaviourism as a descriptive tool provides a unified account of the nature and development of user behaviour over time. It draws attention to the dialectical process by which awareness, learning, and interaction simultaneously shape and are shaped by automation, and looks beyond a human inside the AV, but rather imagines a user responding to stimuli by the system (e.g. DMS), internalising feedback (on correct behaviour or use) through repetition. As it has been argued that to behave is to control perception [14] on the nature of things or systems. The use of reinforcement learning in designing in-vehicle monitoring systems is imperative on compelling users to behave correctly, through learning-induced by rewards and punishments. For example, the Super-cruise monitoring system incorporates some form of learning with the use of rewards and punishments as means to train correct behaviour. For instance, author [17] aimed to observe if the alerting strategy (as a form of punishment) would help "shape drivers' timesharing patterns, reducing instances in which the driver's gaze was off-road for extended periods of time." Reference [5] illustrates a horse's behaviour as "initially based on instinct but through training that relies heavily on learned response to stimuli that are reinforced with positive or negative feedback". Reference [5] suggested that what facilitate the learning process is said to be the positive interaction between the trainer and appropriate motivators. To support this, some experts suggested the use of 'punishment-rewards' as ways of training and shaping user behaviour. Even though [5] favoured removing punishments and encouraging voluntary learning, they did infer punishment or fear as a fathomable technique. For example, the use of discomfort in order to make the agent move away from the unpleasant stimulus and move in an appropriate direction to remove the discomfort [5]. Experts mentioned incorporating punishments (such as when taking your eyes of the road, the system coasts or disables) and rewards (when keeping your eyes on the road, the system continues to operate and you have more leeway) as forms of covertly training the user how to behave.

For example, [17] noted system cues and alerts compliance as a way of training the user to keep their eyes on the road and as an incentive to use efficiently. Thus, they noted that if users fail to obey system cues and alerts (experience a stage 3 alert), the system would disable for a period of time (a form of punishment), but if used effectively, the system continues to operate (a form of reward). Moreover, the punishment must be applied immediately after disobedience, as a disciplinary technique. Therefore, the aim is to use positive and negative reinforces (rewards and punishments motivators), as users learn to associate specific behaviours with desirable or undesirable performance. For instance, instilling a sense of discomfort through punishments and instilling a sense of comfort through rewards may increase consideration for restraint design and adaptive safety systems to optimally reduce risks. Furthermore, incorporate various persuasive features that employ game theory to reinforce different types of motivation: intrinsic motivation (personal hand-off elements or non-driving related activities) and extrinsic motivation (socially safe driving).

4.3 In-Vehicle Persuasive Interactive AI Systems

Reference [12] argues that "with continually advancing automation capabilities in vehicles, there is increasing potential for these capabilities to be used not only as stopgaps towards full automation but to enhance humans' manual driving capabilities during this transition phase and beyond." The long-term improvement of user behaviour towards automation may be depended on the learning method presented via system design. The ability for AVs to be equipped with AI to detect user misbehaviours would be key for long-term safety, as this has the advantage of persuading users to behave correctly. We consider a system equipped with persuasive AI abilities, with persuasion described as "the act of causing people to do or believe something", and persuasive technology as "technology that is designed to influence people in their actions or believes" [8]. Thus, intelligent in-vehicle persuasive AI systems equipped with deep learning networks are important to consider. For engineers, this has the dual advantage of massively increasing the speed with which automation can learn the user and the user can learn the automation, and thus advocate for collaborative learning. Author [1] noted that good interface design could result in good experiences with AVs, providing support to users through output channels that provide information about the system state to the user. Moreover, experts noted important to consider a system that provides information about the risk situation and user state to the user. When we think of new ways of interacting or interfacing with AVs, an intrinsic interest is exploring the safety dynamics of in-vehicle persuasive AI towards safe long-term use, using verbal and non-verbal communication to provide regular feedback. A possible approach to designing an in-vehicle persuasive AI system is through demonstrative intelligence, augmented machine learning, simulated safety character, comprehensive knowledge of human factors and social norms, etc. The in-vehicle persuasive AI system can be a pleasant interacting feature that bridges the gap between the human and automation, as well as an affective personified companion. This in some way personifies the AV as a social unit or a social rule of influence and knowledge, a socially intelligent vehicle where rules apply. Thus, we consider anthropomorphism (as the attribution of human-like features) and humanization can be related either to the appearance or to the nature of the object [7]. We also consider the attribution of human character traits as well as emotional acting and complex intellectual capacity [7] in the design. In essence, the system should be able to recognise a distraction or drowsiness as a way to persuade the user to pay attention to the road. Some OEMs have been prototyping with similar technology by integrating advanced vehicle control and safety technologies with cutting-edge AI to enrich performance though in-vehicle interfaces. Moreover, we can expect a future where AI plays a significant role in the delivery of support scenarios, for example, in-vehicle-bots, which means that new solutions for control can emerge and risky behaviours can be solved. Other example, may include AV operation assisted by an AI-powered voice embedded inside a smart speaker such as Amazon's Echo or Alexa (in a hologram form), Gatebox (a cartoon persona encased in a glass tube), a socially cognitive robot (displayed on a tablet or screen) or even a design similar to ElliQ (AI-driven contraption). These are interesting technological concepts to consider for AVs, which can be used as a safety assistant co-driver/co-pilot. Thus, as virtual assistants proliferate, it would prove interesting for OEMs to experiment with different forms of AI learning strategies. Personal factors such as personality and pre-existing motivation may affect the effectiveness of certain persuasive technology features and users' adoption decisions. Thus, integrating gamified persuasive technology features as motivational elements, especially when they are distracted. Some experts noted the need for systems that draw on human feelings and emotions, a condition of feeling machine learning. Hence, how humans learn and how automation learns through AI, both apply. The latter becomes interesting when we consider the 'social' and 'affective' within machine learning.

Author [24] urges that with "rapidly increasing system functionalities, there is an urgent need to consider appropriate HMIs to communicate system states via auditory, visual, and haptic elements and also to let the user interact with these systems through the same modalities." They also reasoned that, "this proliferation of functionality and interaction possibilities raises the necessity for developing robust methods to evaluate such interfaces" [24]. AVs should be equipped with sophisticated self-modifying software that can adapt itself to different users' way of thinking (as thoughts become actions) as well as improved modelling of, and interaction with the real world. As part of the long-term safety design philosophy, automation should be capable of coordination, cooperation and collaboration with users in different interactive tasks. Thus, a fundamental interest is exploring co-risk cues and dynamics as a way to enhance safety in phases of AV operation, which means designing risk identifiers. One possible approach towards designing a risk-identification HMI (rHMI) would be via AI that monitors and communicates user, environment and AV states. Author [1] proposed a framework, which categorises HMI into subsections, based on the relationship between humans and AVs. They suggested separating the range of HMIs into explicit task-related strategies, and argue it to be valuable for understanding the underlying cognitive elements imposed by each [1]. In considering an in-vehicle AI rHMI, we consider or refer to dynamic HMI (dHMI) as the case with reference [1], which is used to recognise the need for a "multi-actor" communication channel.

Author [28] stressed the need for better defined, systematic, and design guidelines for suitable in-vehicle HMIs. A novel concept here is the introduction of an in-vehicle AI rHMI, an interactive system for mediating risky behaviour. Nonetheless, a challenge with

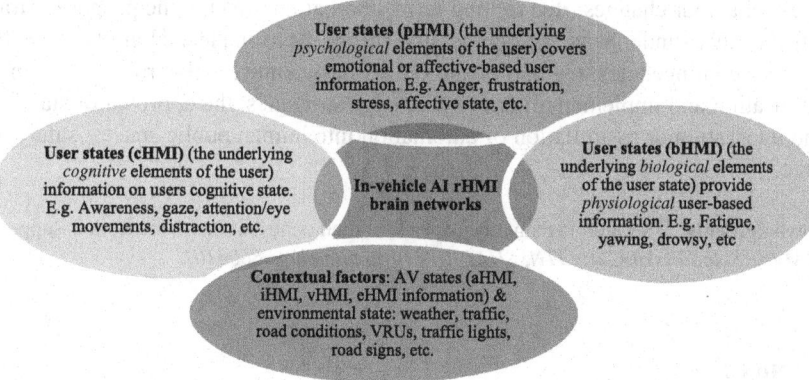

Fig. 4. In-vehicle AI rHMI for estimating users' risk probability.

designing an AI rHMI is with condition that it provides accurate and suitable information through an integrated holistic "multi-actor" communication channel, for example, cHMI, pHMI, and bHMI user states (Fig. 4). This information can be communicated in conjunction with the dHMI by [1], divided the various internally presented messages by the AV. As information based on aHMI (automated system: mode awareness), iHMI (vehicle's infotainment system), vHMI (vehicle-based: tyre pressure, fuel level/engine), and eHMI (external vehicle state to VRUs). We also need to consider other influencing factors, such as static infrastructure (road type, traffic rules), automation system (functional scope), dynamic factors (weather, lighting conditions), and users (personality/age/experience/gender) [1]. Somewhat similar to [1], this concept entails output channels that also provide information about the user state to the operator (e.g., via displays and auditory signals), input channels to receive the operator input (e.g., via buttons, steering wheel, and pedals), and a dialog logic to specify the relationships among input, output, and context parameters.

5   Concluding Remarks

In essence, humans' ways on learning to interact with and use automation in a smart or sustainable way should be considered, as well as how an automated system acts and reacts to human stimuli. It is important to train users concerning automation's reliability, mechanisms governing its behaviour, and intended use, as this may help maintain long-term safety and risk-free behaviours. There also needs to be careful evaluation of any personification on the interaction to ensure appropriate behaviours. Nevertheless, like any other innovative technology, it takes several seasons to build the needed credibility for users to become accustomed to it. Thus, these learning strategies are useful in engineering and generating information for support, on the topic of mobilising awareness, and assembling robust knowledge, and even in cases of renting and buying a second-hand vehicle. In conclusion, human factor issues need to be evaluated with solutionism that stand the test of time. Given the uncertainties of automation (as an unruly technology) and user behaviour changes (due to long-term automation effects), the proposed strategies offer a way to understand and democratise the means to enhanced long-term HAI/U. Given these contingencies, regulators and OEMs have a more active role to play in the standardisation and implementation of knowledge strategies, the approval of standards and the integration or assimilation of automation into human public spaces, safely and risk-free.

Acknowledgments. The authors thank the experts for their contributions. This work is supported by the *Marie Skłodowska-Curie ITN, SHAPE-IT [grant number: 860410].*

References

1. Bengler, K., Rettenmaier, M., Fritz, N., Feierle, A.: From HMI to HMIs: towards an hmi framework for automated driving. Information **11**, 61 (2020)
2. Carroll, J.M., Carrithers, C.: Training wheels in a user interface. Commun. ACM **27**(8), 800–806 (1984)
3. Cabrall, C.D., Eriksson, A., Dreger, F., Happee, R., de Winter, J.: How to keep drivers engaged while supervising driving automation? A literature survey and categorisation of six solution areas. Theor. Issues Ergon. Sci. **20**(3), 332–365 (2019)
4. Endsley, M.R., Kiris, E.O.: The out-of-the-loop performance problem and level of control in automation. Hum. Factors: J. Hum. Factors Ergon. Soc. **37**(2), 381–394 (1995)
5. Flemisch, F.O., Adams, C.A., Conway, S.R., Goodrich, K.H., Palmer, M.T., Schutte, P.C.: The H-metaphor as a guideline for vehicle automation and interaction (Report No. NASA/TM—2003–212672). Hampton: NASA, Langley Research Center (2003)
6. Flemisch, F., Kelsch, J., Löper, C., Schieben, A., Schindler, J.: Automation spectrum, inner/outer compatibility and other potentially useful human factors concepts for assistance and automation. In: de Waard, D., Flemisch, F.O., Lorenz, B., Oberheid, H., Brookhuis, K.A. (eds.) Human Factors for assistance and automation, pp. 1–16 (2008). Shaker Publishing, Maastricht, the Netherlands (2008)
7. Haeuslschmid, R., Buelow, M., Pfleging, B., Butz, A.: Supporting trust in autonomous driving. IUI 2017, March 13–16, Limassol, Cyprus. ACM (2017)
8. Hock, P., Kraus, J., Walch, M., Lang, N., Baumann, M.: Elaborating feedback strategies for maintaining automation in highly automated driving. In: Proceedings of the 8th International Conference on Automotive User Interfaces and Interactive Vehicular Applications (AutomotiveUI 2016), October 24–26, 2016, Ann Arbor, MI, USA (2016)
9. Lee, J.D., See, K.A.: Trust in automation: designing for appropriate reliance. Hum. Factors: J. Hum. Factors Ergon. Soc. **46**(1), 50–80 (2004)
10. Lee, J.D., Sanquist, T.F.: Augmenting the operator function model with cognitive operations: assessing the cognitive demands of technological innovation in ship navigation. IEEE Trans. Syst. Man Cybern. Syst. Hum. **30**, 273–285 (2000)
11. Manser, M.P., et al.: Driver training research and guidelines for automated vehicle technology. Performing Organization Report No. Report 01–004 (2019)
12. Mirnig, A.G., et al.: Workshop on exploring interfaces for enhanced automation assistance for improving manual-driving abilities. In: 13th International Conference on Automotive User Interfaces and Interactive Vehicular Applications. New York, USA, pp. 178–181 (2021)
13. Parasuraman, R., Riley, V.: Humans and automation: use, misuse, disuse, abuse. Hum. Factors **39**, 230–253 (1997)

14. Powers, W.T.: Feedback: beyond behaviorism: stimulus-response laws are wholly predictable within a control-system model of behavioral organization. Science **179**(4071), 351–356 (1973)
15. Robertson, C.L., Petros, T.V., Schumacher, P.M., McHorse, C.A., Ulrich, J.M.: Evaluating the effectiveness of FITS training. University of North Dakota (2006)
16. Russell, D.: Looking beyond the interface: activity theory and distributed learning. In: Lea, M., Nicoll, K. (eds.) Distributed learning, pp. 64–82. Routledge Falmer, NY (2001)
17. Llaneras, R.E., Cannon, B.R., Green, C.A.: Strategies to assist drivers in remaining attentive while under partially automated driving: verification of human–machine interface concepts. Transp. Res. Rec. **2663**(1), 20–26 (2017)
18. Stilgoe, J.: Machine learning, social learning and the governance of self-driving cars. Soc. Stud. Sci. 2018 **48**(1), 25–56 (2018)
19. Wynne, B.: Unruly technology: practical rules, impractical discourses and public understanding. Soc. Stud. Sci. **18**(1), 147–167 (1988)
20. Shalev-Shwartz, S., Shammah, S., Shashua, A.: On a formal model of safe and scalable self-driving cars (2017). arXiv preprint arXiv:1708.06374
21. Strömberg, H., et al.: Designing for social experiences with and within autonomous vehicles – exploring methodological directions. Design Sci. **4**, E13 (2018). https://doi.org/10.1017/dsj.2018.9
22. Endsley, M.R.: Designing for situation awareness in complex systems. In: Proceedings of the 2nd International Workshop on Symbiosis of Humans, Artefacts and Environment, pp. 1–14 (2001)
23. Blackmore, B.S., Fountas, S., Have, H.: System requirements for a small autonomous tractor. Agricultural Engineering International: the CIGR Journal of Scientific Research and Development. PM 04 001 (2004)
24. Forster, Y., Hergeth, S., Naujoks, F., Krems, J., Keinath, A.: User education in automated driving: owner's manual and interactive tutorial support mental model formation and human-automation interaction. Information **10**(4), 143 (2019)
25. Rodionova, A., Alvarez, I., Elli, M.S., Oboril, F., Quast, J., Mangharam, R.: How safe is safe enough? Automatic Safety Constraints Boundary Estimation for Decision-Making in Automated Vehicles. IEEE Intelligent Vehicles Symposium, Las Vegas, USA (2020)
26. Every, J.L., Barickman, F., Martin, J., Rao, S., Schnelle, S., Weng, B.: A novel method to evaluate the safety of highly automated vehicles. In: International Technical Conference on the Enhanced Safety of Vehicles, NHTSA, Detroit, Michigan (2017)
27. Zimmerman, J., Forlizzi, J.: Research through design in HCI. In: Olson, J., Kellogg, W. (eds.) Ways of Knowing in HCI. Springer, New York, NY (2014)
28. Carsten, O., Martens, M.H.: How can humans understand their automated cars? HMI principles, problems and solutions. Cogn. Technol. Work **21**(1), 3–20 (2018). https://doi.org/10.1007/s10111-018-0484-0
29. Davis, S., Wiedenbeck, S.: The effect of interaction style and training method on end user learning of software packages. Interact. Comput. **11**(2), 147–172 (1998)
30. Amershi, S.: Designing for effective end-user interaction with machine learning. In: Proceedings of the 24th annual ACM Symposium Adjunct on User Interface Software and Technology, pp. 47–50 (2011)

SAFERent: Design of a Driver Training Application for Adaptive Cruise Control (ACC)

Molly Mersinger[✉], Shivani Patel, Jenna Korentsides, Elaine Choy, Stephen Woods, Barbara Chaparro, and Alex Chaparro

Embry-Riddle Aeronautical University, Daytona Beach, FL 32114, USA
mersingm@my.erau.edu

Abstract. Research shows that drivers rarely consult the vehicle owner's manual to learn to operate advanced safety features, like Adaptive Cruise Control (ACC). Students from a United States university participated in a national competition, called EcoCAR Mobility Challenge (sponsored by the Department of Energy), to develop an engaging solution that is efficient and effective in educating drivers about the operation of ACC while adhering to user-centered design methodology. Students used an iterative design process and collected quantitative and qualitative data to evaluate the strengths and weaknesses of a mobile application. Iterations of the prototype included an interactive guide, video components, remote connection to the vehicle, and ultimately a solution that could reside on the vehicle's infotainment system. This paper details the process and evaluation outcomes and serves as a model for user-centered design of tutorials for Advanced Driver Assistance Systems (ADAS).

Keywords: Adaptive Cruise Control · Driver interface · Usability Testing · Iterative Design · HCI design and evaluation

1 Introduction

Advanced Driver Assistance Systems (ADAS) are important in promoting safety for drivers on the road. These features include adaptive cruise control (ACC), lane keeping assistance, blind spot warning, etc. They are designed to automate various aspects of driving to minimize the mental workload of the driver. For example, ACC is an advanced version of cruise control, where the vehicle can maintain a predetermined distance behind a lead vehicle and adapt its speed based on the speed of the lead car. ADAS features are an example of Level 2 Automation – Partial Driving Automation, as defined by the Society of Automotive Engineers (SAE) J3106 standards [1]. Vehicles equipped with ACC are able to control speed and some aspects of steering, but still require constant driver monitoring (Fig. 1). Generally, higher levels of automation require less driver intervention.

While automation offers benefits to drivers, there is a lack of standardization across manufacturers in naming conventions and the operations of ACC. According to the Automobile Association of America [3], vehicle manufacturers refer to ACC by 20 unique

H. Krömker (Ed.): HCII 2023, LNCS 14048, pp. 112–128, 2023.
https://doi.org/10.1007/978-3-031-35678-0_7

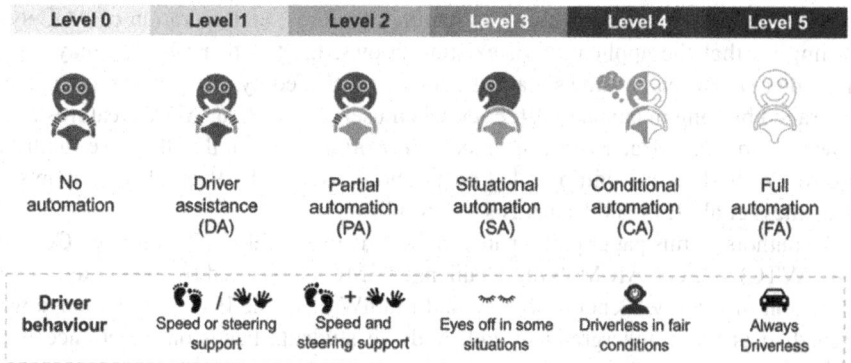

Fig. 1. SAE J3106 – Levels of Driving Automation (https://www.idtechex.com/en/research-rep
ort/autonomous-cars-robotaxis-and-sensors-2022-2042/832)

names, while lane keeping assistance and blind spot warning each have 19 different
names. The lack of consistency in naming conventions poses a problem for drivers that
own and/or rent vehicles with ADAS. For vehicle owners, there are many ways to learn
how to properly engage and use ACC, but they vary in effectiveness and time commitment. A vehicle owner may rely on automotive dealership employees or search the
internet to learn about specific features. However, the veracity of the information cannot
be determined, and the information may not be comprehensive enough to allow use of
the feature to its full capability, possibly decreasing safety. Studies show that drivers
report rarely consulting their vehicle owner's manual partly due to its length [4]. Additionally, studies show that even when user's read the owner's manual, it is not always
an efficient learning tool for understanding the system [23]. A study of 370 participants
[22] reported that 67% claimed to have read the owner's manual, yet 72% were unaware
of system limitations. Similarly, Llaneras [24] found that 90% of a 150 participant pool
reported having read all or parts of the owner's manual, yet 13–57% of the pool answered
questions about system limitations incorrectly. One reason why users may not read the
vehicle owner's manual to learn about new features is because adults generally avoid
situations that require new learning [5]. Second, manuals are written with technical jar-
gon that is unfamiliar to the average reader and difficult to interpret [6, 7] which may
account for users' poor retention of the information. This is an issue because new safety
features, like ADAS, are more sophisticated with novel controls and modes of operation
that drivers are likely to be unfamiliar with. For a manual to be more useful, it "should
not just describe the features of a system, [it] should help people get things done" [6].
Beggiato & Krems [25] proposed an in-vehicle tutoring system that teaches users about
vehicle capabilities and limitations and informs them of system updates.

Additionally, the Production Paradox and the Assimilation Paradox typically apply
to people learning new systems. The Production Paradox states that people are focused
on the end goal of completing their desired task [5]. Learners want to start their task(s)
quickly, but vehicle owner's manuals often emphasize and are organized around descriptions of buttons or features rather than the task(s) the user aims to accomplish. The
Assimilation Paradox states that users apply previous knowledge to new situations [5].

However, the lack of standardization of naming, features, and operation of ACC systems implies that the application of existing knowledge to other vehicles may come with some risk. Research shows that learners are motivated by self-exploration and are discouraged by long manuals [8, 9]. Even when users learn about ADAS features from the owner's manual, dealerships, or videos, they often mention that they use the trial-and-error method as an additional learning source, as seen by 95% of participants in both Jenness et al., [22] and Llaneras's [24] studies.

The authors of this paper participated in the Advanced Vehicle Technology Competition (AVTC)'s "EcoCAR Mobility Challenge (EMC) sponsored by the United States Department of Energy, General Motors, and MathWorks. The EcoCAR challenge was designed to provide undergraduate and graduate students hands-on experience with working in the automotive industry and tackle current day energy and mobility challenges. Each EcoCAR competition is completed in four one-year segments. The current challenge being discussed is called the EcoCAR Mobility Challenge (EMC) that spanned the years 2018–2022 and included 11 participating universities across the U.S. and Canada. The focus of EMC was to improve energy, efficiency, and safety for the 2019 Chevrolet Blazer. General Motors provided each university a 2019 Chevrolet Blazer for testing and engineering design as part of the competition. A central focus of the competition was mobility-as-a-service (MaaS), meaning consumers typically do not own the vehicles they use for travel. Therefore, the target users of the solution are frequent automotive renters or car sharers [11]. The ERAU EcoCAR Human Machine Interaction (HMI) sub-team was responsible for developing a quick learning/education tool for drivers to learn about the advanced driver assistance technology equipped in the vehicle. For vehicle renters, time is limited. The solution must be convenient so they can arrive at their destination as quickly as possible. To mitigate this issue, the solution also must be user-friendly and effective in teaching important information relating to ACC. A minimalist instruction design framework was employed to encourage users through a guided exploration of learning while leveraging the users' likelihood to use prior knowledge in new situations [10]. Since adults may resist investing time in learning new information, a minimalist approach can be beneficial. "The key idea in the minimalist approach is to present the smallest possible obstacle to learners' efforts," [5]. This paper describes the application and challenges of a user-centered methodology to develop a prototype Driver Training Application intended to assist drivers in learning about the features of ACC in a 2019 Chevrolet Blazer.

2 Design and Prototyping

Investigation of the problem and design of the solution occurred in three iterative phases. In Phase 1, the goal was to understand the target audience and competition requirements through a literature review, competitive analysis, a user needs research survey, and a hierarchical task analysis (HTA). In Phase 2, the goal was to develop a low-fidelity prototype, conduct user testing, and incorporate the findings to inform the design of a higher fidelity prototype. Researchers conducted a heuristic evaluation, information architecture evaluation, and formative usability testing to identify necessary improvements to the prototype. Lastly, for Phase 3, the goal was to improve the usability of the high-fidelity prototype; researchers conducted summative usability testing and gathered subject-matter expert feedback to inform the final iteration of the design. Figure 2 shows a schematic of the design methodology used to develop the prototype.

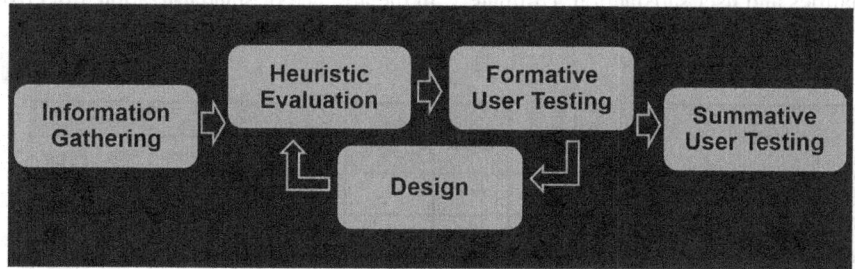

Fig. 2. The cycle of user-testing for prototype design

2.1 Phase 1 Information Gathering

In Phase 1, the competition goal from August 2020 through May 2021 was to develop a low-fidelity prototype that would teach users about and encourage the use of ACC. When presented with this challenge, the researchers focused on information gathering by conducting a brief literature review, competitive analysis, a user needs research survey, and an HTA. These testing methods were chosen in accordance with the Human Factors workflow for product design.

Competitive Analysis. Since ADAS feature design and functionality vary according to manufacturer, drivers are unable to generalize operation knowledge across all vehicles. Therefore, researchers conducted a competitive analysis to gather information on current variations of ACC-like features across different manufacturers. Researchers analyzed vehicle owner's manuals and official videos of eight vehicles from different manufacturers. The eight vehicles were chosen because they were similar to the 2019 Chevrolet Blazer and comparably priced. ACC-related features that were compared include the steering wheel buttons, driver information center (DIC) icons, type of notification feedback or alert (e.g., auditory, visual, or haptic), and type of vehicle sensors (e.g., camera or radar). The final competitive analysis included the Chevrolet Blazer, the Nissan Murano, the Hyundai Santa Fe, the Ford Edge, the Honda Passport, the Subaru Forester, the Jeep Grand Cherokee, and the Toyota Rav 4. These vehicles were chosen for comparison since they occupy roughly the same market segment as the Chevrolet Blazer.

Results from the competitive analysis demonstrated that there was very little standardization of ACC systems, gap distance systems, and related symbology used on DIC across vehicles (Fig. 3). However, compared to the other vehicle types evaluated, the Chevrolet Blazer provided more types of feedback (auditory, visual, and tactile) to the driver, and is one of only three vehicles that uses both radar and camera sensors. Being unfamiliar with vehicle features can also be dangerous and increase crash risk since the driver may not be aware of all the safety features, where they may be located, or how they function [12]. Additionally, ACC is a complex feature, so it is important for users to understand the operational limits of the system. While the Chevrolet Blazer offers more advanced safety features compared to other vehicles, the lack of standardization, both system-wide and industry-wide, may decrease understanding of the system's functionalities and user satisfaction. Ultimately, in the absence of standardization, effective learning tools are crucial for safe use of ADAS.

Make & Model	ACC Name	ACC Indicator	Gap Adjustment Indicator	ACC Feedback			Sensors	
				Auditory	Visual	Haptic	Camera	Radar
Chevy Blazer	Adaptive Cruise Control (ACC)	🚗	🚗	✓	✓	✓	✓	✓
Nissan Murano	Intelligent Cruise Control (ICC)	🚗	🚗	✓	✓			✓
Hyundai Santa Fe	Smart Cruise Control (SCC)	🚗	🚗	✓	✓			✓
Ford Edge	Adaptive Cruise Control (ACC)	ON OFF	🚗		✓			✓
Honda Passport	Adaptive Cruise Control (ACC)	🚗	🚗		✓		✓	✓
Subaru Forester	Adaptive Cruise Control (ACC)	🚗	🚗	✓	✓		✓	
Jeep Grand Cherokee	Adaptive Cruise Control (ACC)	🚗	🚗		✓		✓	✓
Toyota Rav 4	Dynamic Radar Cruise Control	🚗	🚗		✓			✓

Fig. 3. Competitive Analysis Results

User Needs Research Survey. Next, the researchers sought to better understand the target U.S. driving population and their learning preferences when presented with new automotive features. Specifically, the researchers inquired about the drivers' familiarity with advanced safety features, preferences on learning how to use advanced safety features, and vehicle renting habits. A user needs research survey was developed on Qualtrics and included 15 questions across three overarching sections: demographic information, preferences when owning a vehicle with ADAS, and preferences when renting a vehicle with ADAS [13]. Participants were recruited using Amazon Mechanical Turk (MTurk) [14]. To participate in the study, participants had to be 18 years of age or older, have a valid U.S. driver's license, and have over a 90% response rate on MTurk. The 90% response rate inclusion criteria reduces the likelihood of having participants who submit incomplete data. Data was collected from 956 participants with an age range of 18–79 years (M = 37.5, SD = 11.9) (Fig. 4).

When completing the survey, participants were asked to select *all* the learning methods that they used to learn about ADAS. Findings from the user needs research survey showed that for both vehicle owners and renters, the preferred mode of learning for the use ADAS features was by teaching oneself (i.e., trial and error). The next most preferred mode of learning was to consult family and friends or read the vehicle owner's manual. These results concur with the findings of Jenness et al., [22], Llaneras [24], and Viktorová & Šucha [23]. When asked how comfortable users were with using non-corrective and corrective driving features, researchers found that drivers were more comfortable using non-corrective

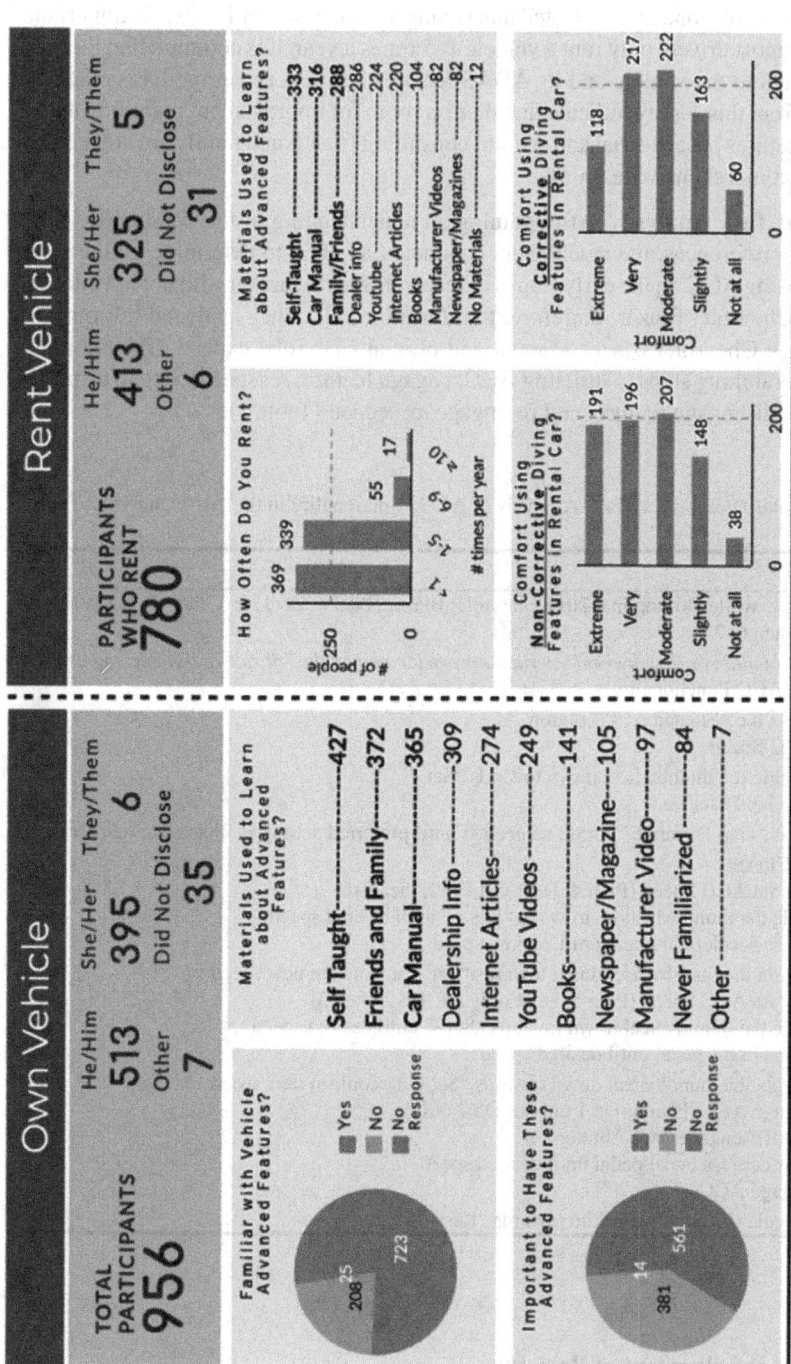

Fig. 4. Results of User Needs Research Survey

vehicle features (e.g., backup camera) over using corrective features (e.g., ACC). How-ever, most (57%) respondents indicated that having advanced safety features is important. Lastly, since most drivers only rent a vehicle 1–5 times a year, it is common that they are unable to adapt to the vehicle and its ADAS features before a newer model is released. The results from this survey indicate that there is room for improvement regarding learn-ing ADAS features because trial and error or consulting friends and family are not reliable methods for comprehensive learning.

Hierarchical Task Analysis. After gaining an understanding of the competitive mar-ket and user preferences, the researchers conducted an HTA to determine all the steps involved in using ACC. In the early stages of research, the researchers did not have access to the 2019 Chevrolet Blazer; therefore, to inform the HTA, they analyzed information from the 2019 Chevrolet Blazer Manual and official Chevrolet experts. Table 2 high-lights six overarching steps to utilizing ACC: engage feature, set speed, set gap distance, adjust speed, disengage feature, and re-engage as needed (Table 1).

Table 1. Tabular Hierarchical Task Analysis of ACC Functionality in the 2019 Chevrolet Blazer

Plan (0-7)
0.   Use ACC while Driving in a 2019 Chevrolet Blazer (Plan 0: Do 1, 2, 3, then 4 or 5 in any order, then 6, 7)
Note: All button or thumbwheel interactions are located on the left side of the steering wheel
1.   Enable ACC Functionality
1.1 Press the "Engage ACC" button
2.   Set ACC Speed
2.1 Toggle the thumbwheel down towards "Set -"
3.   Adjust Gap Distance
3.1 Press "Gap Distance" button as needed until preferred setting is illuminated on Driver Instrument Cluster
4.   Accelerate ACC Speed (Plan 4: Do 1 OR Do 2, then 3)
4.1   Toggle the thumbwheel up towards "Res +" until desired speed
4.2   Depress accelerator pedal until desired speed
4.3 Toggle the thumbwheel down towards "Set -" to confirm new speed
5.   Decelerate ACC Speed (Plan 5: Do 1 only OR Do 2, then 3)
5.1   Toggle the thumbwheel down towards "Set -" until desired speed
5.2   Depress brake pedal until desired speed
5.3 Toggle the thumbwheel down towards "Set -" to confirm new speed
6.   Disengage ACC (Plan 6: Do 1 only or Do 2 only)
6.1   Press "Disengage ACC" button
6.2   Tap or depress brake pedal until desired speed
7.   Reengage ACC
7.1 Toggle the thumbwheel up towards "Res +"

2.2   Phase 2 Development of Prototype

Initial Prototype Design. After gaining an understanding of ACC, its variations across different manufacturers, user preferences when learning about these features, and all the functions needed to use it, the researchers developed the first iteration of the prototype. There were four main learning objectives to achieve with the prototype design: (1) teach users about the purpose of ACC, (2) how it functions, (3) how to interpret relevant buttons on the steering wheel and icons on the DIC, and (4) highlight critical safety precautions.

The researchers developed a low-fidelity, two-part prototype system, called SAFERent, which included a video tutorial and an interactive guide. The video tutorial provided a 4-min 30-s overview of ACC. It taught users about the controls, important safety precautions, and the meaning of steering wheel buttons, as well as DIC icons (Fig. 5). The researchers utilized comedic sound bites as an approach to maintain engagement with

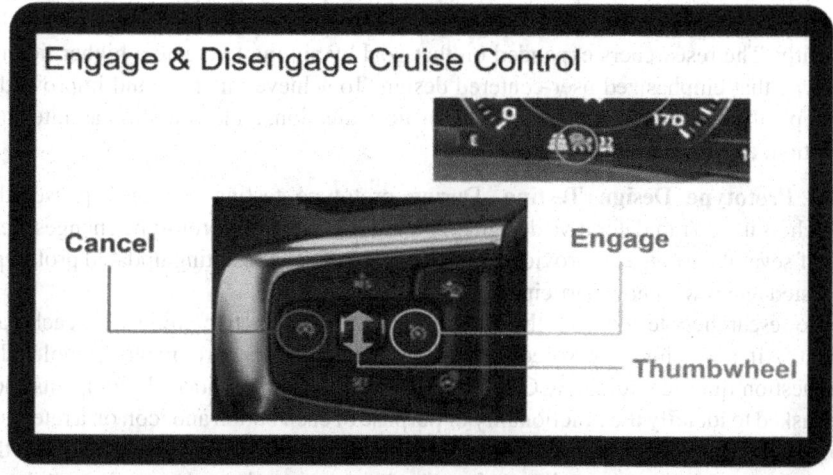

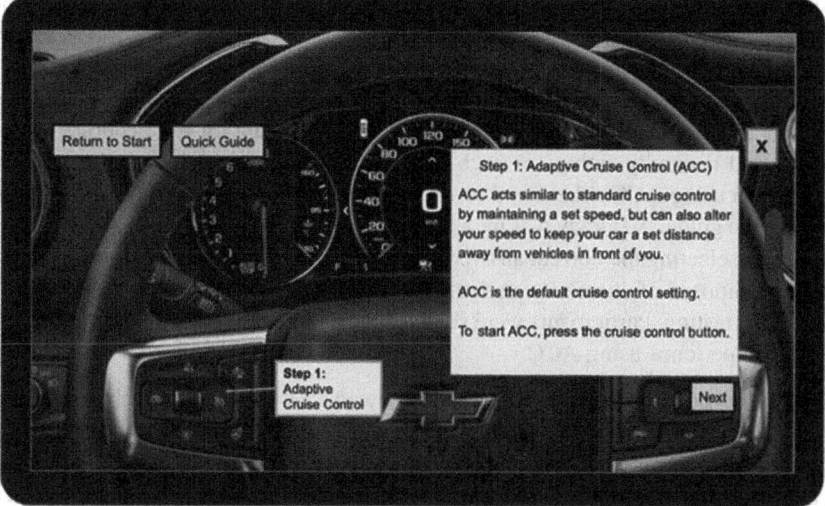

Fig. 5. SAFERent Initial Prototype Video Tutorial (top) and Interactive Guide (bottom)

the media. At this stage of the research, the 2019 Chevrolet Blazer was not available to the students, otherwise they would have explored the efficiency of using sounds and other auditory components from the vehicle. The second part of the prototype system was the interactive guide, developed on Google Slides, which functioned similarly to a mobile application and served as a refresher that allowed the user to interact with the features they wanted to review (Fig. 5) [15]. Next, the researchers began initial usability testing of this prototype.

2.3 Phase 2 Research and Usability Testing

Phase 2 started at the beginning of a new academic year with new competition challenges. From Fall 2021 through Spring 2022, the competition goal, as required by EMC organizers, was to develop a standalone prototype that a user can interact with independently. The researchers expanded on that goal by aiming to create a higher fidelity prototype that emphasized user-centered design. To achieve this goal and improve the prototype, the researchers conducted a heuristic evaluation, an information architecture evaluation, and formative usability testing.

Initial Prototype Design Testing. During prototype testing for each phase, the researchers used a rapid iterative design process. This meant that prototype changes were made if several participants provided similar feedback. The resulting updated prototype was tested again with new participants.

The researchers tested both the video tutorial and interactive guide with each participant. After watching the video tutorial on a smartphone, participants completed a five-question quiz regarding ACC capabilities featured in the video. Participants then were tasked to identify the functionality or purpose of each button and icon on a retention map (Fig. 6). The researchers gathered data on the accuracy of performance for both the quiz and retention map, as well as subjective information about the user's confidence in using ACC, likelihood to watch the video, and perceived effectiveness of the video tutorial.

Participants were then presented with the interactive guide on the same smartphone. Participants navigated through the interactive guide to complete six tasks while vocalizing their thought processes. The researchers gathered data on the accuracy of performance and difficulty rating for each task, subjective information about the user's confidence in using ACC, likelihood to use the interactive guide, and perceived effectiveness of the guide. Accuracy was measured based on the number of errors participants made before selecting the correct answer when presented with a task. A total of 18 participants, composed of university students and faculty at Embry-Riddle, were recruited for prototype testing. Participants were selected via a screener survey to ensure they had little to no experience using ACC.

Initial Prototype Iterations. Quiz results showed that users that watched the video tutorial had the most difficulty understanding the gap distance functionality, as well as cautions and warnings of the operational limits of ACC. The retention map results indicated that users thought the motion of "toggling up" the thumbwheel on the steering wheel increased their gap distance rather than speed. To mitigate these issues, the researchers updated the video tutorial by eliminating the static images of buttons, and

Fig. 6. Retention Map to Assess User's Knowledge of ACC

instead, showing a video of actual button pressing or toggling of switches on the steering wheel. To add emphasis to the safety precautions, the researchers improved the audio quality and added more relevant imagery. Lastly, the researchers changed the comedic approach from sound bites to themed imagery since the audio was reported to hinder learning.

For the interactive guide, most participants reported that they enjoyed this mode of learning, while others indicated that they would not use it if offered to do when renting a vehicle. The research team began brainstorming incentivization to encourage use of the prototype in a real-world environment. Participants also provided feedback that adding a "quick reference" tool, serving as a legend for all buttons and icons, would be helpful in refreshing knowledge quickly. After developing a Quick Guide (Fig. 7), participants performed better on tasks and rated the tasks as easier to complete.

Heuristic Evaluation. The researchers conducted an evaluation using Nielsen's 10 Usability Heuristics for User Interface Design because this seminal inspection method includes best practices to create an interface that is useful for many users [16]. For example, it is important that the system matches user expectations and for the user to feel in control when interacting with the system. The research team, composed of seven graduate-level Human Factors students, completed individual heuristic evaluations, with guidance from Nielsen's 10 and additional mobile-specific questions developed by Gómez and colleagues [17]. The mobile-specific questions expanded on Nielsen's heuristics by considering other aspects of design, including but not limited to, the crowding of buttons on mobile interfaces, and clarity on types of interactions (e.g., swipe, pinch). Overall, the most impactful results related to two heuristics: the Visibility of System Status and Error Prevention.

Visibility of System Status is a heuristic highlighting the importance for a user to know what is happening with the system and have a clear mental model of where they

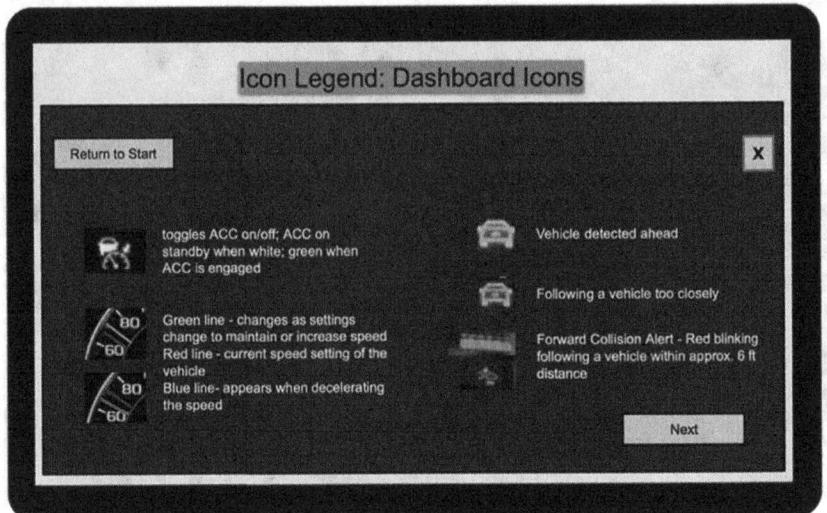

Fig. 7. First Iteration of Quick Guide

are in the navigation system [16]. Results indicated that users did not know the length of the ACC Tour and how much of it they had completed. To give users feedback and an indication of tour length, a progress bar was added (Table 2).

Error Prevention is a heuristic highlighting the importance for a user to receive feedback when making an erroneous action [16]. One area of the prototype that was problematic for users was the proximity of the hamburger menu and home button. Users would accidentally tap the wrong option because the buttons were too close together (Table 2). To resolve this issue, both buttons were enlarged and their spacing increased. The "Test Yourself" section of the prototype, where users could check their knowledge of ACC, was another problematic area since users did not originally receive feedback for selecting the wrong button on the steering wheel. To resolve this issue, users would be notified with a positive message if they selected the correct response or an encouraging message to try again if they had selected an incorrect response (Table 2).

Information Architecture Evaluation. To assess whether the refinements made to the prototype facilitated intuitive navigation, researchers conducted first-click testing and tree testing through Optimal Workshop [18]. In the first-click test, participants were given multiple tasks regarding where they would find information about ACC within the prototype. then, they would indicate their response by clicking on a particular button on the home screen. To avoid learning biases, participants were not given feedback on where their selection would navigate them. The results from this study helped the researchers determine if navigation of the prototype was intuitive and if the users generally employed the shortest path when searching for information of interest.

The tree testing study was implemented to determine if users could navigate using the correct path when searching for specific information about ACC features within the prototype. Users completed a series of search tasks and clicked through a series of categories until they felt they were at the location where they would find the desired

Table 2. Heuristic Evaluation Findings – Issues and Solutions

Nielson Heuristic	Issue Description	Font size and style
Visibility of System Status	Users cannot identify how far they are within the tutorial	- Adding a progress bar visible at the top of every page
Error Prevention	Users mis-tapped between the hamburger menu icon and home button icon	- Enlarging the hamburger menu and home button icons - Increasing the distance between the two icons
Help Users Recognize, Diagnose, and Recover from Errors	Test Yourself questions did not notify the user of correct/incorrect responses	- Adding pop-up feedback messages including "Try Again" and "Great Job"

information. Findings from both the first-click and tree testing studies demonstrated mostly positive results in the information architecture of the features within the prototype. However, results also indicated that the categories were not completely self-explanatory to every user. Some participants were not fully sure what to expect within each category. These results were used to inform updates to make navigation and feature descriptions more intuitive for the user.

Formative Usability Testing. After evaluating the information architecture, researchers conducted formative usability testing to gather feedback on various components of the prototype and determine its effectiveness for users. Nine participants were asked to explore the prototype as if they were offered this service from a rental company and told to speak aloud on all of their thoughts and actions. There were three main areas of interest assessed through formative usability testing: the overall navigation, the Test Your Knowledge section, and the ACC Tour section. Recorded comments from the participants were sorted into one of four categories: issues, positives, suggestions, or comments.

There were 72 comments addressing issues with the prototype from the nine participants of the summative usability testing. A common issue cited in the comments was that the prototype design, as a whole, was too verbose. To address this issue without removing important information, researchers created an overview screen that explained the ACC features at a high level. For example, it showcases a brief explanation of what ACC is, important precautions to consider when using ACC, and incentives for using ACC. This provided the opportunity to alleviate some of the text-heavy content on other pages of the prototype. Another finding from formative usability testing was that many users were not aware that the buttons on the steering wheel in the ACC Tour section were clickable. To resolve this issue, researchers added a cursor icon, as that is used in other interfaces to suggest interactivity (Table 3). Lastly, when addressing overall impression of the application, many users were confused about the meaning of the "i" icon button (e.g., more information). Users were unaware that this led them to an overview of DIC icons and steering wheel buttons. When exploring this page, users commented that the information presented was valuable, but that the relationship between the icon and referent information was unclear, therefore the icon was revised to resemble an existing one found on the vehicle's dashboard (Table 3).

Table 3. Formative Testing Findings – Issues and Solutions

Prototype Issue	Description of Issue	Image of Change
"Too wordy"	Pages within the app were filled with dense text	Addition of an overview screen
Confusing symbology of DIC symbol	Icons used to represent the dashboard icon page was not intuitive to users	
Unclear button functionality	Users did not deem the steering wheel buttons to be tappable within the prototype	

2.4 Phase 2 Prototype Results

The phase 2, higher fidelity prototype, was created on Adobe XD and combined both the interactive guide and video elements to mimic the look and feel of a mobile application – now named the Driver Training Application [19]. This iteration included significant changes due to results from heuristic evaluations, an information architecture evaluation, and formative usability testing. many of the changes involved modifications to the overall aesthetic layout, presentation of information, type of information displayed (e.g., text or video), and navigation, in an effort to streamline the content and user experience. Utilizing the existing prototype's core information, a new color scheme and font style were implemented. Additionally, there were alterations to icons that represented new sections on the Driver Training Application. This was a way to produce a clear portrayal of the materials being presented.

Furthermore, content was analyzed, and wording was modified to ensure readability that accommodated drivers of varying experience levels. Selected video tutorials were recreated to match the professional aesthetic of the revamped Driver Training Application. Finally, the entirety of the application was resized to fit an 8-inch tablet, similar in size to the infotainment system of the Chevrolet Blazer. The extra spacing provided by the larger device increased target touch sizes for icons and improved visual appeal.

2.5 Phase 3 Research and Usability

For phase 3, the researchers expanded on the standalone, high-fidelity prototype goal by migrating the Driver Training Application to more sophisticated software that could include logic that would allow it to adapt to user inputs. To further enhance usability, the researchers conducted summative usability testing and gathered feedback from subject-matter experts with Human Factors and Engineering expertise.

Summative Testing. Summative usability testing involved a formal, systematic evaluation of effectiveness for the Driver Training Application, as a whole, in teaching ACC features to users. In total, nine users were recruited for summative usability testing with 50% of recruited participants being from age groups older than the average college student. Participants of the study were randomly assigned to one of three experimental groups: control group, brief overview group, and interactive guide group. Regardless of the study group, participants sat in a driving simulator seat with a Chevrolet steering wheel mounted in front of them to simulate the interior of a Chevrolet Blazer. All participants were then asked to read aloud a background scenario where they were a traveler renting a new vehicle with advanced safety features, including ACC. Then, depending on the study group, participants were presented with the corresponding learning materials to gain knowledge about ACC.

Control group participants were only presented with a short description of ACC, brief overview group participants were presented with several overview screens of the prototype explaining ACC, and participants within the interactive guide group were presented the entire prototype. After reviewing their learning materials, all participants were asked to complete a questionnaire assessing their confidence using ACC. They were then presented with six tasks that required them to identify steering wheel buttons or DIC icons that they would view or select to complete the task. After each task, participants were asked to rate the difficulty of the task.

After making changes to the Driver Training Application from summative usability testing results, researchers pursued feedback from subject-matter experts to ensure a wide range of demographic feedback for the application. Subject-matter experts were asked to look through the prototype and express their impressions on strengths and weaknesses of the application, as well as possible recommendations for improvements. This feedback was used to create the final iteration of the Driver Training Application.

2.6   Phase 3 Prototype Results

The final prototype was recreated in Axure, an interface design software that is more sophisticated than Adobe XD [20]. The sections of the application include the following elements: Start screen, Home screen (Fig. 8), 'Adaptive Cruise Control (ACC) At a Glance', Hamburger/Slide Menu, 'Adaptive Cruise Control (ACC) Tour' (Fig. 8), 'Dashboard Icons & Steering Wheel Buttons', 'Test Yourself', and 'My Rewards'.

Additionally, videos were implemented into the final application so that users were provided a combination of auditory and visual methods for learning. Incentivized screens were also added to the Driver Training Application that would allow users to redeem rewards, such as a $40 gas gift card and/or a 15% discount on their current rental, upon completion of specific features labeled with a ribbon on the home page. These incentives were chosen after users expressed that rewards such as gas gift cards, reward points, or discounts would help encourage them to utilize the designed prototype. Overall, the final high-fidelity application catered to a range of learning styles, and the intuitions of all users of any experience level.

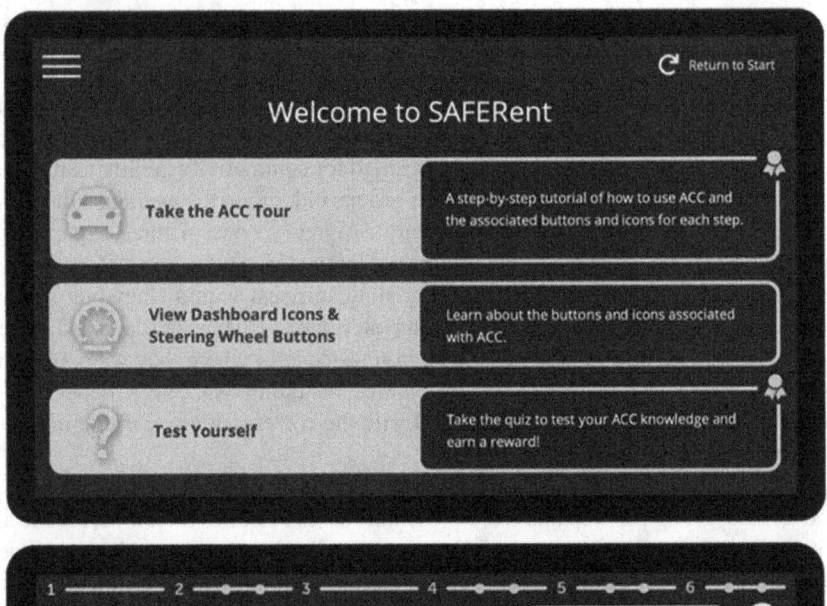

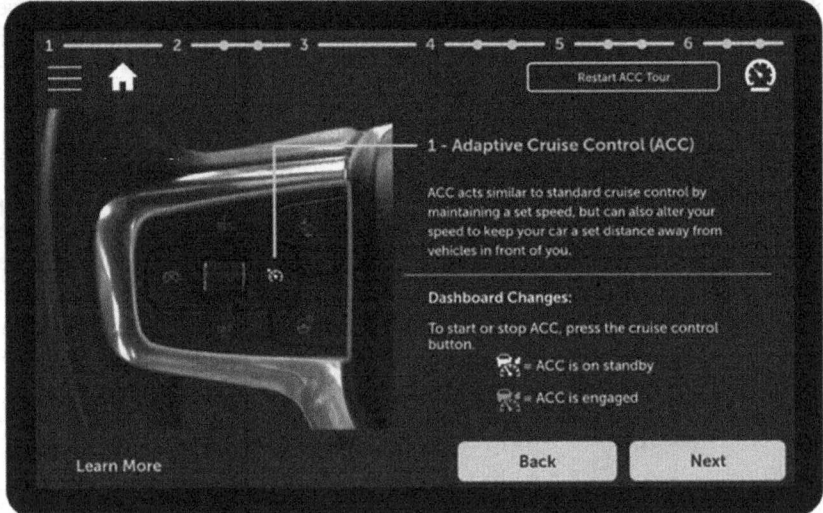

Fig. 8. Final Prototype Home Screen (top) and Adaptive Cruise Control (ACC) Tour Screen (bottom)

3 Discussion

Lack of standardization in ADAS features across manufactures is problematic and could jeopardize driver safety. Although it is important to understand how to operate these features, learning them has proven to be difficult and time-consuming for users. This issue continues to elevate in severity with increasing complexity of newer vehicles coming to the market. Implementing an application, like SAFERent, that consolidates teachings of safety features on all makes and models of vehicles would help mitigate the issue. Usability testing with an iterative design process has been a successful method for gathering and implementing feedback from users when developing such a prototype. The SAFERent application that resulted from this study allows drivers to maximize safety and comfort on the road by informing them of how to use ACC, the limitations of the system, and by providing them with the ability to test/review their knowledge. The results from user testing of the SAFERent application showed that, on average, users explored the prototype in 4 min and 21 s, satisfying the goal of creating a learning too for users to be able to be review quickly to keep engaged. SAFERent was also successful in supporting multiple modes of learning through the use of short videos, a visual written tutorial, and with an interactive "Test Yourself" section. Lastly, SAFERent matched user expectations for interface layout as all participants were able to correctly navigate throughout the prototype for all 6 summative usability testing tasks, with 80% of participants doing so in the most efficient paths possible. These results supported the achieved goal of developing an efficient, effective, and satisfying prototype.

Limitations and Future Research. Participation in usability testing was limited to college students, as well as a few older adult individuals. College students are not representative to the general population and therefore this posed a limitation to our overall results. It is important to take different age ranges into account when creating instructions, as older adults approach manual instructions differently than younger generations. Older adults are more likely to read portions of a manual for a new product, but they can often feel overwhelmed when encountering confusing and frustrating information [21]. Future research is needed in order to incorporate older adults and other special populations.

There was limited access to the 2019 Chevrolet Blazer for the duration of testing, in all phases. Therefore, the testing was completed in a laboratory setting, where participants sat in a mock-up or test-bed of a vehicle seat. As this is not representative of the actual use case scenario (i.e., sitting in the driver seat of a vehicle, it posed another limitation to our findings. Future research should focus on testing this application in real-life settings. The use of a functional vehicle would increase the fidelity of user testing by facilitating a seamless integration between the driver, application, and ADAS features.

The development of the SAFERent prototype enables future ADAS training research to be conducted. Future research should also be conducted to expand the services provided by the Driver Training Application so that it can become an all-inclusive vehicle rental application. It could provide services, like vehicle rentals, vehicle access (e.g., lock and unlock), feature training, and a global positioning system (GPS). Creating an all-inclusive application would provide additional convenience for drivers, especially renters. Additionally, continuing advancements would ensure that SAFERent can be up-to-date with the annual release of new vehicles, promoting sustainable development.

Acknowledgements. We would like to acknowledge the EcoCAR Mobility Challenge (EMC) sponsored by the United States Department of Energy, General Motors, and MathWorks.

References

1. Shuttleworth, J.: SAE Standards News: J3016 automated-driving graphic update (2019). https://www.sae.org/news/2019/01/sae-updates-j3016-automated-driving-graphic

2. Jeff, J.: Autonomous Cars, Robotaxis & Sensors 2022–2042 (2022). https://www.idtechex.com/en/research-report/autonomous-cars-robotaxis-and-sensors-2022-2042/832
3. American Automobile Association. Advanced Driver Assistance Technology Names (2019). https://www.aaa.com/AAA/common/AAR/files/ADAS-Technology-Names-Research-Report.pdf
4. Mehlenbacher, B., Wogalter, M.S., Laughery, K.R.: On the reading of product owner's manuals: perceptions and product complexity. In: Proceedings of the Human Factors and Ergonomics Society Annual Meeting, vol. 46, no. 6, pp. 730–734 (2002). https://doi.org/10.1177/154193120204600610
5. Carroll, J.M., Rosson, M.B.: Paradox of the active user. In: Interfacing Thought: Cognitive Aspects of Human-Computer Interaction, pp. 80–111 (1987)
6. Rettig, M.: Nobody reads documentation. Commun. ACM **34**(7), 19–24 (1991)
7. Oviedo-Trespalacios, O., Tichon, J., Briant, O.: Is a flick-through enough? A content analysis of Advanced Driver Assistance Systems (ADAS) user manuals. PLoS ONE **16**, 6 (2021)
8. Brockmann, R.J.: The why, where and how of minimalism. In: Proceedings of the 8th Annual International Conference on Systems Documentation, pp. 111–119 (1990)
9. Carroll, J.M.: The Nurnberg Funnel: Designing Minimalist Instruction for Practical Computer Skill. MIT press (1990)
10. Carroll, J.M.: Creating minimalist instruction. Int. J. Des. Learn. **5**, 2 (2014)
11. Advanced Vehicle Technology Competitions. EcoCAR Mobility Challenge (2022). https://avtcseries.org/ecocar-mobility-challenge/
12. Evans, L.: Human Behavior and Traffic Safety, 1st ed. Springer, New York (2012)
13. Qualtrics. Qualtrics XM (2022). https://www.qualtrics.com/
14. Amazon Mechanical Turk. Amazon Mechanical Turk (2022). https://www.mturk.com/
15. Google. Google Slides (2022). https://www.google.com/slides/about/
16. Jakob Nielsen. 10 Usability Heuristics for User Interface Design (1994). https://www.nngroup.com/articles/ten-usability-heuristics/
17. Gomez, R.Y., Caballero, D.C., Sevillano, J.-L.: Heuristic evaluation on mobile interfaces: a new checklist. Sci. World J. (2014)
18. Optimal Workshop. Optimal Workshop (2022). https://www.optimalworkshop.com/
19. Adobe. Adobe XD (2022). https://www.adobe.com/products/xd.html
20. Axure Software Solutions. Axure RP 10 (2022). https://www.axure.com/
21. Tsai, W.-C., Rogers, W.A., Lee, C.-F.: Older adults' motivations, patterns, and improvised strategies of using product manuals. Int. J. Des. **6**(2), 55–65 (2012)
22. Jenness, J. W., Lerner, N. D., Mazor, S., Osberg, J. S., & Tefft, B. C. (2008). Use of advanced in-vehicle technology by young and older early adopters. Survey results on adaptive cruise control systems. Report No. DOT HS 810 917. Washington, DC: National Highway Traffic Safety Administration
23. Viktorová, L., Šucha, M.: Learning about advanced driver assistance systems–the case of ACC and FCW in a sample of Czech drivers. Transport. Res. F: Traffic Psychol. Behav. **65**, 576–583 (2019)
24. Llaneras, R.E.: Exploratory study of early adopters, safety-related driving with advanced technologies. Report No. DOT HS 809 972. Washington, DC: National Highway Transportation Safety Administration (2006)
25. Beggiato, M., Krems, J.F.: Transport. Res. Part F Traffic Psychol. Behav. 47–57 (2013)

Cooperative and Intelligent Transport Systems

Who's in Charge of Charging? Investigating Human-Machine-Cooperation in Smart Charging of Electric Vehicles

Meike E. Kühne[1], Christiane B. Wiebel-Herboth[2] , Patricia Wollstadt[2] ,
André Calero Valdez[1] , and Thomas Franke[1(✉)]

[1] University of Lübeck, Ratzeburger Allee 160, 23562 Lübeck, Germany
{kuehne,calerovaldez,franke}@imis.uni-luebeck.de
[2] Honda Research Institute Europe, Carl-Legien-Str. 30, 63073 Offenbach am Main,
Germany
{christiane.wiebel,patricia.wollstadt}@honda-ri.de

Abstract. We investigated the effect of varying the level of cooperation in a smart charging agent (SCA) on user perception and behavior. Our study involved manipulating the SCA's cooperativeness by varying its degree of automation and the amount of information sharing with the user and measuring effects on changes in user behavior, perceived goal alignment, the user's awareness of the SCA's information processing, and perceived cooperativeness. Our hypothesis that a lower degree of automation of the SCA would increase human-agent cooperation was not supported by our results. Instead, participants in the high-automation condition chose a later charging endpoint more often, implying greater cooperation. Our hypothesis that a higher amount of information shared by the SCA would increase human-agent cooperation was only partially confirmed. Cooperation led to a more positive user experience, but the correlation was only moderate to strong. The study shows the limitations of using the degree of automation as a sole measure of human-machine cooperation and highlights the need to explore other operationalizations of human-machine cooperation. Further research is needed to explore other scenarios and variations in the information provided to the user to better understand human-machine cooperation in the context of smart charging.

Keywords: Human-Machine-Cooperation ·
Human-Machine-Interaction · Smart Charging · Battery Electric
Vehicles · Demand Side Management

1 Introduction

The transition to utilizing renewable energy sources, such as wind and solar power, is crucial for achieving a sustainable energy system. However, the integration of these sources into the power grid poses significant challenges due to

H. Krömker (Ed.): HCII 2023, LNCS 14048, pp. 131–143, 2023.
https://doi.org/10.1007/978-3-031-35678-0_8

their inherently variable nature. To address these challenges and ensure a stable power grid, it is necessary to implement measures such as decentral energy generation and storage, as well as flexibly adjusting energy demand, also known as demand-side management. These measures will not only stabilize the power grid but also enable greater penetration of renewable energy into the energy mix. The idea of demand-side management specifically is to alleviate strain on the power grid by incentivizing large energy consumers to shift consumption to periods of increased renewable energy availability and decreased energy demand [1,6]. Furthermore, the immediate use of renewable energies at their time of availability cannot only lead to more efficient use of energy resources at the energy system level but also at the consumer level [20].

An energy consumer unit that is of particular interest for demand side management is the battery electric vehicle (BEV) since it has a large storage unit as well as a high energy demand that can easily be flexibilized through approaches like smart charging. With smart charging, consumers can adjust the charging of their BEV in real time to electricity prices and availability. This process is typically controlled by an automated system (agent), which relies on user preferences and other information, such as energy availability, to optimize the process (e.g., [18]). Therefore, smart charging cannot only lead to a stabilization of the power grid and a reduction in CO_2 emissions but also in lower costs for the consumer [11].

So how do we design a smart charging agent (SCA) that encourages people to regulate their charging behavior to ensure an efficient allocation of energy resources? We suggest that a cooperative approach to system design might help support the effective joint regulation of energy and other resources such as time, information, or comfort. Prior work shows that human-machine-cooperation not only copes with the shortcomings of many current automation approaches but also enables greater flexibility in the shared action and, all in all, enhances the joint performance of the human and the system [2,16,25].

2   Background

Comparative-cognition research has demonstrated a unique motivation for humans to collaborate. Tomasello and Vaish [26] state that "human social interaction and organization are fundamentally cooperative in ways that the social interaction and organization of other great apes simply are not" (p. 239). In recent years, several authors tried to utilize this inherent motivation of humans to cooperate to facilitate human-machine interaction (HMI). The construct of human-machine-cooperation has been discussed for many different use cases, such as driver assistance [17,19,27] or human-robot interaction [9]. However, there is no full consensus on how to define cooperation, and different authors propose different operationalizations of cooperation for different contexts [2,15,28,30]. Moreover, while the effects of cooperative automation design have been investigated in many human-technology contexts, as of writing, we are not aware of any study that has investigated a cooperative approach in the domain of smart charging.

2.1 Formalizing Cooperation

There is a multitude of theories and models on cooperation from different disciplines such as social psychology [13], philosophy [3], and human factors [4,5,7,10,14]. In the following, we conceptualize cooperation based on the similarities between these theories.

How to Design a Cooperative SCA? Klein et al. [14] propose four requirements for successful cooperation between a human and an automated system: An agreement to work together, a common ground, mutual directability, and mutual predictability. For this study, we manipulate the latter two aspects, *directability and predictability*, as we assume the other two prerequisites are already fulfilled by the design of the SCA.

Previous work in the context of automated driving has often operationalized directability by the degree of automation of the system. Several studies find that participants report a higher feeling of safety, pleasure, and trust when interacting with a cooperative system with which they share the task of driving compared to a higher automated system [17,27]. On the other hand, research in artificial intelligence shows that the perceived predictability of the system can be promoted through additional information on the inner workings of the system, which also promotes trust [12,21].

How to Measure Successful Cooperation Between Humans and the SCA? We propose measuring the effects of our SCA manipulation as follows. First, we assess the perceived quality (*perception*) of the human-SCA interaction with respect to the following aspects: (1) goal alignment, (2) system understanding, and (3) perceived cooperativeness. Second, we observe induced behavioral changes in the human charging patterns (*behavioral change*). In addition, we measure user experience as a function of the user's perceived cooperativeness of the interaction.

Goal alignment. It has been proposed that the need for cooperation between two or more agents arises from an interdependence that results from shared goals or overlapping intentions [3–5,10,13,14,26]. Hence, to measure cooperation on a motivational (intentional) level, we assess users' perceived goal alignment with the SCA.

Understanding of the system. Multiple theories emphasize the importance of shared representation, knowledge, and beliefs between the interaction partners as well as some degree of mutual predictability and reliability in cooperation [3,5,7, 10,13,14]. To measure cooperation on a cognitive level, we assess the users' level of awareness and understanding of the system and its information processing.

Perceived cooperativeness. We measure if the interaction with the SCA was perceived as cooperative by the user. To that end, we developed a five-item-scale to measure perceived cooperativeness specifically.

Behavioral change towards a shared goal. Last, according to Klein et al. [14], joint action entails "one or more participants relaxing some shorter-term local goals in order to permit more global and long-term goals to be addressed" (p. 6). For the context of smart charging, this can be understood as users giving up some flexibility and shifting their charging window to times of high energy availability to permit demand side management. Here, we measure cooperation on a behavioral level by assessing whether users shift their charging window upon request.

2.2 Hypotheses

In this study, we set out to examine the following three hypotheses:

- **H1:** A lower degree of automation should increase human-agent-cooperation (behavioral change and perception).
- **H2:** Increasing the amount of shared information should increase human-agent-cooperation (behavioral change and perception).
- **H3:** An increase in perceived human-agent-cooperation should lead to a better general user experience.

3 Method

3.1 Sample

Participants were recruited through mailing lists and social media. The experiment was conducted via the online survey platform LimeSurvey. All participants were required to speak German fluently. After excluding those participants who did not complete the questionnaire, those who took an unusually long time to complete the questionnaire, and those who showed no variance in their responses on any of the pages of the questionnaire, the final sample consisted of 91 participants (29% female) with an average age of $M = 42.3$ ($SD = 12.5$). In the sample, 76% had driven a BEV before, and 53% had prior experience with driving a BEV on a regular basis.

3.2 Experimental Design and Scenario

We conducted an experiment with a 2×2 between-subject design. Participants were presented with scenarios that included interactions with an SCA that varied in the degree of automation (low vs. high) and the amount of information sharing (low vs. high). In the beginning, participants were instructed to imagine themselves in the situation described in Fig. 1.

Next, participants were presented with a message from the SCA that asked them to prolong their charging window to increase the share of self-produced power. The message differed based on which experimental group participants were in.

Imagine the following scenario: You live in the greater Lübeck area in a single-family house. You own a PV system, a private wallbox and a battery electric vehicle. To charge your car at home, you use a smart charging agent. The charging agent supports you in ensuring that your car is charged with as much electricity as possible from your own PV system during each charging session. This is the most cost effective and climate friendly way to charge your car.

To do this, each time you plug in your car, you specify in an app when you need the car again and how much range you need for the next trip. Your charging agent uses this information, along with predicted production data from your PV system, to find an optimal charging window.

You plug in your car for charging on Friday evening after work with a remaining range of 15 km. The next day, you plan to leave in the afternoon at 2 p.m. because you have arranged to meet friends at Timmendorfer Strand. You plan a range of 120 km including a buffer for the trip.

Fig. 1. Introduction to the scenario presented to every participant.

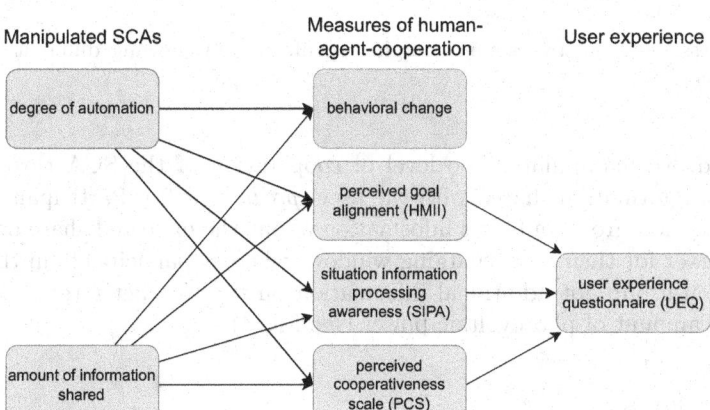

Fig. 2. Research design. Manipulated variables are shown in orange, measures variables are shown in green (behavioral variable) and blue (subjective variables). (Color figure online)

3.3 Experimental Manipulation

For an overview of manipulated and measured variables see Fig. 2.

First, we manipulated the level of cooperation of the SCA through the system's degree of automation (*directability*). The degree of automation was altered according to the levels of automation of decision and action selection by Parasuraman [22]. Participants in the low-level condition got a pop-up message from a level three agent that suggested two alternative charging endpoints while participants in the high automation condition interacted with a level six agent that automatically prolonged the charging window of the user and gave them a restricted time to veto (see Fig. 3).

Message of the SCA in the **low automation** condition (Level 3 according to Parasuraman et al., 2000)

Message of the SCA in the **high automation** condition (Level 6 according to Parasuraman et al, 2000)

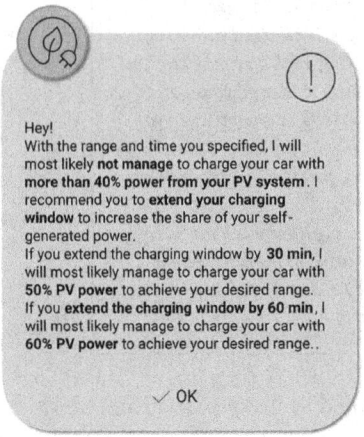

Hey!
With the range and time you specified, I will most likely **not manage** to charge your car with **more than 40% power from your PV system** . I recommend you to **extend your charging window** to increase the share of your self-generated power.
If you extend the charging window by **30 min**, I will most likely manage to charge your car with **50% PV power** to achieve your desired range.
If you **extend the charging window by 60 min**, I will most likely manage to charge your car with **60% PV power** to achieve your desired range..

✓ OK

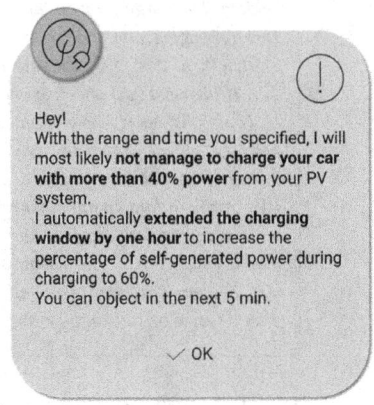

Hey!
With the range and time you specified, I will most likely **not manage to charge your car with more than 40% power** from your PV system.
I automatically **extended the charging window by one hour** to increase the percentage of self-generated power during charging to 60%.
You can object in the next 5 min.

✓ OK

Fig. 3. Messages from two smart charging agents (SCA) showing different levels of automation.

Second, we manipulated the level of cooperation of the SCA through the amount of information shared with the user (*predictability*). Participants in the low-information group only got information about the expected share of photovoltaic power for their next charging window, whereas participants in the high-information group got additional information on the weather forecast and the projected amount of photovoltaic power (see Fig. 4).

3.4   Measures

To measure the effects of our independent variables on human-agent-cooperation, we assessed the following metrics based on our formalization of cooperation described in Sect. 2.1.

To assess participants' perceived goal alignment, we administered the conflict subscale of the Human-Machine-Interaction-Interdependence Questionnaire (HMII, [29]). A low score on this scale would indicate a high perceived goal alignment. Further, we measured participants' understanding of the system through the Subjective Information Processing Awareness Scale (SIPA, [24]). SIPA describes the experience of being enabled by a system to perceive, understand and predict its information processing. A low score on this scale would indicate a low awareness and understanding of the system and its information processing. Last, to quantify behavioral changes, we measured whether participants prolonged their charging window.

Information all subjects received (**low and high information** condition)

Additional information only subjects in the **high information** condition received

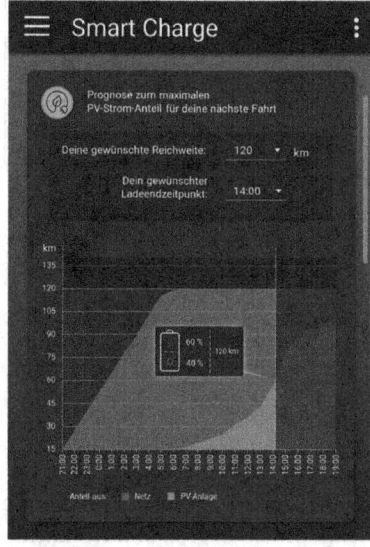 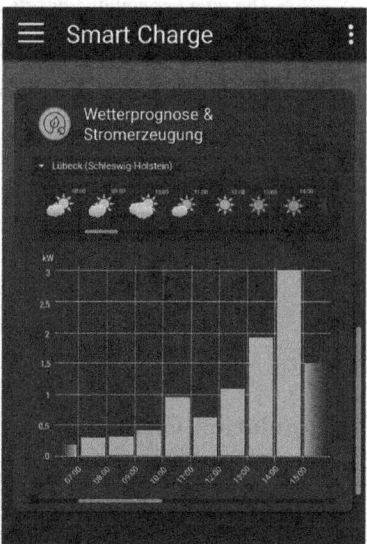

Fig. 4. Manipulated interface providing different levels of information from the smart charging agent (SCA).

As an additional measure for human-agent cooperation, we measured the perceived cooperativeness by a self-developed scale. The items were generated in a workshop with six students. As a basis, participants of this workshop got an introduction to various theories on cooperation and joint action research as well as examples of everyday interactions with automated systems. The scale uses a 6-point Likert response scale. A translation of the German items can be found in Fig. 5.

Additionally, we used the User Experience Questionnaire (UEQ, [23]) to measure participants' user experience through perceived attractiveness, pragmatic quality, and hedonic quality of the agent. We further surveyed the power and information certainty (human to system) sub-scales from the HMII [29] to check if our manipulation of the degree of automation and amount of shared information was successful.

1. *The agent supports me in achieving my goals.*

2. *The agent communicates its decisions and actions with me.*

3. *The agent helps me make better decisions.*

4. *The agent's actions are tailored to my personal circumstances.*

5. *The agent is cooperative.*

Fig. 5. The items of the perceived cooperativeness scale.

4 Results

4.1 Manipulation Check

Participants in the low-automation condition reported significantly higher perceived power in the situation ($U = 1452$, $p_{\text{one-tailed}} < .001$) with a medium effect size $r = .37$. We conclude that the manipulation of the degree of automation was successful. For the variation of the amount of information shared, there was no significant difference in the reported information certainty (human to system) between the low and high amount of given information condition ($U = 984$, $p_{\text{one-tailed}} = .369$). Therefore, results regarding the difference in the amount of given information should be regarded with caution.

4.2 Regression Analysis

To examine our first and second hypotheses, we conducted four different regression analyses for each of our dependent measures.

The first analysis was performed for the variable *behavioral change* measured by the chosen charging endpoint. Since the charging endpoint variable was not normally distributed, we transformed this measure into a dichotomous variable on whether participants choose to prolong the charging window (1) or not (0). We performed a binomial logistic regression (see Table 1) on the respective data. We found that a higher degree of automation led to a significantly higher probability of changing the behavior, i.e., adjusting the charging window.

For the other three measures, we calculated linear regression analysis as the data fulfilled all statistical requirements. We did not find an effect of the experimental manipulations on *perceived goal alignment*, as measured by the HMII's conflict subscale ($R^2 = .016$, $p = .483$, Table. 2, *Conflict regression*), nor on the *perceived cooperativeness* ($R^2 = .013$, $p = .564$, Table. 2. *Perceived cooperativeness regression*). The regression model for the variable system understanding (*SIPA*) showed an effect for the amount of shared information ($R^2 = .068$, $t = 2.48$, $p = .046$), Table. 2, *SIPA regression*, such that a higher amount of information led to a better system understanding.

Table 1. Results of the binomial logistic regression (OR: odds ratio).

Parameter	Estimate	Standard error	z-value	p-value (two-tailed)	Standardized OR	95 % CI for OR
Intercept	−3.09	1.12	−2.77	.006	1.68	[0.99, 2.98]
Degree of automation	2.80	0.55	5.09	<.001	4.08	[2.45, 7.31]
Amount of information	−0.39	0.53	−0.73	.465	0.82	[0.48, 1.39]

Table 2. Linear regression results for conflict, SIPA, and perceived cooperativeness.

Parameter	Standardized beta coefficient	Standard error	t-value	p-value (one-tailed)
Conflict regression				
Degree of automation	0.122	0.12	1.16	.125
Amount of information	0.045	0.12	0.43	.665
SIPA regression				
Degree of automation	−0.035	0.19	−0.35	.367
Amount of information	0.256	0.19	2.48	.008
Perceived cooperativeness regression				
Degree of automation	−0.084	0.19	−0.79	.215
Amount of information	0.072	0.19	0.68	.250

4.3   Correlations with Reported User Experience

In the UEQ, participants overall reported a slightly positive attractiveness ($M = 0.98$, $SD = 1.07$), hedonic ($M = 0.79$, $SD = 0.89$), and pragmatic quality for the SCA independent of the experimental condition ($M = 1.22$, $SD = 0.87$). To examine our third hypothesis, we calculated correlations between our four measures of cooperation and the three subscales of the UEQ.

For the variables charging endpoint (*behavioral change*) and the HMIII conflict subscale (*perceived goal alignment*), Kendall's Tau was calculated. For *SIPA* and *perceived cooperativeness*, Pearson correlation coefficients are reported. All p-values are one-tailed. Results are shown in Table 3.

5   Discussion

We investigated the effect of varying the level of cooperation in a smart charging agent (SCA) on user perception and behavior. We manipulated the SCA's cooperativeness by varying its degree of automation and the amount of information sharing with the user and measured effects on changes in user behavior, perceived goal alignment, the user's awareness of the SCA's information processing, and perceived cooperativeness, measured by a five-item scale developed for the present study.

Table 3. Correlations between measures of cooperation and reported user experience.

	Charging time	Conflict	SIPA	Perceived Cooperativeness
Attractiveness	.05	-.05	.38***	.69***
Pragmatic quality	.04	-.03	.57***	.61***
Hedonic quality	.04	-.15*	.28**	.64***

$*p < .05, **p < .01, ***p < .001$

5.1   H1: A Low Degree of Automation Increases Cooperation

We hypothesized that a lower degree of automation of the SCA would increase human-agent cooperation. However, in three out of four analyses, the degree of automation had no significant effect on our measures of cooperation. For our behavioral variable, participants in the high automation condition chose a later charging endpoint more often than participants in the low automation condition. In other words, participants behaved more often cooperatively in the high-automation condition, which is contrary to our hypothesis.

5.2   H2: A High Amount of Information Shared Increases Cooperation

We hypothesized that a higher amount of information shared by the SCA should increase human-agent cooperation. However, in three out of four analyses, the amount of shared information had no significant effect on our measures of cooperation. There was a small positive effect on the amount of information shared on reported SIPA. Our second hypothesis was, therefore, only partially confirmed.

5.3   H3: Cooperation Leads to a More Positive User Experience

Two of the four cooperation measures correlated moderately to strongly with the three subscales of the UEQ (SIPA and perceived cooperativeness, see Table 3), indicating that higher cooperation (perception measures) led to a more positive user experience. There was a small negative correlation between our measure of perceived goal alignment and the hedonic quality of the SCA. None of the UEQ subscales correlated with our variable of cooperative behavior, i.e., the chosen charging endpoint. Our third hypothesis was, therefore, only partially confirmed.

5.4   Theoretical and Practical Implications

Cooperation is often operationalized by the degree of automation of a system, which is usually attributed to positive effects (such as higher reported pleasure and trust [17,27]. However, we could not replicate equivalent effects in our study. Most literature on human-automation cooperation focuses on use cases in which systems become more automated (such as manual driving developing towards autonomous driving), and users are expected to give up control in the future. This cannot be applied as easily to the use case of human-agent cooperation in smart charging, as described here. In the present scenario, users arguably *keep* more responsibility because they remain involved in the decision process also in the highly automated condition.

Additionally, charging your car might satisfy less hedonic needs compared to driving your car [8]. Thus, users might not want to stay in the loop and share the task as much compared to other use cases.

5.5   Conclusion and Outlook

We present a first study on human-machine cooperation in the context of smart charging. We developed a theoretically driven concept for designing different levels of cooperation in an SCA and for investigating its effect on the user. Contrary to our hypothesis, we did not find a positive effect of lower automation on our measures of cooperation (perception and behavior change). Instead, a high degree of automation even led to a higher probability for the user to shift the charging window, which we interpreted as a higher willingness to cooperate. We discussed that one potential reason for this result might be that the degree of automation as defined by Parasuraman [22] may have different implications in the context of smart charging compared to autonomous driving. It may be conceivable that human-machine cooperation on eye level has requirements that are different from those found in automation, and new concepts and approaches are required. Thus, future work should explore other operationalizations of human-machine cooperation within this context, focusing, for example, on shared responsibility [16].

Furthermore, the degree of information shared by the agent did hardly affect cooperation measures. The manipulation might not have had the intended effect overall. Here, a stronger variation in the provided information to the user might yield a higher effect. Another potential limitation might arise from the between-subject design since participants had no comparison for their assessment of the SCA. In addition, the study displayed only one use case (meeting with friends). A scenario with a more urgent appointment could have led to different user reactions. Taken together, we provide new insights for future research on how to design an SCA for human-machine cooperation successfully.

References

1. Anwar, M.B., et al.: Assessing the value of electric vehicle managed charging: a review of methodologies and results. Energy Environ. Sci. **15**, 466–498 (2022)
2. Bengler, K., Zimmermann, M., Bortot, D., Kienle, M., Damböck, D.: Interaction Principles for Cooperative Human-Machine Systems. Inf. Technol. **54**(4), 157–164 (2012)
3. Bratman, M.E.: Shared cooperative activity. Philos. Rev. **101**(2), 327–341 (1992)

4. Bütepage, J., Kragic, D.: Human-robot collaboration: from psychology to social robotics. ArXiv preprint arXiv:1705.10146 [cs.RO] (2017)

5. Castelfranchi, C., Falcone, R.: Principles of trust for MAS: Cognitive anatomy, social importance, and quantification. In: Proceedings International Conference on Multi Agent Systems. pp. 72–79. IEEE (1998)

6. Finn, P., Fitzpatrick, C., Connolly, D.: Demand side management of electric car charging: benefits for consumer and grid. Energy **42**(1), 358–363 (2012)

7. Flemisch, F., Baltzer, M.C.A., Sadeghian, S., Meyer, R., Hernández, D.L., Baier, R.: Making HSI More Intelligent: human systems exploration versus experiment for the integration of humans and artificial cognitive systems. In: Karwowski, W., Ahram, T. (eds.) IHSI 2019. AISC, vol. 903, pp. 563–569. Springer, Cham (2019). https://doi.org/10.1007/978-3-030-11051-2_85

8. Frison, A.K., Wintersberger, P., Riener, A., Schartmüller, C.: Driving hotzenplotz: a hybrid interface for vehicle control aiming to maximize pleasure in highway driving. In: Proceedings of the 9th International Conference on Automotive user Interfaces and Interactive Vehicular Applications. pp. 236–244 (2017)

9. Gienger, M., et al.: Human-robot cooperative object manipulation with contact changes. In: 2018 IEEE/RSJ International Conference on Intelligent Robots and Systems (IROS). pp. 1354–1360. IEEE (2018)

10. Hoc, J.M.: Towards a cognitive approach to human-machine cooperation in dynamic situations. Int. J. Hum. Comput. Stud. **54**(4), 509–540 (2001)

11. Huber, J., Jung, D., Schaule, E., Weinhardt, C.: Goal framing in smart charging - Increasing BEV users' charging flexibility with digital nudges. In: 27th European Conference on Information Systems-Information Systems for a Sharing Society, ECIS 2019. pp. 1–16 (2020)

12. Jacovi, A., Marasović, A., Miller, T., Goldberg, Y.: Formalizing trust in artificial intelligence: Prerequisites, causes and goals of human trust in AI. In: Proceedings of the 2021 ACM Conference on Fairness, Accountability, and Transparency. pp. 624–635 (2021)

13. Kelley, H.H., Thibaut, J.W.: Interpersonal relations: A theory of interdependence. Wiley, New York, NY (1978)

14. Klein, G., Feltovich, P.J., Bradshaw, J.M., Woods, D.D.: Common ground and coordination in joint activity. Organ. Simul. **53**, 139–184 (2005)

15. Kraft, A.K., Maag, C., Baumann, M.: How to support cooperative driving by HMI design? Transp. Res. Interdis. Perspect. **3**, 100064 (2019)

16. Krüger, M., Wiebel, C.B., Wersing, H.: From tools towards cooperative assistants. In: Proceedings of the 5th International Conference on Human Agent Interaction (HAI '17) pp. 287–294 (2017)

17. Kuramochi, H., Utsumi, A., Ikeda, T., Kato, Y.O., Nagasawa, I., Takahashi, K.: Effect of human-machine cooperation on driving comfort in highly automated steering maneuvers. In: Proceedings of the 11th International Conference on Automotive User Interfaces and Interactive Vehicular Applications: Adjunct Proceedings. pp. 151–155 (2019)

18. Limmer, S., Rodemann, T.: Peak load reduction through dynamic pricing for electric vehicle charging. Int. J. Electr. Power Energy Syst. **113**, 117–128 (2019)

19. Maag, C., Kraft, A.K., Neukum, A., Baumann, M.: Supporting cooperative driving behaviour by technology-HMI solution, acceptance by drivers and effects on workload and driving behaviour. Transp. Res. Part F Traffic Psychol. Behav. **84**, 139–154 (2022)

20. Öhrlund, I., Stikvoort, B., Schultzberg, M., Bartusch, C.: Rising with the sun? encouraging solar electricity self-consumption among apartment owners in Sweden. Energy Res. Soc. Sci. **64**, 101424 (2020)

21. Papenmeier, A., Kern, D., Englebienne, G., Seifert, C.: It's complicated: The relationship between user trust, model accuracy and explanations in AI. ACM Trans. Comput. Hum. Interact. (TOCHI) **29**(4), 1–33 (2022)

22. Parasuraman, R., Sheridan, T.B., Wickens, C.D.: A model for types and levels of human interaction with automation. IEEE Trans. Syst. Man Cybern. Part A Syst. Hum. **30**(3), 286–297 (2000)

23. Schrepp, M., Hinderks, A., Thomaschewski, J.: Design and evaluation of a short version of the user experience questionnaire (UEQ-S). Int. J. Interact. Multimedia Artificial Intel. **4**(6), 103–108 (2017)

24. Schrills, T.P.P., Kargl, S., Bickel, M., Franke, T.: Perceive, understand & predict-empirical indication for facets in subjective information processing awareness. PsyArXiv (2022). https://doi.org/10.31234/osf.io/3n95u

25. Sendhoff, B., Wersing, H.: Cooperative intelligence-a humane perspective. In: 2020 IEEE International Conference on Human-Machine Systems (ICHMS). pp. 1–6 (2020)

26. Tomasello, M., Vaish, A.: Origins of human cooperation and morality. Annual Rev. Psychol. **64**(1), 231–255 (2013)

27. Walch, M., Woide, M., Mühl, K., Baumann, M., Weber, M.: Cooperative overtaking: Overcoming automated vehicles' obstructed sensor range via driver help. In: Proceedings of the 11th international conference on automotive user interfaces and interactive vehicular applications. pp. 144–155 (2019)

28. Wang, C., Usai, M., Li, J., Baumann, M., Flemisch, F.: Workshop on human-vehicle-environment cooperation in automated driving: The next stage of a classic topic. In: 13th International Conference on Automotive User Interfaces and Interactive Vehicular Applications. pp. 200–203 (2021)

29. Woide, M., Stiegemeier, D., Pfattheicher, S., Baumann, M.: Measuring driver-vehicle cooperation: development and validation of the Human-Machine-Interaction-Interdependence Questionnaire (HMII). Trans. Res. Part F Traffic Psychol. Behav. **83**, 424–439 (2021)

30. Wollstadt, P., Krüger, M.: Quantifying cooperation between artificial agents using synergistic information. In: 2022 IEEE Symposium Series on Computational Intelligence (SSCI). pp. 1044–1051. IEEE (2022)

Leaders or Team-Mates: Exploring the Role-Based Relationship Between Multiple Intelligent Agents in Driving Scenarios

Research on the Role-Based Relationship Between Multiple Intelligent Agents in Driving Scenarios

Shuo Li, Xiang Yuan$^{(\boxtimes)}$, Xinyuan Zhao, and Shirao Yang

Hunan University, Changsha, Hunan, China
{leisure,yuanx,s2008w0568}@hnu.edu.cn

Abstract. Intelligent agents (IAs) are increasingly used in vehicles and associated services (e.g. navigation, entertainment) to enhance user experience, as IAs were applied to the car and turned the vehicle into a service platform under the rapid development of the intellectualized and connected vehicle. However, various IAs may be employed by other services and devices. In the case of in-vehicle cross-device interaction, when users interact simultaneously with multiple services or devices, the actions and decisions of one IA may conflict with those of others. This paper presents a role-based relationship framework to resolve potential conflicts between different IAs in the driving scenarios. The article discusses four types of IA relationships: Partnership, Representative, Subordinate, and Co-embodiment. To examine people's perceptions and attitudes towards different types of relationships, we apply an evaluation system and conduct user studies (N = 30). In two scenarios (Navigation Plan & Music Switching), Participants are required to engage in conversations with IAs based on various types of relationships. Data analysis and user interviews show that Partnership is gaining popularity in leisure and entertainment settings. Moreover, Representative is more effective in efficiency-oriented use cases. In addition, the research on driver's attention behavior suggests that Representatives can convince the driver to focus on the road more efficiently in navigation scenarios than in music settings. After evaluating the different role-based relationships of IAs, design recommendations for user interactions with multiple IAs in driving scenarios are offered.

Keywords: Intelligent agents · Multiple agent interaction · Driving Scenarios · Role-based relationship First Section

1 Introduction

Intelligent driving agents are widely used in driving scenarios, and previous research on the characteristics of in-vehicle agents (IVAs) has provided very valuable insights [20, 52]. These studies inspire further discussion on how to optimize IVAs for the driving environment and what additional considerations should be taken into account. In real

H. Krömker (Ed.): HCII 2023, LNCS 14048, pp. 144–165, 2023.
https://doi.org/10.1007/978-3-031-35678-0_9

driving scenarios, there are often two IAs coexisting: a personal assistant represented by a cell phone and a driving agent represented by a car system. Since the two IAs may belong to different service providers, the IAs maintain their own uniqueness, which may lead to different task decision outcomes for the two IAs. Devising mechanisms to encourage autonomous agents to collaborate with each other is the central theme of multiagent systems [35], and to improve efficiency and ensure driving experience, IAs will share resources, exchange data, or perform operations collectively to bring decision results into an agreement among themselves.

Agents are often viewed as multimodal interfaces that provide useful information rather than as social partners with whom humans can build relationships [37]. But any form of interaction, including human-to-human interactions, can expand human horizons by presenting multilateral viewpoints. However, when humans view agents as mere tools for providing information, they cannot assume the agent's point of view. In human-computer interaction (HCI), humans first need to recognize that agent has different and meaningful viewpoints, and one of the mental positions that recognize this fact is known as the intentional stance. Thus, by designing social interactions and expressing information exchange between IAs, users can increase their confidence in IAs and change the user's perception of their capabilities. Currently, most relevant studies focus on the impact of a single IA feature on the driving experience, without considering the cross-platform and cross-context user experience issues when two IAs exist. To achieve this optimization, different IAs should use a criterion to mediate relationships, positions, and behaviors when cross-platform interactions occur. This raises the question of how the interactions between IAs should be designed to induce the user's intentional stance and enhance the driving experience during the collaboration.

In the driving scenario, Yoshiike et al. [19]. Developed a system called MAWARI, which consists of three social robots, and the multiparty conversation format reduces the driver's attention problem of overuse of the system and reduces the psychological load on the driver compared to the traditional one-to-one communication-based approach directly targeting the driver. When we reach SAE level 5 for automated vehicles (AVs), people will be free from driving tasks in AVs, and exploring the simultaneous presence of more than two IVAs and drivers enhances the user experience. Research on this topic is still in its early stages, but there are already teams [23] presenting futuristic situations of multiple IAs interacting with each other in the form of videos. We attempt to address the issue of cooperation between heterogeneous agents with different desires and principles by character-building relationships between the interiors of multiple intelligent systems.

This paper conducted a comprehensive investigation based on research in the field of mobile human-computer interaction. From the current literature, four relationships between multiple IAs were created and compared. They can answer some questions about the existence of role relationships: what are the manifestations of role-based relationships among multiple IAs and how do such role-based relationships affect the user experience? Second, at the application level, basic design strategies and guiding design principles are proposed for the different expressions of role-based relationships of IAs in driving scenarios. We help enterprises to develop IAs interaction design specifications and support the improvement of their product strategies. The aims of this paper are: (1) propose a framework on roles-based relationships to address how different IA's in a driving scenario behave during the collaboration process (2) compare people's attitudes toward different types of relationships and apply an evaluation system, (3) highlight design principles to help enterprises develop multiple IAs' interaction design specifications and support the improvement of their IAs platform technologies and product strategies.

2 Related Works

With the rising use of multi-agent technology in a variety of industries [33, 53], the social relationship between humans and intelligent agents (IAs) has also been widely studied. Katsunori et al. studied how IAs collaborated to persuade users from the standpoint of the balance theory and confirmed that the social relationship between two agents and a human user influences the effectiveness of persuasion. The notion of an intelligent system as a human collaborator is becoming more prevalent in interaction design [2], and in a collaborative partnership, IAs will aid humans in completing tasks more quickly. In a real-world driving scenario, however, there are likely to be more than two IAs for different devices, and their interaction is fundamentally distinct from that of a single IVA. Additionally, the interaction process has shifted from "one person to one device" to "one person to several intelligence." In this context, a number of research groups [19, 23, 40] have proposed research on challenges such as numerous agent priorities, negotiation with one another, and how to conduct autonomy to allow drivers to make decisions when conflicts emerge.

2.1 Design Characteristics of IAs in a Driving Scenario

Previous research investigated the properties of IAs in vehicles. Various topics, including IAs' forms [17, 22, 24], linguistic styles [43, 49], and variances in timbre [9, 18]. The deployment of such technologies supports the driver in doing driving or non-driving-related duties, hence enhancing road safety for both manual and conditional autonomous driving by preventing driver distraction [47]. For example, the study by Manhua Wang et al. [44] compared the differences between conversational and informative language; the research team confirmed that more confident voice reminders result in faster reaction times for drivers than less secure voices [49]; and other studies have investigated the linguistic characteristics of in-vehicle agents, such as voice age [18], and voice gender [9]. Braun et al. demonstrated [4] that voice assistant roles correspond with user personalities. They created various components for in-vehicle assistants in order to investigate the favorable effects of voice assistant customization on trust and likeability. Yoshiike et al. developed the MAWARI [19] system, which consists of three social robots engaging in multi-party dialogues to prevent driver misuse of the system compared to typical one-to-one communication-based systems that target drivers directly. The attention issue minimizes the driver's mental workload. As the reception technology for information grows more intelligent and complicated, the communication channel is presented in a more natural and fluid manner [28, 32].

2.2   Social Relationships Between IAs

According to some sociological research, the role symbolizes the agent's conduct inside a specific agent group, and the role model consists of obligations, goals, permissions, rights, etc. [29]. Multi-agent systems have been strongly influenced by related studies in behavioral ecology [51], sociology [54], and psychology [8]. In a multi-agent system, roles limit rivalry between agents in certain activities, hence it is vital to consider the issue of role relations [55]. Confucian ethics is a paradigm based on rules [46, 48], hence it is used to develop robot ethics by assigning robots various social roles to fulfill their obligations in their interactions with humans. Williams et al. [50] explain how Confucian Role Ethics might inspire Role-theoretic approaches to moral reasoning based on robot-oriented substitutes for Confucian Cardinal Relationships (e.g., supervisor-subordinate, adept-novice, teammate-teammate, and friend-friend). Ruchen Wen et al. [45] investigated norm violation responses based on role-based relational norms and elucidated the distinctions and characteristics of various robot roles. In these investigations, roles are assigned to control behavior and define responsibilities. We also consider linkages based on roles in natural systems. The notion of animal communities by Anderson and Franks [56] applies equally to humans and multi-intelligence systems. Roles allow teams to split their responsibilities. In interactions between humans and multi-intelligent systems, intelligent entities will either collaborate or compete to aid humans in performing tasks. In Yoshimasa's research [30], for instance, observing the interaction between multiple agents (colleague agent and instructor agent) can determine whether the person's intentional stance toward the agent can be encouraged and maintained; Venus & Mars and Recommendation Battlers are additional examples of retrieving or recommending Web information by multiple agents cooperatively or competitively, respectively [21].

According to past studies, the number of agents with social identities may vary in certain settings. For instance, in a given context, numerous agents with distinct social identities exist on separate smart devices. People are able to communicate with one another. Agents with social identities interact; it is also feasible that all devices are managed by one agent with social identities, similar to the one-to-many notion presented by Luria's research [26]; users primarily engage with this agent to control the equipment of other agents. We contend that the number of intelligent agent answers has a significant impact on how humans comprehend multi-agent relationships. On another level, role-based relationships in multiple agents' systems have also been widely discussed. It was mentioned above that interaction between multiple agents may induce intentional stances toward agents. People estimate the behavior of interacting partners through behavioral models, emotional aspects, and the partner's decision-making strategies [30]. The social interaction between two agents is usually an equal and cooperative relationship [57], and there is also a subordinate relationship to assist people in completing tasks [40]. The differences in information exchange brought about by the hierarchical relationship between agents may affect the behavior of humans interacting with them, which is an important factor in user experience. According to the hierarchical relationship and number of responses of the device, we split it into four categories: Partnership, Co-embodiment, Representative, and Subordinate.

Partnership. Hayashi studied the impact of user experience in the four communication modes of "passive-passive social-interaction-interactive social" in the paper [16]. Users can freely interact with the robots in the interactive social mode, which facilitates communication between many robots used to present the content of the information. In the social model, robots exchange user information through social discourse and direct contact.

Co-embodiment. Reig et al. proposed an intelligent agent concept: "Re-embodiment" [26], which described the migration of a kind of intelligent software among different robot subjects [14, 41]. Luria et al. studied the collaboration and insights of multiple IAs in task succession. They explored the driver's perception when the smartphone and in-vehicle agents appeared together in the autonomous driving scenario. This sort of manifestation is known as co-embodiment.

Representative. Tan et al. [39, 40] combined previous research and set up the communication modes based on different logic such as "silent- explicit -reciting" and "representative-direct-social" for robots. Fraune [11, 13] made some adjustments based on the research model and added new dimensions. The transparent mode refers to the process of information exchange between robots, but it will indicate that the information has been successfully communicated. The recitation mode means that the data is directly displayed through voice, and the information is shared. In the usual way, there is no direct communication or interaction between robots, but a robot communicates with users on behalf of other robots.

Subordinate. The current intelligent home assistant is a more fundamental and widely existing form in which an agent controls various other agents in the environment. In this case, users mainly communicate with this agent. This solution mode has been highly market-oriented and applied to many products, such as Alex audio [6]. This relationship can be summarized as a subordinate relationship, which is a relationship in which a high-level agent manages a low-privileged agent (Table 1).

Table 1. Multi-device interaction model framework

Response number of devices	Hierarchical	
	Equality	Inequality
Single	Co-embodiment	Representative
Multiple	Partnership	Subordinate

Through summarizing and sorting out previous studies, it is found that although many research teams have described the communication mode between multiple devices from different perspectives when multiple agents coexist, there is no framework to explain the relationship between them. Therefore, we hope to create a relationship to define the behavior between them. By summarizing the previous studies, it is found that the number of equipment responses may be the factor that affects driver distraction [19]; According to the division of cooperative relationships [34] in sociology, it can be concluded that hierarchical relationships (equal or subordinate) may become a factor affecting multi-agent systems. We construct a multi-agent role-based relationship model based on these two dimensions in driving scenarios. In this work, we ask two key questions: (1) How should the role-based relationships between IAs behave? (2) What is the user's preference for the role-based relationship between IAs in different scenarios? In response to these questions, we put forward the following assumptions:

Hypothesis 1: In the Navigation Planning scenario, the Subordinate has better driving performance than the Partnership; in the Music Switching scenario, the Partnership has better driving performance.

Hypothesis 2: Partnership and Co-embodiment have higher social attributes and are suitable for entertainment-oriented scenarios; Representative and Subordinate are more ideal for efficiency-oriented systems.

3 Material

3.1 Feature Design of Intelligent Agents

To maintain a decent interaction flow, the intelligent agent must display nonverbal cues of the present discussion state [1]. These signals may assist users' cognitive involvement during interactions, allowing them to concentrate on ongoing tasks [58]. Therefore, while building intelligent entities, non-verbal emotional expressions must be considered in addition to changes in timbre. To prevent insufficient feedback on the agent's status from influencing the interaction process. We refer to the research on sentiment analysis [27] and expand the expressions of mobile phone IAs based on the research of [59]. These include inactivity, regular operation, blinking, speaking, and left and proper rotation. A range of distinct expressions, such as moving in and out, mind control, information transmission, and dialogue, are designed. To prevent the potential impact of timbre on the role-based connection, we employ a somewhat flat female voice.

To distinguish the roles of different agents, the IVA's expression design is distinct from the mobile phone intelligent agent's expression design. For the expressions of car IAs, we first refer to the current commercial vehicle IAs, such as NIO Nomi, XPeng Motors X mart OS, and Baidu Apollo; secondly, in Michael Braun et al. studies [3, 47], we investigate the interface design of car dialogue. Certain expressions in the car, such as entering and exiting, mind control, information transmission, and maintaining contact with the mobile intelligent agent, are necessary to ensure that the participants notice continuity between them.

3.2 Multiple Intelligent Agents' Relationship Constructions

One of the approaches for influencing someone via numerous agents is overhead communication [42]. The technique demonstrates indirect contact through multiple agents and is frequently employed in online buying scenarios [38]. In the driving environment, Yoshiike et al. confirmed that the form of multi-party conversation reduces the driver's attention problem to excessive use of the system; under the premise of cooperation between intelligent systems and humans, between devices, we believe that the number of responses devices may become a factor that affects driving; Under the premise of cooperation between intelligent systems and humans, the class relationship between devices may also affect the driving experience. For example, in some sociological studies, the definition of collectivism (subordinate relationship, peer relationship). According to the characteristics of the role relationship, we divide the four role relationships according to the two variables of the number of responding devices and the hierarchical relationship of dialogue agents. Each role relationship has a special significance, as seen in Fig. 1's comparison of the four role-based relationships (Table 2).

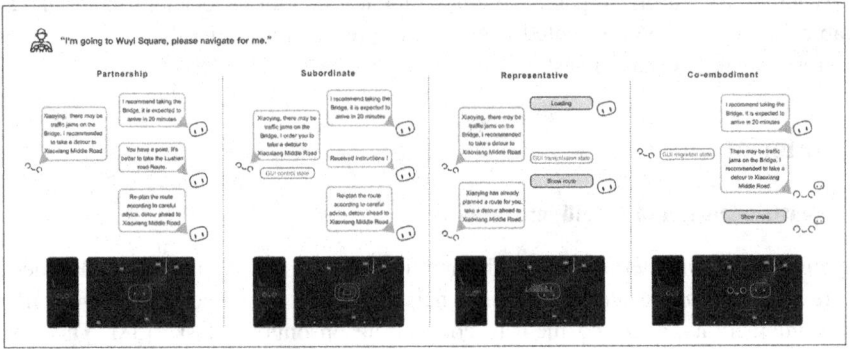

Fig. 1. Dialogue script

- Partnership means IAs with social identities in mobile phones and car machines. They exchange information through natural language, and people can communicate with IAs; mobile smart agents (Xiaoxin) Advise the intelligent agent of the car (Xiaoying) to change the strategy through communication and dialogue.
- Representative: It means that both mobile phones and cars have IAs with social identities. They exchange information through electronic signals, and people can only communicate with the intelligent agent group's agents. The mobile phone intelligent agent (Xiaoxin) changes the strategy of the car-machine agent (Xiaoying) by transmitting electronic signals, and Xiaoxin replaces Xiaoying to end the task.
- Subordinate: It means that among multiple devices, each device has an intelligent agent with social identity, but there is only one intelligent agent with control authority in various devices, and this intelligent agent controls others in a fixed agent with natural language capabilities. Equipment and people can communicate with any intelligent agent, and the intelligent agent with the control right can intervene. When a match occurs in this mode, the mobile phone intelligent agent (Xiaoxin) orders the car-machine intelligent agent (Xiaoying) to change the original strategy. Xiaoxin gave a speech to end the task.
- Co-embodiment: It means that among multiple devices, each device has an intelligent agent with a social identity, and the mobile phone intelligent agent (Xiaoxin) can be transferred between different devices, the intelligent agent will migrate according to the needs; the mobile intelligent agent will migrate to the car and change the default policy of the vehicle, and Xiaoxin will end the task by speaking on the car.

Table 2. Styles available in the Word template

Relationships	Interactive object	Interactive form
Partnership	Mobile phone & The dashboard	Conversational information exchange between IAs
Representative	Mobile phone	Mobile phone intelligent agent speaks instead of vehicle intelligent agent
Subordinate	Mobile phone & The dashboard	Mobile phone intelligent agent to control car and machine, intelligent agent
Co-embodiment	The dashboard	Mobile smart agents migrate to the car-machine system

3.3 Prototype Design

The prototype is divided into the mobile phone terminal and the car terminal. The car-machine terminal adopts the map interface as the main view. The feedback page for route planning and music switching is designed according to the scenario; the mobile terminal displays dynamic expressions. The Graphical User Interface (GUI) provides information transmission, listening, and receiving instructions during driving. Different dialogue forms and language styles [24] and animation effects are used to express other relationships between agents [7, 10]. The prototype is made by Protopie v 6.2.0, AE, Figma. The display screen is a three-channel display mode with a total resolution of 5760×1080 (long) $\times$ three connected horizontal displays; Logitech's driving simulator is used to make the experiment more realistic; iPad pro is used to simulate the car's central control, iPhone11 is used as a test phone.

4   Method

This work concerns the role-based relationship among multiple agents in the driving context. The experimental design was an offline study between 2×4 subjects. The independent variables were relationships (Partnership, Subordinates, Representative, Co-embodiment) and control variables (Navigation Planning, Music Switching), resulting in 8 situations. Participants were randomly assigned to these conditions to account for sequence effects [25] during the experiment. This paper deals with the process of information exchange between multiple IAs. We sequentially design equivalent and representative prototypes to equally compare the four role relationship concepts.

4.1   Setting

This experiment employs a two-factor mixed design. In-group variable role-based relationship between agents (Partnership versus Subordinate versus Representative versus Co-embodiment) is the first independent variable. The second independent variable, a between-group variable, was the driving task (Navigation planning versus Music switching). Each participant was expected to engage in four role-based partnerships in two different settings. In a simulated driving environment, the problems drove autonomously while adhering to traffic laws, but they also had to make turns. The quantitative examination of pertinent experiment questions is conducted mostly with the aid of the Likert scale. Some tweaks were made to the rankings to make them more acceptable for the assessment of this study, and some less relevant questions were eliminated to prevent participants from overproducing throughout the experiment and filling out the questionnaire. Investigate tiredness. The Robot Social Qualities Scale [5, 12] was used to determine whether the agent displayed good social attributes in the conversation mode; the DALI Driving Load Scale [31], a refined version of NASA-TLX [15], was used to determine the cause of driving stress and was applied to driving activities. In our study, we utilized five elements, with the exception of interference factors, which are most applicable in natural driving environments (Fig. 2).

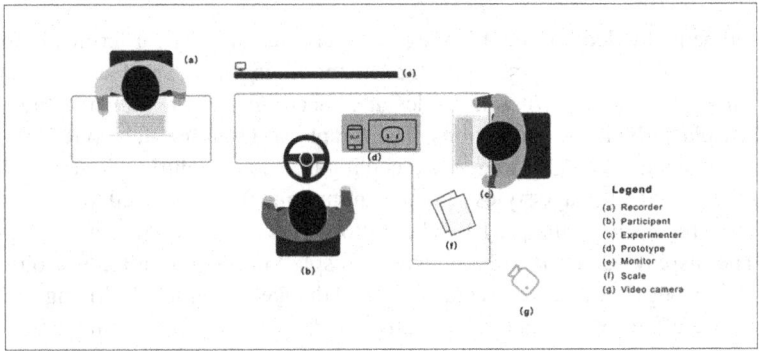

Fig. 2. Experimental scenario

4.2 Pre-experiment

Pre-experiments must be conducted to test the experimental environment and materials and ensure their validity. A total of 10 subjects were recruited to participate in the pre-experiment. The pre-experiment is divided into two parts. The first part is to test whether there are significant differences in the interaction forms of the role-based relational and whether the subjects can perceive the information in the simulated scenario. After the pre-experiment verification, the main feedback from the issues was "the difference between Representative and Subordinate is not obvious enough," "the voices between the two agents are too close to be distinguished," etc. These problems were all carried out before the formal experiment. Adjustment. The second part is to screen further and clarify the scenarios. We first listed a list of driving and non-driving tasks, which were further strained by the users who participated in the pre-experiment, and the scenarios were sorted according to the frequency of use. In the end, two scenarios of navigation planning and music switching were selected.

4.3 Participants

The experimental study used social media to recruit 33 participants. According to the statistics of the filled-in participant information form, the ages were concentrated in the 21–40 years old; 16 participants were males, 17 participants were females; 28 of them were participants Familiar with the intelligent voice agent or had a long-time use experience; ensure that all participants had a driving license. A total of 30 valid interview data were collected in the experiment.

4.4 Procedure

The experiment process is divided into the practice, formal, and interview stages. First, the experimenter briefly introduced the laboratory's functions and the simulation system's driving operation mode to the subjects and told them the general content of the experiment. Under the guidance of the experimenter, the subjects will first conduct a simulated driving practice on the same road setting as the setting in the formal investigation. During this process, the subjects were familiar with the operation of the simulator, adapted to the road conditions in advance, and listened to the specific guidance of the experimental tasks. In the formal experiment, subjects had to experience four different role relationships in two scenarios, navigation planning, and music switching, respectively. The completion order of the eight experimental groups has been balanced. Each time the subjects experienced a relationship, the issues were required to fill out a scale to score the driving experience. It takes about 5 min to complete the driving of one experimental group. Rest periods were arranged in the middle of the experiment. Subjects were asked to wear Tobii glasses two during the investigation to collect eye movement behavioral data. In the interview stage, the experimenter conducted semi-structured interviews with the subjects. The subjects ranked the role-based relationships in different scenarios according to their feelings and expressed their thoughts on the characteristics of each relationship (Fig. 3).

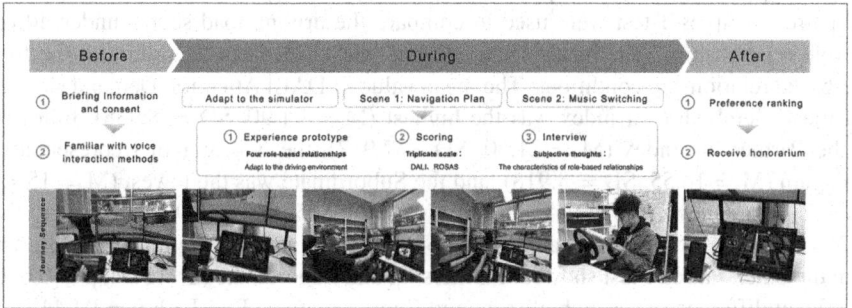

Fig. 3. Experimental flowchart

5   Results

The results mainly come from five aspects: mental workload, eye movement analysis, RoSAS analysis, and interview results. The eye movement response was largely explained from the behavioral level of the subjects, and the subjective psychological load and RoSAS analysis were mainly defined from the psychological level. As a supplement, the interview results verified the acceptance and rationality of the role-based relationship of multiple agents from the side.

A total of 30 valid interviews were conducted in the experimental study, and the overall rating data of the scale showed good reliability and validity. The alpha reliability coefficient of the driving load scale data was 0.841, close to the early Lucey[36] small sample study of 0.8552. The data of the three dimensions of the social attribute scale analyzed by the same method showed excellent reliability: $\alpha = 0.912$, $\alpha = 0.900$, $\alpha = 0.923$. Preliminary data mean-variance analysis indicated that some demographic attributes of the participants (e.g., gender and age) did not significantly affect the data's variability. Significant differences between different conditions are indicated in the figure, and the degree is indicated ($p < 0.05$, marked with*) ($p < 0.01$, marked with **) ($p < 0.001$, observed with ***).

5.1   Driving Activity Load Index (DALI)

DALI (Driving Activity Load Index) is a SWAT technique, and one of the main advantages is that it can identify the source of the driver's workload. It includes six predefined factors: attention, interference, situation stress, visual, auditory, and temporal demands. In this study, we used five elements, excluding interference factors, as this factor is most appropriate when used in natural driving environments. We judge the difference in DALI scores from two scenarios; each DALI factor is calculated from the subjects' scores according to work.

Mental Workload. To test whether the role-based relationship between agents in the navigation planning scenario impacts the driver's psychological load, one-way ANOVA and paired samples T-test were used to compare the driving load scores under different role relationships. The mean value analysis was carried out according to the four role-based relationship conditions. The mean value of DALI Attention Demand showed that the Co-embodiment index was the highest (M = 21.90, SD = 8.938), followed by the Partnership index (M = 21.60, SD = 7.917), the indicator of Representative was again (M = 16.55, SD = 8.918), and the Subordinate was the lowest (M = 15.15, SD = 7.110). One-way ANOVA showed: F(3) = 3, 505, p = 0.019(p < 0.05), $\eta 2 = 0.123$, so it was considered that at least one of the four groups was significantly different from the other. Paired t-test showed: Partnership and Subordinate (T = 3.84, p = 0.016, significant difference); Co-embodiment and Representative (T = 3.12, p = 0.044, significant difference); Co-embodiment and Subordinate (T = 3.57, p = 0.012, significant difference), these data show that Partnership requires more attention than Subordinate, Co-embodiment than Representative and Subordinate in remembering presented information. In terms of Auditory Demand, one-way ANOVA showed that F(3) = 0.189, p = 0.903(p > 0.05), $\eta 2 = 0.007$, the difference between the four relationships was not significant, which indicated that this kind of driving In terms of Visual Demand, F(3) = 0.625, p = 0.601(p > 0.05), η 2 = 0.024, among which the difference between Partnership and Representative is relatively higher. High but not significant (T = 2.20, p = 0.232, non-significant), and the differences among several other relationships were also not significant, which may be because the Participants did not pay more to the intelligent agent in the driving environment. Visual effort; in terms of Temporal Demand, F(3) = 1.911, p = 0.135(p > 0.05), η 2 = 0.070, the difference between the four relationships is not significant; in terms of Situation Stress, F(3) = 0.032, p = 0.992(p > 0.05), η 2 = 0.012, which indicates that the difference between the four role relationships is not significant, which may be because the experimental environment is a simulated driving rather than a more realistic environment (Fig. 4).

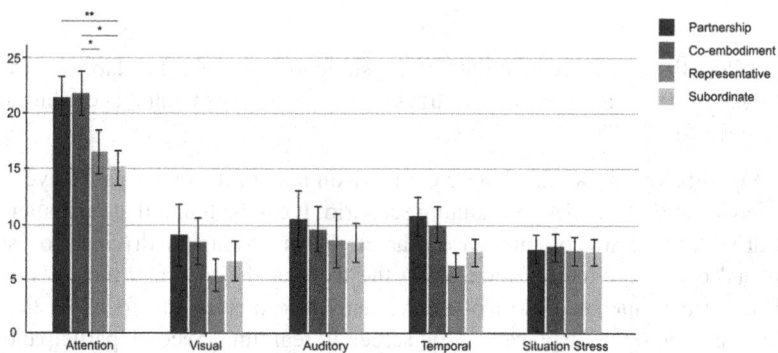

Fig. 4. The results of the DALI factors under the role-based relationship in the Navigation Planning scenario.

Analysis of Music Switching scenario: In terms of Attention Demands, the indicators of Co-embodiment are the highest (M = 17.70, SD = 8.640), the hands under the Subordinate are the next (M = 14.40, SD = 7.214), and the arrows under the Representative is again (M = 14.30, SD = 8.670), and the indicators under the Partnership is the lowest (M = 13.55, SD = 7.075), which indicated that the Partnership required less attention than the other three relationships in the Music Switching scenario. From the one-way ANOVA, it can be concluded that F(3) = 1.084, p = 0.361 (p > 0.05), η 2 = 0.041, so there is no significant difference among the four groups of data. In terms of Auditory Demand, F(3) = 0.217, p = 0.884 (p > 0.05), η 2 = 0.009, there is no significant difference among the four groups of data; in terms of Visual Demand, the differences between Partnership and Co-embodiment are relatively high, But it was not significant (T = −1.976, p = 0.419, non-significant), and there was no significant difference among the four groups of data; in terms of Temporal Demand, F(3) = 0.472, p = 0.703 (p > 0.05), η 2 = 0.018, there is no significant difference among the four groups of data; in terms of Situation Stress, F(3) = 0.488, p = 0.692 (p > 0.05), η 2 = 0.019, there is no significant difference among the four groups of data (Fig. 5).

Attention Behavior of the Driver. To explore the eye movement behavior characteristics of each information display mode in the navigation context, Tobbi glasses two was

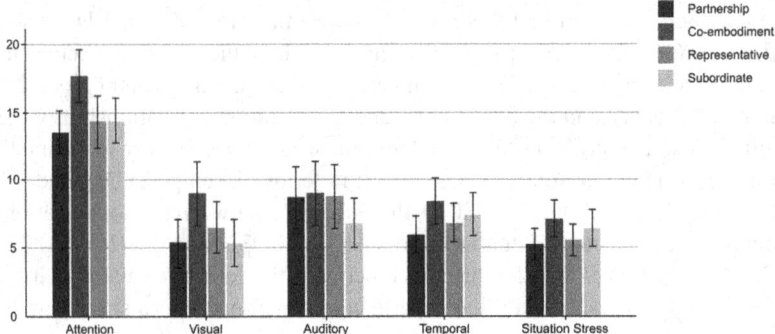

Fig. 5. The results of the DALI factors under the role-based relationship in the Music Switching scenario.

used to collect the eye movement data of the subjects, and Tobii Pro lab was used for data analysis. Gaze heatmaps and eye-tracking maps were generated according to the period of the issues.

The Fig. 6 below shows the feature comparison map of the participant's eye movement behavior in the Navigation Planning scenario. It can be found that no matter what kind of driving, there are common gaze characteristics. During the driving process, the participants' eyes were mainly focused on the front of the road and the car's central control; the driver allocated most of their attention to this area to focus on the road conditions and the route displayed on the screen in real time. Second, participants will jump between the mobile phone and the car's central control to obtain road condition information from the multi-party conversation. Eye movement heatmaps for Partnership, Co-embodiment, Representative, and Subordinate are presented separately. After comparing the thermal images of different groups, it was found that the Partnership and Co-embodiment relationships were more frequently concentrated on the mobile phone and the central control device of the car than on the other two groups. Therefore, it can be judged that in these two relationships, participants must allocate additional attention. Under the Representative, the gaze points of the eyes will be relatively concentrated on the road surface, and less is given to the two devices, so it can be judged that the participants are more focused on the driving behavior.

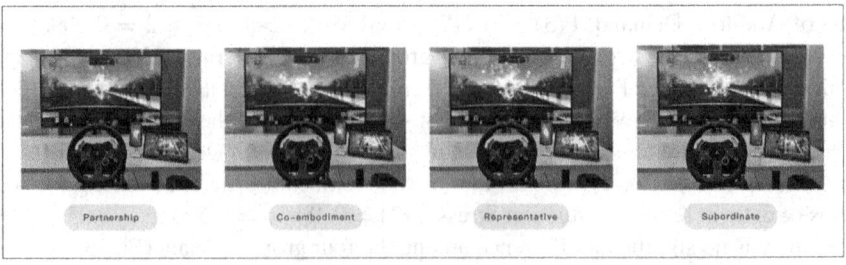

Fig. 6. Eye movement thermogram in the Navigation Planning scenario.

Figure 7 below compares the participants' eye movement behavior in the Music Switching scenario. As a non-driving task, the participants' load is relatively small so they will focus more on the road. In this scenario, the user's jump between two devices showed a more significant difference, with the participant in the Partnership being less distracted between the devices. In contrast, the participants in the Co-embodiment and Subordinate showed more significant differences; attention was switched between the two devices more frequently. This may be because, in entertainment-oriented scenarios, people see intelligent systems as a level relationship, so an equal and intimate form of a dialogue between intelligent systems will feel more natural and focus more on driving behavior.

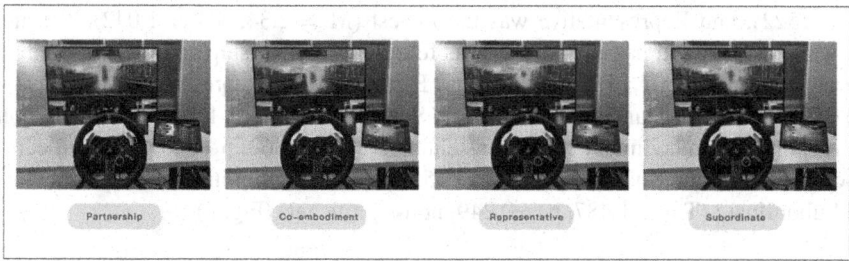

Fig. 7. Eye movement thermogram in the Music Switching scenario.

5.2 The Robotic Social Attributes Scale Index (RoSAS)

RoSAS is how people think about robots using an evaluation dimension, detecting whether agents exhibit good social attributes in communication patterns. We want to know whether the change in the role-based relationship between agents will have different perceptions of warmth, competence, and discomfort.

Competence Judgments. The figure below shows the mean difference between the four role-based relationships under the competence dimension. The indicator of Partnership is the highest (M = 5.36, SD = 1.064), followed by the indicator of Co-embodiment (M = 5.02, SD = 1.848), the indicator of Representative was again (M = 4.48, SD = 1.218), and Subordinate was the lowest (M = 4.82, SD = 1.044). Through one-way ANOVA and paired T-test, it was found that Partnership and Co-embodiment (T = 2.376, p = 0.128, significant difference); partnership and agency relationship (T = 2.211, p = 0.020, non-significant); Partnership and Subordinate (T = 2.834, p = 0.016, significant); Co-embodiment and Representative (T = 0.087, p = 0.420, non-significant); Co-embodiment and Subordinate (T = 0.363, p = 0.370, non-significant); Representative and Subordinate (T = 0.207, p = 0.928, non-significant).

Warmth Judgments. The figure below shows the mean difference between the four roles-based relationships under the warmth dimension. The indicator of Partnership is the highest (M = 5.18, SD = 1.004), followed by the indicator of Representative (M = 4.80, SD = 1.278). The Co-embodiment indicator was again (M = 4.54, SD = 1.297), with the lowest Subordinate (M = 3.84, SD = 1.131). Through one-way ANOVA and paired T-test, it was found that the Partnership and Co-embodiment (T = 9.115, p = 0.007, highly significant); the Partnership and Representative (T = 7.464, p = 0.110, non-significant); Partnership and Subordinate (T = 2.834, p = 0.000, highly significant); Co-embodiment and Representative (T = −1.198, p = 0.273, non-significant); Co-embodiment and Subordinate (T = 0.994, p = 0.003, highly significant); Representative and Subordinate (T = 2.142, p = 0.000, highly significant).

Discomfort Judgments. The figure below shows the mean difference between the different role-based relationships in the discomfort dimension. The index of Co-embodiment is the highest (M = 3.40, SD = 1.485), followed by the index of Sub-ordinate (M = 2.66, SD = 1.437), the indicator of Partnership was again (M = 2.64, SD = 1522), and Representative was the lowest (M = 2.34, SD = 1.042). Through one-way ANOVA and paired T-test, it was found that Partnership and Co-embodiment (T = −4.806, p = 0.007, highly significant); Partnership and Representative (T = 0.444, p = 0.280, non-significant); Partnership and Subordinate (T = −1.148, p = 0.943, non-significant); Co-embodiment and Representative (T = 4.862, p = 0.000, significant); Co-embodiment and Subordinate (T = 3.355, p = 0.008, significant); Representative and Subordinate (T = −1.487, p = 0.249, non-significant) (Fig. 8).

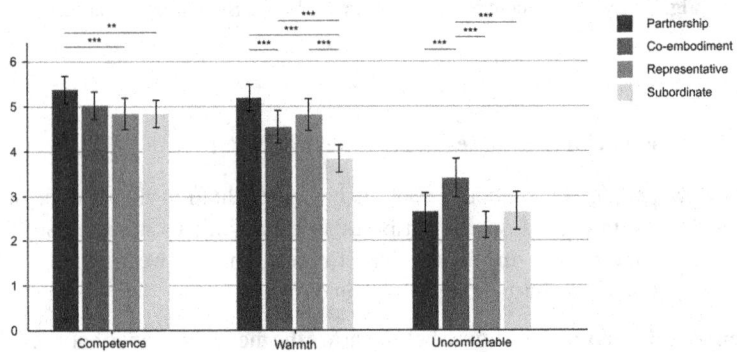

Fig. 8. Bar graph of RoSAS of role-based relationships

5.3   Analysis of Interview

To compare the user experience differences between the four role-based relationships in different scenarios, we conducted a multivariate analysis of variance. From the study of variance test results, it can be concluded that the significance of the role-based relation-ship is p = 0.000; the significance level of the scenario is P = 0.017. The importance of the interaction between the two factors is also less than 0.05, indicating that the exchange of the two factors significantly impacts user experience perception.

The four kinds of role-based relationships are sorted by preference in two different scenarios, and the results are shown in the figure. It can be seen from the figure that the user's choice for the role-based relationship is other in different scenarios: in the Navigation Planning scenario, 40% of the users prefer Representative, and 36.67% of the users rank the Co-embodiment last; In the Music Switching scenario, 46.67% of users ranked Partnership first, and 40% of users ranked Subordinate last. In short, users' preferences for role-based relationships also vary in different scenarios. Partnership and Representative are generally more popular, while Subordinate and Co-embodiment have typically little difference and are ranked lower (Fig. 9).

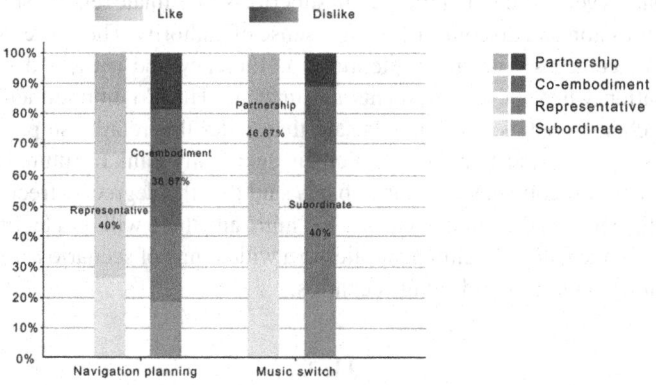

Fig. 9. Preference ranking percentage chart

6 Discussion

The results of data analysis are studied from the perspective of role-based relationships. From the corpus of users, it can be found that users have different views on different communication modes. Based on these analyses, we can start from the characteristics of different communication modes and apply different communication modes to appropriate scenarios to improve user experience. The comparison of different communication modes is shown in Table 3.

The findings for different communication modes seem somewhat different from the expected impressions, and hypotheses 1 and 2 are partially supported. In Navigation Planning scenarios, users prefer Representative and Subordination; in Music Switching scenarios, users prefer Partnership. The different role-based relationships between agents should be considered in conjunction with specific driving tasks and design requirements. Participants believe that Partnership shows more warmth than other relationships, and it is also interesting to apply it to agents. Still, subject to complex interaction logic and agent relationships, users generally believe this relationship seems redundant and unnecessary in efficiency-based scenarios (Navigation Plan). Therefore, Partnership is suitable for scenarios emphasizing attractive and emotional attributes, such as music switching, or some settings emphasizing easy-to-understand and natural social interaction. The Representative correlates with the experimental results of the representation pattern in Tan's study. Some participants said that their natural language when communicating with people and machine behavior when communicating with machines showed a sense of delay, but because a device replaced the speech throughout the whole process, it made the participants feel clearer, which was consistent with the expected impression. Is biased. Users said that the Representative is more concise and efficient, and they can pay more attention to the road conditions during driving. Some participants also noted that the Representative is suitable for scenarios that need to express multiple social identities without the spatial conditions for voice communication, such as remote control.

The Subordinate also deviates from our expectations. Some users said that although the Subordinate is very efficient, the apparent superior-subordinate relationship will bring a sense of oppression and discomfort. Bring a sense of authority. Therefore, Subordinate is suitable for scenarios with great attention to efficiency and trust, and it is safer to choose affiliation when the requirements are unclear. The Co-embodiment has novel and exciting characteristics. Participants said that under this relationship, the integrity of the devices is more robust, and the participants generally think of future autonomous driving scenarios. Of course, some participants said that the degree of freedom of form that can shuttle between different devices is too high, and there will be a feeling of losing control. Overall, the Co-embodiment applies to a wide range of scenarios, and there will be more exploration in future driving scenarios.

Table 3 .

Relationships	Advantage	Disadvantage	Features	Suitable scenario
Partnership	The highest social attributes	Not concise enough, slightly verbose	Kind and natural	Natural social, leisure, and entertainment scenario
Co-embodiment	Strong integrity	Confusion	Flexible and fun	Autonomous driving scenario
Representative	Can be connected remotely	Strong sense of delay	Concise	Scenarios that focus on efficiency and driving safety
Subordinate	Strong execution and high efficiency	Strong sense of oppression	Efficient and boring	Efficiency-oriented scenarios

After completing the experiment, we sent the Big Five Personality Scale to the participants in the experiment, hoping to make discoveries on the preference for the relationship between personality and role. According to previous studies, extraversion is the most easily perceived personality trait in oral language, and agreeableness represents an individual's attitude toward others. Therefore, the research on the relationship between personality and attitude-voice focuses on examining the extraversion in the Big Five. Extraversion and Agreeableness are two traits, so we also choose these two traits as the essential reference in this study. The questionnaire feedback shows that among the 30 users who participated in the test, the proportion of high Extraversion users is about

57%, the balance of high agreeable users is about 60%, the proportion of high openness users is about 47%, and the ratio of high Conscientiousness users is about It is 63%. The balance of high Neuroticism users is about 13%. After analysis, it is found that there is no apparent relationship between personality traits and role relationships. This may be because the participants will pay more attention to driving in the driving scenario, so when sorting preferences, they will prioritize completion efficiency, whether driving is distracted, etc. factor.

7 Conclusion and Future Work

This paper proposes a framework for roles-based relationships in driving environments and discusses the differences in user preferences for different relationships. We designed an experiment to test our hypothesis by comparing Partnership, Co-embodiment, Representative, and Subordinate through a simulated driving environment. In the current study, we compared the four role-based relationships regarding psychological conformity and investigated the differences in driving load and attention behavior among different role-based relationships. We evaluated our proposed framework using the DALI questionnaire, trend analysis of experimentally collected eye gaze data, and the Supervisor Impression Questionnaire. The results of DALI show that the Representative is prominent in driving scenarios. Its concise communication mode requires less eye-gazing behavior and is more suitable for task-oriented methods. It is also demonstrated that Partnerships have higher social attributes and need less eye-gazing behavior in entertainment-oriented scenarios, which is ideal for naturally social, casual, and entertaining systems. Therefore, in different scenarios, users show distinct preferences for role-based relationships. Trend analysis shows that our proposed role-based relationship framework is expected to play a design guiding role in multiple IAs designs in driving scenarios. More scenarios, including driving takeover and autonomous driving, will be considered in future research.

Acknowledgments. This research is supported by the National Key Research and Development Program of China (No. 2021YFF0900600).

References

1. Andrist, S., Tan, X.Z., Gleicher, M., Mutlu, B.: Conversational gaze aversion for human-like robots. In: Proceedings of the 2014 ACM/IEEE International Conference on Human-robot Interaction, pp. 25–32. Association for Computing Machinery (2014). https://doi.org/10.1145/2559636.2559666
2. Bellamy, R.K.E., Andrist, S., Bickmore, T., Churchill, E.F., Erickson, T.: Human-agent collaboration: can an agent be a partner? In: Proceedings of the 2017 CHI Conference Extended Abstracts on Human Factors in Computing Systems, pp. 1289–1294. Association for Computing Machinery (2017). https://doi.org/10.1145/3027063.3051138
3. Braun, M., Broy, N., Pfleging, B., Alt, F.: A design space for conversational in-vehicle information systems. In: Proceedings of the 19th International Conference on Human-Computer Interaction with Mobile Devices and Services, pp. 1–8. Association for Computing Machinery (2017). https://doi.org/10.1145/3098279.3122122

4. Braun, M., Mainz, A., Chadowitz, R., Pfleging, B., Alt, F.: At Your Service: Designing Voice Assistant Personalities to Improve Automotive User Interfaces (2019). https://doi.org/10.1145/3290605.3300270

5. Carpinella, C., Wyman, A., Perez, M., Stroessner, S.: The Robotic Social Attributes Scale (RoSAS): development and validation, 254–262 (2017). https://doi.org/10.1145/2909824.3020208

6. Chen, X., et al.: Chat with smart conversational agents: how to evaluate chat experience in smart home. In: Proceedings of the 21st International Conference on Human-Computer Interaction with Mobile Devices and Services (MobileHCI 2019), New York, NY, USA, pp. 1–6. Association for Computing Machinery (2019). https://doi.org/10.1145/3338286.3344408

7. Clark, L., et al.: The State of Speech in HCI: Trends, Themes and Challenges. Interact. Comput. 31(4), 349–371 (2019). https://doi.org/10.1093/iwc/iwz016

8. Darling, N., Hamilton, S., Toyokawa, T., Matsuda, S.: Naturally occurring mentoring in Japan and the United States: social roles and correlates. Am. J. Community Psychol. 30(2), 245–270 (2002). https://doi.org/10.1023/A:1014684928461

9. Dong, J., Lawson, E., Olsen, J., Jeon, M.: Female voice agents in fully autonomous vehicles are not only more likeable and comfortable, but also more competent. Proc. Hum. Factors Ergon. Soc. Annu. Meet. 64(1), 1033–1037 (2020). https://doi.org/10.1177/1071181320641248

10. Doyle, P.R., Edwards, J., Dumbleton, O., Clark, L., Cowan, B.R.: Mapping perceptions of humanness in intelligent personal assistant interaction. In: Proceedings of the 21st International Conference on Human-Computer Interaction with Mobile Devices and Services (MobileHCI 2019), New York, NY, USA, pp. 1–12. Association for Computing Machinery (2019). https://doi.org/10.1145/3338286.3340116

11. Fraune, M.R., Šabanović, S.: Robot gossip: effects of mode of robot communication on human perceptions of robots. In: Proceedings of the 2014 ACM/IEEE international conference on Human-robot interaction (HRI 2014), New York, NY, USA, pp. 160–161. Association for Computing Machinery (2014). https://doi.org/10.1145/2559636.2559832

12. Fraune, M.R., Šabanović, S., Smith, E.R., Nishiwaki, Y., Okada, M.: Threatening flocks and mindful snowflakes: how group entitativity affects perceptions of robots. In: 2017 12th ACM/IEEE International Conference on Human-Robot Interaction, HRI, pp. 205–213 (2017)

13. Fraune, M., Sabanovic, S.: Negative attitudes toward minimalistic robots with intragroup communication styles. In: Proceedings - IEEE International Workshop on Robot and Human Interactive Communication, vol. 2014 (2014)

14. Guisewite, A.: Leveraging robot embodiment to facilitate trust and smoothness. The Robotics Institute Carnegie Mellon University. https://www.ri.cmu.edu/publications/leveraging-robot-embodiment-to-facilitate-trust-and-smoothness/

15. Hart, S.G., Staveland, L.E.: Development of NASA-TLX (Task Load Index): results of empirical and theoretical research. In: Hancock, P.A., Meshkati, N. (eds.) Advances in Psychology, vol. 52, pp. 139–183. North-Holland (1988)

16. Hayashi, K., et al.: Humanoid robots as a passive-social medium: a field experiment at a train station. In: Proceedings of the ACM/IEEE International Conference on Human-robot interaction, pp. 137–144. Association for Computing Machinery (2007). https://doi.org/10.1145/1228716.1228735

17. Hock, P., Kraus, J., Walch, M., Lang, N., Baumann, M.: Elaborating feedback strategies for maintaining automation in highly automated driving. In: Proceedings of the 8th International Conference on Automotive User Interfaces and Interactive Vehicular Applications, pp. 105–112. ACM (2016). https://doi.org/10.1145/3003715.3005414

18. Jonsson, I.-M., Zajicek, M., Harris, H., Nass, C.: In-car speech based information systems for older adults. In: Intelligent Transportation Society of America - 12th World Congress on Intelligent Transport Systems 2005, vol. 4, pp. 2113–2124 (2009)

19. Karatas, N., Yoshikawa, S., Okada, M.: NAMIDA: sociable driving agents with multi-party conversation. In: Proceedings of the Fourth International Conference on Human Agent Interaction, pp. 35–42 (ACM, 2016). https://doi.org/10.1145/2974804.2974811

20. Karatas, N., et al.: Sociable driving agents to maintain driver's attention in autonomous driving. In: 2017 26th IEEE International Symposium on Robot and Human Interactive Communication (RO-MAN), 143–149, IEEE (2017). https://doi.org/10.1109/ROMAN.2017.817 2293

21. Kitamura, Y.: Web information integration using multiple character agents. In: Prendinger, H., Ishizuka, M. (eds.) Life-Like Characters, pp. 295–315. Springer, Heidelberg (2004). https://doi.org/10.1007/978-3-662-08373-4_13

22. Kraus, J.M., et al.: Human after all: effects of mere presence and social interaction of a humanoid robot as a co-driver in automated driving. In: Adjunct Proceedings of the 8th International Conference on Automotive User Interfaces and Interactive Vehicular Applications, pp. 129–134. ACM (2016). https://doi.org/10.1145/3004323.3004338

23. Lee, S.C., et al.: To Go or Not To Go? That is the Question”: When In-Vehicle Agents Argue with Each Other. In: 13th International Conference on Automotive User Interfaces and Interactive Vehicular Applications, pp. 223–224. Association for Computing Machinery (2021). https://doi.org/10.1145/3473682.3481876

24. Lee, S.C., Sanghavi, H., Ko, S., Jeon, M.: Autonomous driving with an agent: speech style and embodiment. In: Proceedings of the 11th International Conference on Automotive User Interfaces and Interactive Vehicular Applications: Adjunct Proceedings, pp. 209–214. ACM (2019). https://doi.org/10.1145/3349263.3351515

25. Li, Y., Hollender, N., Held, T.: Task Sequence Effects in Usability Tests (2013)

26. Luria, M., et al.: Re-Embodiment and Co-Embodiment: Exploration of social presence for robots and conversational agents. In: Proceedings of the 2019 on Designing Interactive Systems Conference, pp. 633–644. ACM (2019). https://doi.org/10.1145/3322276.3322340

27. Ma, X., Forlizzi, J., Dow, S.: Guidelines for Depicting Emotions in Storyboard Scenarios. undefined (2012)

28. Maciej, J., Vollrath, M.: Comparison of manual vs. speech-based interaction with in-vehicle information systems. Acc. Anal. Prevent. **41**, 924–930 (2009)

29. Mao, X., Yu, E.: Organizational and social concepts in agent oriented software engineering. In: Odell, J., Giorgini, P., Müller, J.P. (eds.) AOSE 2004. LNCS, vol. 3382, pp. 1–15. Springer, Heidelberg (2005). https://doi.org/10.1007/978-3-540-30578-1_1

30. Ohmoto, Y., Karasaki, J., Nishida, T.: Inducing and maintaining the intentional stance by showing interactions between multiple agents. In: Proceedings of the 18th International Conference on Intelligent Virtual Agents, pp. 203–210. ACM (2018). https://doi.org/10.1145/326 7851.3267886

31. Pauzié, A., Manzano, J., Dapzol, N.: Driver's Behavior and Workload Assessment for New In-Vehicle Technologies Design. 9

32. Peissner, M., Doebler, V.: Can voice interaction help reducing the level of distraction and prevent accidents? Meta-Study on Driver Distraction and Voice Interaction (2011). https://www.semanticscholar.org/paper/Can-voice-interaction-help-reducing-the-level-of-on-Peissner-Doebler/5454baad8cb1d0c2702f1a3b020018a4b7d1956b

33. Prajod, P., Al Owayyed, M., Rietveld, T., van der Steeg, J.-J., Broekens, J.: The effect of virtual agent warmth on human-agent negotiation. In: Proceedings of the 18th International Conference on Autonomous Agents and MultiAgent Systems 71–76. International Foundation for Autonomous Agents and Multiagent Systems (2019)

34. Romani, L.: Japanese views on superior-subordinate relationship in Swedish-Japanese collaboration. In: Proceedings of the 3rd International Conference on Intercultural Collaboration, pp. 231–234. Association for Computing Machinery (2010). https://doi.org/10.1145/1841853.1841895

35. Santos, F.P.: Social norms of cooperation in multiagent systems. In: Proceedings of the 16th Conference on Autonomous Agents and MultiAgent Systems 1859–1860 (International Foundation for Autonomous Agents and Multiagent Systems (2017)

36. Sauro, J., Lewis, J.R.: Quantifying the User Experience: Practical Statistics for User Research. Morgan Kaufmann (2016)

37. Shneiderman, B., Maes, P.: Direct manipulation vs. interface agents. Interactions **4**, 42–61 (1997)

38. Suzuki, S.V., Yamada, S.: Persuasion through overheard communication by life-like agents. In: Proceedings. IEEE/WIC/ACM International Conference on Intelligent Agent Technology, 2004. (IAT 2004), pp. 225–231 (2004). https://doi.org/10.1109/IAT.2004.1342948

39. Tan, X. Z., Luria, M., Steinfeld, A., Forlizzi, J.: Charting sequential person transfers between devices, agents, and robots. In: Proceedings of the 2021 ACM/IEEE International Conference on Human-Robot Interaction, pp. 43–52. ACM (2021). https://doi.org/10.1145/3434073.3444654

40. Tan, X.Z., Reig, S., Carter, E.J., Steinfeld, A.: From one to another: how robot-robot interaction affects users' perceptions following a transition between robots. In: 2019 14th ACM/IEEE International Conference on Human-Robot Interaction (HRI), pp. 114–122. IEEE (2019). https://doi.org/10.1109/HRI.2019.8673304

41. Tejwani, R., Moreno, F., Jeong, S., Won Park, H., Breazeal, C.: Migratable AI: effect of identity and information migration on users' perception of conversational AI agents. In: 2020 29th IEEE International Conference on Robot and Human Interactive Communication (RO-MAN), pp. 877–884. IEEE (2020). https://doi.org/10.1109/RO-MAN47096.2020.9223436

42. Walster, E., Festinger, L.: The effectiveness of 'over-heard' persuasive communications. J. Abnorm. Soc. Psychol. **65**, 395–402 (1962)

43. Wang, M., Hock, P., Lee, S.C., Baumann, M., Jeon, M.: Genie vs. Jarvis: characteristics and design considerations of in-vehicle intelligent agents. In: 13th International Conference on Automotive User Interfaces and Interactive Vehicular Applications, pp. 197–199. ACM (2021). https://doi.org/10.1145/3473682.3479720

44. Wang, M., et al.: In-vehicle intelligent agents in fully autonomous driving: the effects of speech style and embodiment together and separately. In: 13th International Conference on Automotive User Interfaces and Interactive Vehicular Applications, pp. 247–254. ACM (2021). https://doi.org/10.1145/3409118.3475142

45. Wen, R., Han, Z., Williams, T.: Teacher, teammate, subordinate, friend: generating norm violation responses grounded in role-based relational norms 10 (2022)

46. Wen, R., Kim, B., Phillips, E., Zhu, Q., Williams, T.: Comparing strategies for robot communication of role-grounded moral norms. In: Companion of the 2021 ACM/IEEE International Conference on Human-Robot Interaction, pp. 323–327. Association for Computing Machinery (2021). https://doi.org/10.1145/3434074.3447185

47. Williams, K., Flores, J.A., Peters, J.: Affective robot influence on driver adherence to safety, cognitive load reduction and sociability. In: Proceedings of the 6th International Conference on Automotive User Interfaces and Interactive Vehicular Applications, pp. 1–8. ACM (2014). https://doi.org/10.1145/2667317.2667342

48. Williams, T., Zhu, Q., Wen, R., de Visser, E.J.: The confucian matador: three defenses against the mechanical bull. In: Companion of the 2020 ACM/IEEE International Conference on Human-Robot Interaction, pp. 25–33. Association for Computing Machinery (2020). https://doi.org/10.1145/3371382.3380740

49. Wong, P.N.Y., Brumby, D., Babu, H., Kobayashi, K.: 'Watch Out!': Semi-Autonomous Vehicles Using Assertive Voices to Grab Distracted Drivers' Attention. 6 (2019). https://doi.org/10.1145/3290607.3312838
50. Zhu, Q., Williams, T., Wen, R. Confucian Robot Ethics. 12
51. Caste and ecology in the social insects. Acta Biotheor **28**, 234–235 (1979)
52. Kraus - 2016 - Human After All Effects of Mere Presence and Soci.pdf
53. Design of Coal Mine Safety Monitoring System Based on Multi-Agent. In: 2021 2nd International Conference on Artificial Intelligence and Information Systems. ACM Other conferences https://dl.acm.org/doi/abs/10.1145/3469213.3470710
54. The ontological properties of social roles in multi-agent systems: definitional dependence, powers and roles playing roles. SpringerLink. https://doi.org/10.1007/s10506-007-9030-8
55. Multi-agent role allocation: issues, approaches, and multiple perspectives: Autonomous Agents and Multi-Agent Systems: Vol 22, No 2. https://dl.acm.org/doi/10.1007/s10458-010-9127-4
56. Teams in animal societies. Behavioral Ecology. Oxford Academic. https://academic.oup.com/beheco/article/12/5/534/311666
57. Karatas 等。 - 2016 - NAMIDA Sociable Driving Agents with Multiparty Co.pdf
58. Pay attention!. Proceedings of the SIGCHI Conference on Human Factors in Computing Systems. https://dl.acm.org/doi/10.1145/2207676.2207679
59. Designing Emotional Expressions of Conversational States for Voice Assistants. In: Extended Abstracts of the 2018 CHI Conference on Human Factors in Computing Systems. https://dl.acm.org/doi/10.1145/3170427.3188560

Nudging the Safe Zone: Design and Assessment of HMI Strategies Based on Intelligent Driver State Monitoring Systems

Roberta Presta[1]([⊠]) [iD], Flavia De Simone[1] [iD], Chiara Tancredi[1] [iD],
and Silvia Chiesa[2] [iD]

[1] Suor Orsola Benincasa University, Naples, Italy
roberta.presta@unisob.na.it, flavia.desimone@docenti.unisob.na.it,
chiara.tancredi@studenti.unisob.na.it
[2] RE:Lab s.r.l., Reggio Emilia, Italy
silvia.chiesa@re-lab.it

Abstract. Dangerous driver behavior can arise from different factors: distraction, sleepiness, and emotional states like anger, anxiety, boredom, or happiness. The Driver Monitoring Systems (DMS) collect data on driver behavior and emotional states, which can help design safer driving systems. Human-machine interfaces (HMIs) can leverage the detection of altered states and foster a safe driving style. To this end, we presents two visual HMI prototypes designed to assist drivers in countering distraction conditions and emotional states of too high or too low activation. The HMI prototypes combine voice assistance, ambient lighting, and visual displays. The HMI visual strategies are designed to indicate the dangerous conditions to the driver and to provide the driver with additional information about the type of dangerous state detected. This work provides details on the design and of the methodology applied to evaluate the two HMI prototypes and presents the results of a user assessment with 26 participants, showing insights into user attitudes and helping to identify future design directions.

Keywords: Driver Monitoring System · Human Machine Interface · User test · Nudge · Emotion

1 Introduction

Most traffic accidents are caused by dangerous driver behavior: human error is responsible for more than 90% of all motor vehicle accidents [1]. The driver's

This study is part of the NextPerception project that has received funding from the European Union Horizon 2020, $ECSEL - 2019 - 2 - RIA$ Joint Undertaking (Grant Agreement Number 876487). The authors would like to thank Luca Tramarin for his help in the implementation of the study.

H. Krömker (Ed.): HCII 2023, LNCS 14048, pp. 166–185, 2023.
https://doi.org/10.1007/978-3-031-35678-0_10

state is therefore critical: distraction, sleepiness, and altered mental and physical conditions can lead drivers to misjudge the road situation and put themselves and others at risk.

Recently in the field of human-machine interaction there has been a focus on emotional conditions that can negatively affect driving [3,15,22,32]. Emotions such as anger, anxiety or boredom can increase the frequency of aggressive behavior, lead to overlooking important details or lower attention levels [3,15,22,32]. Even overly positive emotions, such as too much happiness, can lead to dangerous driving because the driver underestimates dangerous situations [15].

Drivers may not always be impeccable, but it is reasonable to think that pointing out their unsafe driving status may lead drivers to take action on their status by stopping or trying to refocus. Imagining in-vehicle systems for monitoring and communicating unsafe states can therefore make a difference. Technological advances have led to new development and research possibilities in the automotive field. The ability to collect numerous data on driver state and behavior offered by Driver Monitoring Systems (DMS) can help to design increasingly safe and intelligent driving systems, nudging the driver in a secure state. Indeed, by collecting monitoring data, such as facial expressions, body temperature, tone of voice, heart rate, and movement style, emotions can be detected, and driving states classified [7,25].

Designing human-machine interfaces (HMIs) that can leverage the detection of altered states and then foster the driver to a safe driving style is the next step in the automotive industry. For a long time, the design of HMIs depended first on the limitations and then on the breadth of technological capabilities [10], but it is nowadays outstanding that it has to be taken into account at first that the driver can handle just a limited amount of information to avoid dangerous driving situations [10]. Thus there has been a shift from technology-centered approaches to human-centered approaches, trying to design interfaces that follow the criteria of usability, non-distraction, and acceptance, and involving drivers in HMI tests to identifiy promising and effective solutions [10].

In this paper we present two visual HMI prototypes built on top of advanced DMS and the user tests conducted to compare them. The work hereing presented is part of European project NextPerception[1], which began in 2020 and will be completed by 2023. Composed of more than 40 European partners, the project aims to improve safety critical domains leveraging systems based on distributed architecture of sensors and artificial intelligence components that perform complex human monitoring functions, such as those peculiar to the Driver Monitoring System. NextPerception Driver Monitoring System can detect the so-called Driver Complex State (DCS), i.e., a set of state indicators dealing with cognitive distraction, visual distraction, emotion type and arousal of drivers, allowing this way to understand how to best help them in dangerous driving situations [7]. The HMI prototypes are designed considering recent research efforts in the field and combine vocal assistant and ambient lighting to help the driver cope with dangerous situations in partial autonomous driving scenarios.

---

[1] https://www.nextperception.eu.

Given the different nuances that a driver's state can take, there are several initiatives that the HMI can take to help the driver regain a driving-conscious state. In combination with a voice assistant and ambient lighting, two strategies have been created for the visual part of the HMI on the car's dashboard: a minimalist one, geared toward indicating only the dangerous condition due to a dangerous state of the driver detected by the DMS, and a more articulate one, geared toward providing the driver with more information about the type of dangerous state detected.

The objective of this study is twofold. On the one hand, we propose the design an HMI strategy aimed at helping the driver in avoiding or countering distraction conditions, and/or states of too high activation, such as anger, stress, or anxiety, as well as too low ones such as drowsiness. We provide details about the functioning and about the rationale behind the HMI strategies we defined according to the possible detected states leveraging a vocal assistant, ambient lighting and eventually music, as well as visuals. In particular, two different visual concepts have been designed: the first one is aimed at providing the driver with further information about the quality of the dangerous state that have been detected by the DMS, thus explaining why the HMI is acting in a certain way, while the second one is only aimed at visually informing them that a dangerous state have been detected. On the other, we evaluate the designed HMI strategies. The evaluation proposed in this paper aims to answer the following research questions:

- RQ1: what is people's attitude toward an in-vehicle HMI strategy that seeks to keep the driver in a safe state?
- RQ2: what are the user experience differences that can be detected between the two HMI proposals?

Indeed, understanding these issues is helpful in identifying solution strategies on which to focus future design efforts. To this aim we conducted a user study leveraging a driving simulator and present the results in the following.

The rest of the paper is organized as follows. In the next section, the reader is provided with the state-of-the-art features about the Driver Monitoring System solutions realized in the context of European project NextPerception and about the enabled user scenarios. Then, we present the design of two driver state-aware HMI strategies aimed at keeping the driver in a safe state. In Sect. 4, the methodology applied for the evaluation of the HMI prototypes is presented, followed by the presentation of the results of the user-based experimentation in Sect. 5 and their discussion in Sect. 6. Finally, final remarks and design driving directions are identified in Sect. 7.

2   NextPerception DMS and Enabled User Scenarios

Driver monitoring systems are aimed at timely understanding different aspects of the driver's psychophysiological state for the sake of the driving task safety, both in manual and in partially autonomous driving, where the role of the human is the one of being a reliable fallback [17, 19]. Indeed, the state information provided by such monitoring systems can be leveraged to adapt the vehicle human machine interface to better communicate with the drivers and to help in keeping them in the loop effectively, such as by recalling their visual attention on the road when needed [14], or by regulating emotional conditions that have a negative impact on the driving performance [3, 32]. Visual distraction and also cognitive distraction, defined as the detour of the subject's attention from the primary task of driving to secondary tasks or the driver's mind [16], are both undesired state condition since pave the way for a geopardized situation awareness of the driver during the driving task. Emotions such as anger, anxiety, and euphoria as well, have been proved to have negative effects on the driver behavior and judgement of driving situations [3, 8, 15]. Indeed, considering the Russel's circumplex of emotions [28], it has been widely recognized that the safe zone, corresponding to the most suitable condition for driving, is the one having a medium arousal and a positive valence for the emotional components of the driver state. [2, 5]

Nowadays, driving monitoring systems can benefit from the availability of a plethora of heterogeneous sensors, each one capturing specific cues useful to build, by means of distributed artificial intelligence components, a comprehensive picture of the current driver's state [14, 17, 19].

NextPerception Driver Monitoring System (DMS) is aimed at detecting cognitive distraction, visual distraction, emotion type and arousal of drivers to understand how to help them at best in safety-critical situations [7]. To this aim, the solution developed in the project leverage different driver's data captured a set of in-car sensors to have a complete picture of the so-called driver complex state. Focusing on the combined measurement of cognitive, behavioral, and emotional factors collected through unobtrusive sensors, such a DMS is able to fuse data and obtain a $0 - 100$ fitness to drive index estimating the driver's ability to have control of the vehicle [20]. This information about the state of the driver, combined with an estimation of the external driving environment is exploited by a Decision Support System (DSS) in charge of determining the most appropriate action to support the user: this action ranging from starting a "state recovering" strategy by means of the HMI, for example for calming down or refocusing the driver, to take over the control of the vehicle from the driver to the automation for the sake of safety. This is indeed one of the driving scenarios that is envisioned in the partially automated driving. At the inner core of this system, however, there is the DMS with its capability of understanding the driver state, according to their behavioral and psychophysiological signals.

To facilitate the design of the driving scenarios enabled by this cutting edge technology, a set of hypothetical scenarios have been developed: they show the kind of help the vehicle automation can provide to the driver, thanks to the monitoring function of the DMS.

The user stories considered in this study are those of Michael and Peter, two ordinary people who use partially autonomous vehicles during ordinary journeys, and who for different reasons find themselves in unsuitable conditions for driving: thanks to the detection capabilities of the state of the driver, by means of the

DMS, the vehicle can therefore decide to take control or to give appropriate suggestions to the driver aimed at increasing the safety level of the driver-vehicle system:

- **Peter** (distraction-focused use case): Peter is driving into town to pick up his daughter from kindergarten. He is in a great hurry and is agitated since he is running late. While driving, he receives an important message from his boss that he must respond to immediately. He then starts composing the reply using his cell phone and driving badly. The DMS detects that Peter is too agitated and distracted to drive, and so the voice assistant urges Peter to focus. However, Peter continues to be too distracted, and so the voice assistant asks him if he wants to activate autonomous driving. After Peter's affirmative answer, the automation takes control of the vehicle.
- **Michael** (drowsiness-focused use case): Michael is facing a long overnight journey on a highway road in his truck, and suddenly he begins to feel tired. Using sensors, the DMS detects Michael's distraction. The vocal assistant is activated and prompts Michael to focus and to put on his favorite rhythm playlist, while enhancing the internal ambient lighting. Michael does not respond to the voice assistant's suggestion, however, because he is falling asleep. The DMS detects the dangerous situation and, consequently, the vocal assistant announces that autonomous driving is activated.

3   Designing the DMS-Enabled Human Machine Interface Strategy

As presented in Sect. 2, according to the DSS decision, the HMI is called to:

- implement a recovery strategy to convey the Fitness to Drive index (FtD) in a safe range
- allow the passage of the driving control from the driver to the vehicle automation (and from the vehicle automation to the driver)
- signal the implementation of a recovery maneuver in case of inability to drive both of the vehicle automation and of the driver because of their usafe state

Since vocal interaction (and the mirrored visual interface) appears to be a promising solution to tackle the in-car interaction with the vehicle intelligence [21], all the dialogues between the automation and the users in the HMI use cases are implemented in that way.

The recovery strategy is applied in case dangerous circumstances are detected, such as prolonged visual and/or cognitive distraction, unsafe emotional conditions detected, and the DSS judges there is enough time to try to recall the driver in a safe state.

This is the case where, for example, the driver might have to take back control of the vehicle, but they are too emotionally upset to do that; or, for example, when the driver is driving erratically because committed to secondary tasks such as texting for urgent communications.

Recovery strategies can include warnings to refocus on the primary task of driving, as well as emotion regulation approaches based on ambient lighting [13], music [3], breathing exercises [32], and so on, to lead the emotion component of the driver complex state in the so-called safe zone of a medium positive valence and arousal [3]. State of the art research work has shown that there is no one-fits-all approach when coming to the emotion regulation, and that personalization shall be taken into account to match users' personalities and preferences [2,23,31].

While reactive HMI solutions of that kind are applied in a corrective way, preventive solutions should be taken into consideration to foster the application of safe driving behavior (i.e., focused, wise, and emotionally controlled) in general. Gamification strategies [24] and digital nudges [6] are starting to become explored in the HMI research for the automotive field in that way [9].

One of the challenges for the HMI is effectively communicating to the user the why of the HMI adaptive behavior, i.e., the motivation because the in-vehicle interaction is acting in a specific way [18]. The user indeed must understand why the HMI is acting in a certain way according to his state as detected by the DMS to develop trust in the system and foster its acceptance.

Based on these considerations, we decided to implement the visual strategies of the HMI by exploiting the instrument panel (a 20 cm × 35 cm display behind the steering wheel). We also integrated a series of intervention strategies based on music, ambient lighting (implemented by Philips Hue5 LED light stripes), and vocal interaction.

3.1 Recovery Strategies

Recovery strategies are applied in case dangerous circumstances are detected, such as prolonged visual and/or cognitive distraction, unsafe emotional conditions detected, and the DSS judges there is enough time to try to recall the driver in a safe state. More precisely, recovery strategies are activated when the Fitness to Drive index is below the safety threshold and the driving situation is considered safe enough to try to activate a state recovery strategy. Recovery strategies have to be declined into different cases according to the monitored driver state that has been judged by the DMS as dangerous with a consequent Fitness to Drive index below the safety threshold.

Dangerous driver states are identified in the following cases:

- *distraction*: visual distraction (when the subject is not looking at the road and then could miss important events) OR cognitive distraction (when subject's attention is diverted from the primary task of driving towards in-car secondary tasks or mind wandering). In both cases, the HMI should try to recall the subject's attention to the primary task.
- *high activation*: this could be the case of (i) high arousal and positive emotion (for example, euphoria), or (ii) high arousal and negative emotion, as in the situation of anger and anxiety, or fear. It is important in that case to convey the activation level in the middle level, where the safe zone is identified, thus applying in both cases a relaxing strategy.

- *low activation*: This could be symptomatic of an extreme calm (positive valence) or boredom (negative valence) and also of drowsiness and of fatigue. Similarly to the precedent case, the arousal should be brought back to the safe zone, thus increased to allow an acceptable level of activation of the subject. In that case, that means to implement an activating strategy.
- *distraction* and *high activation*.
- *distraction* and *low activation*.

Given this rationale, three main HMI recovery strategies have been identified:

- *refocusing*: the HMI is aimed at call back the driver's attention on driving The refocusing strategy applies whenever visual distraction and/or cognitive distraction are detected, thus even in combination with altered emotional states. The proposed implementation strategy involves both the visual and audio HMI channel to warn the driver of their unsafe behavior. A vocal interaction warning message is accompanied by a proper representation of the issue on the instrument panel.
- *relaxing*: the HMI is aimed at lowering the too high activation level of the subject The relaxing strategy applies whenever too high levels of arousal are detected. The proposed implementation strategy involves a visual representation of the issue on the instrumental panel and a concurrent vocal interaction prompting the playing of a relaxing playlist, which is assumed to be personalized by the driver in a previous configuration moment. Moreover, the ambient lighting increases progressively the intensity of the blue light, since it has been proved to have positive effects in relaxing people.
- *activating*: the HMI is aimed at increasing the too low activation level of the subject The stimulating strategy applies whenever too low levels of arousal are detected. The proposed implementation strategy involves a visual representation of the issue on the instrumental panel and a concurrent vocal interaction prompting the playing of the driver's favorite power-up playlist, which is assumed to be personalized by the driver in a previous configuration moment. In addition, the ambient lighting increases progressively the intensity of the orange light, since it should have positive effects in increasing the activation level of people.

The application of each strategy is associated with the aforementioned driver complex state cases as depicted in Table 1.

The detected valence of the emotion felt by the driver is exploited to have the vocal interaction empathizing with the driver and explaining to them why the HMI is starting acting in a certain way. As an example (case 2.1 in Table 1), when the driver is feeling anger, or frustration, or anxiety or fear, the vocal interaction, in the recovery strategy message, would say *"Hey, you seem a little bit too agitated...What about your relaxing playlist?"*, while when the driver is experiencing euphoria or is particularly elated (case 2.2 in Table 1), the vocal interaction would say *"Hey, you seem a little bit too up... What about your relaxing playlist?"*. By using the phrase "you seem a little bit too up" or "you seem a

Table 1. HMI recovery strategies associated with different driver complex states.

Case n	Distraction	Arousal	Valence	HMI strategy
1	present	normal	any	refocusing
2	absent	high	2.1 - positive 2.2 - negative	relaxing
3	absent	low	3.1 - positive 3.2 - negative	activating
4	present	high	any	refocusing + activating
5	present	low	any	refocusing + activating

little bit too agitated", the vehicle intelligence is, on the one hand, demonstrating empathy with the driver since it understands what is going on with their state, and, on the other hand, is explaining why the car HMI is adapting in a certain way, in the specific case by changing the ambient lighting and proposing a relaxing or empowering playlist. In case of low activation (case 3.1 and case 3.2), the message of the vocal interaction would say respectively *"Hey, you seem a little bit tired... What about your power-up playlist?"* and *"Hey, you seem a little bit sluggish... What about your power-up playlist?"* respectively for negative and positive valence. Finally, when both distraction and extreme levels of activation are detected, the vocal interaction would simply say *"Hey, you seem a little bit too distracted"*, followed by the proposal of the playlist to be used to respectively relax or stimulate the driver (*"What about your relaxing playlist to help you focus on driving?"* and *"What about your power-up playlist to help you focus on driving?"*). This simplification has been chosen to let the driver reflect primarily on their inattention, and then the emotion regulation by music is presented by the vehicle intelligence as a means towards the primary goal of refocusing.

3.2   Visual HMIs: Warning vs Continuous

During the design of the HMI, it was decided to explore two different concepts related to the visual communication of the information that the DMS detected about the driver's state. Therefore, two different visual HMIs were designed: a minimalist one, called "warning", oriented to indicate only the dangerous condition due to the driver's dangerous state, and a more articulate one, called "continuous", oriented to provide the driver with more information about the type of dangerous state detected.

The visual concept of the HMI has been implemented on the instrument panel, and the space exploited is that of a semicircle to be displayed through the steering wheel of the driving simulator. For both HMIs, the space was divided into two sections:

- the top section is dedicated to ordinary driving information (driving part). This has remained graphically unchanged between the two HMIs;

– the bottom section is dedicated to explaining to the driver the logic of adapting the HMI to the driver's state (driver state part). This part has varied between the two HMIs depending on the logic followed.

The warning HMI was designed following the concept of simplicity and immediacy, deciding to communicate with the driver only when his driving status is no longer safe. This concept, in graphical terms, has been represented by warning lights (represented by an exclamation mark in a triangle) accompanied by the words *"Dangerous driver status"*, which appear in the center of the section dedicated to the driver's status at the moment when the fitness to drive is below the safety threshold (Fig. 1).

The continuous HMI, on the other hand, was designed following theories related to digital nudge [6] and Quantified Self applied in automotive [2,30]. The concept behind it is that the driver, by visualizing and self-monitoring their driving state, may be able to better understand why and to what extent their driving state is unsafe and, with this increased awareness, be able to return to a safe zone. Given these goals, the user is given a simplified view of the aspects of the driver's complex state that are measured through the DMS: level of distraction (merging visual and cognitive distraction detection), level of activation, and subsequent fitness to drive in the vehicle. These three aspects were then rendered graphically in the continuous HMI in the following way (Fig. 2):

– on the left was placed the *distraction* indicator that constantly indicates to the driver the distraction level[2] and also the limit beyond which distraction level is high and therefore dangerous: in this case, the indicator turns orange and the indicative light comes on;
– on the right was placed the *activation* indicator that continuously maps the user's arousal level[3], and whose safety zone is within the two limit indicators (i.e., a medium level of activation, to stay in the "safe" emotional zone). When the activation is too high or too low, the indicator turns orange and the related light comes on;
– in the center, on the other hand, is a numerical *fitness to drive* indicator that shows, in percentage, whether or not the user is fit to drive. When the threshold falls below the limit, a warning light (exclamation mark with triangle) appears in conjunction with one or both status indicators, and the percentage turns orange.

4   Methodology

The two HMI concepts designed as described in the previous section need to be tested with real users to see if they can be considered promising solutions.

---

[2] The level of distraction is calculated as the maximum between the percentage of visual distraction events detected in the observation window and the percentage of cognitive distraction events detected in the observation windows.

[3] The level of arousal is calculated as the average value of the N numerical estimations of the arousal performed by the arousal classifier in the observation window.

To this aim we designed an experimental evaluation leveraging a driving simulator to understand what are the effects they could play in a driving environment from a user experience perspective. Namely, we want to assess and compare the acceptance features, the user experience and the comprehensibility of the two proposed solutions to nudge the driver in a safe state, by identifying pros and cons, and to provide a comparison gauge for future design evaluations.

To this aim we designed an experiment where participants would have been divided into two groups: a first group testing the continuous HMI and a second group testing the warning one. Within each group, each participant would have tested only one user scenario, namely or the Michael's one (drowsiness focused) or the Peter's one (distraction focused), realizing this way a $2x2$ between study with two independent variables: the HMI to be tested (warning vs continuous) and the scenario (Peter vs Michael).

Each participant, after having filled an entry questionnaire, is asked to drive on an urban or highway road (according to the assigned scenario) for three minutes. This allows the participant to familiarize with the driving simulator environment.

After the driving trial, the participant listens to a presentation of the correspondent user scenario performed by the experimenter: in this phase, the experimenter leverages the illustration of the scenarios by means of vignettes, like the one shown in Figs. 1, 2. The participant is then asked to empathize with the protagonist of the story, and is told that they are going to drive in the same context of the protagonist, where the on-board HMI will perform as described in principle in the vignettes. The participant is told that after this second driving trial they would be asked to fill-in an exit questionnaire to assess the user experience. The driving simulation is conceived to let the participant live in first person the envisioned in-car scenario and to assess in a more conscious way the interaction with the HMI prototype.

The driving trial lasts 5 min. A recorded narrator's voice reminds the participant what the protagonist of each story is doing in the various minutes of the guide. For example, the narrator's voice in Peter's scenario starts by telling: *"Peter drives in a hurry because he is late to pick up his daughter from kindergarten. Meanwhile, he chats with his boss about an urgent business deal."*; then, the HMI behaves according to the mentioned driver state, namely by activating the vocal assistant saying *"Peter, you seem distracted...Pay attention to driving!"* and modifying the visual interface corresponding to the prototype of the participant group (warning vs continuous). The HMI interaction is then simulated and is played according to the user scenarios, not according to the real participant state: the participant is aware of that, since it has been explained before the test.

To consider the participants characteristics and to assess the dependent variables of interest for the study, an entry questionnaire and an exit questionnaire followed by a de-brief interview have been designed as described in the following subsections.

4.1 Entry Questionnaire

The entry questionnaire was aimed to understand what kind of people joined the experimental session from the perspectives of attitude towards technology, personality, and familiarity with the DMS and its sensors. Because of that, we designed a questionnaire leveraging preferably standard tools, and ad-hoc questions only when needed.

The sections considered in the entry questionnaire were:

1. demographics (age, gender identity, level of education, years of driving experience);
2. a short validated version of the big 5 questionnaire to identify personality traits [12];
3. the ATI questionnaire to assess the openness towards the usage of technology [11];
4. the familiarity towards (i) driving experience in cars equipped with driver monitoring systems; (ii) knowledge of driving monitoring systems in general. These items were evaluated by asking participants their level of familiarity on a 7 point Likert scale ranging from *1 - Very Low* to *7 - Very High*.

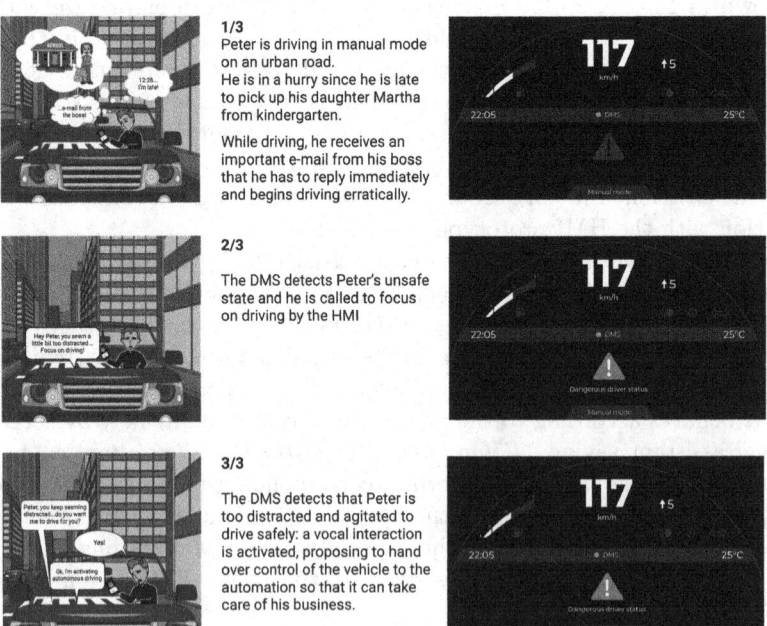

Fig. 1. Peter's scenario side by side with the warning HMI.

4.2   Exit Questionnaire and Interview

The exit questionnaire followed the same design principle of the entry question-naire. We leveraged preferably standard questionnaires to assess the constructs of interest for the study and ad-hoc questions only when needed. Namely, after the driving simulation, we proposed to participants the following:

- to assess their user experience according to the UEQ-short form questionnaire [29];
- to assess the usability of the proposed HMI approach according to the SUS questionnaire [4];
- the perceived usefulness, trust towards the system, perceived behavioral con-trol, willingness to pay, willingness to buy, endorsement and behavioral inten-tion according to the questionnaire we found in [27] to assess the acceptance towards driver monitoring systems;
- finally, to check the perceptibility of the interface, we designed an ad-hoc quiz of four questions. They checked if the participants were aware of the changes and positions of distraction and activation indicators as well as of the danger indicator during the driving simulations.

Written open comments were registered as well, to check what worked and especially what did not according to the participant's judgment.

The compilation of the exit questionnaire was followed by an exit interview of the experimenter asking a free comment about the overall experience, and more specific questions about their opinion of the effectiveness of the recovery strategy applied in each case (what about the vocal assistant / visual interface / ambient lighting or music). The alternative interface is shown on a paper prototype to ask for rapid feedback about the alternative visual strategy.

5   Results

Participants were offered compensation in the form of Amazon shopping vouch-ers as an incentive to participate in the experiment. 32 participants joined the experiment. Because of the driving simulation sickness that was reported by par-ticipants after the driving trials, we had to remove the trials of 6 people from our sample. The 26 considered participants (13 Females, 13 Males), are aged $24 - 58$ years (M = 34.9, SD = 10.98), all of them with at least 3 years holding a driving license. The 26 participants are distributed as in the Table 2. The two groups had similar characteristics in terms of gender (C: 4 Females, 7 Males; W: 9 Females, 6 Males), age (C: M=36.4; W: M=33.8) and obtained similar scores in the $Big5$ questionnaire.

At the beginning of the test, participants were asked to answer a list of questions related to their relationship and familiarity with technology and driver monitoring systems. Generally, participants obtained high values regarding their openness towards the usage of technology (M = 4.34, range $1 - 6$) even if their familiarity and experience with driving monitoring systems were not very high (familiarity mean = 3.12; experience mean = 1.7; range $1 - 7$).

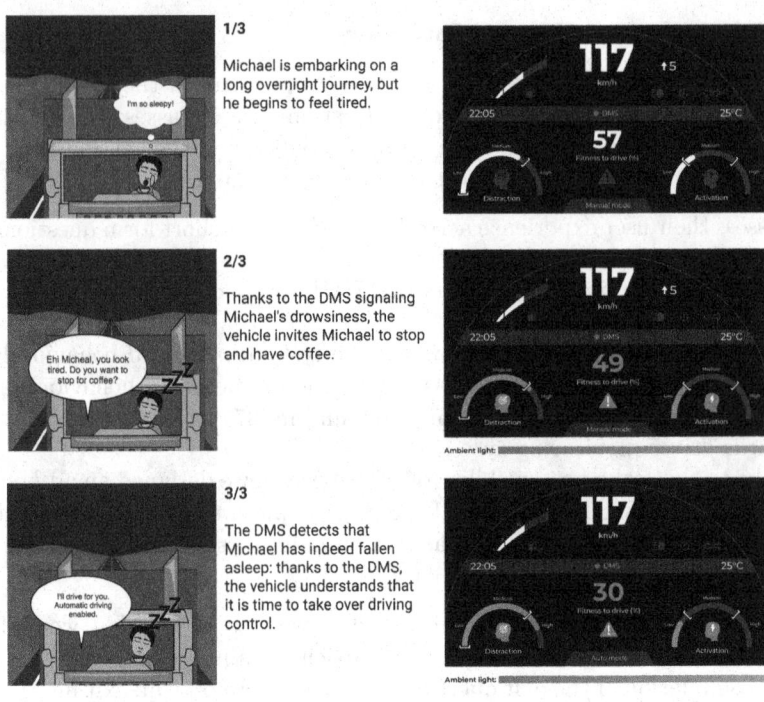

Fig. 2. Michael's scenario side by side with the continuous HMI

Table 2. Distribution of the 26 participants per groups.

	Continuous HMI (Group C)	Warning HMI (Group W)
Peter's scenario	4	7
Michael's scenario	7	8

The level of acceptance has been investigated focusing on different aspects presented in [27] as important acceptance predictors besides usability, that was considered by means of the SUS questionnaire. Participants were asked to express with a 7-point scale their opinion regarding the perceived utility of the HMIs, their level of trust, the perceived behavioural control, the level of endorsement and the behavioural intention. All these factors identified a global level of acceptance. Both interfaces obtained high value considering the average of such factors (C: M = 5.84; W: M = 5.68) and no significant differences between HMIs has been found.

In confirmation of the good acceptance of the two systems, the willingness to buy for both HMIs is high. Both the continuous and warning interfaces obtained a score of 5.8 (range of the scale from 1 to 7). In Tables 3 and 4 are reported the willingness to pay of participants in case they have to buy a new car (most frequented classes: 250€ – 500€ for prototype C and 751€ – 1000€ for prototype

W) and if they have to add this new feature on their current vehicle (most frequented classes: 501€ – 750€ for both).

Table 3. Willingness to pay the feature on a new vehicle.

	less than 250€	250€ – 500€	501€ – 750€	751€ – 1000€	1001€ – 1250€	1251€ – 1500€	more than 1500€
Continuous	1	6	2	0	1	1	0
Warning	1	4	3	6	0	0	1

Table 4. Willingness to pay the feature on the current vehicle.

	less than 250€	250€ – 500€	501€ – 750€	751€ – 1000€	1001€ – 1250€	1251€ – 1500€	more than 1500€
Continuous	2	3	4	0	0	2	0
Warning	2	4	5	4	0	0	0

For the usability aspects, participants were asked to evaluate the systems using the System Usability Scale (SUS). Generally, participants reported they would like to use both systems frequently and reported that the systems were easy to use. The warning system obtained a higher score (SUS = 81.67) than the continuous system (SUS = 80.45), but both conditions obtained quite high values and both systems obtained results above the normative average (68).

We also assessed the overall user experience using the User Experience Questionnaire - Short form (UEQ-S). The range of this questionnaire is from -3 to $+3$ and the average of all values obtained are positive. The general value of the user experience is statistically higher ($F=5.6$; $p < .05$) for the warning interface (M=1.16) than for the continuous interface (M=1.09). Although this general result, for specific subscales the Continuous HMI obtained higher values as for example for the stimulation level (C: M=1.05; W: M=0.9), the perspicuity level (C: M=1.66; W: M= 1.53) and statistically higher for the attractiveness (C: M=1.2; W: M= 1.08 $F=7.17$; $p < .05$). On the other hand, the warning HMI is considered better for other values as the novelty (W: M=0.88; C: M= 0.48), the efficiency (W: M=1.13; C: M= 0.91) and the dependability (W: M=1.40; C: M= 1.25).

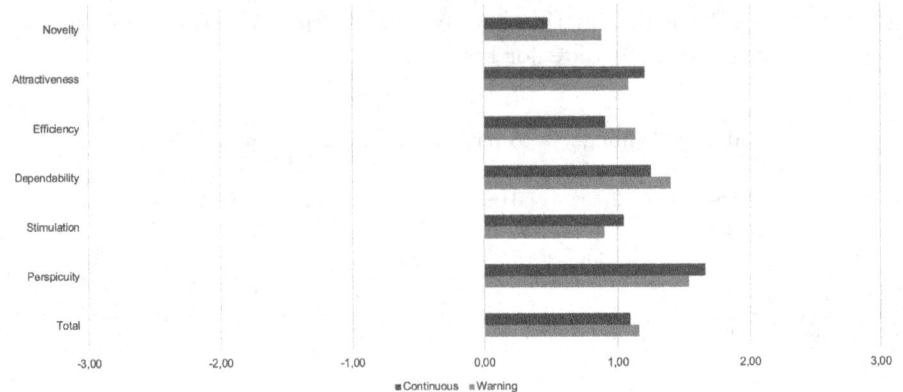

Fig. 3. UEQ-S results.

At the end of the exit questionnaire, the participants answered a quiz about the visual HMIs. The scores obtained in both interfaces are quite low (W: M=1.36; C: M= 1.07, Range 0-4) and detailed messages, as for example changes of the distraction level, activation level or fitness to drive indicator have been missed.

5.1   Analysis of the Open Comments and of the Interviews

Open comments of the exit questionnaire have been analysed according to a thematic analysis performed by two of the authors. Codes have been identified independently and then, after a merging phase, the final set has been defined. Agreement has been achieved by discussion for the identification of each code in the participants' open comments.

The most recurrent topic is about the perceived usefulness of the experienced system. 16 people judged the system as useful: 8 of them tested the continuous HMI while the other 8 the warning one. Interestingly, 10 of them tested the drowsiness-focused scenario and the remaining 6 the distraction-focused one. 5 times the visual interfaces was mentioned to go unnoticed: 4 times of them was the case of the warning interface, where the visual contribution was less articulated than the case of the continuous interface. The lighting solution was appreciated 5 times, 4 of them in the warning HMI case, followed by the effectiveness of the vocal assistant (3 mentions, 2 for the warning HMI). Other recurrent topics are the definition of the HMI as "interesting" (2 times for the warning HMI and 2 for the continuous one), "simple" (3 mentions, all for the warning interface) and as "direct" (2 mentions, all for the continuous interface), well distributed among the scenarios in all cases.

Besides recurrent topics, there are two groups of topics occurring just one time for the continuous interface and for the warning one. The continuous HMI was defined once as "to be improved", "distracting" and it was argued that the activation indicator could be not considered as interesting by drivers in general.

Positive comments were also released, defining the continuous HMI as "optimal", and that it should be distributed at a convenient price to allow for a wide distribution (both cases for Michael's scenario). The warning HMI raised on the other hand a comment saying that more information is felt as needed to understand the driver state and that more driving trials would be needed to thoroughly understand the functioning behind the warning strategy. One participant testing the warning HMI in the drowsiness case commented that, imagining to have that system in place, the driver could get used to driving even in non-optimal conditions and that the problem could be the one of forgetting to activate the DMS while continuing to rely on it.

We collected also feedback by means of ex-post interviews, performed after the participants answered the exit questionnaire, i.e., after the open comments summarized above. In this context, after a general feedback about the overall experience, participants were provided with a paper representation of the HMI they didn't test. When comparing the two different HMI strategies, some reflect that it is a matter of personality to appreciate visual HMI (3) and that the visual representation should be chosen according to personal preferences (2). Some are afraid of the mirror effect in the continuous one (4), while others appreciate it a lot (4) and find it more engaging (one finds it similar to a game) and helpful in self-regulation (2). In contrast, regarding the warning HMI, some prefer it for simplicity (3), but others emphasize the need for more information (3). Some shortcomings or improvements to reason about at the HMI strategy level come up, i.e. a fair number of people who said that visual HMI went unnoticed (6) or not very noticeable (4) (equally distributed between the two HMIs), the lack of the warning sound coinciding with any warning of a driver's dangerous state (4). Some fear deresponsibilization of the driver due to the system and look with suspicion at it for that reason (2).

6   Discussion

Participants did show a positive attitude about the idea of a system that supports the driver in detecting and intervening in dangerous driving states: the acceptance of both HMIs stood at a very high mean value (5.84 for continuous and 5.68 for warning), indicating that this type of initiative is well regarded, as also confirmed by the results obtained in willingness to buy and willingness to pay.

Both HMIs in terms of usability were rated very positively. An overall preference in terms of user experience was noted toward the warning HMI, with a higher attractiveness value, however, for the continuous one. We can hypothesize that the interactivity and completeness of information of the continuous HMI aroused more interest than the warning HMI, which in the overall driving experience, however, was found to be preferable. This result is in line with the open comments that showed the need for more information when they drove with the warning interface, but also with the comments that pointed out that the continuous interface was distracting. Considerations of visual appreciation related to personal preference and of a system that alternates between the two views are in line with the study [2].

In general, the lighting and voice assistant solutions were appreciated and found to be effective, often sufficient to communicate the driver's status regardless of the visual HMI: the low quiz results are in line with the comments about the HMI not being noticeable, probably also due to the lack of an alarm sound when a dangerous state was detected, in addition to the voice assistant.

The drowsiness-focused scenario where both HMIs were tested raised particular interest in the participants. In the open comments, those who tried that scenario felt the urgency of commenting that the system is very useful (more times than the distraction-focused one) and that it should be widespread. This is in line with the scenario assessment performed in a similar study from the authors in [26].

7   Conclusion and Future Work

In this study, we presented the design of two HMIs designed to assist drivers in countering distraction conditions and emotional states of too high or too low activation detected by the DMS. The designed prototypes leverage the DMS detected information about the driver state and were based both on ambient lighting and vocal assistant, but differ in the visual strategy used for communicating the driver their monitored state: the warning one only provides information about the detection of a dangerous condition, while the continuous one provides further detail about the fitness to drive, distraction and activation level. The two different ways of displaying the driver state have been tested with 26 participants in two different scenarios, one drowsiness-based and another one distraction-based.

The test were aimed at understanding different aspects of the overall user experience, acceptance and usability of the two solutions. In general, we found a high acceptance of driver state support systems, despite few doubts arising from the fear that the driver will become deresponsibilized in partial autonomous driving scenarios thanks to such a system. In particular, ambient lighting and voice assistant were highly appreciated. They seem to be predominant with respect to the visual HMI strategy, putting the visual display solutions in the background, as verified also by means of the quiz part of the exit questionnaire.

About the visual HMI prototypes, the results were in general positive for both options, with significant differences only for the overall user experience and usability that registered a preference for the warning HMI. In light of the obtained results, the HMI strategy should show the warning about the low fitness to drive monitored by the DMS accompained by a warning sound, together with the vocal assistant intervention and ambient lighting. By the way, considering the user experience sub-scales of and open comments coming from both the exit questionnaire and from the interviews, it seems that the most convenient choice would be to consider the two HMIs as two different display options, selectable according to the driver's personality or preferences, keeping the visual warning as the default one.

To understand the effectiveness of keeping the driver in a safe zone, the use of such a system should be considered on longer time intervals, as used on a daily basis and in the long term, to see if the increased knowledge of the system would lead to a change in the preference of a visual strategy instead the other one, or to a change on the level of acceptance of the system in general. Therefore, we would like to direct future research to test such solutions in a more extended way, not only by tying them to personalization, but also to long-term experimentation.

References

1. NHTSA: 2016 fatal motor vehicle crashes: Overview. traffic safety facts: research note. Report No. DOT HS 812 456 (2017)
2. Braun, M., Schubert, J., Pfleging, B., Alt, F.: Improving driver emotions with affective strategies. Multimodal Technol. Interac. **3**(1), 21 (2019)
3. Braun, M., Weber, F., Alt, F.: Affective automotive user interfaces-reviewing the state of driver affect research and emotion regulation in the car. ACM Comput. Surv. (CSUR) **54**(7), 1–26 (2021)
4. Brooke, J., et al.: SUS-a quick and dirty usability scale. Usability evaluation in industry **189**(194), 4–7 (1996)
5. Cai, H., Lin, Y.: Modeling of operators' emotion and task performance in a virtual driving environment. Int. J. Hum Comput Stud. **69**(9), 571–586 (2011)
6. Caraban, A., Karapanos, E., Gonçalves, D., Campos, P.: 23 ways to nudge: a review of technology-mediated nudging in human-computer interaction. In: Proceedings of the 2019 CHI conference on human factors in computing systems, pp. 1–15 (2019)
7. Davoli, L., et al.: On driver behavior recognition for increased safety: a roadmap. Safety **6**(4), 55 (2020)
8. De Simone, F., Presta, R.: A song can do that: an emotion induction study for the development of intelligent emotion-aware systems. In: Arai, K. (eds) Intelligent Systems and Applications. IntelliSys 2022. Lecture Notes in Networks and Systems, vol. 543, pp. 363–377. Springer, Cham (2022). https://doi.org/10.1007/978-3-031-16078-3_24
9. Di Lena, P., Mirri, S., Prandi, C., Salomoni, P., Delnevo, G.: In-vehicle human machine interface: an approach to enhance eco-driving behaviors. In: Proceedings of the 2017 ACM Workshop on Interacting With Smart Objects, pp. 7–12 (2017)
10. François, M., Osiurak, F., Fort, A., Crave, P., Navarro, J.: Automotive HMI design and participatory user involvement: review and perspectives. Ergonomics **60**(4), 541–552 (2017)
11. Franke, T., Attig, C., Wessel, D.: A personal resource for technology interaction: development and validation of the affinity for technology interaction (ati) scale. Int. J. Human-Comput. Interac. **35**(6), 456–467 (2019)
12. Guido, G., Peluso, A.M., Capestro, M., Miglietta, M.: An Italian version of the 10-item big five inventory: An application to hedonic and utilitarian shopping values. Personality Individ. Differ. **76**, 135–140 (2015)
13. Hassib, M., Braun, M., Pfleging, B., Alt, F.: Detecting and influencing driver emotions using psycho-physiological sensors and ambient light. In: Lamas, D., Loizides, F., Nacke, L., Petrie, H., Winckler, M., Zaphiris, P. (eds.) INTERACT 2019. LNCS, vol. 11746, pp. 721–742. Springer, Cham (2019). https://doi.org/10.1007/978-3-030-29381-9_43

14. Horberry, T., et al.: Human-centered design for an in-vehicle truck driver fatigue and distraction warning system. IEEE Trans. Intell. Transp. Syst. **PP**, 1–10 (2021)
15. Jeon, M.: Emotions and affect in human factors and human-computer interaction: taxonomy, theories, approaches, and methods. Emotions and affect in human factors and human-computer interaction, pp. 3–26 (2017)
16. Kaber, D.B., Liang, Y., Zhang, Y., Rogers, M.L., Gangakhedkar, S.: Driver performance effects of simultaneous visual and cognitive distraction and adaptation behavior. Transport. Res. F: Traffic Psychol. Behav. **15**(5), 491–501 (2012)
17. Koesdwiady, A., Soua, R., Karray, F., Kamel, M.S.: Recent trends in driver safety monitoring systems: state of the art and challenges. IEEE Trans. Veh. Technol. **66**(6), 4550–4563 (2016)
18. Koo, J., Kwac, J., Ju, W., Steinert, M., Leifer, L., Nass, C.: Why did my car just do that? explaining semi-autonomous driving actions to improve driver understanding, trust, and performance. Int. J. Inter. Design Manufact. (IJIDeM) **9**, 269–275 (2015)
19. Manstetten, D., et al.: The evolution of driver monitoring systems: a shortened story on past, current and future approaches how cars acquire knowledge about the driver's state. In: 22nd International Conference on Human-Computer Interaction with Mobile Devices and Services, pp. 1–6 (2020)
20. Andruccioli, M., Mengozzi, M., Presta, R., Mirri, S., Girau, R.: Arousal effects on fitness-to-drive assessment: algorithms and experiments. In: 2023 IEEE 20th Annual Consumer Communications & Networking Conference (CCNC). IEEE (2023)
21. Meck, A.M., Precht, L.: How to design the perfect prompt: a linguistic approach to prompt design in automotive voice assistants-an exploratory study. In: 13th International Conference on Automotive User Interfaces and Interactive Vehicular Applications, pp. 237–246 (2021)
22. Neta, M., Cantelon, J., Haga, Z., Mahoney, C.R., Taylor, H.A., Davis, F.C.: The impact of uncertain threat on affective bias: Individual differences in response to ambiguity. Emotion **17**(8), 1137 (2017)
23. Oehl, M., Lienhop, M., Ihme, K.: Mitigating frustration in the car: which emotion regulation strategies might work for different age groups? In: Stephanidis, C., Antona, M., Ntoa, S. (eds.) HCII 2021. CCIS, vol. 1421, pp. 273–280. Springer, Cham (2021). https://doi.org/10.1007/978-3-030-78645-8_34
24. Pinder, C., Vermeulen, J., Cowan, B.R., Beale, R.: Digital behaviour change interventions to break and form habits. ACM Trans. Comput.-Human Inter. (TOCHI) **25**(3), 1–66 (2018)
25. Presta, R., Chiesa, S., Tancredi, C.: Driver monitoring systems to increase road safety. Human Body Interaction, p. 247 (2022)
26. Presta, R., De Simone, F., Mancuso, L., Chiesa, S., Montanari, R.: Would i consent if it monitors me better? a technology acceptance comparison of bci-based and unobtrusive driver monitoring systems. In: 2022 IEEE International Conference on Metrology for Extended Reality, Artificial Intelligence and Neural Engineering (MetroXRAINE), pp. 545–550. IEEE (2022)
27. Rahman, M.M., Strawderman, L., Lesch, M.F., Horrey, W.J., Babski-Reeves, K., Garrison, T.: Modelling driver acceptance of driver support systems. Accident Anal. Preven.ion **121**, 134–147 (2018)
28. Russell, J.A.: A circumplex model of affect. J. Pers. Soc. Psychol. **39**(6), 1161 (1980)
29. Schrepp, M., Hinderks, A., Thomaschewski, J.: Design and evaluation of a short version of the user experience questionnaire (ueq-s). Int. J. Inter. Multimedia Artif. Intell. **4**(6), 103–108 (2017)

30. Swan, M.: Connected car: quantified self becomes quantified car. J. Sens. Actuator Netw. **4**(1), 2–29 (2015)
31. Wadley, G., Smith, W., Koval, P., Gross, J.J.: Digital emotion regulation. Curr. Dir. Psychol. Sci. **29**(4), 412–418 (2020)
32. Zepf, S., Hernandez, J., Schmitt, A., Minker, W., Picard, R.W.: Driver emotion recognition for intelligent vehicles: a survey. ACM Comput. Surv. (CSUR) **53**(3), 1–30 (2020)

Evaluation Study of Intelligent Vehicle Virtual Personal Assistant

Jianqiong Pu[⊠], Liping Li, Qihao Huang, Wen Jiang, Xiaojun Luo, and Jifang Wang

Apollo Intelligent Connectivity (Beijing) Technology Co., Ltd., Beijing 100000, China
pujianqiong@baidu.com

Abstract. With the outburst of the intelligent vehicle market, vehicle Virtual Personal Assistant (VPA), as an important interactive interface of drivers and smart vehicles, will be more and more common in our daily life. Since VPA is the carrier of Voice User Interaction, previous VPA research topics mainly focus on the field of voice interaction and personification, such as personification, emotionalization, gender cognition preference, and role cognition preference. However, there is a growing demand to comprehensively evaluate vehicle VPA experience, concentrating on the vehicle application field. In this study, we constructed a vehicle VPA experience evaluation model and verified its validity and reliability.

There are 3 phases included in this study: (a) First of all, through interviews with 20 real users, we extracted users' perception elements of vehicle VPA, summarized the factors affecting the experience of vehicle VPA, and converted them into the initial dimensions of the evaluation model, included *Sensory Comfort, Depth Assistance, Natural Interaction, Pleasant Emotions* 4 modules. (b)Secondly, we invited 10 vehicle VPA designers to discuss and adjust the model based on professionals suggestions. (c) Thirdly, to verify the adjusted model, 364 vehicle VPA users (including NIO(ET/ES/EC), XPENG(P5/P7), Li One, Lynk&Co 09) were invited to participate in an online questionnaire. Reliability analysis, validity analysis, and factor analysis are carried out to confirm the quality of the questionnaire and to complete the structural adjustment of the model. The final output of the vehicle VPA evaluation model included 5 first-level indicators, 7 s-level indicators, and 45 third-level indicators. First-level indicators include *Sensory, Resource, Interaction, Emotion, Design* modules. Among third-level indicators, high influence (weight) ones are Voice of VPA coverage span, Comprehension Accuracy, Recognition Accuracy, Sense of Trust, and Satisfaction of Resource Content.

This vehicle VPA experience evaluation model study is grounded on real driver interviews and questionnaires, reflecting vehicle VPA daily users' demands, and comprehensively evaluating *Sensory, Resource, Interaction, Emotion, Design,* and 5 modules. It can be scientific guidelines for vehicle VPA design and product diagnosis.

Keywords: VPA · Intelligent Vehicle · Evaluation Model · User Experience · Factor Analysis

H. Krömker (Ed.): HCII 2023, LNCS 14048, pp. 186–199, 2023.
https://doi.org/10.1007/978-3-031-35678-0_11

1   Introduction

1.1   Research Background

With the development of science and technology, our society has gradually entered the era of the industrial revolution dominated by artificial intelligence, expanding from the "Internet Plus" stage to the "Intelligent Plus" stage, which has changed our life dramatically. The most explicit change was that the Virtual Personal Assistant (VPA), through carrying various smart products (smartphones, smart cars, smart speakers, smart TV, etc.), entered our lives and played an increasingly important role in our lives and work quietly. Concurrently, with the improvement of intelligent voice interaction, the positioning of VPA in the user's mind was not only a useful tool, but also a warm companion, and an interesting configuration that showed personality.

In the field of automobile cockpit, with the popularization of networking, vehicle VPA has gradually become one of the important optional parts of vehicles. According to a configuration survey conducted by Automotive Data of China in 2021, consumers had strong demand for vehicle VPA, ranking fifth among the top 10 must-install prospective configurations. The data showed that consumers' preference for vehicle VPA in the form of a holographic images and a physical robots is 43% and 35.8% respectively; Concurrently In the relevant research of Baidu Apollo Design Center, pointed out that "Gen Z (people born between 1995 and 2009) generally show their personality through the interior and exterior decoration of automobile, as well as vehicle VPA."

Vehicle VPA generally uses voice recognition as the human-vehicle interaction channel, which can provide navigation access, telephone calls, music playback, vehicle function control, and other services. Globally, the market of vehicle VPA is increasing fast. Amazon's vehicle VPA product named Alexa is currently in the leading market. Alexa has reached cooperation with Hyundai BMW, Benz, Ford, Lexus, Volkswagen, Toyota, and other vehicles; In addition, influential VPA service providers include Google, Microsoft, Apple, Baidu, Alibaba, Samsung, Panasonic, Bosch, and other companies.

1.2   Introduction to the Research Status of Vehicle VPA

What does not match the booming development of vehicle VPA in the automobile cockpit industry is that the research on vehicle VPA is not sufficient. Since vehicle VPA was the carrier of Voice User Interaction, vehicle VPA research topics such as personification, emotionalization, gender recognition preference, role recognition preference, are usually focused on the field of voice interaction and personification. For example, there were sufficient studies pointed out that "human VPA is more useful, durable, and beautiful than non-human VPA", and "female voice VPA is more attractive than male voice". In actual scenarios, vehicle VPA is often closely connected with the resources of the Internet and functions, and the user perception of these contents will also affect the evaluation of vehicle VPA; There are no mature scale and research on quality characteristics for the evaluation of experience quality on vehicle VPA.

1.3 Research Value and Objectives of Vehicle VPA

This study would focus on the vehicle VPA carrying the representative models of China's emerging new energy vehicles (NIO(ET/ES/EC), XPENG(P5/P7), Li One, Lynk&Co 09, etc.) Through rigorous and reasonable methods, we constructed a vehicle VPA experience evaluation model and verified its validity and reliability. Which created product measurement tools, deepened the understanding of relevant designers on the quality characteristics of vehicle VPA, and improved user experience quality of vehicle VPA.

2   Research Methodology

2.1   Phase I

Through desktop research and user interviews, the influential factors of users' perception and experience of the vehicle VPA are extracted, and the prototype of vehicle VPA evaluation model framework is constructed.

Participant. A total of 20 users were interviewed, who covered 4 Li One owners, 5 NIO (ES6/ES8) owners, 5 XPENG P7 owners, 2 LEAPMOTOR (T03 or S01) owners, 4 Roewe RX5 owners. The distribution of their gender, age, income(Family, Monthly), marriage, and childbearing are as follows Table 1.

Table 1. Demographic information distribution of participants

Demographic information		Proportion of users(N = 20)
Gender	Male	65%
	Female	35%
Age	20–30 years old (included)	30%
	30–40 years old (included)	65%
	40–50 years old (included)	5%
Income	20000–30000 Yuan (RMB,included)	35%
	30000–40000 Yuan (RMB,included)	25%
	40000–50000 Yuan (RMB,included)	40%
Marriage and Childbearing	Unmarried	10%
	Married without children	25%
	Married with children	65%

Method of Analysis. Through in-depth interviews, the real needs of vehicle owners were disassembled, and their subjective perception factors for the vehicle VPA were extracted, this information was analyzed and summarized in text and then converted into an evaluation model. The principle of constructing the evaluation model followed the user experience design level theory: 1) Visceral; 2) Behavioral; 3) Reflective.

Research Results. Through analysis and induction, we found that the main factors affecting the user experience of vehicle VPA are *user status, sensory preferences, function*

realization, emotional experience, and *interaction effects*. According to the analysis of this information, it was transformed into an evaluation model, which was composed of four modules: *Sensory Comfort, Depth Assistance, Natural Interaction, and Pleasant Emotions*. As shown in the figure below, we call this model V1.0 Table 2.

Table 2. Content of model V1.0

First-level indicators	Second-level indicator	Third-level indicators
Sensory Comfort	Auditory experience	The voice quality of VPA is comfortable and natural
		The voice speed of VPA well situated
		The tone and intonation of VPA are emotional
	Visual experience	The interface of VPA looks good
		The layout of VPA is reasonable
		The clothing and accessories of VPA are satisfactory
		The expression of VPA is natural and interesting
		The movements and posture of VPA are harmonious
		The expression, action, and command of VPA are harmonious
		The color matching of VPA is reasonable
	Multi-sensor consistency experience	VPA lip shape and dialogue are synchronized
		The audio-visual experience of VPA is unified
Depth Assistance	Abundant Function	Voice of VPA Coverage Span
		Voice of VPA Coverage Depth
		Satisfaction of Resource Content
	Abundant image resources	The image of VPA is abundant
		The clothing and accessories of VPA are abundant
		The expression of VPA is abundant
		The action and posture of VPA are abundant
	Abundant voice resources	The voice quality of VPA is abundant

(*continued*)

Table 2. (*continued*)

First-level indicators	Second-level indicator	Third-level indicators
		The voice speed regulation of VPA is abundant
	Abundant Personality orientation	The personality orientation of VPA is abundant
		The Interactive script of VPA is abundant
Natural Interaction	Efficient Voice interaction experience	Wake up Easily
		Recognition Accuracy
		Comprehension Accuracy
		Feedback Quickly
		Feedback Effectively
	Intelligent Voice interaction experience	Multi-region Speech Recognition
		Voiceprint Recognition
		Full Duplex natural dialogue
		Active Voice Interaction
Pleasant Emotions	Emotional gain	Sense of Interest
		Sense of Trust
		Sense of Individualistic
	Emotional identity	Personified Identity

2.2 Phase II

According to the actual form of the product, the evaluation model also needs to introduce professional suggestions. We invited the relevant personnel of vehicle VPA design to have a group discussion on the V1.0 model and produce the V2.0 model after adjustment.

Participants. There were 10 participants in the discussion, covering 2 user experience researchers, 5 designers, and 3 technicians; Their ages are between 25 and 40 years old, including 4 males and 6 females. They had been engaged in vehicle VPA design for more than half a year.

Method of Analysis. Focusing on the interaction scenario and business model of VPA in the cockpit, we had a group discussion on the model of Phase 1, which mainly focused on two topics: 1) which dimensions were not applicable and should be eliminated; 2) Which dimensions should be added.

Based on the above discussion, we eliminated the dimensions with low importance at present, such as V*oiceprint Recognition* and *personality orientation of VPA*. Concurrently, in combination with the current focus in the field of intelligent cockpit design, new dimensions such as *Innovativeness* and *Design Integration* had been added, and the definition and location of some dimensions had been adjusted and corrected.

Research Results. According to the above discussion and adjustment, we outputted model V2.0 as follows Table 3.

Table 3. Content of model V2.0

First-level indicators	Second-level indicator	Third-level indicators
Sensory Comfort	Auditory experience	The voice of VPA is coherent and fluent
		The tone of VPA is fluctuating and comfortable
		The voice of VPA is comfortable and natural
		The voice speed of VPA well situated
		The tone and intonation of VPA are emotional
	Visual experience	The interface of VPA looks good(Focus on position)
		The interface of VPA looks good(Focus on size)
		The layout of VPA is reasonable
		The clothing and accessories of VPA are satisfactory
		The expression of VPA is natural and interesting
		The movements and posture of VPA are harmonious
		The expression, action, and command of VPA are harmonious
		The color matching of VPA is reasonable
	Audio-visual integration	VPA lip shape and dialogue are synchronized
		The audio-visual experience of VPA is unified
Depth Assistance	Abundant Function	Voice of VPA Coverage Span
		Voice of VPA Coverage Depth
		Satisfaction of Resource Content
	Abundant image resources	The image of VPA is abundant
		The clothing and accessories of VPA are abundant
		The expression of VPA is abundant
		The action and posture of VPA are abundant
	Abundant voice resources	The voice quality of VPA is abundant
		The voice speed regulation of VPA is abundant
	Abundant Personality orientation	The personality orientation of VPA is abundant
		The Interactive script of VPA is abundant

(continued)

Table 3. (*continued*)

First-level indicators	Second-level indicator	Third-level indicators
Natural Interaction	Efficient Voice interaction experience	Wake-up Rate
		Wake-up Speed
		Wake up Easily
		Recognition Accuracy
		Comprehension Accuracy
		Feedback Quickly
		Feedback Effectively
	Intelligent Voice interaction experience	Multi-region Speech Recognition
		Keep Listening
		Continuous Dialogue
		Active Voice Interaction
Pleasant Emotions	Emotional gain	Emotional Expression (Focus on body)
		Emotional Expression (Focus on content)
		Emotional Broadcasting (Focus on voice)
		Emotion and broadcast match accurately
	Emotional identity	Personified Identity
		Sense of Trust
		Sense of Individualistic
	Design gain	Innovativeness
		Design Integration
		Design Style
		Matching of design and brand

2.3   Phase III

Through an online quantitative survey of large samples, model V2.0 was quantitatively verified, the structure adjustment was completed, and model V3.0 was outputted.

Participants. Through Baidu Map APP and Lynk&Co 09 vehicle information operation platform, we collected 364 valid real users data (Focus on China's emerging new energy vehicle owners data represented by NIO(ET/ES/EC), XPENG(P5/P7), Li One, Lynk&Co 09). Among them, there were 99 data of NIO(ET/ES/EC) owners, 91 data of Li One owners, 85 data of XPENG(P5/P7) owners and, 89 data of Lynk&Co 09 owners.

The distribution of their gender, age, educational level, residence, marriage and childbearing were as follows Table 4.

Method of Analysis
Questionnaire Design. The questionnaire was designed with the model V2.0 as the core framework. The questionnaire scored 10 points, with 1 representing "very dissatisfied" and 10 representing "very satisfied". In addition, users' sociological information (such

Table 4. Demographic information distribution of participants

Demographic information		Proportion of users (N = 364)
Gender	Male	82.97%
	Female	17.03%
Age	19–25 years old	10.71%
	26–35 years old (included)	41.48%
	36–45 years old (included)	39.84%
	More than 46 years old	7.97%
Educational level	Junior High Schools (Below)	1.65%
	High Schools (and equivalent)	12.36%
	College	23.63%
	Undergraduate	48.90%
	Postgraduate	13.46%
Residence	Township/rural	4.00%
	Small cities/Counties	17.82%
	Capital city/Municipality	46.91%
	Big Cities(Beijing/Shanghai, etc.)	30.55%
Marriage and childbearing	Unmarried (no partner)	14.84%
	Unmarried (partnered)	9.34%
	Married without children	9.62%
	Married with children	66.21%

as gender, age, education level, residence, marriage, and childbearing) would be added for later analysis.

Questionnaire Reliability. To test the reliability of the questionnaire, we used the consistency coefficient - Cronbach's Alpha (α) to test the internal consistency of each module of the questionnaire. The reliability of each module in the questionnaire was analyzed by SPSS, and the results were as follows Table 5.

It is generally believed that when α is above 0.8, it means that the reliability of the scale is very good; When α Between 0.7 and 0.8, the scale is acceptable; When α lower than 0.6, it means that the scale should be revised, but still has some value; When α lower than 0.5, it means that the scale needs to be redesigned. As shown in the table above, the α of each module is greater than 0.8, indicating that the internal consistency of the questionnaire is good and it is stable.

Questionnaire Validity. The validity of the questionnaire is used to measure whether the item design is reasonable. We verified the corresponding relationship between variables and items through exploratory factor analysis. After factor analysis, when the corresponding relationship between factors and items was consistent with expectations

Table 5. Reliability statistics of each module of the questionnaire

The module of Questionnaire (Second-level indicator)	Number of items	Cronbach's Alpha(α))
Auditory experience	5	0.951
Visual experience	8	0.961
Audio-visual integration	2	0.898
Abundant Function	3	0.906
Abundant image resources	4	0.97
Abundant voice resources	2	0.822
Abundant Personality orientation	2	0.939
Efficient Voice interaction experience	6	0.96
Intelligent Voice interaction experience	4	0.934
Emotional gain	4	0.95
Emotional identity	3	0.937
Design gain	4	0.948

basically, it indicated that the questionnaire has a good level of validity. For this reason, we use the following indicators:

$KMO > 0.6$, *usually indicating that the validity of the scale is reasonable (KMO, Kaiser-Meyer-Olkin).*

Characteristic root value (Rotated) > 1, usually indicating that the validity of the scale is reasonable.

P (Bartlett's test of sphericity) < 0.01, usually indicating that the validity of the scale is reasonable.

Cumulative variance interpretation rate > 50%, usually indicating that the validity of the scale is reasonable.

This study analyzed the overall questionnaire through SPSS exploratory factor analysis, and the results were as follows Table 6.

Table 6. Validity statistics of each module of the questionnaire

KMO	0.979
Characteristic root value (Rotated)	1.775
P	0.00
Cumulative variance interpretation rate	84.68%

As shown in the table above, the validity indicators of this questionnaire met the requirements, and we can carry out subsequent analysis.

Exploratory Factor Analysis. In exploratory factor analysis, we used principal component extraction analysis and the maximum variance method to obtain the factor load coefficient of each item. By comparing the factor load coefficients, we deleted the unrealistic items and adjusted the attribution of the items. In this study, SPSS software was used for analysis and several attempts were made to classify factors. Finally, the total number of extracted factors was 7. Meanwhile, KMO = 0.979, $P = 0.00$, and the cumulative variance explained rate was 84.68%.

Research Results. Through the above analysis, we have completed the structural adjustment of model V2.0. According to the data results and the judgment based on the actual logic, we have changed the attribution of dimensions. The model structure has been adjusted from 4–12-47 to 5–7-45 (5 first-level indicators, 7 s-level indicators, and 45 third-level indicators). Concurrently, due to the abnormal factor load coefficients, the two dimensions of "The layout of VPA is reasonable "and "The voice speed of VPA well situated " have been deleted. Besides, the connotations of these two dimensions overlap in other dimensions. After analysis and adjustment, our outputted model V3.0 was as follows Table 7.

Table 7. Content of model V3.0

First-level indicators	Second-level indicator	Third-level indicators
Sensory	Auditory experience	The voice of VPA is coherent and fluent
		The tone of VPA is fluctuating and comfortable
		The voice of VPA is comfortable and natural
		The tone and intonation of VPA are emotional
	Visual experience	The clothing and accessories of VPA are satisfactory
		The expression of VPA is natural and interesting
		The movements and posture of VPA are harmonious
		The expression, action, and command of VPA are harmonious
		The color matching of VPA is reasonable
		VPA lip shape and dialogue are synchronized

(*continued*)

Table 7. (*continued*)

First-level indicators	Second-level indicator	Third-level indicators
		The audio-visual experience of VPA is unified
Resources	Abundant Function	Voice of VPA Coverage Span
		Voice of VPA Coverage Depth
		Satisfaction of Resource Content
	Abundant resources	The image of VPA is abundant
		The clothing and accessories of VPA are abundant
		The expression of VPA is abundant
		The action and posture of VPA are abundant
		The voice quality of VPA is abundant
		The voice speed regulation of VPA is abundant
		The personality orientation of VPA is abundant
		The Interactive script of VPA is abundant
Interaction	Voice interaction experience	Wake-up Rate
		Wake-up Speed
		Wake up Easily
		Recognition Accuracy
		Comprehension Accuracy
		Feedback Quickly
		Feedback Effectively
		Multi-region Speech Recognition
		Keep Listening
		Continuous Dialogue
		Active Voice Interaction
Emotional	Emotional experience	Emotional Expression (Focus on body)
		Emotional Expression (Focus on content)
		Emotional Broadcasting (Focus on voice)
		Emotion and broadcast match accurately

(*continued*)

Table 7. (*continued*)

First-level indicators	Second-level indicator	Third-level indicators
		Personified Identity
		Sense of Trust
		Sense of Individualistic
Design	Design experience	Innovativeness
		Design Integration
		Design Style
		Matching of design and brand
		The interface of VPA looks good(Focus on position)
		The interface of VPA looks good(Focus on size)

After adjustment, the classification of the model was more clear and more intuitive, consistent with the actual logic, and easy to understand. To reflect the differences in the importance of each dimension, we invited 15 vehicle VPA experts (including 8 designers, 5 product managers, and 2 marketing personnel; 7 males and 8 females, who were engaged in the design of vehicle VPA more than 1 year)to assign weight to the model. According to expert data, based on AHP (Analytic Hierarchy Process,) calculation, we found that the five most important dimensions that affect the vehicle VPA user experience were *Voice of VPA Coverage Span (weight value 13.02%), Comprehension Accuracy (weight value 10.24%), Recognition Accuracy (weight 7.59%), Sense of Trust (weight 5.61%), Satisfaction of Resource Content(weight 5.52%).*

3 Summary and Discussion

3.1 Summary

Pathway for this study: 1) By focusing on domestic Intelligent vehicle desktop research and in-depth interviews, extracted the perception information of Chinese vehicle owners and constructed a vehicle VPA evaluation model; 2) Discussed and optimized the model in combination with the actual product form; 3) Then, through large-scale data (collection of China's emerging new energy vehicle owners' data) to quantify and complete the adjustment of the evaluation model, and to confirm the weight of the evaluation model through expert empowerment scoring.

On the whole, the model construction method was rigorous and the process was clear. We have obtained a vehicle VPA evaluation model focusing on real vehicle owners, which included five modules: *Sensory, Resources, Interaction, Emotional, and Design.* The weight calculation extracts the dimensions that highly affect the user experience as *Voice of VPA coverage span, Comprehension Accuracy, Recognition Accuracy, Sense of*

Trust, and Satisfaction of Resource Content, which provided powerful tool support for the evaluation and design of the subsequent vehicle VPA.

3.2 Design Inspiration

Vehicle VPA is an important interface between drivers and intelligent vehicles, and it will become more and more common in our life. According to the index system of the model, we found that the vehicle VPA was not a single visual image in the user's mind, but also consists of voice capabilities, functions, resources, and other modules. Therefore, in future designs, automobile manufacturers and automobile digital service providers can evaluate its experience from five aspects as *Sensory, Resources, Interaction, Emotional, and Design.*

According to the weight of each dimension, we found that *Comprehension Accuracy, Recognition Accuracy* would greatly affect people's evaluation of vehicle VPA. Based on the actual situation of vehicle users, if the VPA cannot understand the user's words and give users the expected feedback, then its beautiful image is unattractive. For example, although the overall image of Li One's VPA was very simple, it had won praise from users by its high comprehension accuracy, high recognition accuracy and humanistic feedback.

Concurrently, many studies have shown that VPA with human images was more trusted by users. For example, relevant experiments on VPA trust showed that in the driving situation, people, robots, and abstract images were the most suitable VPA images, with the highest degree of perceived trust, while animal and mechanical images perception trust was lowest. The XPENG'S VPA in this study often appeared as a racing driver, coupled with its very individual actions, so the satisfaction of trust is relatively high, which inspired automobile manufacturers in terms of VPA image design.

Furthermore, in the driving situation, VPA is usually associated with many functions (vehicle control, navigation, etc.) and network resources (multimedia, life services, etc.) The experience of functions and resources will also be migrated to the VPA experience in the user's mind. In this model, *Voice of VPA coverage span* and *Satisfaction of Resource Content* had higher weight, indicating that they had a higher weight in users' evaluation of VPA. Therefore, the preferable VPA experience needs to consider its integration with vehicle functions, as well as the richness of resources, whether resources meet user needs and the integration of resources and vehicle VPA.

3.3 Discussion

The purpose of this study is to explore the evaluation models, which suitable for emerging intelligent vehicle VPA based on Chinese vehicle owners. It is hoped that it is reasonable, representative, and can help business positioning problems. In the study, we found that different user groups have different expectations for vehicle VPA, for example, young users and groups living Tier 1 cities have low satisfaction with vehicle VPA. They paid more attention to the dimensions such as voice interaction experience and emotional experience. The next research will focus on different types of vehicle VPA user groups, and explore their differences in vehicle VPA interaction and consumption patterns through in-depth, qualitative, and exploratory research methods. Thereby, providing more deep and detailed insights into design and product marketing. This suggested information will be used to improve user experience and business efficiency of vehicle VPA.

References

1. Automotive Data of China, Prospective Vehicle Configuration. https://mp.weixin.qq.com/s/5FHteIX3gmeUy6aO8tCw0Q. Accessed 20 Sept 2022
2. iResearch, Research Report on China's Intelligent Voice Industry. https://www.iresearch.com.cn/Detail/report?id=3526&isfree=0. Accessed 12 Oct 2022
3. Euronet Alliance. Research on Intelligent Function Module Series of National Automobile - Voice Interaction. https://www.iyiou.com/research/20220318982. Accessed 09 Oct 2022
4. Hu, K.: "prototype" empowerment_ The "personality" setting of virtual anchors from the perspective of psychology. Broadcasting Realm (5), 32–37 (2020)
5. Chen, S.Y.: Research on vehicle robots anthropomorphic design. Res. Expl. 137–139 (2022)
6. Fang, H.: Research on voice assistant personality design based on car emotion (2020)
7. Mo, Y.F.: Design and research of on-boardhuman-computerinter action system based on situational awareness (2020)
8. Shi, L.X., Song, S.: Research on the Service Quality Scale of Virtual Personal Assistant. http://www.paper.edu.cn. Accessed 25 Aug 20225
9. Wei, Q.Y., Li, Y.T., Yao, Y.: Research on the user experience evaluation index system of mobile library. J. Natl. Lib. Chin. (119), 21–31 (2018)
10. Niu, L.: Assistant or partner? Research on users' preferences of intelligent voice products --From the perspective of gender stereotyoes (2021)
11. Zhang, B.J.: Analytic Hierarchy Process and Its Application Cases. Beijing Electronic Industry Press (2014)
12. Huang, Q.H., Mao, G.Y.: A Deep dive into the China's gen z: how they use and what they expect for their cars. In: Krömker, H. (ed.) HCII 2022, LNCS, vol. 13335, pp. 167–183 (2022). https://doi.org/10.1007/978-3-031-04987-3_11
13. Saad, U., Afzal, U., El-Issawi, A., et al.: A model to measure QoE for virtual personal assistant. Multimedia Tools Appl. (2018)

Demonstrating a V2X Enabled System for Transition of Control and Minimum Risk Manoeuvre When Leaving the Operational Design Domain

Joschua Schulte-Tigges[1]([envelope])[iD], Dominik Matheis[2]([envelope])[iD], Michael Reke[1]([envelope])[iD], Thomas Walter[2]([envelope])[iD], and Daniel Kaszner[2]([envelope])[iD]

[1] FH Aachen University of Applied Sciences, Aachen, Germany
{schulte-tigges,reke}@fh-aachen.de
[2] Hyundai Motor Europe Technical Center GmbH, Ruesselsheim am Main, Germany
{dmatheis,twalter,dkaszner}@hyundai-europe.com

Abstract. Modern implementations of driver assistance systems are evolving from a pure driver assistance to a independently acting automation system. Still these systems are not covering the full vehicle usage range, also called operational design domain, which require the human driver as fall-back mechanism. Transition of control and potential minimum risk manoeuvres are currently research topics and will bridge the gap until full autonomous vehicles are available. The authors showed in a demonstration that the transition of control mechanisms can be further improved by usage of communication technology. Receiving the incident type and position information by usage of standardised vehicle to everything (V2X) messages can improve the driver safety and comfort level. The connected and automated vehicle's software framework can take this information to plan areas where the driver should take back control by initiating a transition of control which can be followed by a minimum risk manoeuvre in case of an unresponsive driver. This transition of control has been implemented in a test vehicle and was presented to the public during the IEEE IV2022 (IEEE Intelligent Vehicle Symposium) in Aachen, Germany.

Keywords: V2X · Transiton of Control · Minimum Risk Manoeuvre · Operational Design Domain · Connected Automated Vehicle · Infrastructure Assisted Driving · Vehicle HMI

1 Introduction

A modern car is equipped with a variety of safety and comfort features. Most of the safety and comfort features from the last years are Advanced Driver Assistance Systems (ADAS). Based on sensor information, these systems support the driver (up to SAE Level 2). To achieve level 3 or higher levels of the SAE standardized levels of automated driving [8], the control capabilities of those cars

H. Krömker (Ed.): HCII 2023, LNCS 14048, pp. 200–210, 2023.
https://doi.org/10.1007/978-3-031-35678-0_12

have to increase. The biggest problem is not following a vehicle while keeping inside of a lane. More of a problem are sudden and unexpected events that can occur, like construction sites or accidents. One approach to handle such events are wireless transmitted location information by the infrastructure to the vehicle via a wireless interface. This is called a vehicle to infrastructure communication, I2V, which is a subset of V2X (Vehicle to everything). In recent years, researchers and industries have been working and are still working on standardisation of V2X services [5]. First V2X enabled vehicles have been introduced in different markets using standardized message sets. Future automated cars should be able to cooperate with each other, via V2X communication, which additionally could be used to guide it. To enable the V2X supported vehicle control, a Framework to collect sensor data and vehicle behaviour planning is required. The University of Applied Science Aachen, in cooperation with the Hyundai Motor Europe Technical Center (HMETC), built up two vehicles to build a common platform for automated driving research and development. The underlying software structure, is based on the architecture proposed in [14]. The joint work presented in this paper is part of the European Hi-Drive project [1], where a set of different cross border V2X scenarios are developed and tested by various project partners.

The scope of the Hi-Drive Project is to advance the state of the art of automated driving technologies, by testing and evaluating demonstrators for high automation functions. The first generation of automated vehicles must manoeuvre in complex and challenging environments of mixed traffic. Such vehicles will be able to drive within their specified operational design domain (ODD) [9]. But what is to do when the vehicle is about to leave the operational design domain it is designed for? If a Level 3 automated vehicle leaves its specified operational design domain, the driver of the vehicle should take back the control over the car. This process is called Transition of Control (ToC), where the car will first initiate a Takeover Request (ToR). In the proposed demonstrator the vehicle should alert the driver for a defined time period of 10 s (e.g. [11]). In this 10 s, the car ensures to be able to further drive automated. If the driver does not react, the vehicle has to ensure that it is capable to come to a safe stop. Making a safe stop can be e.g. to drive to an emergency lane. This process is called Minimum Risk Manoeuvre (MRM) as proposed in an earlier proof of concept by [6].

A demonstrator was developed to show how V2X can help automated vehicles with this transition by informing the vehicle upfront about a traffic situation. The vehicle receives the information via its V2X interface, so the vehicle's software can decide if it will leave its ODD. The architecture of the base automated driving software, developed for the demonstrator, was extended to make decisions after a V2X advice was received. The standardised ITS message for Decentralised Environmental Notifications (DEN) [3] was used to populate information about upcoming traffic situations like traffic jams, accidents etc. Further, the software had to extract the information from the V2X message and control the manoeuvres of the vehicle, e.g., making lane changes and stopping.

2 Related Work

Assisting Connected Automated Vehicles (CAV) by supporting infrastructure, when leaving their ODD, was part of the research in the European TransAID ('Transition Areas for Infrastructure-assisted Driving') project. The TransAID deliverable D7.2 [16] describes different use cases in which the infrastructure supported CAVs to avoid a ToR or inform the CAV that it is about to leave its ODD. Use case 4.1 presented earlier demonstrators that were evaluated by [6] and is a foundation of the demonstrator described in this paper. The main focus here was to evaluate if the triggering of ToRs can be distributed, to tackle scenarios where multiple CAVs are about to leave their ODD and trigger ToRs at the same time. By providing a designated safe spot for every car, the team of [6] were able to show a safer execution of the MRM.

Based on the description by [10], a transition, in the context of automated driving functionality, is defined as *'The process of changing from one static state of driving to another static state'*. Hence, a state of driving can be an ADAS function (like ACC) and another higher state can be a more advanced automated driving function with less driver control. In this paper, the ToC process will just transition between automated driving and human/driver control, like [6,7,12, 15,17] using the ToC process to transition between automated driving as either on or off. The whole process of executing a ToC has to be modelled first, where [12] proposed a state machine modelling scheme for. By using simulation, the team behind [12] evaluated the effects of the ToC and MRM process in mixed traffic work zones.

Where [6,15,17] also show a prototype demonstrator of a CAV that is about to leave its ODD and therefore does a ToC. Their demonstrators focused on the evaluation of the feasibility of infrastructure aided transition areas. In this setup the infrastructure advises the vehicle to make a ToC instead of the CAV deciding by itself. That is where the here stated demonstrator distinguishes from others, the infrastructure does transmit a notification and the Automated Driving Software does have a module to decide how to handle in the upcoming situation.

3 Demonstrator Setup

The demonstrator was developed and evaluated on a proving ground near Darmstadt Griesheim. It was publicly shown on the proving ground Aldenhoven Testing Center (ATC) near Aachen during the IEEE IV'22 [2]. The demonstrator shows a highway driving scenario, where the proving ground segments were chosen to have at least two or more driving lanes and an emergency lane, to emulate a real highway.

In the beginning of the demonstration, the driver drives manually. When the CAV is ready to drive automated the HMI (see sect. 3.1 for further details) indicates that the automated driving function is available (step 1 in Fig. 1). The driver enables the system and the CAV starts driving automated (step 2

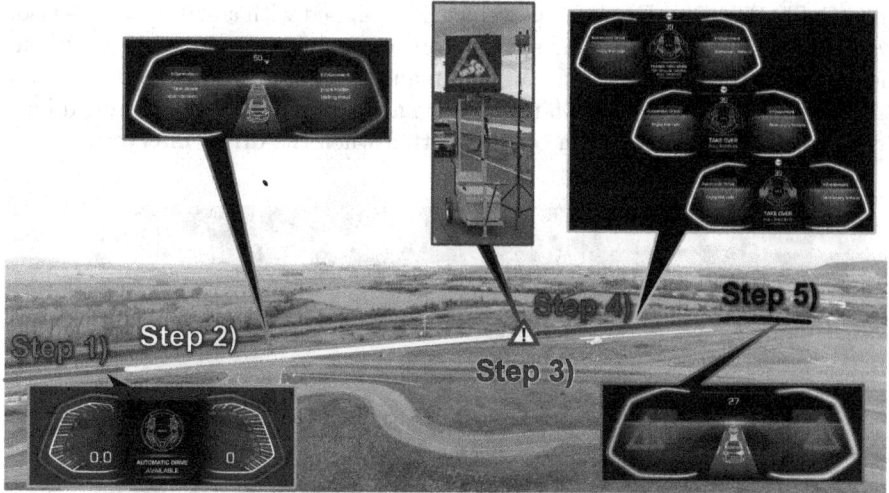

Fig. 1. Showing the demonstrator setup on the Aldenhoven Testing Center, with the different steps in the demonstrated scenario.

in Fig. 1). The CAV is driving automated continuously and will receive a notification message (DENM) via V2X from the road side unit (RSU) as it drives towards the variable message sign (VMS), where the RSU is mounted on. The vehicle's manoeuvre deciding module processes the information from the DENM, which contains a warning of a stationary vehicle (step 3 in Fig. 1, VMS and RSU). Arriving at the event position of the DENM based on the information transmitted to the CAV, the automated driving software (ADSW) identifies that the CAV will leave its ODD and starts the ToC (step 4 in Fig. 1). For ten seconds, the HMI will keep alerting the driver with a ToR in three cascaded more and more obtrusive audiovisual steps (step 4 in Fig. 1). In case, the driver does not respond to the ToR within the ten seconds, the ToC fails and the CAV will start to perform a MRM (step 5 in Fig. 1). The ADSW checks in the map, if an emergency lane is available, if not it will initiate the MRM right away by slowing down and stopping within the actual driving lane. If an emergency lane is available the CAV checks for a free spot without any occupying traffic. It starts to change to the emergency lane and to slowly stop safely at the free spot. The CAVs remains in this safe position with hazard warning lights turned on until the driver takes over again.

3.1 Vehicle

The vehicle (shown in Fig. 2) used in the demonstration was retrofitted with sensors and other hardware, to cover the use cases during the Hi-Drive project, and is totally separated from KIA's serial development. The vehicle is equipped with the required control units and sensors for automated driving. For localisation, the vehicle uses a precise GNSS unit including an inertial measurement unit (OxTS RT3000). Further, the vehicle is equipped with a 360°C radar sensor system utilising multiple radar sensors to sense information about its environment. Installed on the roof is a V2X antenna connected to an On-Board-Unit (OBU) to receive ETSI ITS-G5 messages. The prototype can be controlled by a drive-by-wire system, that can be overwritten when the driver intervenes.

Fig. 2. Vehicle used as automated driving platform (KIA Niro).

The original instrument cluster was replaced, and a new cluster was developed for prototype use cases within the project. The new instrument cluster was developed to give the driver information from the automated driving software. For example, the cluster shows other detected vehicles and visualises lane changes. To activate the automated driving functions, the steering wheel pedal shifters of the car were repurposed to function as activation buttons. Figure 3 shows the driver pulling both pedals, after the cluster signalled, that automated driving is available. For three seconds the cluster shows a visual indication of the activating progress.

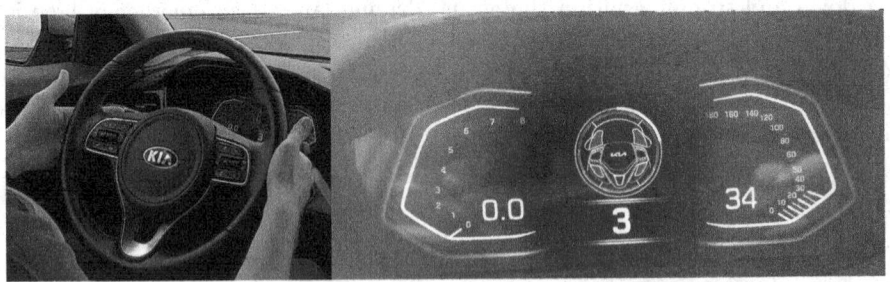

Fig. 3. Driver activates Automated Driving Function.

3.2   Software Architecture

The automated vehicle is controlled by the software architecture developed by a research team of the University of Applied Sciences, Aachen [14]. It was extended to make decisions and handle different manoeuvres. A software module here called, manoeuvre decider, is a state machine that was used to design the decision-making for different scenarios. This manoeuvre decider creates a sub-goal. A sub-goal consisting of three parameters: a specific driving lane, a specific speed and information for the driver. Based on the sub-goal another software module called motion planner has to make manoeuvres to reach the specified lane and the specified speed, e.g. lane changes or speed adaptations.

Figure 4 shows a simplified representation of the software architecture.

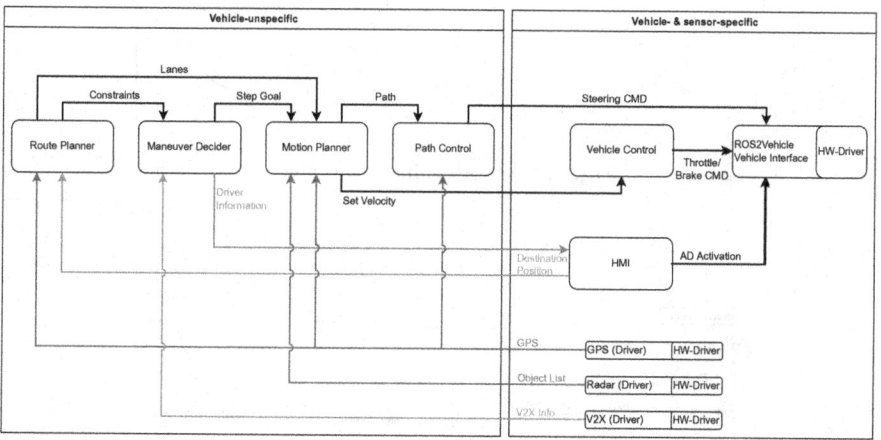

Fig. 4. Automated Driving Software Architecture.

The user interacts with the HMI, where he can select a destination and a maximum speed. Once a destination is selected, the route planning module generates a route to the destination, containing road segments that can consist of multiple lanes and constraints like speed limits. For the route planning, the map format Lanelet2 [13] is used.

When messages are received via V2X, the information is passed to the manoeuvre decider module. A small snipped of the underlying state machine is shown in Fig. 5, where every scenario is a state machine on its own and some of them can trigger a MRM. In the demonstrator the CAV receives a DENM warning about a stationary vehicle. Figure 6 shows the state machine for this stationary vehicle scenario. When receiving the warning about a stationary vehicle, the CAV first should slow down to a predefined safe speed. According to the DENM, if the message contains information about the location of the stationary vehicle, the CAV should change to a designated lane where it can continue driving safely and should keep that lane until the event ends. If the message does not contain information about the stationary vehicle's position, the CAV is about to leave its ODD, will start a ToC and will therefore execute a ToR. In the case the driver does not take over control during 10 s, the state transitions to the MRM.

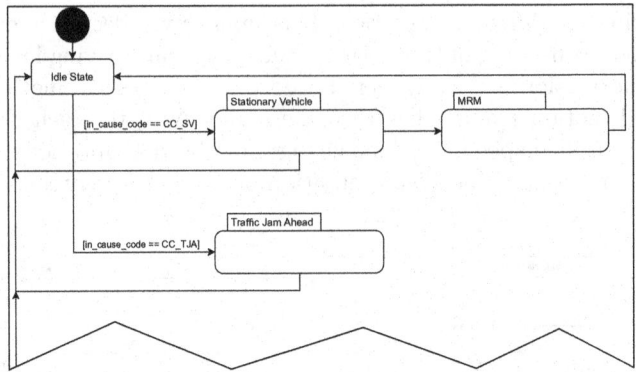

Fig. 5. State Machine Overview

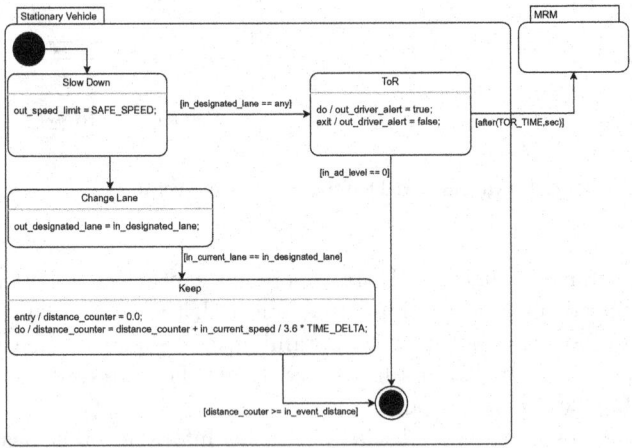

Fig. 6. Stationary Vehicle Sequence

Figure 7 shows the state machine for the MRM. First, the CAV should again reduce its velocity. If no emergency lane is available, the CAV should stop immediately, else the CAV should first change to the emergency lane before stopping. The CAV should remain stationary until the driver takes over.

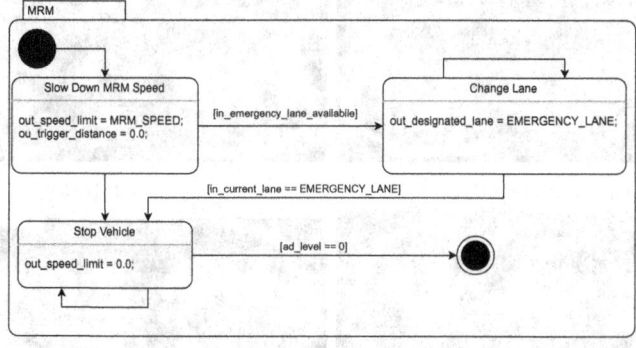

Fig. 7. MRM Sequence

Each state in the manoeuvre decider module creates a sub-goal in the ADS The motion planner module has to execute those sub-goals. The motion planner module has information about the surrounding environment of the CAV in the form of an object list. Sub-goals are executed within the environmental restrictions, for example a lane change should only be executed when a gap exists that is big enough to make the lane change safely. The motion planner module creates a path and a set speed for the vehicle controller.

Two controllers execute the planned path and velocity-set-values. The lateral controlling from [4] is used to calculate steering commands, while the longitudinal controller calculates throttle and brake commands, that are passed to the drive by wire module by the vehicle interface.

4 Evaluation and Discussion

The demonstrator was validated and demonstrated in a real world environment on a highway like proving ground, at the Aldenhoven Testing Center (ATC) near Aachen during the IEEE IV'22 [2]. It was able to show that V2X can help if a vehicle is about to leave its ODD. Further, the demonstrator proved that the software architecture was able to handle the scenario and to stop the vehicle safely when the driver does not take over.

As seen in Sect. 3.1 the cluster instrument was retrofitted to visualise information from the automated driving software. For the takeover request, three different phases were implemented, each with a duration of about 3.3 s. Figure 8(a-c) shows the three different phases for the ToR. During the ToR a audible beeping sound started that got more intrusive for each phase. Figure 8(d) shows the cluster instrument when the vehicle executing an MRM, intentionally without asking the driver to take over any more.

During the presentation guest were allowed in the vehicle (two guests per ride, a ride consisting of two times the described scenario). A safety driver was driving, the guests were not allowed to drive the car, hence they were not able to perform a ToC themselves. Impressions of the guest were mostly positive, they

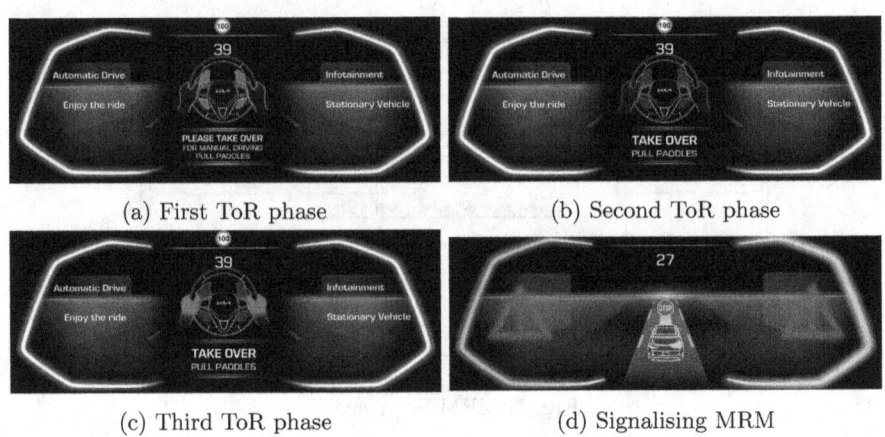

(a) First ToR phase   (b) Second ToR phase

(c) Third ToR phase   (d) Signalising MRM

Fig. 8. Cluster instrument visuals for ToR and MRM

described the ride quality as very good and the smooth lane change and stopping were often acknowledged in particular. Guest further gave the impression to be comfortable at any time trusting the system in what it was doing. This paper should point out, if a vehicle has already left its ODD it is too late for a safe transition, so a vehicle should be able to decide upfront. This paper showed a possible solution by transmitting notifications via V2X to the CAV upfront and let the vehicle decide how to handle the situation.

5   Conclusion

This paper described how V2X technology can help automated vehicles when leaving their intended operational design domain. A use case was presented in which a CAV receives a decentralised notification message containing information about a stationary vehicle, causing the automated vehicle to leave its ODD. The authors of this paper recommend the following strategy to avoid further vehicle incidents: A transition of control back to the driver should be initiated by an automated vehicle when leaving the operational design domain. In case of an unresponsive driver (no control take over), the vehicle should start a minimum risk manoeuvre. The presented software architecture supports the described use case, by e.g. taking decisions and controlling the vehicle until the full stop. The demonstrator was also equipped with a human machine interface to show the current system state to the driver.

After receiving the V2X messages the CAV triggered a takeover request and the driver was informed to take over control by audiovisual notifications for a time period of ten seconds. The demonstrator showed that in case that the transition of control failed, the CAV was able to execute a minimum risk manoeuvre consisting of a lane change to the emergency lane and a full stop.

The authors received a lot of positive feedback during the demonstration from drivers and guests. Visitors agreed to the approach of a warning cascade, which still allows the driver to intervene before entering the phase of the minimum risk manoeuvre. Further research on the warning timings and the way how the driver will be informed about the take over request, are subject of future research. A haptic acknowledgement (e.g. pressing a button) of the driver to show full awareness is mandatory.

Acknowledgement. The Hi-Drive project has received funding from the European Union's Horizon 2020 research and innovation programme under grant agreement No 101006664.

References

1. Addressing challenges toward the deployment of higher automation hi-drive project h2020 cordis european commission. https://cordis.europa.eu/project/id/101006664. Accessed 20 Oct 2022
2. Vehicles for the demonstration day - ieee intelligent vehicle 2022. https://iv2022.com/sponsoring/vehicles-for-the-demonstration-day. Accessed 02 June 2023
3. Intelligent transport systems (its); vehicular communications; basic set of applications; part 3: Specifications of decentralized environmental notification basic service. ETSI 1.2.1 (2014)
4. Chajan, E., Schulte-Tigges, J., Reke, M., Ferrein, A., Matheis, D., Walter, T.: Gpu based model-predictive path control for self-driving vehicles. In: 2021 IEEE Intelligent Vehicles Symposium (IV21). IEEE (2021)
5. Chen, L., Englund, C.: Cooperative its - eu standards to accelerate cooperative mobility. In: 2014 International Conference on Connected Vehicles and Expo (ICCVE). pp. 681–686 (2014). https://doi.org/10.1109/ICCVE.2014.7297636
6. Coll-Perales, B., et al.: Prototyping and evaluation of infrastructure-assisted transition of control for cooperative automated vehicles. IEEE Transactions on Intelligent Transportation Systems pp. 1–17 (2021). https://doi.org/10.1109/TITS.2021.3061085
7. Correa, A., et al.: Management of transitions of control in mixed traffic with automated vehicles. In: 2018 16th International Conference on Intelligent Transportation Systems Telecommunications (ITST). pp. 1–7. IEEE (2018)
8. Driving, A.: Levels of driving automation are defined in new sae international standard j3016: 2014. Warrendale, PA, USA, SAE International (2014)
9. Driving, A.: Taxonomy & definitions for operational design domain (odd) for driving automation systems j3259. SAE International (2021)
10. Lu, Z., de Winter, J.C.: A review and framework of control authority transitions in automated driving. Procedia Manuf. **3**, 2510–2517 (2015)
11. Melcher, V., Rauh, S., Diederichs, F., Widlroither, H., Bauer, W.: Take-over requests for automated driving. Procedia Manuf. **3**, 2867–2873 (2015)
12. Mintsis, E., et al.: Joint deployment of infrastructure-assisted traffic management and cooperative driving around work zones. In: 2020 IEEE 23rd International Conference on Intelligent Transportation Systems (ITSC). pp. 1–8. IEEE (2020)

13. Poggenhans, F., et al.: Lanelet2: a high-definition map framework for the future of automated driving. In: 2018 21st International Conference on Intelligent Transportation Systems (ITSC), Maui, HI, USA(2018), https://www.mrt.kit.edu/z/publ/download/2018/Poggenhans2018Lanelet2.pdf

14. Reke, M., et al.: A self-driving car architecture in ros2. In: 2020 International SAUPEC/RobMech/PRASA Conference. pp. 1–6 (2020). https://doi.org/10.1109/SAUPEC/RobMech/PRASA48453.2020.9041020

15. Schindler, J., Coll-Perales, B., Zhang, X., Rondinone, M., Thandavarayan, G.: Infrastructure-supported cooperative automated driving in transition areas. In: 2020 IEEE Vehicular Networking Conference (VNC). pp. 1–8. IEEE (2020)

16. Schindler, J., et al.: Transaid deliverable 7.2: System prototype demonstration (iteration 2) (2021)

17. Schindler, J., Markowski, R., Wesemeyer, D., Coll-Perales, B., Böker, C., Khan, S.: Infrastructure supported automated driving in transition areas-a prototypic implementation. In: 2020 IEEE 3rd Connected and Automated Vehicles Symposium (CAVS). pp. 1–6. IEEE (2020)

Research on Sound-Guided Design of Lane Departure Scenarios in Intelligent Car Cockpit

Yu Wang[1,2](✉), Jie Wu[2], and Hanfu He[2]

[1] School of Pop Music and Dance, Shanghai Institute of Visual Arts, Shanghai, China
asterwangyu@126.com
[2] College of Design and Innovation of Tongji University, Shanghai, China

Abstract. With the advancement of science and technology, the cabin of a car is becoming increasingly intelligent. The design of audible, intelligible, conductive and useful sounds in the audio-visual design of various interfaces in the cabin of a car, to enable driving safety and experience, has become a new design issue in the context of future smart mobility. Through literature research and analysis, this paper summarises the human factors parameters and design methods for the design of vehicle guidance sound. This study is based on the lane departure warning scenario of the car's intelligent cockpit and investigates the sound design in this scenario. We conducted a case study of the warning cues of four mainstream car brands in this scenario and explored the design method of the cueing sound in the lane departure warning scenario.

Keywords: Sound Design · Intelligent Car Cockpit · Lane Departure Warning

1 Introduction

With the progress and development of science and technology, the car has become the norm in people's lives and the cabin of the car is becoming increasingly intelligent. The intelligent cockpit gives the driver a new type of interactive driving experience. With the development of various types of interactive interfaces in the intelligent cockpit and the evolution of automated driving technologies, the driver often needs multimodal sensory information rather than single-channel sensory information to quickly receive and understand the information at hand and to improve driving performance and safety in human-machine driving situations in order to achieve a rational distribution of attention.

The literature suggests that the auditory system can compensate for obscured objects that the visual system cannot detect, objects that are not visible in the dark and objects that are out of sight. The auditory senses are the second most important senses after the visual senses, being more sensitive than the visual senses at the level of the time-domain dimension, and therefore the auditory senses have an advantage when people are confronted with the integration of audiovisual information [1]. The combination of the audio-visual channels allows one to selectively ignore the transmission of information through the visual channel. In certain situations, people can choose to transmit the necessary information only through the auditory channel [2].

H. Krömker (Ed.): HCII 2023, LNCS 14048, pp. 211–224, 2023.
https://doi.org/10.1007/978-3-031-35678-0_13

Therefore, according to the information integration and processing characteristics of human beings, the incorporation of auditory interaction elements in the various warning scenarios of the intelligent cockpit of the car, and the design of information through the sound channel to complement the information transmission through the visual channel, allows the user to reduce the perceptual load on the visual channel and increase his trust in the driving system. The integration of interactive information from both the audio and visual channels is more effective in alerting drivers to signals of danger. In situations where the level of danger is high, we can design the driver to be alerted by both audio and visual channels.

Some studies have shown that auditory warning signals have an effect on driver behavior and reactions during unforeseen events, and that designing auditory warnings with a high sense of urgency is effective for driver safety judgement and feedback [3]. Properly designed in-cockpit warning tones can help drivers determine the current vehicle status and road conditions accurately and quickly, reducing the likelihood of driving accidents [4, 5]. However, poorly designed in-cockpit warning tones can have a startling effect on the driver, causing a range of negative driving effects. Depending on the sound characteristics provided by the in-cockpit audible warning system, the driver's behavior and driving performance may vary. How to design audible, intelligible, informative and useful sound in the audio-visual design of the diverse interfaces of the car's intelligent cockpit to empower driving safety and experience has become a new design issue in the context of future smart mobility.

Among the application scenarios for intelligent cockpit automated assisted driving technology, the lane departure warning scenario is one that is used more frequently by drivers, with a higher degree of urgency in this scenario. Different brands of cars have also designed different warning guidance tones based on this application scenario to assist users in making safe and efficient driving behavior judgements, but this has also created new problems and needs. There is an urgent need to address the issue of how to enable drivers to receive and understand the meaning of the warning tones in this scenario more effectively with minimal learning costs, to improve their driving safety performance and to summarise the design principles of the warning tones in lane departure scenarios.

This paper defines the problem of guided sound in automotive intelligent cockpits and investigates the principles of sound information processing in multimodal sensory interaction. Based on the lane departure warning scenario, an auditory experience design strategy is explored that incorporates both visual and auditory information. The human factors parameters and relevant cases of the guiding sound are analysed, and a set of guiding sound design principles and methods for the lane departure warning scenario in the automotive intelligent cockpit are designed and summarised.

2 Materials and Methods

There are two main types of guidance tones in a smart cockpit, the first is a tone used to convey UI interaction information, such as switch on/off. The second type is feedback related to the driver's driving of the vehicle, e.g. driving hazard sounds.

Lane Departure Warning (LDW) is an advanced driver assistance system (ADAS) that is available in many newer vehicles. It warns drivers when they are unconsciously leaving the lane their vehicle is travelling in, for example by providing visual information via a colour change of the lane lines on the HMI or HUD, or by emitting a warning sound in the cockpit, or by vibrations from the seat belt and steering wheel.

LDW can reduce some single-vehicle sideswipes and head-on collisions. When these types of crashes do occur, LDW can reduce the incidence of injury. LDW is becoming more common in the automotive industry. Some OEMs are even making it a standard feature on many models.

Lane departure warning systems are generally visual devices that calculate the predicted moment of lane departure by looking at lane line markings and alert the driver when an unintended lane departure is about to occur, thus avoiding false warnings due to subtle lateral lane position changes. Starting with simple lane line detection video, LDW has evolved into sophisticated lane marking recognition and lane boundary projection systems that warn drivers if a vehicle's trajectory is about to deviate from the lane. While most LDWS use video technology, sound, infrared, LIDAR, magnetic and electronic mapping technologies are also being researched and developed. LDW gives an audible, visual or acoustic warning when a car crosses or approaches a lane boundary and the driver fails to activate the turn signal.

In Lane Departure Context (LDW) scenarios, we usually design guide tones to deliver the warning message. Figure 1 shows a model of the processing of auditory information.

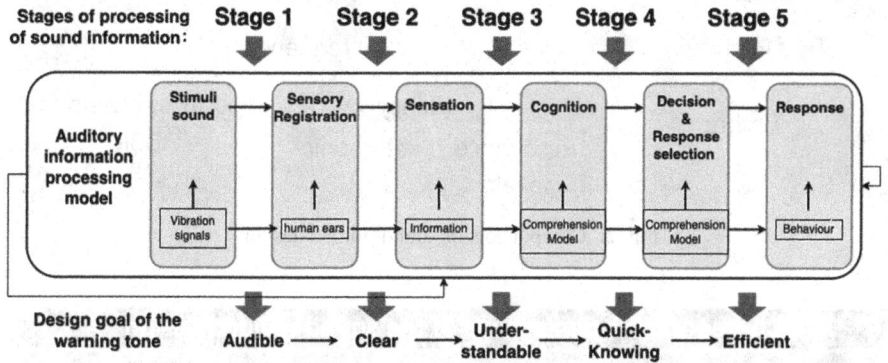

Fig. 1. Auditory information processing model (from[1])

According to the literature, poorly designed timbre characteristics, loudness and their delivery mechanisms may lead to cognitive overload, perceptual frustration and loss of primary attention [3]. During auditory perceptual processing, individuals are unable to consciously perceive and understand all sensory stimuli simultaneously and have developed the ability to separate the flow of information, i.e. audiovisual asynchronous processing.

Therefore, taking into account the five stages of sound information processing and the perceptual characteristics of sound, we set the objectives for the design of interactive sound in the automotive cockpit as: audibility, intelligibility, intelligibility and usefulness. In the different scenarios of an intelligent cockpit, we need to design sounds that can be received and heard clearly by the driver (i.e. audibility), so that the driver can quickly understand the information conveyed by the sound with minimal learning costs (i.e. intelligibility) and, later, so that the driver can make quick judgements and actions (i.e. intelligibility and usefulness), while at the same time, we need to consider the auditory comfort of the audible sound. *Simultaneously*, we also need to consider the auditory comfort of the guidance sound.

In the warning scenario of lane departure in a car smart cockpit, we use the sound of the Warning category for guidance [4].

The warning and intelligibility (the functional semantics of the sound) of the guide sound is related to the constituent elements of the sound and the form of its arrangement. Figure 2 and Fig. 3 show the details.

a) Interactive sounds with higher fundamental frequencies and higher loudness are more likely to alert audiences and thus get their attention more quickly.
b) Sounds with shorter duration and higher cyclic frequency convey more urgent functional semantics, thus prompting the audience to quickly avoid danger.

Components of sound-guided design			
Psychological elements	Physical elements		
Loudness	Sound intensity		
Pitch	Frequency		
Timbre	Frequency		Loudness envelope ADSR
	Resonance peak	harmonic	

Fig. 2. Components of sound-guided design

Choreographic dimensions of the sound-guided design					
Time Domain			Frequency domain		Space Domain
duration	Loudness	Number of cycles	Pitch	Timbre	Sound Panning

Fig. 3. Choreographic dimensions of the sound-guided design

We have divided the design of the guidance sound for the lane departure warning scenario in the vehicle intelligent cockpit into the following levels.

The first level is the sound physical property level, the human factors parameters in this level mainly include: frequency; Sound intensity; sound panning (spatial location of the sound).

The second level is the interaction mode of the sound. The design factors of this level include the triggering mode of the sound in the collision warning scenario and the urgency of the sound, where the triggering mode mainly includes the trigger time and the trigger conditions.

The third level is the sound design method level, where the design factors include the morphological characteristics of the sound signal, the duration of the sound, the number of cycles and the degree of urgency.

We have studied the international standard literature published by NHTSA, ISO and other organisations [6–9], and summarised the design methods of the guidance sound in lane departure warning scenarios from the physical and psychological aspects of the vehicle intelligent cockpit.

3   Results

We have studied the literature on international standards published by organisations such as NHTSA and ISO at three levels: the physical properties of sound, how sound interacts and how sound is designed, and the following are the results obtained.

3.1   Design of the Physical Properties of the Guide Sound

Frequency

1) The recommended frequency range for in-vehicle auditory signals is 200 Hz to 8000 Hz and for tonal signals the predominantly audible frequency component should be between 400 Hz and 2000 Hz, avoiding exceeding 2000 Hz in order to be inclusive to accommodate elderly populations or the hearing impaired.
2) The audible warning fundamental frequency should be at least in the range of 500–1500 Hz.
3) When using frequency as a code to differentiate warning levels, the chosen fundamental frequency should be widely distributed in the range 200 Hz–3000 Hz and not concentrated in only a small range.
4) The frequencies selected should be those least affected by ambient noise. If frequencies are used to code the warning level, the imminent collision avoidance warning will have the highest fundamental frequency.
5) If a single sound is used for a collision avoidance warning, a composite sound should be used instead of a pure sine waveform.
6) A broadband signal or a mixture of narrowband signals with clearly separated centre frequencies should be used to improve signal position detection and driver attention direction.

Sound Intensity

1) The recommended audible sound pressure level range for all driving conditions is between 50 and 90 dB(A), with an ideal minimum acoustic signal of 75 dB(A) depending on the parameters set for the playback side of the vehicle.

2) At the driver's ear, the default intensity value of the sound warning should be at least 20 dB but not more than 30 dB above the threshold for shielding based on ambient noise under relatively noisy operating conditions.

3) The onset speed of the sound or tone used in the collision prevention warning should be fast enough to alert the driver. However, the onset speed should not be too fast, as this could cause a severe startle effect. It is recommended that the onset speed of the sound is greater than 1 dB/ms but less than 10 dB/ms. To avoid the startle effect of the sound, the steep increase in sound should be less than 30 dB increments over 0.5 s.

4) For critical warning sounds, the time from the start of the sound head to the maximum loudness should be less than 30 ms. This is mainly to way the sound head is blurred, resulting in a poor audible warning.

5) In addition, sound intensity should not be used to differentiate the level of urgency of a warning due to poor perception and judgement of sound intensity.

Sound Panning

1) Sound Panning is a human factor parameter that reflects the spatial physical properties of sound, i.e. information about the orientation, distance and proximity of the sound. It is known from the Simon effect that when stimulus and response are spatially compatible (on the same side), the subject responds faster than in the case of incompatibility. Therefore the auditory stimulus has to be spatially compatible with the indicated spatial hazard information. If there is a hazard on the right and the orientation information of the tone points to the right, the driver will respond faster.

2) Where appropriate, the warning should elicit a directional response, causing the driver to look in the direction of the hazard. The source of the auditory warning should be consistent with the direction of the hazard.

3.2   Interactive Design of the Guide Sound

Trigger time

1) The brain's integration of stimuli is delayed by about 30–50 ms in comparison to the fast auditory conduction system, due to the electrochemical sensing system used in the retina of the visual system.

2) Sound appears before the image and the response to the target stimulus is fast; sound appears simultaneously with the image or with a delay in the appearance of the sound, and the response time to the target stimulus is prolonged.

3) In order to synchronise the integration of audio and visual information, and to speed up the response to the target stimulus, the sound information needs to precede the visual information by more than 50 ms.

4) Under close proximity conditions of the hazard source, the propagation time difference between audio and visual signals can be ignored, while as the distance increases,

the audio collision warning signal needs to be presented before the visual signal in order to form audio-visual integration.

Trigger Condition

The combination of audible triggering conditions for collision warnings with visual cues in the intelligent cockpit or the design of triggering modes associated with TTC levels can go some way to improving driver performance in collision warning scenarios. The TTC collision time is the value of the distance between the two vehicles/relative vehicle speed.

3.3 Methodology for the Design of Guide Sound

Urgency

1) ISO 16352–2005 uses the following signal words as criteria for communicating the intensity of danger: 'DANGER', 'WARNING' and 'CAUTION' respectively.
2) Both DANGER and WARNING are high emergency levels, while CAUTION is generally a low emergency level.
3) Very urgent collision warnings must be presented through at least two modalities: visual & auditory or visual & tactile modalities; the auditory channel is the recommended channel to use in emergency collision warnings.

Sound Format Design

1) ISO 11429–1996 classifies the characteristics of auditory signals: Sweeps, Bursts (quick-pulses sound), Alternating sounds and Short sounds are the four sound patterns that express a hazardous situation.
2) Sweeps, bursts and Alternating sounds express a higher degree of hazard urgency, while Short sounds express a lower degree of hazard urgency.
3) In a collision warning scenario, the Sweep, Bursts and Alternating sound patterns are more appropriate as they convey a higher level of warning.
4) Pulse format and pulse interval time have a significant impact on the perceived urgency. An emergency warning should have a higher signal repetition rate, higher intensity and higher fundamental frequency than a warning alert.

Sound Duration

1) Temporal attributes are one of the most important features of tone signals. They can draw attention to themselves and can also be used as discriminatory factors and emergency cues.

2) It is recommended that intermittent beeps or chirps are repeated at a frequency of 1 to 8 tones per second (i.e. single tones with a temporal value between 125 ms and 1000 ms) for warning alarms in vehicles.

Cycle Times

1) Use intermittent or continuous long repetitive tone signals only in exceptional circumstances (until the driver or passenger takes appropriate action).
2) These include situations where very important information affecting the safety of vehicle occupants or the ability to drive the vehicle is communicated.

4   Case Study

We have collected the audible signals from four car brands BMW, Tesla, SERES and NIO in the lane departure warning scenario through test drives, and analysed them in terms of sound signal characteristics and spectrum. By comparing the principles of sound design in the lane departure warning scenario with those summarised in the previous section, we summarise the commonalities and discuss the shortcomings of the sound design in the lane departure warning scenario for these four vehicle brands.

The recorded audio was noise reduced and analysed by Sonic Visualizer software. The parameters analysed included the time value of the sound, the sound format, the frequency and the number of cycles. We use T1 for single syllable time values, T2 for syllable interval time values, T3 for loop interval time values and Td for the total duration of the warning tone.

Figure 4 shows the sound signal waveform (left) and spectrum (right) of the Tesla LDW and Fig. 5 shows the sound design parameter table of the Tesla LDW.

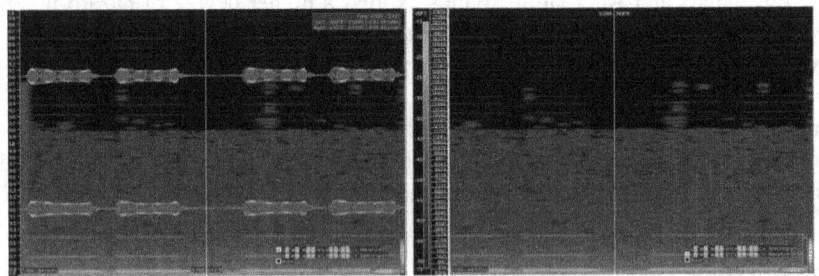

Fig. 4. Tesla LDW sound signal waveform (Left) and spectrogram (Right)

As can be seen from Fig. 4 and Fig. 5, the Tesla LDW sound uses a short monosyllabic sound with a sound format of 'Burts'. The Tesla LDW sound presents a non-variable pitch, uniformly repetitive rhythm with no variation in intensity. The Tesla LDW sound is designed to cycle through the number of cycles until a measure is taken, without designing the orientation of the sound. In terms of urgency, the Tesla LDW scenes are warned through a combination of visual and auditory channels, using only one tone.

Tesla			
Frequency	**Duration**	**Cycle Times**	**Sound panning**
Fundamental frequency range: 473-3186Hz	T1: 218ms T2: 45ms T3: 168ms monosyllable number: 2 Td: 481ms	Cycle until action is taken	Not designed
Urgency	**Urgency Design**	**Sound Format**	
Danger: Visual (Central Control) + Sound	One danger tone only	Bursts Set of 2 sound pulses (monosyllables) Pulse period: 218ms + 45ms = 263ms (0.263s) Pulse frequency: 3.8Hz	

Fig. 5. Tesla LDW sound design parameter table

The Tesla LDW's sound has a pulse frequency of 3.8 Hz, which is slightly below the recommended value and may cause a slight lack of urgency in the listening experience. Overall, the Tesla LDW's sound matches the design human factors data summarised in the previous section to an average degree.

Figure 6 shows the sound signal waveform (left) and the sound spectrum (right) of the NIO LDW and Fig. 7 shows the sound design parameter table of the NIO LDW.

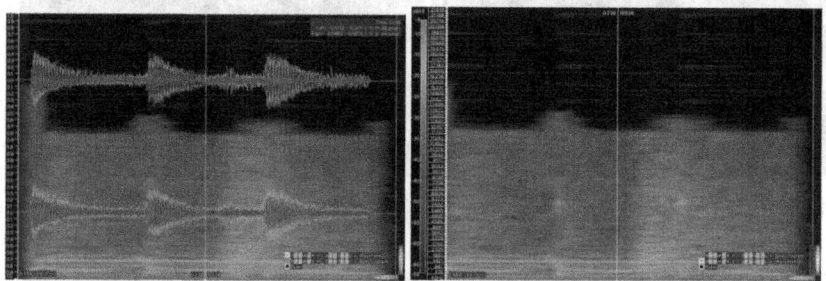

Fig. 6. NIO LDW sound signal waveform (Left) and spectrogram (Right)

As can be seen from Fig. 6 and Fig. 7, the NIO LDW sound is a short monophonic compound with the sound format 'Burts', but the number of sound pulses is only 3, which is below the recommended minimum of 5 and may result in a less urgent sound. The fundamental frequency of the NIO LDW sound is not within the recommended range, but slightly above the recommended fundamental frequency range. In terms of the number of cycles, NIO LDW sound is designed to be played only once. The spatial orientation of the sound is also not designed for NIO. In addition, in terms of the design of the urgency level, the warning of the NIO LDW scene is achieved through a combination of visual and auditory channels, using only one alert tone. Overall, the NIO LDW sound matches the design human factors data summarised in the previous section to an average degree.

NIO			
Frequency	**Duration**	**Cycle Times**	**Sound panning**
Fundamental frequency range: 775-1636Hz	T1: 106ms T2: 44ms monosyllable number: 3 Td: 406ms	1 time	Not designed
Urgency	**Urgency Design**	**Sound Format**	
Danger: Visual (Central Control) + Sound	One danger tone only	Bursts Set of 3 sound pulses (monosyllables) Pulse period: 106ms + 44ms = 150ms (0.155s) Pulse frequency: 6.6Hz	

Fig. 7. NIO LDW sound design parameter table

Figure 8 shows the sound signal waveform (left) and the sound spectrum (right) of the BMW LDW under high deviation. Figure 9 shows the sound signal waveform (left) and the sound spectrum (right) of the BMW LDW under mid deviation. Figure 10 shows the sound design parameter table of the BMW LDW.

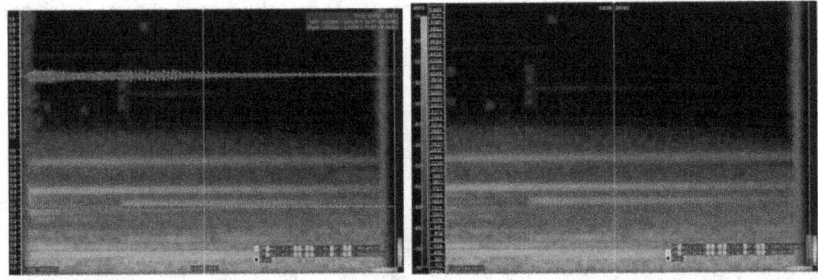

Fig. 8. BMW LDW sound signal waveform (Left) and spectrogram (Right) under high deviation

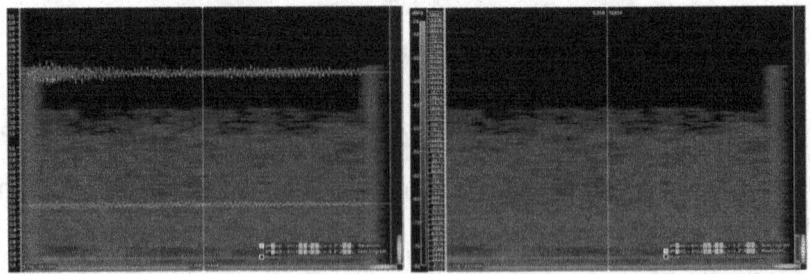

Fig. 9. BMW LDW sound signal waveform (Left) and spectrogram (Right) under mid deviation

As can be seen from Fig. 8, Fig. 9 and Fig. 10, BMW divides the warning guidance sound into two scenarios in the lane departure warning scenario, the high departure

BMW			
Frequency	**Duration**	**Cycle Times**	**Sound panning**
1.High deviation Fundamental frequency range: 473-2325Hz **2.Moderate deviation** Fundamental frequency range: 473-2583Hz	**1.High deviation** T1：325ms+905ms T2：0ms T3：464ms monosyllable number：2 Td：1230ms **2.Moderate deviation** T1：661ms Td：661ms	**1.High deviation** Cycle until action is taken **2.Moderate deviation** 1 time	Not designed
Urgency	**Urgency Design**	**Sound Format**	
Danger: Visual (Central Control) + Sound	3 different levels of beeping. 1.Mild deviation: no audible alert 2.Moderate deviation: monosyllabic alert 3.High deviation : visual (central control + sound) Two-syllable prompts that cycle until action is taken.	**1.High deviation:** Alternating Step sequence of 2 different pitches (bisyllabic) Syllable time:325ms+905ms **2. Moderate deviation:** Short sound 1 syllable Syllable time: 661ms	

Fig. 10. BMW LDW sound design parameter table

scenario and the moderate departure scenario. In the high deviation scenario, the BMW LDW sound uses a step sequence of 2 different pitches, i.e. a two-syllable sound with varying intensity and a sound format of 'Alternating', the fundamental frequency of the sound is not in the recommended range and is slightly thin on the ears. In terms of the number of cycles designed for highly deviant scenes, the BMW LDW sound is played in a loop until a measure is taken.

In medium deviation scenarios, the BMW LDW sound uses a short monosyllabic compound sound with a sound format of 'Short sound'. The fundamental frequency of the sound is also out of the recommended range. In terms of the number of cycles, the BMW LDW sound is designed to be played only once, which can serve as a cue in moderate deviation scenarios, but is not very urgent.

For the spatial orientation information of the sound, neither the height nor the medium deviation scenarios have been designed by BMW. For the degree of urgency design, the warning for the BMW LDW scenario is achieved through a combination of visual and auditory channels. Overall, BMW LDW sound matches the design human factors data summarised in the previous section to a high degree.

Figure 11 shows the sound signal waveform (left) and spectrum (right) of the SERES LDW and Fig. 12 shows the sound design parameter table of the SERES LDW.

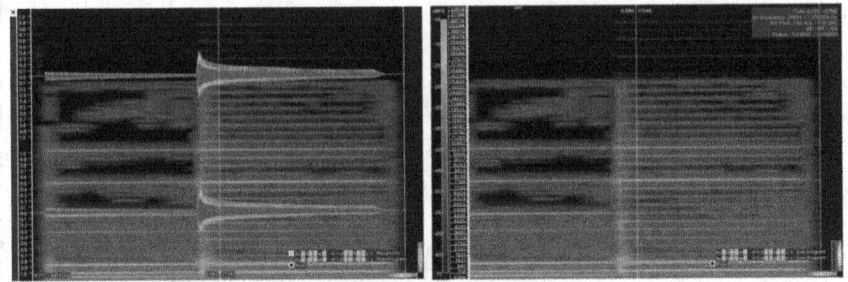

Fig. 11. SERES LDW sound signal waveform (Left) and spectrogram (Right)

SERES			
Frequency	**Duration**	**Cycle Times**	**Sound panning**
Fundamental frequency range: 200-1500Hz	T1: 550ms+550ms T2: 0ms monosyllable number: 2 Td: 1.1s	1 time	Not designed
Urgency	**Urgency Design**	**Sound Format**	
Danger: Visual (Central Control) + Sound	One danger tone only	Alternating Step sequence of 2 different pitches (bisyllabic) Syllable time :550ms+550ms	

Fig. 12. SERES LDW sound design parameter table

As can be seen from Fig. 11 and Fig. 12, the SERES LDW sound uses a step sequence of 2 different pitches, i.e. a bisyllabic sound, with variations in intensity, the second sound is louder, the sound format is 'Alternating', the fundamental frequency of the sound is not in the recommended.The sound design of the SERES LDW has only 1 cycle in terms of the number of cycles and no design for the orientation of the sound, which may result in poor warning. In terms of urgency design, the SERES LDW scenario's warning is achieved through a combination of visual and auditory channels, using only one alert tone. Overall, the sound for SERES LDW matches the design human factors data summarised in the previous section to an average degree.

5   Discussion and Conclusion

We conducted a Case Study on four mainstream brands of cars and found that not all brands follow the relevant international standards for the design of warning sounds, and that some sound designs may be created by designers based on their own experience and subjective awareness. In general, in the lane departure warning scenario, the design of the guidance sounds of the four car brands did not correspond to the human factors constraints of the international standards, and the sound design was generally consistent with the human factors parameters recommended by the international standards.

In terms of Frequency, the fundamental frequencies of all four vehicle brands are not within the recommended frequency range of international standards, which may have an impact on the audible urgency of the sound in lane departure warning scenarios.

On the basis of Sound Format, two brands of car chose 'Bursts Sound' and two brands chose 'Alternating'. The relatively more focused choice of categories for Sound Format allows users to reduce the perceived barriers to in-cabin sound when using cross-brand vehicles, reducing learning costs and adaptation time for users. Only BMW divides the warning guidance sound into two scenarios in the lane departure warning scenario, a high departure scenario and a medium departure scenario, which may be a more efficient driving experience for the driver. In terms of the number of sound cycles for LDW, both NIO and SERES are designed for one sound cycle, which does not comply with the provisions of the relevant international standards and may result in a poor sound warning.

According to Sound Panning, none of the four vehicle brands have designed the orientation of the warning sound. Through research into relevant international standards, we have found that the driver's avoidance response is accelerated when the orientation of the sound of the warning tone is consistent with the spatial hazard information indicated. Therefore, it is possible to include the orientation of the warning sound in the lane departure scenario of the car's intelligent cockpit.

In terms of the degree of urgency design, the warning for the LDW scenarios of all four car brands is achieved through a combination of visual and auditory channels. This shows that the information in the visual channel is also very important for the user in the lane departure warning scenario.

During the Case study, due to the limited digital audio equipment used to collect the sound footage, we were unable to make an accurate determination of the trigger time (audio-visual synchronous or audio-visual asynchronous) and trigger conditions for the four LDW warning sounds for the time being. We will investigate this issue in more depth in our subsequent studies.

Funding Information. Chenguang Project of Shanghai Municipal Education Commission and Shanghai Education Development Foundation. Grant/Award Number: 19CGB06.

Key Laboratory of Intelligent Processing Technology for Digital Music (Zhejiang Conservatory of Music), Ministry of Culture and Tourism. Grant/Award Number: 2022DMKLC 011.

References

1. Ge, S.L., et al.: Current status of research on audiovisual cross-modal interactions. Beijing Biomed. Eng. **30**(4), 431–434 (2011)
2. Niu, D., et al.: A study on trust enhancement of unmanned vehicles based on multi-channel information. Packaging Eng. Art Ed. **41**(6), 81–85 (2020)

3. Zaki, N.I.M., et al.: Auditory alert for in-vehicle safety technologies: a review. J. Soc. Autom. Eng. Malaysia 5(1), 88–102 (2021)
4. Delle Monache, S., Rocchesso, D.: Bauhaus legacy in research through design: the case of basic sonic interaction design. Int. J. Des. 8(3) (2014)
5. Rocchesso, D., Polotti, P., Monache, S.D.: Designing continuous sonic interaction. Int. J. Des. 3(3) (2009)
6. NHTSA DOT HS 808 535: In-Vehicle Crash Avoidance Warning Systems: Human Factors Considerations
7. ISO 11429–1996: Ergonomics System of auditory and visual danger and information signals
8. ISO 15006–2011: Road vehicles - Ergonomic aspects of transport information and control systems -Specifications for in-vehicle auditory presentation
9. ISO TR 16352–2005: Road vehicles - Ergonomic aspects of in-vehicle presentation for transport information and control systems - Warning systems

Customized Intelligent Vehicle Based on Innovation Diffusion Theory - Research on Modification Service Design

Ziyi Zheng and Hongyu Ma[✉]

Wuhan University of Technology, Wuhan 430000, China
375378355@qq.com

Abstract. This research obtains the functional requirements by mining the needs of the group of intelligent car owners, analyzing the relationship between user needs and functional requirements, and based on the innovation diffusion theory, using the online and offline diffusion channels, designs a service system that not only meets the needs of the modified car owners, but also has good usability. The innovation diffusion model was used as a research method for the availability of customized intelligent vehicle modification services. Collect the user needs of refitters through research, and analyze the user demand elements and their importance of refitting APP by using the diffusion group division method; Analyze and dissect the user needs to obtain the corresponding functional requirements, take the functional requirements as the design basis, combine the needs of the strong diffusion group, and design the information architecture, interaction prototype and visual interface of the modified service APP to optimize the contacts in the service system, and use the O2O business model to enhance the diffusion efficiency. It provides a solution for the optimization service system of China's intelligent vehicle modification industry.

Keywords: Intelligent vehicle · Innovation diffusion theory · O2O business model

1 Introduction

Automobile modification refers to the modification of the external shape, internal shape and mechanical properties of the prototype vehicle produced by the automobile manufacturer according to the needs of the vehicle owner, mainly including body modification and power modification.

2 Current Situation of Refitting Industry

In recent years, the user group of smart cars has gradually shown a younger trend. The refitting industry of smart cars has developed rapidly and has become an important part of the after-sales service of modern cars. The auto refitting industry has developed from the initial professional racing refitting to more and more customized smart car refitting.

According to the current situation, China's automobile modification industry is still at the initial stage of development. In terms of specific modifications, most people do not have a comprehensive understanding of the automobile modification industry. They think that it is only a simple repair, and their understanding is relatively one-sided. Common automobile modification projects include light modification, appearance modification, sound modification, internal modification, etc. Among them, the internal modification includes engine intake, engine ignition system modification, oil supply system hardware and software modification, etc. These modification projects require relatively high technology level, and the general automobile modification department can not achieve such high technology. In addition, consumers can not fully understand the automobile modification industry, and the automobile modification companies also have their own deficiencies in development, resulting in the automobile modification industry in China still staying at some relatively low level of maintenance. It often costs a lot of money to refit cars, and these funds can not be reasonably budgeted before refitting. Due to different brands, prices will vary greatly.

With regard to refitting services, China's automobile service industry has just started and is basically still in the manual operation stage. Even in developed cities, the automobile service industry has only partially changed from the manual operation stage to the current information-based technology service. Whether in hardware or software, the gap between China's automobile service industry and developed countries is still large. The existing problems can be summed up in the following three points:

1. Weak service concept. With the automobile consumption market entering the era of personal family, the characteristics of automobile consumption demand have changed greatly. For example, the price elasticity of demand is large, the demand for quality assurance is fine, and the after-sales service requirements are high, but the basic cultural quality of the personnel engaged in the automobile service industry in China, especially the cleaning, maintenance and repair personnel, is relatively low, resulting in the poor service experience of some car owners.
2. Lack of intellectual talents. At present, there is a general lack of automotive maintenance technicians with professional knowledge, understanding of principles, familiar instruments, and certain practical experience in automotive after-sales service. Some automotive service projects cannot be launched, and can only provide simple cleaning, paint repair, and repair.
3. Challenges brought by the new four modernizations of the automobile industry. With the progress of the "new four modernizations" (electrification, networking, intelligence and sharing) in the automobile industry and the upgrading of user needs, the automobile service field also ushers in the evolution of the whole field. How to use advanced technologies such as internet and intelligence to provide customers with better service experience is a difficult problem that automobile service enterprises must face.

At present, with the gradual rise of O2O business model in China's market, the automobile maintenance industry is developing in the direction of "Internet plus + automobile maintenance", which has formed a new market pattern. As a link in the automobile third-party service industry chain, automobile refitting urgently needs to build a multi-dimensional driven organic dynamic product service system in order to adapt the intelligent automobile refitting service to the development of new business types. If the contradiction between design and demand in the above issues is defined as innovation and innovation adoption, such issues can be discussed in the theoretical perspective of innovation diffusion. This theory provides a certain basis for the study of the efficiency process and the principle of action accepted in the general society or specific groups. At the same time, the feedback of the receiver on innovation can also effectively guide the direction of subsequent iteration innovation. Therefore, this paper selects the smart car modification industry to explore the feedback of basic innovation patterns according to different user groups. So as to clarify the demand characteristics of specific types of users for different products, and then point out the iterative innovation direction of subsequent products.

3   Relevant Theoretical Research

The innovation diffusion theory was proposed by American scholar Everett Rogers. The audience can feel that some information is spreading in this process. This innovation diffusion includes five stages: understanding, interest, evaluation, experiment and adoption.

3.1   Research on Diffusion Mechanism

American scholar Everett M. Rogers believes that the degree of diffusion of an innovation mainly depends on the understanding and acceptance of innovation methods by the social system, and the imitation of the acceptance of this innovation behavior among individuals in the social system, that is, the social system factor that allows innovation to spread. The diffusion process of innovation diffusion is an "s" curve. From the "s" curve, we can see that in the early stage of innovation diffusion, the number of adopters is small, the development is limited, and the curve of diffusion speed is relatively flat. When the number of adopters increases to 10%–25% of the residents, the innovation diffusion rate reaches a certain standard, and the diffusion speed is rapidly improved. In the process of follow-up development, the vast majority of people have accepted this view. It is difficult to expand the number of adopters, and the diffusion of innovation has reached saturation, and the diffusion speed has gradually decreased. From the perspective of innovation diffusion, the innovation of a thing needs to first meet the following two aspects: rich communication paths, which can attract a large number of adopters, improve the compatibility of innovative content and audience, and consider the audience's requirements and tolerance for innovation.

3.2   Relationship Between Innovation and Diffusion

When innovation is put into the product service system, it will break the original service link, establish new contacts and react on the product service system where it belongs, and become an important driving force to promote its continuous diffusion within the social system. At the same time, changes in the market also have a negative effect on the innovation behavior itself. Diffusion has an impact on the innovation efficiency, and affects the innovation behavior through certain mechanisms such as market feedback.

To sum up, the innovation and diffusion of products present a trend of mutual influence. Product-oriented innovation behavior not only actively changes the environment of innovation diffusion, but also changes due to the constraints of the environment. This change is mainly reflected in the iterative method of innovation and the specific differences ultimately reflected in the products. First of all, innovation needs to take products as the material form and be realized through the adoption and diffusion of social systems. Accordingly, product function upgrading, experience optimization, and even use methods and concepts.

4 Research on the Optimization Path of Customized Intelligent Vehicle Modification Service

First, based on the process of innovation diffusion, a reference system is established based on the diffusion phenomenon and behavior of design innovation within a specific social system, and the diffusion mechanism of service system and method innovation is launched against the innovation in design activities. Explore the innovation methods in the process of new technology diffusion, mainly including the innovation direction, innovation process, innovation gradient, etc. Put forward the ideas and methods to guide industrial product design and service system update under the environment of modern technology and concept diffusion.

4.1 Basic Elements of Smart Car Modification Service

At present, China's smart car modification services are concentrated in offline stores such as auto stores and auto repair and modification plants. Customers with modification needs need to connect with offline stores to handle modification business. Online services mainly include telephone consultation, WeChat group created by individual stores, and other channels. Therefore, at present, the online modification industry has a narrow service range and lacks an integrated platform to select and compare stores and online consultation Operation costs such as appointment are high.

At present, the smart car modification service links are basically: telephone reservation consultation, selecting parts at the store, communicating with the service personnel of the modification store, consulting inventory, ordering parts, understanding relevant modification laws and regulations, price negotiation, determining the time of vehicle collection, vehicle collection, after-sales service, etc.

From the service process contact point, we can see that smart car modification belongs to the traditional service industry. Some online operation processes can be carried out by telephone reservation. However, when it is used, due to its comprehensive nature of service. Users tend to classify them as traditional offline services in terms of concept cognition, which to some extent limits the diffusion of their innovative forms among users. In addition, due to the general idea that the adoption of traditional service systems that rely on offline stores is age sensitive, and is generally applicable to the middle-aged and elderly, who have limited understanding of smart car modification services. Therefore, there is room for improvement in its diffusion efficiency among users.

4.2 Diffusion of Smart Car Modification Service

In recent years, under the comprehensive influence of the new information technology environment such as the mobile internet, the innovation (including core functions, user experience, application scenarios and other aspects) and diffusion of the service system have shown the characteristics of diversification of diffusion methods and diversification of channels. This puts forward new requirements for the existing diffusion description and prediction methods based on the innovation diffusion theory model. Accordingly, the diffusion mechanism of innovation and service, as well as the specific performance of each contact point in the service system, is more complex.

Restricted by the over-reliance of the traditional service industry on the offline service mode, the diffusion efficiency of the smart car modification service has been affected. In the current smart car modification service link, all contacts are roughly divided into three stages: before arrival, when arrival, and when leaving the store. From the perspective of the diffusion channel, the intervention of the mobile internet channel has a significant impact on the product diffusion effect. Without considering the negative impact, Its diffusion efficiency is far ahead of the traditional diffusion channels. The contact points of mobile devices are only telephone reservation and consultation. The core links and contacts of modification services are still offline. Therefore, the diffusion of modification services is restricted by the diffusion channels. At the same time as China's car ownership is increasing, the smart car user group is also gradually showing a younger trend. According to online research and interviews, from the perspective of the diffusing population, the middle-aged car owners in China lack trust and acceptance in the auto modification industry. From the perspective of interviews, most middle-aged car owners are unwilling to invest in auto modification, On the contrary, young smart car owners have a higher acceptance of innovation and better diffusion efficiency. At the same time, young people also have more willingness and demand for modification.

4.3 Investigation on the Modification Needs of Smart Car Owners

Collect the needs of smart car owners for modification services by issuing questionnaires and interviews to smart car owners, and summarize them as follows:

1. Within the legal scope, you can modify the vehicle at will according to your personal preferences, and you can see the real-time 3D modification effect. 2. I hope there are a large number of comprehensive and complete types of convertible models; 3. An online community sharing personal modification plans for public comment and exchange; 4. Relevant refitted parts can be purchased according to the model; 5. I hope you can easily query the nearby "wash, repair and change" stores; 6. I hope you can give feedback on your own experience and follow-up problems in a timely and effective manner. 7 Distribute a questionnaire to understand the user's modification intention. 8. The original factory sells modified products. 9. Officially participate in the modified auto show. 10. Hold a modified auto show. 11. Share modification consultation and answer users' doubts (Fig. 1).

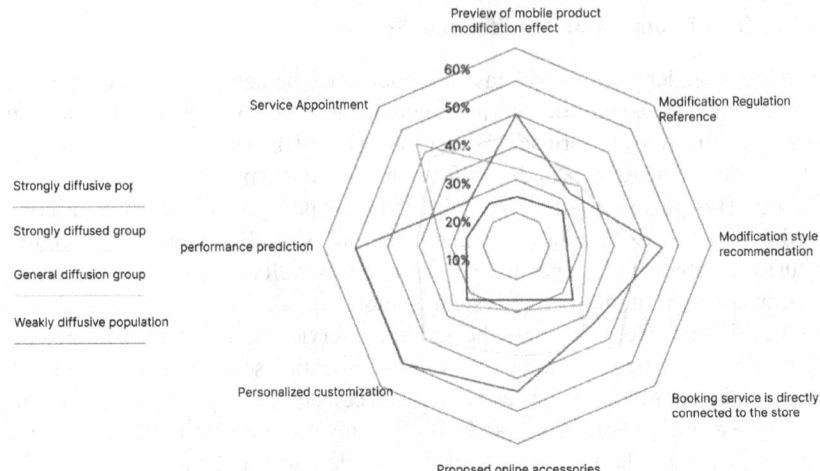

Distribution of potential needs of users of customized
intelligent vehicle modification product service system

Fig. 1. Potential demand distribution of customized intelligent vehicle modification users

At the same time, from the objective social environment, the refitting industry needs to strengthen its propaganda and positive guidance; The government departments need to further strengthen the laws and regulations related to refitted vehicles; The refitting industry needs more professionals engaged in automobile refitting.

Starting from the summary of the above user needs, this study aims to innovate and improve the above pain points from the design perspective.

4.4 Division of Diffusion System Group

The user group is divided according to the individual diffusion efficiency in the innovation diffusion system, and different improvement routes are formulated according to the product adaptability and improvement needs of different groups, and different structural contacts are formed according to the corresponding iteration strategies.

According to the basic division principle of the adoption population based on the innovation diffusion theory, combined with the characteristics and needs of the user groups in the collected samples, the survey population can be roughly divided into innovation adopters, early adopters and early adopters, late adopters and late adopters according to their age characteristics and attitudes towards innovation.

The follow-up design practice will consider the possibility of innovation adoption and market launch, starting from the needs of innovation adopters and early adopters and early adopters.

4.5   Specific Design Optimization Objectives of Modification Service

Starting from the design means and types, the specific design optimization objectives are divided into three main directions: core functions, user experience, and application scenarios. On this basis, the specific innovation path is drawn as follows: mobile end product modification effect preview, service reservation, and performance prediction function innovation; The specific model modification style recommendation, modification regulation reference and personalized customization service innovation provided during the vehicle modification process; The system innovation of online optional parts and reservation service to offline stores.

According to the case product diffusion characteristics and the performance of the key influencing factors in the diffusion process, the corresponding product iterative design strategy was developed in this study. From the three points of product, service chain and product service system, the feasible iteration paths are determined as three:

1. Independent innovation of products - positive planning of product innovation. At present, in the smart car refitting industry, there are industry-related products such as CAR++'s game-based refitting experience app and the matching applet based on the brand car enterprise, and the analysis of similar products is carried out. CAR++ mainly provides online modification scheme preview and modification consultation services, while the matching app under the Porsche brand car enterprise mainly provides the function of car matching effect preview and price display, Through the analysis of the above two modification service products, it is concluded that the main iterative factors of mobile software are modification experience and online selection. By extracting the iterative factors of similar products, a diffusion model is built, and the modification experience and online selection of similar products are introduced into the mechanism category for quantitative analysis.

2. Path to promote innovation - fully consider the characteristics of the distribution path of various elements in the system and use it reasonably to promote innovation and improvement. When O2O mode is adopted in the field of smart car modification, in terms of the online diffusion path, the mobile internet channel has faster diffusion than the offline diffusion channel, shorter time to achieve stability, and relatively ideal diffusion efficiency. As far as the offline dissemination path is concerned, it can persuade people to accept and use innovation more directly and effectively, but it also has the characteristics of slow diffusion and poor diffusion efficiency. Therefore, in the field of customized retrofit service, we can establish a community of refitters through mobile devices, provide lightweight virtual customized retrofit experience to achieve good diffusion efficiency, and connect with the offline industry, expand the third-party service of smart cars in the later stage, and expand the offline dissemination path in this context (Fig. 2).

3. System iterative update - adjust the strategy according to social recognition, consider and anticipate social effects, and form an iterative mechanism. In the system iteration,

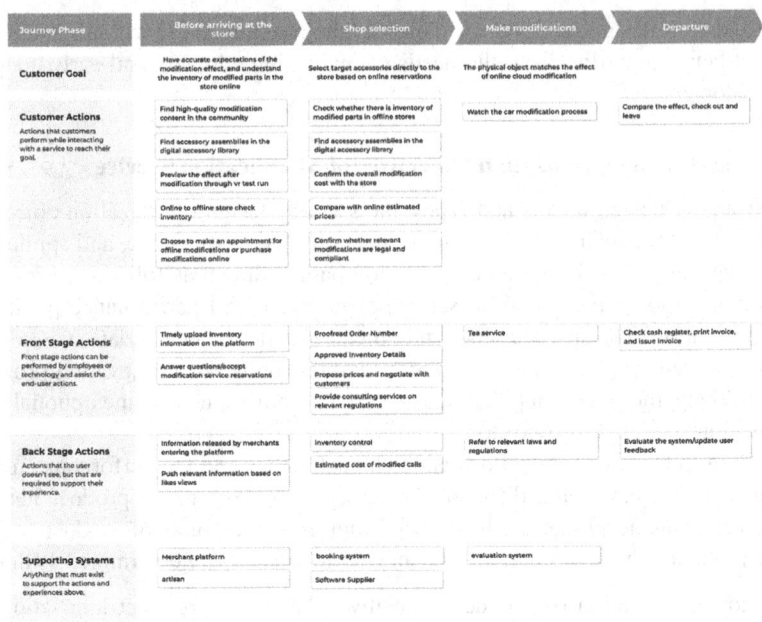

Fig. 2. Blueprint of customized smart car modification service

the social system and culture play a subtle role. In the field of smart car modification, the recognition of modification culture varies greatly among different groups and different social environments. In China, the audience of modification culture has been small for a long time, but the rising trend is obvious. The recognition of modification culture by society is also gradually increasing, In the system iteration, not only the current concept needs to be considered, but also the future use concept needs to be considered in advance and dynamically, thus forming the value identification and community atmosphere.

5 Design Practice of Customized Modification Service App for Smart Cars

5.1 Interactive Prototype

The interaction prototype is the basic framework for modifying the service APP interface, which has strong adjustability and lower cost compared with the visual interface. It will determine the final form of the interface as the blueprint of the APP. The prototype of the refitting service APP is to think about the operation task process on the basis of the information architecture, and consider how to improve and improve the contact points in the service link from the perspective of users, so as to solve the learning cost, time cost and other problems that users pay when refitting the car to a certain extent. Design the low-fidelity interaction prototype of the modified APP by combing the user's operation behavior in each task process (Figs. 3 and 4).

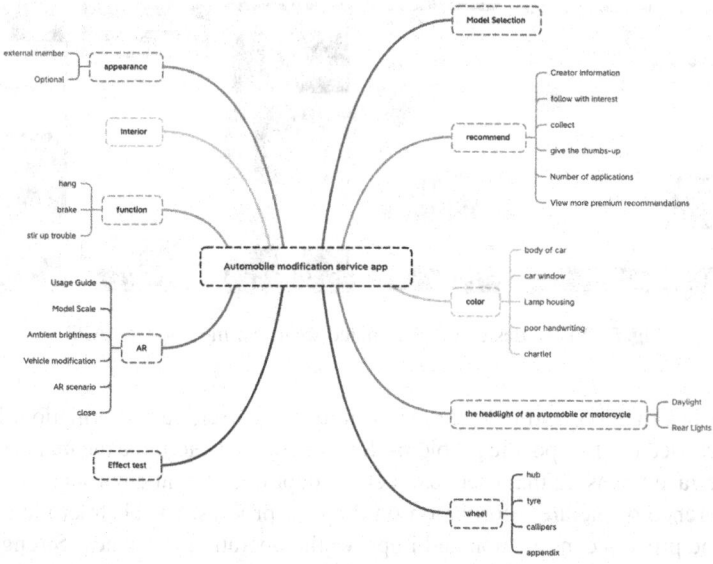

Fig. 3. Customized smart car modification APP information architecture

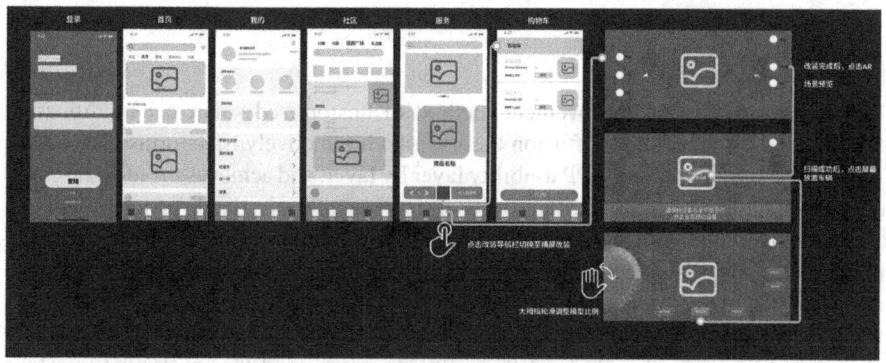

Fig. 4. Low-fidelity prototype of customized smart car modification APP

5.2 Visual Interface

On the basis of modifying the interactive prototype of the APP, the potential users and existing users are considered respectively, and the modifier community is set up to attract potential users. At the same time, the APP can be integrated into the overall service link, and connect users and stores at the same time. Secondly, the product availability of users in three aspects of environment, process and operation is analyzed to improve the user's satisfaction with the use of the product, and the young characteristics of the population with strong diffusion groups are targeted, Positioning the visual style of APP as young, simple, warm and bright. Carry out visual design for the organization and layout of elements such as status bar, content partition, navigation bar, function buttons, etc. in the prototype, highlight important functions and main contents, weaken the level of

Fig. 5. Visual design of customized smart car modification APP

secondary information, and reduce the interference of irrelevant information. Detailed design is carried out for specific problems. See the Fig. 5. Determine the main color tone of the interface to ensure that users can better adapt to different environments; Reduce the interference of picture information on the task process, use abstract identification to guide the process continuation and improve the operation efficiency; Strengthen the visual influence and physical area of function buttons to ensure the effectiveness of functions.

6   Conclusion

This study proposes an improvement method of customized smart car modification service based on innovation diffusion theory, which effectively transforms user needs, functional requirements and APP usability layer by layer, and achieves the ideal research goal at each design stage. In the stage of user demand analysis, the diffusion model can sort the demand qualitatively and get the real demand of the strong diffusion group. In the expansion stage of functional requirements, integrate the requirements importance ranking of strong diffusion groups to help clarify functional requirements. In the design improvement stage, based on the functional requirements, the information architecture is optimized by using the needs of the diffusion group, and the interactive interface and service blueprint are output to provide usability reference for China's smart car modification industry.

Acknowledgment. This paper is supported by "The Fundamental Research Funds for the Central Universities (WUT: 2002VI003–04)".

References

1. Shuqing, Y.: The development and dissemination of Dongyang woodcarving – based on the theoretical model of innovation diffusion. Popular Literat. Art **10**, 191–193 (2022)
2. Gang, M., Shu, C., Yuan, W.: Research on the product design strategy of cervical pillow in the perspective of innovation diffusion. Packag. Eng. **43**(10), 257–264 (2022). https://doi.org/10.19554/j.cnki.1001-3563.2022.10031

3. Daofeng, D.: Promote the development of automobile customization and modification industry in an orderly manner. Auto Horizon **12**, 27–30 (2021)
4. Hu, P.: Research on the diffusion of Didi Travel under the innovation diffusion theory. Huazhong University of Science and Technology (2016)
5. Wei, Z.: Countermeasures for the development of automobile modification industry. Autom. Pract. Technol. **45**(23), 249–251 (2020). https://doi.org/10.16638/j.cnki.1671-7988.2020. 23.082

Research on Personality Traits of In-Vehicle Intelligent Voice Assistants to Enhance Driving Experience

Xinyi Zhou and Yangshuo Zheng[✉]

Wuhan University of Technology, Wuhan 430000, China
zhengyangshuo@163.com

Abstract. Despite the rapidly expanding research on in-car intelligent voice assistants, relatively little attention has been paid to the study of their personality traits and how to improve the user's emotional experience by cutting through from this perspective. In this study, first, an exploratory interview-based study is conducted to identify key traits that elicit driver emotions and five personality traits that can be used in in-car intelligent voice assistants: emotional enrichment, emotional expressiveness, persona congruence, empathic expressiveness, and user impressions. Second, an online survey-based study is conducted to investigate how the above five performance dimensions that focus on the degree of personification of intelligent assistants can be incorporated into smart car voice interaction systems in different scenarios. Finally, the ideal use cases for each personality trait are investigated, as well as how in-car intelligent voice assistants can apply the corresponding trait dimensions according to the current real-time driving situation. This study explores how the five personality trait dimensions can enhance driving users' emotional experience and emotional connection with intelligent voice assistants by adding a pleasant attitude, a sense of pleasure, and can accurately grasp users' emotional state and provide proactive services conducive to emotional appeasement and safe driving in navigation.

Keywords: In-car intelligent voice assistant · Personality traits · Driving emotional experience

1 Car Intelligent Voice Assistant Concept

Voice is the most commonly used interaction method in people's daily life, and with the development of intelligent AI technology, it is also gradually applied in car products. Voice assistants help to free users' hands and eyes during the driving process, thus enhancing driving safety and enhancing the emotional experience of driving users. People increasingly rely on maps to navigate, check road conditions and find nearby points of interest in their daily travels, while users use their hands to control or view the screen while driving, which poses a great safety risk. The emergence of in-car intelligent voice assistant greatly liberates the user's hands and eyes during the driving process and improves driving safety. It not only supports the whole process of voice wake up, but also can quickly and accurately understand user commands and propose effective solutions.

1.1 Voice Assistant Definition

A voice assistant is a human-machine dialogue program embedded in a hardware device or APP software to assist users to use the functions on the host device or program by voice. A complete human-machine dialogue includes the front-end processing of sound signals, the conversion of sound into text for machine processing, and the conversion of textual language into sound waves using speech synthesis technology after the machine generates the language, thus forming a complete human-machine voice interaction.

1.2 Development Trend of In-Car Intelligent Voice Assistant

Smart cars are vehicles that can use ICT (information and communication technology) and artificial intelligence to interact with users autonomously. With innovations in the areas of electric vehicles, mobile connectivity and autonomous driving, smart cars have captured the attention of many interaction designers. The importance of improving the emotional experience of smart cars is reflected in its ability to improve driving pleasure, efficiency and safety.

With the development of artificial intelligence and voice recognition technology, intelligent voice is becoming the main vehicle interaction tool, in the car HMI, the car voice assistant provides auxiliary functions for the driver. As the third living space, i.e. another living space after home and work, intelligent voice assistant is the core content, which helps users to control the whole intelligent cabin and provide services to users through intelligent voice assistant. The development of AI technology and hardware upgrade has given intelligent devices more and more perception channels, and also continuously enhanced the ability to output information through various media. Based on research results in the fields of intelligent remote technology speech, natural language and machine translation, in-car voice assistant technology has been greatly improved in fluency, quality, fidelity and naturalness. Beyond the perception and output layer, the improvement of emotional computing capability has enabled the machine to make a qualitative leap in the cognitive layer.

With the development of technology and hardware upgrades, smart devices are given more and more perception channels, and also continue to enhance the ability to output information through various media. Beyond the perception and output layer, the improvement of emotional computing capability enables machines to make a qualitative leap in the cognitive layer. They understand users far better than before, and are far more articulate than before. Based on design understanding and practice, it is believed that the voice assistant experience is showing three trends in terms of interaction channels and interaction objects: the expression of information services incorporating multi-channel experience, dialogue close to natural human instincts, and the ability to interact with emotions.

Expression of Information Services Converge Multi-Channel Experience. In addition to voice channels, computer technology has expanded interaction channels such as face and spaced gestures, while traditional interaction methods such as touch and knob have their own advantages in terms of operation accuracy, information output efficiency and technical cost. Multi-channel fusion can bring into play the advantages and scenario applicability of different channels to express information services more naturally and maximize efficiency.

The navigation scenario is the most commonly used driving scenario. In the navigation scenario, based on the map speech of touch screen mobile devices, the collaborative output of visual information can effectively compensate for the defects such as invisible and not easily memorized by voice, and improve the user's understanding of the voice interaction experience. The first time the voice interaction capability was introduced, the interaction form of map voice followed the industry's common closed dialogue flow in order to reduce cognitive costs. This form also has limitations with the expansion of voice supported map functions: (1) The form is independent and closed and does not integrate with the scene, which will interfere with navigation and affect driving safety. (2) The original information expression of the map cannot be used, and the results need to be presented separately in the conversation stream. This not only affects the expansion of complex requirements such as road calculation, but also increases the maintenance cost of design and development.

Dialogue is Close to Natural Human Instincts. Continuous dialogue and can be interrupted at any time, in line with the instinctive awareness of daily communication, but most of the current products voice interaction is still not natural enough: to initiate a dialogue, you need to wake up first and then issue commands in a quiet environment, and the main focus is on "One question and one answer". With the emergence of full-duplex wake-up-free voice technology, the user's instructions can be predicted and judged by contextual information, eliminating the intermediate wake-up process and achieving more natural and smooth multi-talk. Before achieving natural dialogue, effective cognitive education is a necessary way to lower the threshold for users to use speech. due to the "invisibility" of voice information, there is the defect of weak discoverability of skills, which makes users often ignore the use. At the same time, the basic research of voice map found that the primary reason for not using voice is that users are not used to the operation method.

In the user's mental model, the concept of voice as a tool determines that users will use it only when they have the intention, which inevitably affects the frequency and regularity of practice required for learning and habit formation. Therefore, Map Voice Skill Center proposes the concept of Xiaodu growth, completing daily tasks and using rewards to help Xiaodu grow, so that users can learn skills and develop habits quickly. The tasks are mainly organized in personalized, hot and level dimensions, for example, in personalization, priority is given to displaying user error-prone instruction tasks and solving the problem of instruction expression through repeated learning.

Ability to Interact Emotionally. Language is a symbol of human intelligence, and users will have an "Empathy" effect on voice products. emotion computing enables products to process relevant data such as human facial expressions, body movements and various psychological parameters such as heartbeat, pulse, brain waves, etc. Through machine learning algorithms such as emotion analysis, and finally calculate the emotional state of a person by combining external environmental information. Then give three-dimensional emotional feedback from the hardware level, GUI and VUI level to realize emotional interaction.

Car intelligent voice assistant active service currently includes three main scenarios, such as safety advice, road conditions and destination services. For example, when the user drives for a long time at high speed, it provides the nearest rest service area; when the road is crowded, it suggests a suitable route; and when the destination is near, it recommends a convenient parking spot. We provide valuable proactive services at key touchpoints during the driving process to improve driving safety and establish a trustworthy emotional connection.

With more sensors to obtain human body-related data in the future, it can accurately grasp the user's emotional state and provide active services in navigation that are conducive to emotional calming and safe driving.

2 Design Research

2.1 Exploratory Study

Objectives and Methods. An exploratory study was conducted to identify personality traits of voice assistants applicable to smart car interactions, with the aim of identifying key traits that elicit driver emotions and personality traits that can be used in in-car intelligent voice assistants.

Sample. The study was conducted on 10 drivers, 7 males and 3 females, aged 23 to 55 years old, who had experience with in-vehicle voice-activated products. all had experience with frequent use of in-vehicle intelligent voice interaction products within three months to one year.

Procedures. The study involved an interview process consisting of two parts. first, contextual interviews were conducted in the participants' cars to allow users to generate design ideas. the mobile app voice search for place names and voice navigation functions were tested. the steps for testing the voice search place name function included: first step: voice input the place name; second step: select the destination based on voice prompts; and third step: start navigation. the voice navigation function test includes voice announcement: route location, full road length, estimated time required, real-time road conditions, monitoring reminders, service station reminders, etc. next, a structured interview was conducted and the scenario should be needs-based. participants were asked to think in advance about what they imagined in-car intelligent voice interaction would be like. seven driving scenarios were included: (1) entering the vehicle (2) waiting for a traffic signal (3) receiving guided navigation (4) refueling the car (5) dozing off (6) in-car entertainment (7) parking. questions in the contextual interviews included, "how would you communicate with the in-car intelligent voice assistant in each situation?" and "how would you like the in-car intelligent voice assistant to respond anthropomorphically?" etc.

Results. For in-car voice interactive navigation, the needs of participants are as follows: (1) basic needs: in-car intelligent voice assistant can meet the navigation function (2) desired needs: in-car intelligent voice assistant can understand the driver's needs and give timely feedback (3) excited needs: in-car intelligent voice assistant can have personality traits and communicate with the driver like a friend or family member without hindrance.

The structured interviews for in-car intelligent voice interaction resulted in the following conclusions: key traits that elicit driver emotions and five personality traits that can be used for in-car intelligent voice assistants: emotional enrichment, emotional expressiveness, persona consistency, empathic expressiveness, and user impressions.

With the advent of natural language interaction, the first human can use the tool according to his or her customary needs, and it is important to create an appropriate "personification" of the intelligent voice product during conversational interaction. These results provide initial insights into how to use in-car intelligent voice assistants for automotive interactions, which may be designed to bring more emotional experiences to users.

2.2 Online Survey

Objectives and Methods. Second, an online survey-based study was conducted to investigate how the above five performance dimensions that focus on the degree of personality of the intelligent assistant can be incorporated into smart car voice interaction systems in different scenarios. This resulted in the desire to elucidate the most desirable use of each personality trait in smart car voice assistants and how each personality trait can be applied to different instances. An online survey was therefore conducted to investigate whether different in-car intelligent voice assistants are appropriate for application scenarios in specific driving situations, and if so, why these scenarios are preferable.

Survey Design. The survey was designed to evaluate five application scenarios using two independent variables (different driving situations, different in-car intelligent voice assistant personality traits). By using fictional scenarios to familiarize users with the new situations in the survey, selective representations of real user scenarios can help unravel the complexities and conflicts that exist in the real world.

By constructing a short narrative covering all aspects of car use, including task-oriented situations (e.g., refueling, listening to music), specific contexts (e.g., first meeting, return), or covering a period of use (e.g., adaptation), it is easier to imagine being in these scenarios. Questions include "Does the assistant need to empathize with the driver's negative emotions when the vehicle is in an undesirable driving situation?", "How should the assistant respond when a human presents a need that cannot be met by the current intelligent assistant?" and "How do you want the assistant to successfully identify the user's emotional state?" etc.

Five representative driving situations were selected based on previous studies: (1) adapting to the car (2) listening to music (3) refueling (4) parking the car (5) returning to the car. In selecting these situations, both realistic and common situations (e.g., parking) were considered, rather than rarely encountered situations (e.g., dropping the car).

By creating different usage scenarios, these scenarios were specifically explained and expressed in a familiar representation so that participants could easily understand and immerse themselves in them. Testers used scales to indicate their preferences for personality traits of in-car intelligent voice assistants in different application scenarios.

Respondents were also asked to assess the usefulness of each scenario. These scores were used to understand why scenario-specific personality traits were preferred.

Results. The test results showed that the most popular personality trait was empathic expressiveness, while the next most preferred was emotional expressiveness and emotional enrichment. This was closely followed by Personality Consistency. The user impression personality traits of the in-car intelligent voice assistant were the least important.

3   How In-Car intelligent Voice Assistant Personality Traits Enhance the Driving Experience

3.1   Emotional Enrichment

Joy and sorrow examine the emotional richness of intelligent assistants, most of the voice assistants nowadays are a tool-based product, on top of which personality is attached. Take Gaode voice navigation as an example, the experience of using this product is undoubtedly done pleasantly, and the interaction with them is full of fun: "There is a sharp turn ahead, do you understand? It's a sharp turn, probably the sharpest sharp turn in the Eastern Hemisphere". By using the voices of celebrities and internet celebrities or using jingles that people are familiar with and orchestrated for the current driving scenario, the driving fun is enhanced and the driver's emotional experience is enriched.

But in the actual business, it is difficult for us to make a similar design when users are interacting with the assistant. The reason is as follows: users have already managed their expectations for the product when they choose the voice package of Gaode Map. And if it is the first time to use the intelligent assistant users do not have an expectation management. Some times using some jingle or slang, users who don't understand it will have an inexplicable feeling, and sometimes even because of regional and cultural differences may produce offense to certain kinds of groups. Therefore it is safest to try to choose a design that faces the general public.

Designed for the masses, meaning professional, only positive emotions can generally exist, with little negative feedback, and often only the joy of a task being completed. "If the Internet is stuck, does the in-car intelligent voice assistant assistant need to feel the anger in sync with the user?" "If the user makes an unreasonable request, what level of flirtation needs to be used?" These are all questions that will exist in the design. Generally speaking, professional assistants do not have negative emotions and do not flirt freely. Once you choose to design for the masses, often the design is more limited and the safe thing to do is to keep only positive emotions. Moderate is safe, but it seems to have less human light. And Gao De voice navigation processing solution is when the user is familiar with, let the user make a choice, their own management of their own expectations, perhaps a solution.

3.2   Emotional Enrichment

Emotional expressiveness here refers to the expressiveness and infectiousness of the voice assistant in expressing emotions. Assuming that the direction of emotion is joy, anger, sorrow, sadness, fear and shock, how to express and what intensity is appropriate? The computer expressions that can be listed are: text, expressions, voice, sound effects, images, light effects, and even the robot's body movements. The more of these ways are overlaid, the richer their expressiveness. It is important to be able to demonstrate the emotional expressiveness that you can show in order to impress the user on key occasions.

3.3 Personality Consistency

Once the persona of the in-vehicle intelligent voice assistant is defined, its behavior habits, timbre, speech speed, and language content expression should be guaranteed to conform to the consistency. Different people have independent personalities as individuals with very distinct personas. The performance when facing problems, values, language expressions, logical sequence, and the position of interest they are in must be based on the previous persona performance. Although each person is a complex individual, but by and large also fluctuate within a certain range of values, so that it is a reasonable character setting, will not give rise to a sense of dismay. The logic of the intelligent voice assistant, timbre and speech rate towards is based on the same voice model setting, is easier to ensure consistency, and the difficulty is in the presentation of the language content level. "I'm sorry! I can do something else for you" "Hey, I can't do this yet, but I will try to learn to serve my master soon" This is the words expressed by different personalities, if you frequently switch between different language styles, it is easy to have The persona is inconsistent and fails in shaping the personality of the assistant's character.

3.4 Empathic Expressiveness

Emotional intelligence and empathy are high-level competencies - responding to the user's descriptions and corresponding content. "Empathy is a psychological phenomenon in which people actively project their true feelings onto the things they see. "Empathy" is a kind of feeling, standing in the perspective of others to think about the problem, is one of the necessary communication skills. The act of empathy requires feeling and observing first, and then responding. How to successfully identify the user's emotional state? From the machine level, various components are responsible for collecting and various technologies are responsible for analyzing. For example, visual recognition analysis, audio track analysis, text understanding, and even brainwave signal acquisition technology can do to analyze emotions and the corresponding degree. When the user is happy, sad or anxious, how should the voice assistant do empathic feedback?

Users rarely have mood swings, and when they have violent mood swings (ecstasy, anger, sadness), if the assistant can show some empathy and resonate with them with the same frequency of emotions, it can enhance the user's emotional experience. And empathic performance, invariably considers the four dimensions of the previously mentioned capabilities.

3.5 User Impressions

The product is able to manage the psychological expectations of the user and success-fully shape an image. In other words, what kind of brand impression the assistant has in the user's mind. In the past, branding required a lot of effort from various departments such as product, operation, marketing, business, brand, and channel to expose and main-tain. Nowadays, conversational interaction has more opportunities and personalization is easier. At present, the most shipped voice products on the market is the smart speaker, the user interaction with these smart speakers compared to traditional hardware prod-ucts has changed fundamentally, because the anthropomorphic interaction form of voice dialogue, it is easier to attach personality, and then to convey the brand impression. In the actual use of the process, occasionally because of some playful words of the speaker to make themselves laugh out, this use of the process to generate positive emotions, the formation of pleasant memories, and then promote the user's willingness to use, to enhance tolerance and trust.

But in many smart speaker products, the performance of personality traits shaped by much the same, most speakers are still functional, business, service state, once experi-enced it is difficult to leave a deep impression on the user. And the design of the product, if the user experience a period of time, but also can not leave any impression, it is a great failure.

4 Research Conclusion

The purpose of this paper is to improve the in-vehicle voice interaction experience, shape a voice image personality that is more emotional and more in line with user expectations, highlight the humanization and intelligence of the voice assistant, improve the intelligence of the in-vehicle assistant, and assist in optimizing the user driving experience.

After describing what the definition of an in-car intelligent voice assistant is, I hope to think further about, "Emotions need to be rich, how does the assistant handle & apply negative emotions?" "Strong emotional expression is needed, how to grasp the proportion of the assistant's emotions?" "How to do the assistant's persona selection and how to ensure consistency in feedback?" "The empathic performance of the assistant, how to identify as well as feedback?" "How does the product impress the user and shape the brand?". All five personality traits of the in-vehicle intelligent voice assistant are interrelated and yet exist independently. It is easy to make requests, but difficult to have a methodology to apply to specific examples. For example, when the alarm clock wakes up the user, the content could be a cyclic alarm or a repeater voice announcement, or a variety of flirtatious ways to stimulate the user to get up.

Benefiting from the further evolution of voice capabilities, information and services continue to flow around the user rather than the medium. People's demand for natural, emotional and personalized is more prominent than in any previous era, and the voice experience will be more real-time and versatile. However, with the advent of the smart driving era, in-vehicle voice products are developing rapidly, but research on issues related to in-vehicle voice interaction design is still relatively small. In addition to prob-lems such as poor environmental network, improper user operation and environmental noise, existing voice assistants have serious homogenization, rigid language and lack of personality in auditory experience, which cannot bring good emotional experience for users. At present, the majority of products voice interaction is still not natural enough: waking up in a relatively single way, you need to wake up the voice assistant in a quiet environment by voice call and then issue commands. The voice assistant lacks the ability of multi-round interaction, and it is difficult to maintain the state of dialogue with the user after a wakeup, and the dialogue effect is not natural enough. Understanding the user's intention is basically only through the voice, but not through the multimodal information acquisition to identify the user's intention. Although the voice assistant has evolved from "one question and one answer" to "continuous question and answer" and "emotional dialogue", the "emotional dialogue" relies on The "emotional dialogue" relies on keyword recognition technology, and once it is separated from certain specific keywords, it will return to the state of indifferent machine. In-car voice assistant is difficult to understand human emotion and situation, give caring and compassionate response, can not provide immediate emotional support and long-term emotional companionship. Due to the "invisible" nature of voice information, there is a weakness of skill discoverability, resulting in users often ignoring the use. Users often have high expectations of voice interaction, and people will unconsciously compare voice assistants with real people when communicating with language, trying to understand voice assistants with the thinking habits of the human brain, which will inevitably lead to many times users feel that the results of human-computer dialogue do not meet expectations.

This paper focuses on the exploratory research of analyzing the personality traits of in-car voice assistants to enhance user driving experience, and the subsequent step-by-step advancement is needed to improve the interaction fluency and intention understanding degree of in-car voice assistants intelligence.

References

1. Braun, M., et al.: Improving driver emotions with affective strategies. Multimodal Technol. Interact. 3(1), 21 (2019). https://doi.org/10.3390/mti3010021
2. Gruber, D., Aune, A., Koutstaal, W.: Can semi-anthropomorphism influence trust and compliance? In: Proceedings of the Conference on Technology, Mind, and Society, pp. 1–6. ACM, New York (2018). https://doi.org/10.1145/3183654.3183700

Urban Mobility

Usability, Habitability and Cab Performance in Heavy Surface Mining Trucks

Juan A. Castillo-M[1]([✉]) [iD] and Oscar Julian Perdomo[2] [iD]

[1] ErgoMotion-Lab, School of Medicine and Health Sciences, Universidad del Rosario, Bogota, Colombia
`juan.castillom@urosario.edu.co`
[2] School of Medicine and Health Sciences, Universidad del Rosario, Bogota, Colombia

Abstract. A study is presented focusing on the set of variables associated with the link between cognitive aspects, usability and fatigue in the context of the safe operation of heavy vehicles in a surface coal mine, in order to understand how driving performance is affected and the reduction of perceptual activity capacity in the detection of tasks associated with driving. Method: A multi-method approach was used where human activity recognition (HAR) was incorporated to analyze data from different sensor sources to identify activity-related features of a person. The study was developed in a surface mine with the participation of 37 drivers with a mean age of 37.4 years (STDV. 9.6), the study covered a shift design of 12 h duration and in sequence 3Day, 4 Night by three recovery shifts. Systematic observation sessions and data recording were carried out. Results: The various data recorded show that the usability possibilities of the cabins in terms of driver performance is closely related to the influence of aspects of geometry and composition of the cabs on the one hand and on the other hand with the cumulative effects of working time, i.e. it is not only to understand the impact of the 12-h shift and its effects on the modification of metabolic aspects and sleep hygiene of drivers, to explain the problem of driving performance, it must also consider an aspect that influences performance such as sleep debt.

Keywords: Human activity recognition · Inertial sensors · Usability · Fatigue · Use

1 Introduction

1.1 Literature Review

This study focused on the set of variables associated with the link between cognitive aspects, usability and fatigue in the context of the safe operation of heavy vehicles in a surface coal mine, for the construction of the context of the analysis, a literature review was conducted in the domain of fatigue and usability, for this bases were consulted as: Sciencedirect, PubMed, Web of science and Springer Link, a search string was used with terms in the subject of countermeasures for fatigue and for usability of heavy vehicle cabins; this search allowed the location of 1872 articles, for the final analysis 44 articles were selected that met the characteristics required within the research, this review is published in another document.

H. Krömker (Ed.): HCII 2023, LNCS 14048, pp. 247–254, 2023.
https://doi.org/10.1007/978-3-031-35678-0_16

The analysis of the literature allowed establishing that this type of studies have mostly been carried out in European, Asian and North American countries [1]. To a lesser extent, studies were located in Latin America, where the ways of driving heavy vehicles and work are developed in a substantially different way due to the parallels that exist, ranging from the geographical to the cultural.

In the results of the studies analyzed on fatigue, it was found that fatigue affects driving performance as it reduces steering ability and decreases perceptual activity of secondary task detection [2]. Fatigue also caused aberrant behaviors within driving and driving generating an increase in the possibility of events and near-events [3]. This burnout appears as the result of different intersecting elements, among which are working overtime, not having enough breaks, unbalanced schedules, stress, sleep deprivation, lifestyle on personal and social level, economic retribution and workplace environment [4].

1.2 Study on Driver Fatigue

For the study of fatigue, both in Latin America and in other countries that report studies of this type in the field of heavy vehicle driving, the use of subjective (PS), psychological (PP) and performance (PR) tests are among the most frequently cited [1, 5–8], also among the most recommended countermeasures for fatigue control are fatigue warning systems, which are used so that workers can easily recognize when they have symptoms related to drowsiness and fatigue [9], Fatigue Warning Systems (FWSs) [10].

The relationship between driving tasks, fatigue and usability is relevant to driving tasks, a systematic review of 208 articles found that there is a detriment in driving performance when performing a secondary task. Specifically, studies that included cell phone use were 16% more likely to have this performance detriment [11]. Workers report as a "need" to perform certain secondary tasks and also use these actions as measures to stop fatigue such as looking at the cell phone or being distracted by the radio [12]. Many of these tasks are important for the control of equipment such as air conditioning, communications such as the operational radio band. When performing these secondary tasks fatigue is being induced, but the decrease in performance by workers is not recognized, and, in this sense, no actions are taken to counteract fatigue [2].

Concerning the link between fatigue, performance and usability, it was found that in Latin America there is still no exact regulation of working hours vs. rest hours for the transport sector; this is evident in the high number of accidents and accidents recorded in southern countries compared to the figures for countries in the global north where there are stricter regulations [13]. In Latin America, out of 17 countries, only 9 have regulations on daily hours of service and only 5 mention the limitation of the number of consecutive days that must be worked. Only 8 of the 17 countries have policies that establish the duration of the amount of time that can be driven continuously, ranging from 3 to 5.5 h. As a consequence, it can be evidenced that there are still gaps in this type of regulations, in addition to the fact that it is possible to contribute with the recording of information to nurture research that can contribute to the creation of private and public strategies.

1.3 Study Context

Because of the lack of regulation in the countries of the global south, driving workers can work an average of 90 h during a week (12.8 h/day), DVS: 30 h; this generates high rates of tiredness and fatigue since, in addition, they only sleep an average of 5.4 h per day with a DVS of 2.3 h, while the desirable would be 8 h of sleep per day [14]. Under these work circumstances, approximately 62% of workers report that at least once a week they work "nodding off". This establishes the importance of regulating work incentives, since by attempting to meet goals outside the standard objectives, workers may be over-demanding, which can be detrimental to their health [3]. The aforementioned may be accompanied by additional health and operational performance complications, if it is considered that most workers report obesity and high body mass indexes, given the eating habits and sedentary lifestyle characteristic of this type of profession.

In order to better understand the problem of fatigue, use and performance of drivers, a study was designed and developed with experienced drivers (average of 10 years) in heavy vehicles used in open pit mining, in order to understand the relationship between the variables of habitability and usability of cabs, the associated cognitive processes and the links with fatigue, the study was carried out in a controlled and regulated context with agreements on duration, intensity and order of work shifts.

2 Methods and Tools

This study includes human activity recognition (HAR) as an analytical tool, this is a combined aspect approach, which consists of analyzing data from different sensor sources to identify features related to a person's activity [15].

This study was designed with a multi-method approach developed in three parallel scenarios, the first scenario focused on the study of the specific needs of the participating production units, the second scenario concentrated on the records of specific aspects of the drivers' performance in the truck cabs, in the third scenario on the analytics of the data obtained from the literature and the data from the records made with the workers.

In HAR research, wearable sensor based approaches including inertial measurement units (IMU) have been developed, in this approach multi-sensor based systems increase the effectiveness of activity recognition by combining data from multiple sensors, under this approach heavy vehicle drivers were instrumented with measurement devices (Capture U-Vicon for Kinematics; Empatica-E4 wristband for physiological measurements).

2.1 Procedure

Data recording was performed in 12 sessions in three observation cycles in both night and day shifts. In each work session with the operators, the purpose of the study was explained, the systematic observation of the cabin's habitability in natural working conditions was carried out, with analysis of distribution, frequency of use and usefulness of each piece of equipment.

For the instrumentation process, each participant was asked to identify his or her dominant hand, record his or her age and sex, and was given a registration number; this information was stored in a coded file. In the natural conduction operations, monitoring and storage of data was performed for 5-min periods, every 30 min during the operation cycles for each team examined. For the purposes of this study it was considered that driving tasks with path control, loading and unloading, are complex activities that usually combine physical and cognitive characteristics that represent the driver's daily task.

This study was included 37 drivers (31 men, 6 women) of heavy vehicles (Electric Truck, Tractor, Excavator, Motor Grader), with an average age of 37.4 years (STDV. 9.6), with a seniority in the position of driver/operator of 10 years (STDV 6. 5) in heavy equipment; these workers performed driving tasks in 12-h shifts, organized in 3Dx4Nx3R rotation, with an accumulated 84 h of work per 7 days (3 day shifts, followed by 4 night shifts and 3 recovery shifts).

2.2 Ethical Aspects

In accordance with the General Data Protection Regulation (GDPR), written informed consent was obtained from all participants prior to the start of data collection in order to obtain permission for the processing of personal data. All participants received information about the study and the type of data collected prior to any data capture.

They were given the opportunity to continue or withdraw from the study at any time without further questions. This process was carried out following the ethical protocols established by the ErgoMotion-lab laboratory. In addition, the data were stored anonymously, the identity of the participants is not disclosed, nor can this information be obtained from published data.

3 Results

The study began with the application of two questionnaires designed to determine the perception of visual fatigue and fatigue associated with the work activity in the participating drivers, two self-completed questionnaires were used. Two self-completed questionnaires were used; these questionnaires were applied during the operation. The first one, the H. Yoshitake Questionnaire (30 items), of the 36 participants, 15 (41%) reported discomfort or signs of cognitive fatigue. The second examined the aspects of perceived visual fatigue developed by Francois Calais,2017. Which examines the perception of visual fatigue on and off the job, the results indicate that 16 participants (14 males, 2 females), reported feeling visual discomfort, tingling and blurred vision.

In the second stage, a driver profile was constructed by examining the driving style, compliance with standards, road rules and mental adaptation to the driving task. From this analysis it was found that 100% of the participants had a prudent driving style, presented a low risk of non-compliance with standards and road rules, and 4 of them presented a medium risk of mental adaptation to the job. This first part showed the presence of fatigue indexes that would affect the drivers' performance, it also showed the importance of measurements in the different shifts that the drivers work.

Knowing these profiles, the analysis of the activity of the drivers was carried out using recording devices, first of all, an average variation of illumination in the cab during the day equivalent to 780 lx was obtained, with peaks of 1000 lx (glare), in summary, the obtained records of illumination in LUX inside the cabs found values ranging from 250 lx to 700 lx, depending on the time of day and the atmospheric state outside the cab. According to the standard for secondary driving tasks (control of equipment and data reading) these values should be between 300 lx and 700 lx, with an average of 500 lx, when there are values above the average, glare may occur, which makes reading difficult, increases eye strain and may be at the origin of visual fatigue.

The other determining aspect of cab habitability is the ratio of the internal and external temperature at the driver's place of operation. In the morning and night shifts, the average temperature is lower, 29.9 °C and 29.84 °C respectively, while in the afternoon it is 31.17 °C. In these vehicles, the thermal environment in the cab can be adjusted according to the worker's comfort thanks to the air conditioning installed in all the equipment. However, there were cases in which workers made temporary use of this element, establishing hot-cold periods, i.e., they turn off the air conditioning until the heat sensation generates discomfort and turn on the air until the cold sensation also begins to generate it.

Additionally, records of physiological variables such as heart rate, electrodynamic activity of the skin (EDA), accelerations in the driving position were made, finding that in the records made the variation of heart rate, are above 90 Ppm, which indicates the presence of effort, which may be related to the cardiovascular conditions of workers who present these values. The EDA records show peaks that refer to maneuvers inherent to the operation that capture the operator's attention and require control; however, the data shows that a continuous or elevated state of excitation is neither present nor likely to be present, which is at the origin of sustained stress states. This analysis allowed the identification of some drowsiness control strategies associated with cabin temperature management, the peripheral temperature data indicate a decrease in temperature that is associated with periods of wakefulness (Fig. 1).

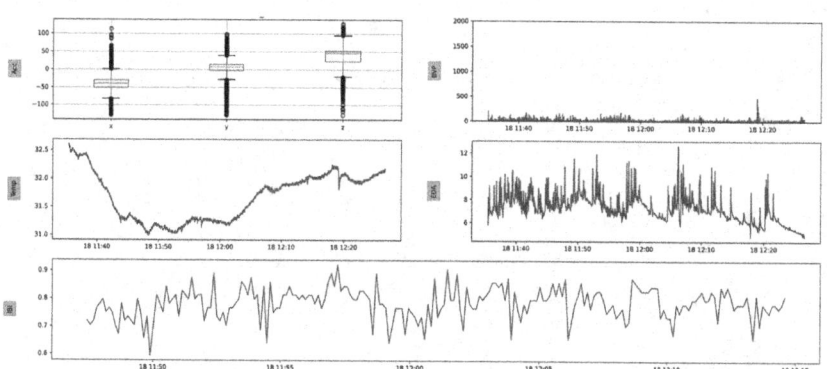

Fig. 1. Shows the records in the late afternoon shift change period.

The summary of graph 1, effectively indicates acceleration changes that are perceived by drivers with modification of control attention, which is manifested with changes in heart rate and electrodermal activation states, in general a variation behavior is observed with stress states that affect the performance of the worker. This also implies the presence of fatigue events associated with numbness or loss of attention.

In this order of ideas, the monitoring of events associated with fatigue in driving tasks, which includes the recording of slow blinking, closing of eyelids for more than 2 s and changes in the focus of attention to control the vehicle's trajectory, it was observed that the highest number of events occurs at night, specifically in the first part of the morning. (Fig. 2) This is associated with an indicator of fatigue that affects the performance of the vehicle operator.

Because of the shift rotation design, the events are distributed throughout the week with different frequencies. The analyzed record indicates that the highest number of events is recorded on Saturday (54 events), followed by Tuesday (47), Friday (36) and Monday (32), these events occur after having accumulated at least 70 h of driving activity, following the sequence of shifts designed. (Fig. 3.)

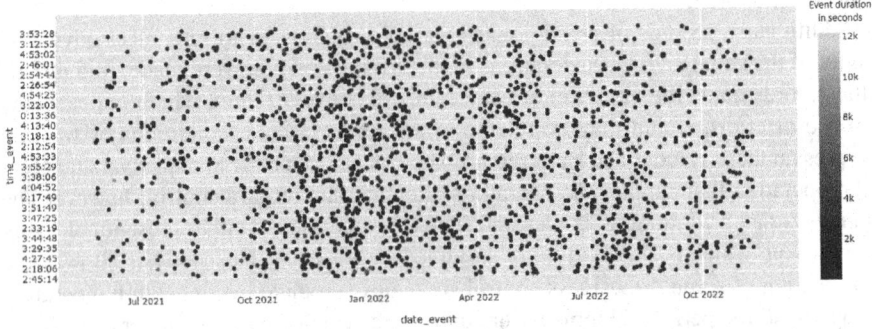

Fig. 2. Record of events over a period of 16 months. 95.4% (1302 events) occurred during the night shift.

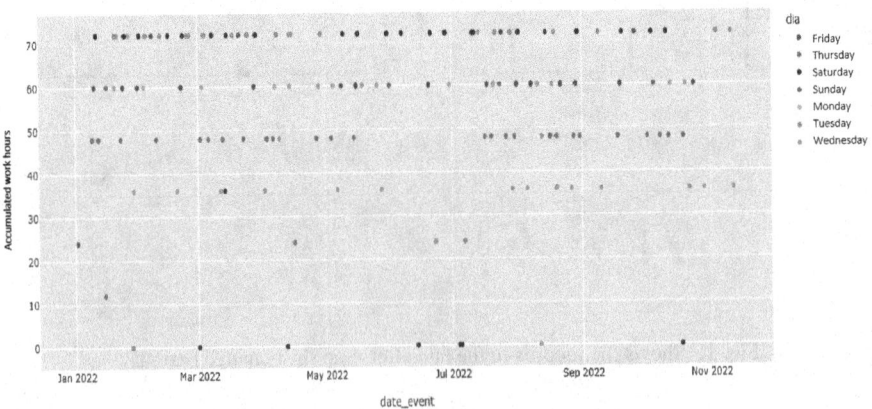

Fig. 3. The points of the diagram represent the events greater than 2 s, where "X" is the date, "Y" is the hours of work accumulated by the worker, the colors represent the days of the week.

The different data recorded show that the problem of cab usability in terms of driver performance is closely related to the influence of aspects of geometry and composition of the cabs on the one hand and on the other hand with the cumulative effects of working time, i.e. it is not only about understanding the impact of the 12-h shift and its effects on the modification of metabolic aspects and sleep hygiene of drivers, to understand the problem of driving performance, must also consider an aspect that influences performance as is the effect of the sleep debt. In fact, during recovery times, workers present changes in behavioral patterns, with intermittent periods of sleep, which is reflected in the records of events and in the number of events.

In fact, shift transitions can involve states of wakefulness in drivers that can reach 23 h in the workers participating in this study, which definitely has effects at the physiological level, concentration and cognitive availability for the effective development of the control tasks of driving heavy vehicles, including the control of body temperature, following instructions and the execution of work plans, in addition to the use of the working cab.

3.1 Contribution to the Industrial Performance

This study helped the company to analyze the relationship between the composition of work shifts, the annual composition of shifts and the working hours calculated for each year for the group of participants, allowing them to identify the intensity of the shifts, the cumulative duration per worker, the patterns of recovery and working time, focusing on the accumulated sleep debt of workers, which would explain the number of events recorded and the quality of life of these workers.

In fact, it was recommended that in this type of shift work organization models, whether 8 or 12 h, it is necessary to examine the worker's sleep debt status before the shift, this helps to understand the problems of use and utilization of the room resources for the worker's performance and for the productivity objectives of the organizations.

References

1. Hashemi Nazari, S.S., Moradi, A., Rahmani, K.: A systematic review of the effect of various interventions on reducing fatigue and sleepiness while driving. Chinese J. Traumatol. - English Ed. **20**(5), 249–258 (2017). https://doi.org/10.1016/j.cjtee.2017.03.005
2. Matthews, G., Desmond, P.A.: Task-induced fatigue states and simulated driving performance. Q. J. Exper. Psychol. Sect. A **55**(2), 659–686 (2002). https://doi.org/10.1080/02724980143000505
3. Gastaldi, M., Rossi, R., Gecchele, G.: Effects of driver task-related fatigue on driving performance. Procedia. Soc. Behav. Sci. **111**, 955–964 (2014). https://doi.org/10.1016/j.sbspro.2014.01.130
4. Aliakbari, M., Moridpour, S.: Managing truck drivers' fatigue: a critical review of the literature and recommended remedies (2016). https://doi.org/10.5281/ZENODO.1126720
5. Ahlström, C., Anund, A., Fors, C., Åkerstedt, T.: Effects of the road environment on the development of driver sleepiness in young male drivers. Accident Analysis and Prevention, 112(September 2017), 127–134 (2018). https://doi.org/10.1016/j.aap.2018.01.012
6. Choi, D., Sato, T., Ando, T., Abe, T., Akamatsu, M., Kitazaki, S.: Effects of cognitive and visual loads on driving performance after take-over request (TOR) in automated driving. Appl. Ergon. **85**(January), 103074 (2020). https://doi.org/10.1016/j.apergo.2020.103074

7. Moreno, C.R.C., Matuzaki, L., Carvalho, F., Alves, R., Pasqua, I., Lorenzi-Filho, G.: Truck drivers sleep-wake time arrangements. Biol. Rhythm. Res. **34**(2), 137–143 (2003). https://doi.org/10.1076/brhm.34.2.137.14487

8. Paper, C., Mar, F.S.: Aportes de la psicología a la investigación de incidentes/accidentes por fatiga: el trabajo de campo en el sector minero (October) (2015)

9. May, J.F., Baldwin, C.L.: Driver fatigue: the importance of identifying causal factors of fatigue when considering detection and countermeasure technologies. Transport. Res. F: Traffic Psychol. Behav. **12**(3), 218–224 (2009). https://doi.org/10.1016/j.trf.2008.11.005

10. Meng, F., et al.: Designing fatigue warning systems: the perspective of professional drivers. Appl. Ergon. **53**, 122–130 (2016). https://doi.org/10.1016/j.apergo.2015.08.003

11. Ferdinand, A.O., Menachemi, N.: Associations between driving performance and engaging in secondary tasks: a systematic review. Am. J. Public Health **104**(3), e39–e48 (2014). https://doi.org/10.2105/AJPH.2013.301750

12. Iseland, T., Johansson, E., Skoog, S., Dåderman, A.M.: An exploratory study of long-haul truck drivers' secondary tasks and reasons for performing them. Accident Analysis and Prevention, 117(December 2017), 154–163 (2018). https://doi.org/10.1016/j.aap.2018.04.010

13. Simonelli, G., et al.: Hours of service regulations for professional drivers in continental Latin America. Sleep Health **4**(5), 472–475 (2018). https://doi.org/10.1016/j.sleh.2018.07.009

14. Castro, J.R., de Gallo, J., Loureiro, H.: Cansancio y somnolencia en conductores de ómnibus y accidentes de carretera en el Perú: estudio cuantitativo. Rev. Panam. Salud Publica **16**(1), 11–18 (2004). https://doi.org/10.1590/s1020-49892004000700002

15. Climent-Pérez, P., Muñoz-Antón, A.M., Poli, A., Spinsante, S., Florez-Revuelta, F.: Dataset of acceleration signals recorded while performing activities of daily living, Data in Brief, **41**, 107896 (2022). ISSN 2352-3409. https://doi.org/10.1016/j.dib.2022.107896

Interaction with Automated Heavy Vehicles Using Gestures and External Interfaces in Underground Mines

Johan Fagerlönn[1]([✉]) [iD], Yanqing Zhang[2] [iD], Lina Orrell[1] [iD], and Hanna Rönntoft[1] [iD]

[1] RISE Research Institutes of Sweden AB, Borås, Sweden
`johan.fagerlonn@ri.se`
[2] Scania AB, Södertälje, Sweden

Abstract. The present study investigated the potential of using gestures to guide and control unmanned automated heavy vehicles in underground mine contexts, as well as the effects of adding external human-machine interfaces (eHMIs) to provide feedback during the gesture interaction. A study with 12 professional operators was conducted in a simulated mine environment, utilizing a Wizard of Oz methodology. The subjects used gestures to guide and control a heavy vehicle in three different scenarios in the mine, and different aspects of the user experience (UX) were assessed. The results support the notion that there is high potential in using gestures when operators stand in close proximity to the vehicles. Moreover, the results suggest that eHMI solutions can enhance the operator's acceptance and feelings of safety. The selected gestures seemed appropriate for the investigated scenarios, which should be valuable insights for practitioners intending to develop and implement gesture interaction for similar applications.

Keywords: Natural user interfaces · Gesture interaction · Interaction design · Autonomous vehicles · Self-driving vehicles

1 Background

In future confined industrial contexts (hubs), unmanned highly automated vehicles (AVs) and human operators may work in shared spaces (mixed traffic environments). Since a driver will no longer be available, new interaction models need to be developed to facilitate useful, satisfying, and safe human-vehicle interactions.

In recent years, gesture interaction has attracted great interest in both academia and industries, such as the gaming and automotive industries. The development of different gesture detection technologies has enabled this type of interaction. However, most studies have focused on the technical development of gesture recognition and the use of small hand gestures [1–3], and in entertainment or relaxing contexts, such as TV watching [4]. A few studies have also investigated gesture control's potential in safety-critical applications [5] and when controlling vehicle movements [3, 6]. However, the suitability of using arm gestures to interact with AVs has been largely unexplored.

© The Author(s), under exclusive license to Springer Nature Switzerland AG 2023
H. Krömker (Ed.): HCII 2023, LNCS 14048, pp. 255–267, 2023.
https://doi.org/10.1007/978-3-031-35678-0_17

The present study aims to investigate the potential of using gestures to interact with AVs in underground mines. The study examines the following main questions:

- What are the general attitudes among human operators toward using gesture interactions to control automated heavy vehicles? Are people comfortable using gestures to control the movements of the vehicles?
- How do human operators experience the selected gestures in terms of understandability, efficiency, and acceptance?
- Would it be valuable to use external human–machine interfaces (eHMI) to provide feedback during gesture interaction?

Procházka et al. [7] wrote that there are several aspects to consider when developers design their control mechanisms based on gestures. Key among them is simplicity; the gesture should neither be long nor complicated. Additionally, the gesture should be natural for the user, such as turning a book page. Furthermore, if possible, the gestures should not be designed by the developer, but by the target user group. Finally, good performance is an important factor; if there is no immediate feedback on whether the gesture was correctly done, the user will become confused.

There are limited opportunities to provide detailed feedback from vehicles to humans. In recent years, however, much research has been conducted into new types of vehicle-mounted eHMIs [8]. These interfaces aim to communicate, for example, intentions and actions of self-driving vehicles (e.g., that it will stop at a pedestrian crossing). By utilizing an eHMI, the vehicle can convey more advanced information and feedback to its environment.

2 Scenarios and Concepts

During the concept development, we focused on generating user stories for future underground mines. We identified three use cases where gesture interaction might be useful. Based on these use cases, we created three scenarios for the study. All these scenarios concern deviations from the normal autonomous operation, where humans provide input and guidance to solve the situation and uphold high levels of production efficiency and safety. Each scenario was implemented in a virtual environment for the study.

- **Stop an autonomous vehicle**: autonomous vehicles should be able to stop automatically when they approach a human. However, the possibility for humans to manually stop a vehicle can be an additional safety solution. For instance, humans may detect some issues (e.g., fire) that require an immediate reaction. The situation may for instance occur when maintenance people are working on a broken machine in a mine ramp. In this scenario, a human operator is standing in a mine ramp. An AV approaches the operator, and the operator stops the vehicle using a gesture.
- **Adjust a vehicle's position**: human operators in underground mines may need to adjust the position of AVs. The situation may for instance occur in complicated traffic situations that the automation can't handle, and where human situation awareness is needed. Opportunities to move the vehicle on-site can also be useful in situations where remote control operators (in a control room) for some reason cannot control the vehicle (e.g., due to a system failure). In this scenario, an AV is standing still in

a narrow part of the ramp and is blocking traffic. The operator uses gestures to guide the vehicle to a wider pocket in the ramp to leave room for other upcoming vehicles. The operator shows where the vehicle should stand, and the vehicle drives there and stops.

- **Guide a vehicle to pass around an obstacle**: In mixed traffic situations, AVs may stop if they don't know how to handle the situation in a safe way. In such situations, human operators may guide the vehicle and allow it to continue the operation. The situation may for instance occur if maintenance people are working on a broken machine in a ramp, and an operator allows the vehicle to drive around the machine and continue. In this scenario, an AV is approaching an operator in a partially blocked ramp. The operator uses gestures to guide the vehicle to drive around the obstacle and continue driving.

We developed interaction model concepts consisting of gestures (see Table 1) and feedback (eHMI) for each scenario. The design of gestures was based on an analysis of current common gestures used in traffic, as well as police and aircraft-marshaling gestures. We also consulted a literature review [1] on the classification of gestures.

Table 1. Description of the gesture concepts and their features according to Aigner et al.'s classification [1].

Scenarios	Stop	Adjust position	Navigate around
Gesture concepts	 Stretch your arm in front of your chest Palm faces towards the vehicle	 Arm bends on one side and one palm faces towards the vehicle Hand goes down and the arm stretches straight	 Arm waves toward the direction that you would like the AV to continuously move in
Features	Semaphoric		
	Static	Single stroke	Dynamic

The eHMI (see Table 2 and Fig. 1) consisted of LED light strips on all sides of the vehicle, which indicated the vehicle's intentions (e.g., that it will stop) or emphasized current vehicle dynamics, to provide feedback for gesture input. The design of the eHMI was not a focus of the study, so a concept developed and tested in other scenarios from another research project was used. However, we adapted the light design to fit the position and length of the specific vehicle.

Fig. 1. A The cabless automated vehicle with eHMI to signal vehicle's intentions in the underground mine simulation.

Table 2. Description of the external signals and light behaviors for the three scenarios.

Scenarios	Stop	Adjust position	Navigate around	
eHMI messages	Intent to stop	Intent to take off	Decelerating	Accelerating
Light behavior				

3   Method

3.1   Participants

Twelve participants took part in the study. Nine of them had experience working in mines, and three of them were tow truck drivers. Their ages ranged from 22 to 51 years (Mean = 36.2, SD = 10.1). Three of the participants (25%) were female. The level of work experience varied among the participants. The average years of experience in different tasks near heavy vehicles was 16.1 years (SD = 10.5).

3.2 Experimental Setup

We conducted the study in a lab with a large cloth display that showed the virtual environment of an underground mine (Fig. 2). We utilized a Wizard of Oz (WoZ) simulation method to move the vehicle in this virtual environment. WoZ is a widely used approach for evaluating user interfaces in various domains, such as robotics [9], mobile applications [10], and automotive industry [11, 12]. This approach is rooted in the idea of simulating a functioning technical system by a human operator – a wizard [13]. It is often used to collect data from users who think they are interacting with an automated system. In our experiment, the wizard, who was controlling the vehicle movements, sat in an adjacent room and was not visible to the test participants. He observed the subject's gestures and controlled the actions of the vehicle.

Fig. 2. A demo of how the participants acted in the study.

3.3 Procedure

Each trial included two rounds: one with eHMI and one without eHMI. The three scenarios were presented in the same order and the controlling gestures were the same in both rounds. The subjects conducted the rounds in a counterbalanced order. A video presented specific gestures for each scenario before the two rounds. Each scenario was introduced to make the subject aware of the context and situation that were to take place. The subject stood about two meters from the cloth, on a marked spot, for the entire scenario, which lasted approximately 0.5 to 1 min. When the subject performed the control gestures, the vehicle reacted with a slight delay (due to the wizard), and in the eHMI round, the feedback from the LED light strip started to change just before or simultaneously as any noticeable action from the vehicle. After each round, the participants answered questions and statements related to their trust in the system, acceptance and efficiency of the interaction, and other aspects of their user experience (UX). The session was concluded with a follow-up interview.

3.4 Measurements

UX was assessed using a combination of quantitative ratings and a short interview. Acceptance of the concepts was evaluated using the method presented by Van der Laan et al. [14], which consists of nine scales with opposite words (e.g., useful – useless). These scales are summarized into two dimensions of acceptance: usability and satisfaction. The usability dimension pertains to whether the solution supported the user in their task (e.g., effectiveness), whereas satisfaction relates more to the emotional aspects of the interaction (e.g., annoyance). The scales range between -2 and + 2. Trust towards the system was assessed using the method presented by [15], which consists of 12 scales focusing on different aspects of trust. The subjects also answered statements regarding whether the interaction felt efficient, natural, collaborative, and appropriate for real-world work. The scales ranged from 1 (do not agree at all) to 7 (completely agree). To gain further insights into the UX, a short interview was conducted with each subject after performing both conditions. In addition to UX, the subjects' behaviors while doing gestures were observed and recorded on video.

4   Results

4.1   User Experience

Table 3 demonstrates the results of the acceptance ratings evaluated using the method proposed by Van der Laan et al. [14]. Both concept variants yielded high mean ratings of usefulness and satisfaction. Paired Student's t-tests revealed no significant differences between conditions at the 0.05 alpha level.

Figure 3 shows the results of the statements related to UX. Overall, the statements yielded high mean scores, indicating that the solutions were perceived as efficient, natural, collaborative, and suitable for real-world work. Additionally, the majority of subjects indicated a preference for gestures, or gestures and eHMI, over a physical control or touch screen. A Paired Student's t-test revealed no significant differences between conditions at the 0.05 alpha level.

4.2   Trust and Concerns

The results of the 12 scales focusing on different aspects of trust in human-machine systems are shown in Fig. 4. In general, the two concepts got about the same mean scores and always on the side of the scales indicating higher levels of trust. Student's t-tests revealed no significant differences between conditions using the 0.05 alpha level.

When asked directly in the interview, all subjects responded that they felt that they had control over the vehicle and that they did not perceive the AV as threatening, which could reflect the high scores.

The statement "I'm wary of the system" had the lowest scores in both concepts. This was raised in interviews of some saying that they would be on their guard regarding the

Table 3. Mean acceptance scores for both conditions, Van der Laan et al. [14].

	Gestures and eHMI	Gestures
Usefulness	**1.5**	**1.5**
Useful - Useless	1.8	1.6
Bad - Good	1.6	1.4
Effective - Superfluous	1.3	1.5
Assisting - Worthless	1.4	1.7
Satisfaction	**1.6**	**1.4**
Pleasant - Unpleasant	1.4	1.3
Nice - Annoying	1.7	1.5
Irritating - Likeable	1.5	1.3
Undesirable - Desirable	1.7	1.5

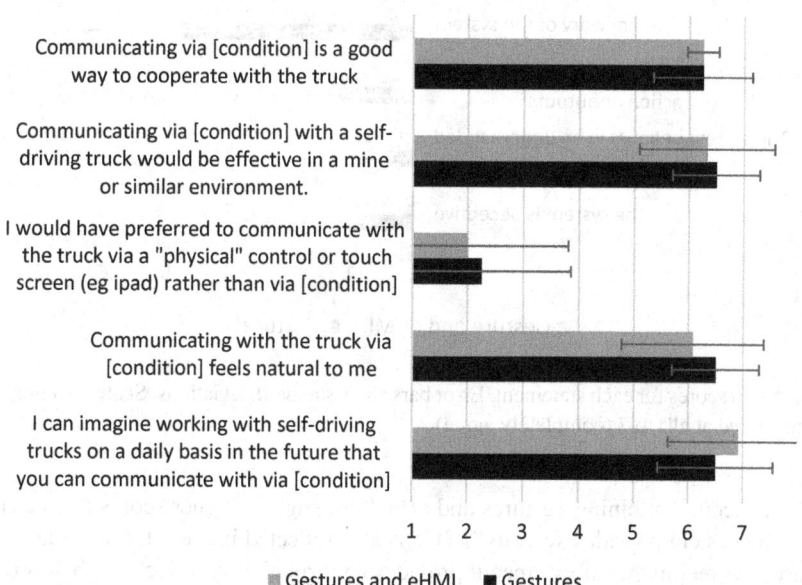

Fig. 3. Mean scores for each statement and condition. Error bars show standard deviations.

vehicle, which would also be the case if there was a driver, for example to be positioned with a certain margin of safety and the possibility to step away if the vehicle does not stop. In other words, these are regular safety precautions in the mining context.

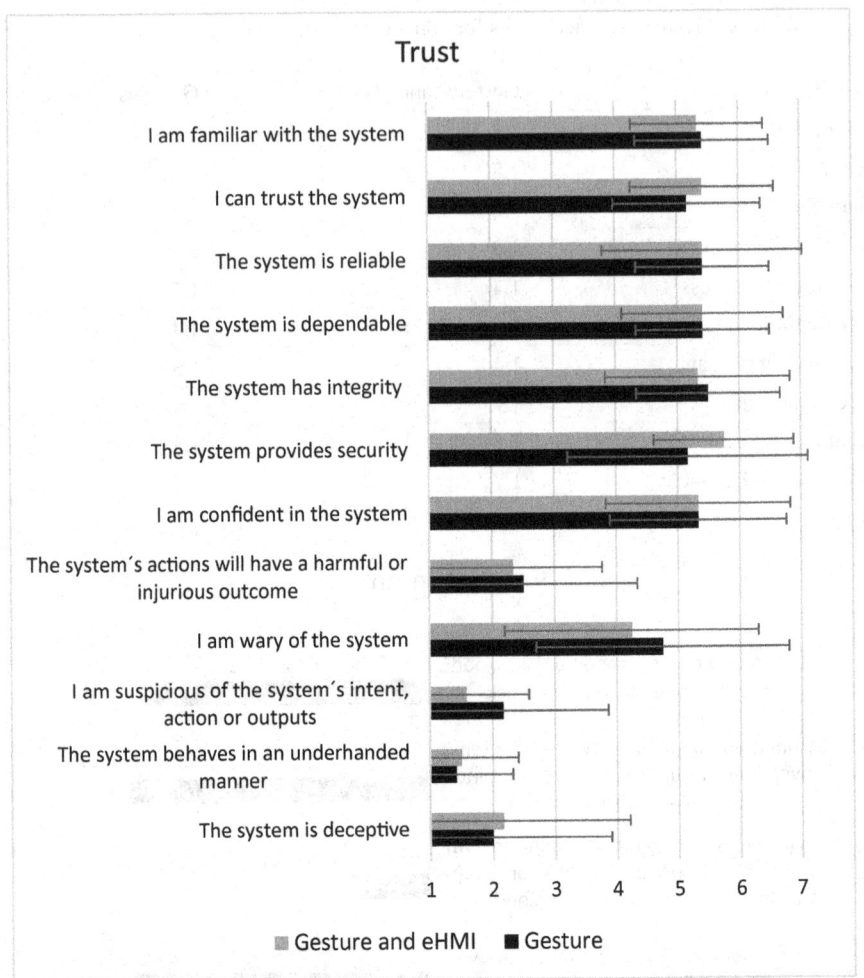

Fig. 4. Mean scores for each statement. Error bars show standard deviations. Scales ranging from 1 (do not agree at all) to 7 (completely agree).

The concept combining gestures and eHMI got slightly higher scores for the statement "The system provides security". This is also reflected in the interview that eHMI increases the feeling of safety since it provides confirmation (see Sect. 4.4 below).

For the statement "I am suspicious of the system's intent, action or outputs" the concept with eHMI also has somewhat higher scores. It was mentioned in the interviews that feedback is important, for example that the AV has received a command or noticed an obstacle. Above all, they thought it was most important that the vehicle indicates that it has registered that you are nearby, and that it provides a direct response.

In the interviews, some participants expressed concerns regarding the technical reliability of the proposed technology (gesture detection and eHMI) in harsh mining environments.

4.3 Gesture Behavior Analysis

Even though all participants watched the same gesture instruction videos, they performed the gestures in slightly different ways, especially in terms of gesture repetition and the number of different gestures performed in a single scenario (see Table 4).

Table 4. A comparison of the test participants' gesture behavior during the three scenarios.

Scenarios	Stop	Adjust position	Navigate around object
Summary	All subjects gestured in a manner very consistent with our original concept	Had the most discrepancies compared to our original concepts	All subjects did the gestures similarly to our original concept, but with different waving times
Performed in the same way by all subjects	Palm facing towards the vehicle	Point to one direction with one arm	Wave with one arm
Variations between subjects	Curve of the elbow	Some pointed several times; some ended with a stop gesture	Differences between subjects in terms of waving frequency and duration

As Table 4 shows, for the Stopping scenario, participants did quite similar to the instructions. There were only slight differences in the curve of the elbow. Some stretched their arms very straight, while others had a bit of bend. This is probably due to different comfort for individuals.

For the Navigate around scenario, all participants did the gesture similarly to our original concept, but with different frequency and duration. Some participants waved one time, but most of them waved several times. Most participants waved until the vehicle came close, while a few waved until the vehicle passed. This shows that while using certain gestures seem to be intuitive for people, the exact way they are performed are different for different individuals. For instance, some participants felt it natural to use the predefined gesture as a single stroke while some preferred using it as a dynamic gesture (waving several times).

Participants did gestures with the most varieties for the Adjust position scenario. Some participants pointed two or three times. This, again, shows that how many times gestures should be performed (single stroke or dynamic) has individual preferences. A few ended with a stop gesture as used in the Stop scenario. In the test, it was mentioned that there was an auto-parking function. A few participants still used the stop gesture.

4.4 External HMI as Feedback

The addition of eHMI did not result in any major differences in ratings related to acceptance or interaction efficiency. This may indicate that external interactions in this context probably rely a lot on vehicle kinematics than light eHMIs.

However, eHMI could still bring value in the sense that it can easily visualize the feedback from the system, once it has recognized the gestures. Our study implies that gesture control should be combined with feedback, 1) to confirm that the system has taken the command or recognized the gestures; 2) to inform users of which actions it is about to take. This could be done through eHMI. In our test, most subjects expressed positivity toward eHMI in the interviews. For instance, 10 subjects (83%) stated that gestures together with eHMI are perceived as safer than gestures alone as you receive confirmation via eHMI, and 9 subjects (75%) expressed that they would prefer to work with eHMI because of the confirmation and since it would feel safer.

The addition of eHMI did not result in any major differences in ratings related to trust, acceptance or interaction efficiency. However, from interviews it appeared that eHMI added other values within UX. For instance, 10 subjects (83%) stated that gestures together with eHMI are perceived as safer than gestures alone as you receive confirmation via eHMI, and 9 subjects (75%) expressed that they would prefer to work with eHMI because of the confirmation and since it would feel safer.

Nine (75%) subjects described the eHMI experience in positive terms, since they got feedback from the vehicle. Seven (58%) participants answered that they felt like they understood the eHMI´s signals. Yet fewer could express how they interpreted the signals when asked to give examples. There are a few mixed comments about the eHMI signals, the response time and some about the fact that there could also be different colors of eHMI to further clarify the meaning of the signals.

5   Discussion

The present study investigated the general attitudes of human operators towards using gestures to guide and control automated heavy vehicles. Specifically, it sought to understand whether people are comfortable with using gestures to control the movements of the vehicles, and how they experienced the selected gestures in terms of understandability, efficiency, and acceptance.

Overall, the participants were positive about using gestures to interact with the vehicle. They found the selected gestures to be natural and logical to perform, and they rated the interactions high in terms of both acceptance and efficiency. The subjective ratings also suggest that the operators found the gesture interaction suitable for real-world use. The selected gestures were considered appropriate for the selected scenarios, which provides valuable insights for practitioners intending to develop and implement gesture interaction for similar applications. In sum, the study indicates that there is good potential for gestures to facilitate interaction and collaboration between human operators and heavy automated vehicles in mines.

The study also explored whether an eHMI solution could provide beneficial feedback during gesture interaction. The eHMI was utilized to confirm the system's received commands and to inform users of its intended actions. However, the results did not demonstrate any considerable differences in ratings of UX or trust. This may indicate that users mainly depend on the vehicle kinematics to receive feedback (the user sees that the vehicle changes its behavior). However, most subjects stated that an eHMI would increase their feeling of safety since it provides confirmation of the interaction, and a majority expressed that they would prefer to have eHMI since it would feel safer. The findings support the notion that eHMIs can be a useful complement during gesture interaction, helping users to feel safer and more comfortable.

While eHMI may provide valuable feedback during gesture interaction, further investigation is needed to determine how to design eHMI for this specific purpose and if necessary to communicate the vehicle's intentions in addition to direct feedback of the received command. For instance, it is necessary to investigate which animations and colors are appropriate. The color should be consistent with the existing communication systems in mines. The color cyan was used for the solution in the present study. In the development and research of driverless vehicles today, cyan is commonly used for external communication between highly automated vehicles and other road users on public roads. However, further investigation is needed to determine if this is the most suitable color for eHMI in the mining industry.

The study investigated the use of gesture interaction in a limited set of scenarios in an underground mine. Considering the positive results found for these three scenarios, the study suggests that gesture interaction may be suitable for situations where emergent actions by humans in a close vicinity of the vehicle are needed. Our study showed that participants considered it useful and satisfying to interact with gesture in the selected scenarios in underground mining contexts:

1. stop an approaching AV when a human is working in a mine
2. guide a stationary AV to move to the side to leave room for other AVs in the mine ramp
3. guide an approaching AV to pass around a big obstacle (e.g., a broken vehicle or machine)

All these scenarios belong to deviations, which might not occur every day. They all need the personnel on site to act towards the vehicle based on their understanding of the situation so that the human beings feel safe, and the operation could continue.

The analysis of gesture behavior revealed variations in how the subjects performed the gestures, despite having watched instructional videos before the scenarios. As we employed a WoZ methodology, these variations did not pose any problems for the study. Nonetheless, the observed variations emphasize the need for future implementations to accommodate how different people naturally perform gesture commands. End users may need to be trained on how to correctly execute the commands, or the machine learning models handling the command recognition need to be able to accommodate the variations in order to understand the user's intentions. The analysis carried out in the study reveals some of the types of variations that can be expected for the specific gestures used in the evaluated scenarios.

Some participants expressed concerns regarding the technical reliability of the technology in harsh mining environments. This highlights that future implementations must offer reliable solutions that can handle dust, vibrations and other challenges to enable smooth interactions with gestures and avoid frustrations. This requires the technology to reach a certain level so that gesture control can function well, and that immediate and reliable feedback can be provided to the user.

The present study focused on scenarios for mining, but gesture control has the potential to be used in other contexts and situations. For instance, future automated vehicles running on public roads might be able to detect and understand the gesture by traffic police and follow their instructions. Thus, more research should be carried out to explore how gesture interaction can be employed to facilitate human-vehicle collaborations.

The results of the study should be interpreted with caution, as it was conducted in a simulated environment. The user experience and feeling of trust may differ when interacting with heavy vehicles in a real context. Additionally, while the study involved actual end-users, the sample size was rather small.

6 Conclusions

In summary, the study shows that there is good potential for gestures to facilitate interaction and collaboration between human operators and heavy automated vehicles in mines. The results also support the potential of combining gestures with eHMI solutions to enhance the operator's acceptance and feelings of safety. The observed variations in gesture performance between subjects highlight the importance of developing future gesture recognition technology that accommodates differently performed gestures. Future researchers should integrate gesture interaction and eHMI and evaluate these interactions using a real vehicle.

Acknowledgments. The work was funded by the Swedish partnership program Strategic Vehicle Research and Innovation (FFI) (Grant no: 2019–05898). It was conducted in a collaboration between Scania, RISE Research Institutes of Sweden, Boliden Mineral, AFRY and Icemakers.

References

1. Aigner, R., et al.: Understanding mid-air hand gestures: a study of human preferences in usage of gesture types for HCI. Microsoft Research TechReport MSR-TR-2012–111, 2, 30. (2012)
2. Xuan, L., Daisong, G., Moli, Z., Jingya, Z., Xingtong, L., Siqi, L.: Comparison on user experience of mid-air gesture interaction and traditional remotes control. In: Proceedings of the Seventh International Symposium of Chinese CHI, pp. 16–22 (2019)
3. Schulte, J., Kocherovsky, M., Paul, N., Pleune, M., Chung, C.J.: Autonomous human-vehicle leader-follower control using deep-learning-driven gesture recognition. Vehicles 4(1), 243–258 (2022)
4. Samimi, N., von der Au, S., Weidner, F., Broll, W.: AR in TV: design and evaluation of mid-air gestures for moderators to control augmented reality applications in TV. In: 20th International Conference on Mobile and Ubiquitous Multimedia, pp. 137–147 (2021)
5. Montebaur, M., Wilhelm, M., Hessler, A., Albayrak, S.: A gesture control system for drones used with special operations forces. In: Companion of the 2020 ACM/IEEE International conference on Human-Robot Interaction, p. 77 (2020)
6. Brynolfsson, N.: Investigating a gesture based interaction model, controlling a truck with the help of gestures. Dissertation (2021). http://urn.kb.se/resolve?urn=urn:nbn:se:umu:diva-191598
7. Procházka, D., Landa, J., Koubek, T., Ondroušek, V.: Mainstreaming gesture based interfaces, Acta Universitatis Agriculturae et Silviculturae Mendelianae Brunensis, vol. 61, no. 7, pp. 2655–2660 (2013). https://doi.org/10.11118/actaun201361072655

8. Debargha, D., et al.: Taming the eHMI jungle: a classification taxonomy to guide, compare, and assess the design principles of automated vehicles' external human-machine interfaces. Transp. Res. Interdiscipl. Persp. 7 (2020)

9. Hoffman, G., Ju, W.: Designing robots with movement in mind. J. Hum. Robot Interact. 3(1), 89–122 (2014)

10. Carter, S., Mankoff, J. Momento: Early-Stage prototyping and evaluation for mobile applications (2005). https://www.eecs.berkeley.edu/Pubs/TechRpts/2005/5224.html

11. Mok, B.K.J., Sirkin, D., Sibi, S., Miller, D, B., Ju, W.: Understanding driver - automated vehicle interactions through wizard of Oz design improvisation. In: Proceedings of the Fourth International Driving Symposium on Human Factors in Driving Assessment, Training, and Vehicle Design (2015)

12. Habibovic, A., Andersson, J., Nilsson, M., Lundgren Malmsten, V., Nilsson, J.: Evaluating interactions with non-existing automated vehicles: three Wizard of oz approaches. In: the Intelligent Vehicles Symposium (IV), IEEE. (2016). https://doi.org/10.1109/IVS.2016.753 5360

13. Steinfeld, A., Jenkins, O.C., Scassellati, B.: The oz of wizard: simulating the human for interaction research. In: The 4th ACM/IEEE International Conference on Human Robot Interaction, pp.101–107 (2009). https://doi.org/10.1145/1514095.1514115

14. Van Der Laan, J.D., Heino, A., De Waard, D.: A simple procedure for the assessment of acceptance of advanced transport telematics. Transp. Res. Part C: Emerg. Technol. 5(1), 1–10 (1997)

15. Jian, J.-Y., Bisantz, A., Drury, C.: Foundations for an empirically determined scale of trust in automated systems. Int. J. Cogn. Ergon. 4(1), 53–71 (2000)

Enhancing Operator Engagement in Safety Critical Control Rooms-Validating Influential Factors and Improving Interview-based Data Collection

Linyi Jin[1]([✉]), Val Mitchell[1], Andrew May[1], and Ning Lu[2]

[1] School of Design and Creative Arts, Loughborough University, Loughborough LE11 3TU, UK
l.jin@lboro.ac.uk
[2] School of Art and Design, Xihua University, Chengdu 610039, China

Abstract. Insufficient operator engagement poses danger to public safety. Identifying and validating the factors that influence Control Room Operator (CRO) engagement and choosing suitable research methods for collecting interview data is a crucial element of informing user experience design (UXD) interventions for enhancing engagement. However, this knowledge has been largely neglected. With the goal of filling this gap, a total of 18 CROs in two motorway control rooms were recruited. They were asked to participate in a two-stage interview to capture the influential factors of operator engagement. The findings corroborate that the factors influencing engagement in non-safety-critical domains still apply in the more safety critical control room context. This helps build a solid theoretical foundation for designing UXD interventions for enhancing operator engagement. In addition, this paper describes some inherent obstacles in using interview-based techniques for collecting the operator working experience -- especially engagement --, and makes suggestions for alleviating these obstacles.

Keywords: Operator engagement · engagement influential factors · interview-based methods

1 Introduction

Operator engagement in highly safety-critical control rooms is closely linked with system safety. However, the vast majority of the literature on enhancing performance has focused on technological improvement e.g., [1–3], human factors e.g., [4, 5] and elimination of risks e.g., [6, 7], whereas user experience (UX) has been largely overlooked. This research contends that it is essential that knowledge about how operators experience their work and engage emotionally should also be considered, when seeking to enhance operator engagement [8–10].

Operator engagement can be defined as a positive, fulfilling, confident, concentrated and even enjoyable experience related to immediate work. Furthermore, an increasing amount of evidence supports the positive link between workforce engagement, well-being and organizational outcomes, such as work performance [11].

H. Krömker (Ed.): HCII 2023, LNCS 14048, pp. 268–281, 2023.
https://doi.org/10.1007/978-3-031-35678-0_18

Many studies argue that individual engagement plays a crucial role in ensuring high operational standards in control rooms, for example, operator engagement can alleviate lapses in vigilance during air traffic control monitoring [12]. Operators experiencing engagement in a railway control room reported a sense of personal growth and increased competence [8]. Similarly, operators who reported positive engagement in a power plant control room said they had a sense of achievement and a more responsible attitude to their work [9]. However, engagement of operators in highly safety-critical control rooms is generally insufficient, such as for CCTV control [13], nuclear power control [14]. Motorway control rooms [10, 15] are not an exception, and this poses a hidden danger to public safety. Effective UXD for enhancing engagement, requires understanding of the factors that influence operator engagement and then choosing suitable research methods for collecting data to contextualize these. This paper describes a study an interview-based study, designed to identify these factors in the motorway control room context.

1.1 Factors that Influence Operator Engagement

Previous research has found engagement to be influenced by organizational factors (psychological meaningfulness, psychological safety, psychological availability [16, 17] and individual factors (personal traits and personalities [18]). Psychological meaningfulness can be as achieved when the work performance of employees brings them emotional, financial or physical benefits, so that they sense that have achieved something that is valuable, worthwhile and significant. Psychological safety allows an employee to behave and react naturally without fearing that this will result in undesired consequences in terms of career prospects or social relationships. Psychological availability enables an employee to believe that (s)he has the physical, emotional and psychological strength to cope with work assignments [16]. These four factors were all identified or validated in environments that were not highly safety-critical, such as the offices of an insurance firm [17], a summer camp and architecture practice [16]. These work contexts and associated tasks are very different from safety critical monitoring domains, undermining confidence that these four factors are relevant to operator engagement. To date, there are no reported studies that validates these influential factors in the actual context of highly safety-critical control rooms.

1.2 Methods for Collecting Data on Operator Engagement

The highly safety-critical control room context brings new challenges to the self-reported methods traditionally used for investigating engagement-related experience, especially interviewing. For example, in the study of operators' emotion [9], no operators said that they were bored when doing monitoring or operational work, and the researchers speculated that maybe this was due to the unacceptability of expressing a sense of boredom when doing this safety-related work. As interviews are frequently used in engagement related studies, e.g., [15, 19]. More work is needed to clarify where attention should be focused when interviewing operators in highly safety-critical control room contexts when the topic may be considered sensitive.

When being interviewed, operators can only describe past experiences of engagement. However, memory is not always reliable [20], and engagement, because it is an immersive and absorbed experience [11], may be difficult to recall. Therefore, a related concern is how to elicit from operators' reliable memories of engagement.

1.3 Aim and Questions

The overarching aim of this research was to evaluate the factors that influence operator engagement, and to identify the concerns needing attention regarding interview-based data collection techniques when collecting data on engagement relating to CRO experience in highly safety-critical control rooms.

The research questions addressed by this study were:

1) Are the factors identified in the literature effective in highly safety-critical contexts?
2) What are potentials of these factors for enhancing operator engagement?
3) What should attention be paid to when interviewing operators about sensitive topics?
4) How to improve these interview-based techniques?

This paper answers these questions by reporting on an empirical study based on two Chinese motorway control rooms (Fig. 1) where operators must remain vigilant as they observe and respond to a continuous stream of data regarding conditions on a motorway and be ready to respond to incidents.

Fig. 1. A Chinese motorway control room investigated in this study.

2   Methodology

The interview was in two stages. Firstly, the three step process [21] of transferring knowledge, was used to inform the participants about engagement. The specific procedures follow the sequence of 'perception, conception and abstraction'. This was to provide the participants with the necessary vocabulary and verbal expression for them to be able to articulate their experience of engagement, and to make sure that participants were able to identify the target experience before describing it, because engagement is a vast concept and hard to articulate [22].

Secondly, after the participants possessed sufficient understanding and vocabulary to describe their work engagement, the interview proceeded to next stage, which was to understand what factors from the perspective of the operators influenced operator engagement. A key premise for asking operators to articulate their engagement experience is that participants are able to recall relevant memories. Because memories are often unreliable [20, 23, 24], and also because engagement is an unconscious and immersive experience, participants will most likely forget detail when recalling past engagement with work tasks. Therefore, this research adopted a systematic approach to enhance recall of engagement. The cognitive interview (CI) method [20] was used to understand engagement in relation to critical incidents. Critical incidents including high engaging tasks and subtasks identified had been identified through review of control room operating procedures, card sorting (CS), fieldwork in the control room and task analysis [15]. The CI method ask respondents about the same target event through different paths to help them recall credible information [20]. This technique is often used to help police interview witnesses to recall details of incidents. After this, an open-ended interview was used to understand more holistically what characteristics influence engagement in the control room context and to build an understanding of ideal engagement from the point of view of the CROs. The interviews were audio recorded transcribed and stored in QSR NVivo software for qualitative analysis, and all data was subject to thematic analysis [25].

3   Methods

3.1   Sample

The recruitment of participants for this study was undertaken in two motorway control rooms in a southwestern China province. The researcher developed two criteria for the selection and recruitment of operators in traffic control rooms:

1. Interviewees should have more than one year's experience and an operator in a control room
2. Operators should be willing to take part in this study.

The gender of the operators was not included in the criteria. A total of 18 participants – thirteen participants in control room A and five participants in control room B – were recruited. All participants were aged between 29 and 55. There are two steps in current study. There were 13 operators in control room A who participated in all aspects of Step one, and 10 operators who participated in all aspects of in Step two. In addition, three operators participated in parts of Step two. In control room B, five operators participated in all aspects of both steps.

3.2   Procedure

Participants were asked to participate in two-step interviews to help them recall their memories related to engagement when performing work tasks. In addition to the recorded comments, the attitudes of the participants during the interviews were also recorded, such as whether they were willing to participate in the interview, and if they were not willing to participate in aspects of the interview, what caused the reluctance.

Step One. At the start of the interviews, participants were informed by the researcher about the definition of work engagement [26] that was also used to help inform participants the concept of engagement. Three concepts: absorption, dedication and vigor [26], are clearly illustrated. Then each participant was asked to explain his or her understanding of operator engagement to the researcher so that the researcher could ensure that their understanding of engagement was identical to the concept specified in this study, while enriching their vocabularies related to operator engagement so that they could explain their relevant experiences in the next step.

Step Two. The interviews progressed in the following three stages to help them recall their experiences related to operator engagement and disengagement.

1. Each operator was asked to rank tasks according to the degree of associated engagement by sorting cards, and she or he was required to describe their typical experience of each task in their own words (Fig. 2).
2. An operator was then asked to describe each step of each task in detail including their experience, thoughts, conditions and behaviors.
3. Each operator was then asked to suggest what kinds of task characteristics could increase or decrease the level of engagement while performing each task, and what kind of task they thought would create the ideal degree of engagement.

All research steps were carried out in the two actual motorway control rooms. Only when the operators were idle could researchers had the opportunity to interview. Sometimes the interview would be interrupted by the operators' sudden work. A complete interview may therefore take more than one day to finish.

Fig. 2. An operator was selecting cards in this study

3.3   Analysis

This study focuses on analyzing the understanding of engagement from the perspectives of the operators. Interviews were recorded, transcribed and imported into NVivo software 11. Codes were assigned to the following major themes that emerged from the data:

1) Engaging tasks and their corresponding characteristics.
2) Engaging parts of tasks and their corresponding characteristics.
3) Unengaging tasks and their corresponding characteristics.
4) The ideal engaging task and its characteristics.
5) Description of the experience of engagement.
6) Operators' feedback and attitudes relating to methods used for data collection.

4   Results

This study presents the characteristics that influence the engagement of operators in the context of Chinese motorway control rooms, and the obstacles CROs encountered during data collection. Influential factors are clustered in two main streams: organizational factors, and individual factors. Obstacles to data collection are also reported.

4.1   Factors that Influence to Operator Engagement

Organizational factors.
Psychological availability. It was found that psychological availability is strongly linked with operator engagement. They could feel engagement when they felt they possessed sufficient resources (e.g., competence and time) to cope with job demands. For example, operator C stated that, *"the ideal state is that when something happens, everyone will work together to complete it, instead of leaving it to one person"*. Another illustration is that two operators reported that an ideal example of a very engaging task was the experience of detecting the cause of a traffic jam by using CCTV cameras along the road. In such tasks, operators were sufficiently competent to successfully overcome difficulties. For instance, operator F reported, *"I checked the camera along the direction of the traffic jam to explore the cause of the traffic jam. This feeling was like finding a watermelon somewhere along a melon vine, or it was like a cat playing with a string ball... I felt interested, serious and had a sense of achievement. My mood was positive, it was difficult to extricate myself, and I felt that the time went by quickly"*. This competent, enthusiastic, meaningful and absorbing CCTV detecting experience is a good example of a high degree of operator engagement. This confirms the influence of psychological availability on operator engagement; a high psychological availability means that even if operators feel that a situation is somewhat difficult, they believe they are competent to overcome the challenges.

Psychological Meaningfulness. In the example above, the CRO also believed that what they were doing is contributing to road safety. Psychological meaningfulness is therefore another important predictor of the degree of operator engagement. Operators preferred to pay more attention to tasks which they thought were important for road safety. In terms of road safety, one example is that Operator G said, *"I pay particular attention to how I can express the information that is to be released on information boards in different locations to guide traffic flow effectively"*. Another example was reported by Operator H who said, *"I think detecting traffic jams during festivals by CCTV or on the TV wall is very engaging. When I had to do this recently, I felt very engaged, because detecting accidents by checking CCTV along the roads can help solving problems that affect many people. And I'm willing to invest effort in doing such tasks"*. Tasks that impacted personal growth were also perceived as meaningful and engaging. CROs reported that they felt much more engaged and significant at work when they met new unfamiliar tasks, because there was much to learn. For example, operator F said, *"Once I had to deal with an accident that I had never encountered before. The road patrol told me that a mud and rock flow had occurred in a tunnel. Because this was a big accident, I dared not report it directly to the superior department, for fear of making mistakes. So, I asked my manager, who went online to enquire about the nature of the accident. It took about an hour. Finally, we found out that it was not a mud and rock flow but a pipe explosion in the tunnel. I felt very committed when dealing with this matter. Although I resolved this situation with the help of others, I learned the importance of checking the facts before passing on information. I understood that, as an operator, I not only need to transmit information, but also that it is very important to use my own judgement."*

In contrast, operators felt unwilling to execute tasks when the purpose of the task was unclear. This happened when some tasks were perceived as dull and repetitive and without challenge, making operators feel very bored. For example, Operator M said, *"watching the CCTV for incidents on roads is very inefficient, because most incidents are not discovered by continuous CCTV surveillance. I feel insignificant and bored when frequently transmitting repeat information... After completing such tasks I couldn't remember anything about what I had been doing."* In line with that, a few operators reported little engagement in tasks which cannot reflect their own values or give them a sense of personal development. These included tasks which don't require any assessment or judgement. As operator E described, *"I'm required to record information and check the correctness of information repeatedly every day, I feel very bored and insignificant when doing tasks that could be completed by a five-year old boy, but I have to force myself to complete them seriously. No mistake is permitted, otherwise I will be held accountable."* This implies in addition to psychological meaningfulness, the fear of being reprimanded or penalized at work could also seriously influence an operator's experience at work.

Psychological Safety. During the interviews, the fear of punishment was mentioned frequently as the motivation for operators to force themselves to conduct their tasks seriously. However, because of this external motivation, they have a sense of being forced to work seriously, and they could not feel positive or fulfilled during the execution of such tasks. Rather, they feel stressed, worried and afraid. For example, operator C said: *"I must be very serious about collecting road information in an emergency, because this step is a prerequisite for subsequently submitting a correct report. I will be punished for an inaccurate report...to be honest, I just want to complete my duty correctly, and I do not feel engaged in this...how can I feel engaged about something that scares me"*. Another example was reported by operator D who said: *"I forced myself to concentrate* on releasing information for the information boards, and guiding motorway patrols to cope with accidents, because I had been punished for mistakes before... The punishment is very serious, including salary deduction and blame from supervisors". The absence of psychological safety has cast a shadow on the operators' engagement, because of the pressure caused by the threat of punishment, operators work seriously, but it is difficult for them to feel operator engagement. These insights confirm that psychological safety is an important predictor of operator engagement in this context.

Individual Factors. One of the conclusions from the interviews is that an operator who expressed a comparatively high sense of responsibility was more likely to feel engagement at work. It was found that a very small number of operators who expressed a very strong sense of responsibility could even find meaning in very repetitive and boring tasks and would force themselves to engage with them. For example, operator P mentioned: *"I think it's possible to find meaning in daily CCTV monitoring in the control room; even if it's very hard I will still insist on checking the CCTV, because I believe it can contribute to our monitoring work".* These responsible operators would like to pay more attention to the job. One example is operator F, who said: *"I would be worried about unfinished tasks after leaving work, and I would sacrifice my own time to finish tasks; for example, once there was a major accident which occupied the road for many days and had not been resolved properly. I was feeling very worried and nervous about it, so that I couldn't even sleep well at night."*

4.2 Obstacles Encountered in Data Collection

Almost every operator said repeatedly that it was difficult to describe the feeling of operator engagement, because they rarely experienced this feeling in their work, and only two operators reported experiences close to the theoretical definition of positive work engagement. Furthermore, one operator reported that the full engagement only happened in his personal hobbies, and eleven operators directly said that they usually felt bored and insignificant at work. Additionally, sixteen operators said that they did not enjoy most of their tasks, but they persisted in carefully completing the tasks because that is what they were expected to do. For example, operator F said, *"I don't like to repeatedly send the same information about the roads to my superiors, because I think that they don't respond to it, but because this is part of my job, I must do it."*

Four operators passively participated in the interviews because they questioned the significance of this study. They thought it was meaningless to ask them about their views on engagement at work. This may have been because they did not believe that any interventions could facilitate an experience of engagement in their work; they might have thought that their work was supposed to be boring. For example, operator A always answered questions passively and said to the researcher from time to time, *"I really don't understand. What's the point of asking me this? I do this because it is my job that's all. I don't expect my job to be very enjoyable."*

Furthermore, operators were very worried that their participation in this study would result in a negative impact on their work or others' opinions of them. This was clearly a concern because every operator repeatedly checked with the researcher about whether the researcher would pass on their comments about their work to their supervisors, and they all emphasized that they should participate anonymously in the study, because they thought that their frank evaluations of the work were often negative and did not meet the expectations of the managers.

What's more, the operators generally complained that the interview process was too cumbersome and repetitive. In the interviews, 16 participants complained many times about why they were asked to answer questions again and again about their experiences in performing their tasks. For example, operator G said: *"Didn't you ask this question yesterday? Why do you have to ask again?"* Three of operators refused to answer questions that seemed similar to what had been asked previously. For example, operator M said: *"You asked me yesterday. Please ask someone else"*.

5 Discussion

The findings from the qualitative study described in this paper supported the conclusion from the literature, that operator engagement in motorway control rooms usually depends on two broad groups of indicators: one is organizational factors, the other is individual factors (e.g., personal characteristics and traits [18]). Influential organizational factors include psychological safety, psychological availability and psychological meaningfulness [16, 17]. Individual factors are employees' personal characteristics and traits that can also influence their engagement at work [18].

In terms of the matters needing attention for interview-based data collection techniques, the participants' enthusiasm for participating in the interviews was generally not high, this is may have been due to three potential obstacles in the interview-based methods used to collect data in the current study: one is Social Desirability, the other is Naive Realism. Firstly, Social Desirability is the tendency of an individual to attempt to ensure that his or her behavior conforms to social norms and standards, and they seek to hide their socially undesirable actions [27]. Secondly, one of the central manifestations of Naive Realism is that people understand the world largely according to their own experience and think that this subjective view of the world is the real world. In addition to that, and repetitive questions asked according to the CI technique [20] also hinder the willingness of operators to participate in interviews.

Influential Factors Impacting CRO Engagement

Psychological availability. Operators who exhibit psychological availability were most likely to feel engagement. This is in line with previous research, operators who seek engagement want to be competent to overcome uncertainties at work (e.g., high equipment failure rates, blind spots, and unpredictable driver and passenger behaviors [8]), and operators who are familiar with work procedures generally show similar experience or behaviors relating to engagement, they have a positive attitude towards their work and perform well [28]. However, when operators cannot control their work situations, they usually have a negative and stressful experience, resulting in incompetence and resignation (not caring) [9], and this can therefore damage work performance. This is similar to the situation in the current research when motorway CROs described an ideal working state as when they have the capacity to complete work or overcome difficulties. As described by operator D, *"I hope there can be a balance in work, not too busy and not too idle.* " The current study also found that a certain degree of work stress and demand can stimulate work performance. This resonates with the work about railway operators [29], which found that an appropriate level of stress can optimize the performance of operators, but excessive or insufficient stress may damage their performance. Therefore, it is recommended that creating "just the right stress and competence requirement" for operators should be further explored in this context.

Psychological Meaningfulness. Psychological meaningfulness is another factor that can significantly influence CRO engagement, the findings in the current study show that operators are willing to carry out the tasks they think are meaningful in an engaged manner. This is consistent with one study based on power plant control rooms where power plant operators said that they generally worked seriously and very hard, because they thought their work was significant to public life [9]. However, the majority of the operators in the two Chinese motorway control rooms stated that they felt their work was often mundane and meaningless. They thought much of their work related to transmitting information repetitively to different organizations. This shows that the lack of psychological meaninfulness in this context is a problem which needs further research to inform the design of interventions that aim at improving engagement.

Psychological Safety. Although it appears that studies that specifically focus on investigating factors that influence CRO engagement were not found in mainstream literature, some studies concerning safety-critical control rooms confirm indirectly the rationality of applying these influential organizational factors in the context of motorway control rooms. Regarding psychological safety, a qualitative study based on a railway control room showed that operators whose managers tolerated mistakes that impacted only system efficiency rather than system safety showed positive attitudes towards making mistakes. They regarded making mistakes at work as a means of learning, and reported that they could experience engagement in their work [8]. However, motorway operators observed in the current study generally suffered from a lack of psychological safety and were always afraid of being punished for mistakes. Even when working conscientiously they rarely felt engaged [15].

Individual factors. Like the conclusions drawn for other work environments, individual factors, e.g., personality traits and characteristics, are also one of the important factors influencing engagement [30]. As in current study, operators who showed a comparatively high sense of responsibility seemed to sometimes experience engagement. For example, among the eighteen operators, operator F was one of only two operators who explicitly mentioned that they had experienced engagement at work. In the interview, she showed a stronger sense of responsibility than most of the other operators; for example, unlike most operators who considered monitoring work as a routine, without the need of initiative, she was one of the few operators who said she would still care about completing unfinished tasks, even after the end of her shift. However, this study only summarizes the influence of personal characteristics on engagement based on the operators' attitudes towards their work as they expressed them during the interviews. A more systematic and in-depth approach may be beneficial. In future research, it is recommended to systematically consider the influence of an operator's personality on his/her work. For example, some widely recognized scales related to engagement, e.g., Core Self-Evaluations (CSE)

[31] and the Five-factor model (FFM) [32], could be used to assess the personality of an operator, so that the impact of the operator's personality on engagement could be determined.

The above information suggest that the factors impacting operator engagement in highly safety-critical control rooms could be largely consistent with those impacting individual engagement in other work environments [30]. Clarifying the factors that can influence operator engagement contributes to identifying the possibilities for influencing operator engagement from a psychological perspective, and it also suggests a promising path for optimizing operator performance and working experience, and for enhancing well-being in the workplace. However, due to the general lack of UXD related research in the field of highly safety-critical domains [9, 28], there exists insufficient guidance on how to collect data on CRO working experiences. Reducing the obstacles to collecting data on operator working experience was therefore also a focus for this research.

Methodological Issues with Interviewing. Interviewing is one of most important and practical methods for collecting experience-related data. The two main obstacles facing interview-based data collection are Social Desirability and Naive Realism. Firstly, Social Desirability may lead to operators of highly safety-critical systems being reluctant to give their true thoughts on some safety related issues. For example, when it came to the reasons that made them feel a sense of disengagement at work, some operators always chose their words carefully, and some of them sometimes denied that they had such experiences, especially when they were unfamiliar with the researcher. Similar situations have been reported in other types of highly safety-critical control rooms e.g., [9].

In order to mitigate these negative effects brought about by Social Desirability when collecting interview data, the researchers in future studies should take time to build rapport [33] with the operators, for example, chatting with operators and trying to build trust with them and reassuring them about confidentiality. Another possible intervention that could alleviate Social Desirability is to use more empathic methods [34] because it is believed that interviewees are more willing to express themselves when they feel understood by others. Future studies could, for example use videos that portray similar situations to operators, such as similar problems encountered at work, to stimulate a more honest and unguarded sharing of information. In addition, generative techniques that aim to encourage the exchange of latent information could also be considered [35], such as asking operators to make a drawing to illustrate their experience, and then encouraging them to describe and explain their drawing.

Secondly, because of Naive Realism, it may be difficult for the motorway operators to describe in detail their specific experiences of engagement at work because they have experienced engagement only rarely. Therefore, it is recommended to expand the sample sizes in future work to address the impact of Naive Realism. This could be achieved by collecting data in other control rooms where operators report experiences of operator engagement, such as the case of operators in a power station control room who said they felt pride and contentment and worked seriously, because they regarded their work as significant to public life [9]. Another case was reported in one study about a railway control room, where the operators said that they felt highly engaged in some tasks, and they had a sense of motivation, concentration, time passing quickly, psychological safety and satisfaction [8]. Furthermore, due to Naive Realism, many participants might be unable to imagine a solution that could improve their engagement, because they think their work should be depressing. Possible alleviations to Naive Realism could be illustrating some achievements that can help improve operators' working conditions and experience successfully. For example, technology that enables motorway operators to effectively identify hazardous road conditions e.g., [1, 3]. This may allow them to see the possibilities for improving what is often tedious and inefficient monitoring work.

In addition to the above two obstacles, the high degree of participation required by current interview-based research methods may make participants feel tired which further hinders data collection. Almost all operators complained that they were repeatedly asked the same questions, perhaps caused by the three step process for transferring knowledge [21] and CI [20]. Indeed, these approaches were effective in helping operators recall related experiences of engagement, because as discussed above, the findings about factors affecting operator engagement collected from operators are supported by many existing related studies. But relatively speaking, CI may be more practical and applicable in situations where the interviewees can easily see the purpose of the interview, such as when the police are questioning witnesses. In this work context however, the largely interview-based technique CI was found to be complicated and laborious because the repetition in this approach greatly reduces the enthusiasm of participants and consumes their patience. In future research, it is therefore recommended to consider research methods that are less time intensive or at least less intrusive, such as observation, or reduce the difficulty of participation by letting operators choose from a list of possible answers instead of requiring them to create their own answers.

Observation-based methods less intrusive, in the sense that they hardly disturb the operator's work. However, the effectiveness of this method as a means to understand engagement largely depends on the researcher having a deep understanding of the tasks and work context. A video-based observation method could alleviate some of, disadvantages as it can be analyzed repeatedly and deeply. However, the approval of filming in the highly-safety critical systems is usually complex, and asking operators to participate in being filmed more difficult than being interviewed. In addition, from a video the wider context of the work could easily be lost, for example, the nature of the incident being dealt with and the role of others in addressing it. Therefore, researchers are advised to have an immediate understanding of CRO work before observation. In general, it is recommended to use multiple methods for collecting engagement. For example, firstly use interviews systematically understand operator tasks and then use observation in the control room context to broaden understanding but also to objectively identify signs of low or high engagement through monitoring of body posture or heart rate e.g., [36]. Further work by the authors (under review) has gone on to explore the use of Machine Learning as a way to augment understanding of engagement using video analysis of working posture.

The choice of methods for collecting data on operator experience in the context of highly safety-critical domains should therefore be in accordance with the conservative culture common to safety critical environments [28].

6 Conclusion

The four influential factors impacting work engagement: psychological safety, psychological availability, psychological meaningfulness and individual factors, have been confirmed as relevant to the CRO context by this study. These provide a solid theoretical basis for guiding future experience design interventions from a theoretical perspective. In addition, some inherent obstacles to using interview-based techniques for collecting CRO work experience, especially engagement, have been described. These can be used in related studies to avoid obstacles in the collection of data on sensitive work experience by improving interviewing techniques. Furthermore, the overall findings of this study indicate that CRO's in this cultural context are likely to experience low levels of engagement. There is therefore great potential to explore the potential to use experience design to enhance work engagement in this context, and consequently enhance the overall performance the safety critical motorway system.

References

1. Sarikan, S.S., Ozbayoglu, A.M.: Anomaly detection in vehicle traffic with image processing and machine learning. Procedia Comput. Sci. **140**, 64–69 (2018)
2. Peppa, M.V., Bell, D., Komar, T., Xiao, W.: Urban traffic flow analysis based on deep learning car detection from CCTV image series. Int. Arch. Photogramm. Remote Sens. Spat. Inf. Sci. - ISPRS Arch. **42**(4), 565–572 (2018)
3. Formosa, N., Quddus, M., Ison, S., Abdel-Aty, M., Yuan, J.: Predicting real-time traffic conflicts using deep learning. Accid. Anal. Prev. **136**(December), 2020 (2019)
4. Reinerman-Jones, L., Matthews, G., Mercado, J.E.: Detection tasks in nuclear power plant operation: vigilance decrement and physiological workload monitoring. Saf. Sci. **88**, 97–107 (2016)
5. Gao, Q., Wang, Y., Song, F., Li, Z., Dong, X.: Mental workload measurement for emergency operating procedures in digital nuclear power plants. Ergonomics **56**(7), 1070–1085 (2013)
6. Zayed, T., Amer, M., Pan, J.: Assessing risk and uncertainty inherent in Chinese highway projects using AHP. Int. J. Proj. Manag. **26**(4), 408–419 (2008)
7. Chang, L.Y.: Exploring contributory factors to highway accidents: a nonparametric multivariate adaptive regression spline approach. J. Transp. Saf. Secur. **9**(4), 419–438 (2017)
8. Smith, P., Blandford, A., Back, J.: Questioning, exploring, narrating and playing in the control room to maintain system safety. Cogn. Technol. Work **11**(4), 279–291 (2009)
9. Schaeffer, J., Lindell, R.: Emotions in design considering user experience for tangible and ambient interaction in control rooms, vol. 22, no. 1, pp. 19–31 (2016)
10. Jin, L., Mitchell, V., May, A.: Understanding engagement in the workplace: studying operators in Chinese traffic control rooms, pp. 653–665, July (2020)
11. Bakker, A.B., Demerouti, E.: Towards a model of work engagement. Career Dev. Int. **13**(3), 209–223 (2008)
12. Pop, V.L., Stearman, E.J., Kazi, S., Durso, F.T.: Using engagement to negate vigilance decrements in the NextGen environment. Int. J. Hum. Comput. Interact. **28**(2), 99–106 (2012)
13. Smith, G.J.D.: Behind the screens: examining constructions of deviance and informal practices among CCTV control room operators in the UK. Surveill. Soc. **2**(2/3), 376–395 (2002)
14. Izsó, L., Antaiovits, M.: An observation method for analyzing operators' routine activity in computerized control rooms. Int. J. Occup. Saf. Ergon. **3**(3–4), 173–189 (1997)

15. Jin, L., Mitchell, V., May, A., Sun, M.: Analysis of the tasks of control room operators within Chinese motorway control rooms, vol. 1, no. 2016, pp. 526–546. Springer International Publishing (2022)
16. Kahn, W.A.: Psychological conditions of personal engagement and disengagement at work. Acad. Manag. J. 33(4), 692–724 (1990)
17. May, D.R., Gilson, R.L., Harter, L.M.: The psychological conditions of meaningfulness, safety and availability and the engagement of the human spirit at work. J. Occup. Organ. Psychol. 77(1), 11–37 (2004)
18. Simpson, M.R.: Engagement at work: A review of the literature. Int. J. Nurs. Stud. 46(7), 1012–1024 (2009)
19. Koskinen, H., Karvonen, H., Tokkonen, H.: User experience targets as design drivers: a case study on the development of a remote crane operator station. In: Proceedings of the 31st European Conference Cognitive Ergonomics, pp. 1–25 (2013)
20. Memon, A., Bull, R.: The cognitive interview: its origins, empirical support, evaluation and practical implications. J. Community Appl. Soc. Psychol. 1(4), 291–307 (1991)
21. Ross, B., Munby, H.: Concept mapping and misconceptions: a study of high-school students' understandings of acids and bases. Int. J. Sci. Educ. 13(1), 11–23 (1991)
22. Burnett, J.R., Lisk, T.C.: The future of employee engagement: real-time monitoring and digital tools for engaging a workforce. Int. Stud. Manag. Organ. 49(1), 108–119 (2019)
23. Edward Geiselman, R., Fisher, R.P., MacKinnon, D.P., Holland, H.L.: Eyewitness memory enhancement in the police interview: cognitive retrieval mnemonics versus hypnosis. J. Appl. Psychol. 70(2), 401–412 (1985). https://doi.org/10.1037/0021-9010.70.2.401
24. Köhnken, G., Milne, R., Memon, A., Bull, R.: The cognitive interview: a meta-analysis. Psychol. Crime Law 5(1–2), 3–27 (1999)
25. Braun, V., Clarke, V.: Using thematic analysis in psychology. Qual. Res. Psychol. 3(2), 77–101 (2006)
26. Schaufeli, W., Salanova, M., González-romá, V., Bakker, A.: The measurement of engagement and burnout: a two sample confirmatory factor analytic approach. J. Happiness Stud. 3(1), 71–92 (2002)
27. Zerbe, W.J., Paulhus, D.L.: Socially desirable responding in organizational behavior: a reconception. Acad. Manag. Rev. 12(2), 250–264 (1987)
28. Savioja, P., Liinasuo, M., Koskinen, H.: User experience: does it matter in complex systems? Cogn. Technol. Work 16(4), 429–449 (2013). https://doi.org/10.1007/s10111-013-0271-x
29. De Felice, F., Petrillo, A.: Methodological approach for performing human reliability and error analysis in railway transportation system. Int. J. Eng. Technol. 3(5), 341–353 (2011)
30. Bakker, A.B., Demerouti, E., Sanz-Vergel, A.I.: Burnout and work engagement: the JDR approach. Annu. Rev. Organ. Psychol. Organ. Behav. 1, 389–411 (2014)
31. Judge, T.A., Erez, A., Bono, J.E., Thoresen, C.J.: The core self-evaluations scale: development of a measure. Pers. Psychol. 56(2), 303–331 (2003)
32. Goldberg, L.R.: Personality processes and individual differences - an alternative description of personality: the big-five factor structure. J. Pers. Soc. Psychol. 59(6), 1216–1229 (1990)
33. Kawulich, B.B.: Participant observation as a data collection method, Forum Qual. Sozialforsch., 6(2), (2005)
34. Stappers, P.J., Visser, F.S.: Bringing participatory techniques to industrial design engineers, DS 43 Proceedings of E PDE 2007, 9th International Conference on Engineering and Product Design Education, no. September, pp. 117–122 (2007)
35. Visser, F.S., Stappers, P.J., van der Lugt, R., Sanders, E.B.-N.: Contextmapping: experiences from practice. CoDesign 1(2), 119–149 (2005)
36. Bustos-López, M., Cruz-Ramírez, N., Guerra-Hernández, A., Sánchez-Morales, L.N., Cruz-Ramos, N.A., Alor-Hernández, G.: Wearables for engagement detection in learning environments: a review. Biosensors 12(7), 1–30 (2022)

"G" Classes for Vehicles Classification According to Size and the Justification of "Fine Mobility"

Sophie Elise Kahnt[1]([⊠]), Jori Milbradt[1], and Carsten Sommer[2]

[1] Chair of Transportation Planning and Traffic Systems, University of Kassel, Kassel, Germany
{sophie.kahnt,jorimilbradt}@uni-kassel.de
[2] Department of Transport Planning and Systems, University of Kassel, Kassel, Germany
c.sommer@uni-kassel.de

Abstract. As a counter to the continuous increase in size of passenger cars and the share of sport utility vehicles in traffic, we focus in this paper on the segment of fine means of transport that can be classified below the classic passenger car/small car. We have called mobility with these fine means of transport "fine mobility". The broad spectrum of means of transport of fine mobility has not yet been perceived as a coherent segment and cannot be captured by any existing vehicle categorization. Therefore, a new classification was developed to define fine mobility, which classifies all road-based means of individual transport according to their (urban) spatial effects into the seven classes XXS to XXL. The classification makes it possible to clearly distinguish fine mobility means of transport - from micro mobiles to mobility aids, bicycles of all kinds to light electric vehicles - from coarse means of transport - from mid-range cars to off-road vehicles - on the basis of the characteristics of the spatial dimension (cuboid around the external dimensions of vehicles) and the turning circle. In this context, the classification builds on various use cases that may condition the promotion of fine mobility through its preference in moving and stationary traffic. The diversity of fine mobility means of transport in terms of their possible uses and transport capacities illustrates the shift potential of many urban as well as city-regional car trips to finer alternatives.

Keywords: fine mobility · vehicle size · vehicle classification · vehicle categorization

1 Fine Mobility: The Relevance of the Size of (Road-Based) Means of Transport

How can the sustainable turnaround in traffic be advanced? The priority areas of action discussed are the avoidance of motorized individual transport (MIT) (including enabling home offices), the shift from MIT to the environmental network, the transition from private car ownership to vehicle and trip sharing (car sharing, carpooling), and the drive system turnaround by switching fossil drives to electric drives. In addition, it is sometimes explored whether and to what extent the digitalization and autonomization of vehicles reduce the environmental impacts [1] and the accident causation risk of private motorized transport [2].

H. Krömker (Ed.): HCII 2023, LNCS 14048, pp. 282–295, 2023.
https://doi.org/10.1007/978-3-031-35678-0_19

Grant-aided by the German Federal Environmental Foundation (Deutsche Bundes-stiftung Umwelt DBU) we look into one important factor of the negative effects of car-oriented traffic that has so far been largely overlooked: the size of the vehicles we use to get around. Even if we share vehicles or drive electrically, it makes a difference for our quality of life and urban life as well as for environmental and climate protection whether we use agile or bulky, small or voluminous vehicles.

There are three parallel negative developments here. First, passenger cars have become larger and heavier from model series to model series over the past decades [3]. Figure 1 shows by way of example how the VW Golf (and an average man) grew in size in the first three decades and extrapolated what the car will look like after another three decades in 2040 if there is no change in the current trend.

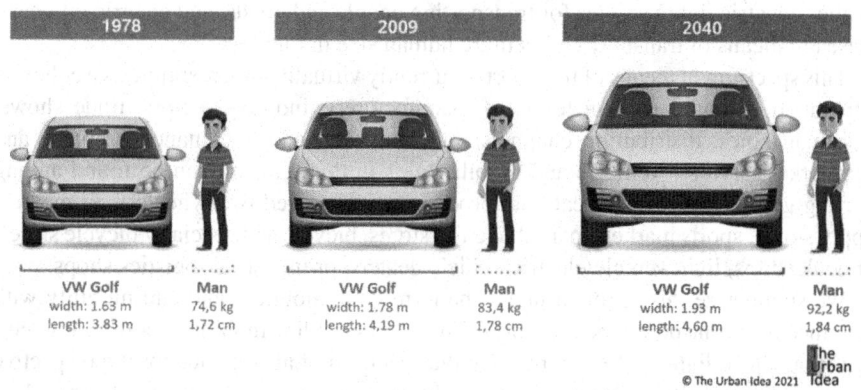

Fig. 1. The size growth of passenger cars using the example of the VW Golf [4].

Secondly, more and more car users are switching from their previous models to larger models (vans, crossovers, SUVs, off-road vehicles). The share of SUVs in new registrations has been rising steadily for years in Europe and is now just under 30% in Germany [5]. Third, more cars are hitting the roads and car density (cars per 1000 inhabitants) continues to rise: In the last decade alone, from 2010 to 2019, it increased by twelve percent in Germany [6]. The Forschungsgesellschaft für Straßen- und Verkehrswesen e. V. (German Road and Traffic Research Association), responsible for setting the technical regulations for traffic engineering, is responding by increasing the size of the design vehicle for passenger cars, so that, among other things, the dimensions of parking spaces are also increasing [7].

At present, the impression is growing that car manufacturers are supporting the trend by thinning out their range of so-called mini and small cars in order to give priority to selling larger vehicles. These generate a higher profit margin. Very small cars are often only presented as trade show vehicles or for image campaigns [8].

In contrast, startups and companies that develop small, lightweight vehicles for fine mobility and bring them to market have a hard time: quite a few had to give up before their vehicle reached the mass market [9]. The German government supports the purchase of battery electric as well as fuel cell vehicles, but not light electric vehicles [10].

Motor vehicles are categorized by different approaches. Among passenger cars, a distinction is made between small, medium and luxury cars (see Sect. 1.3). Since the spread of pedelecs and cargo bikes as well as electric pedal scooters and Segways, bicycles and mini electric vehicles have also been subdivided. However, there is as yet no comprehensive classification of all road-based means of private transport in Germany and Europe.

1.1 Counter Design: Fine Mobility

The fact that, contrary to the development described above, there is a finer way of individual transport is shown by the already existing and currently growing sector below the scale of today's automobiles: fine, small, light and maneuverable means of transport (i.e., mobility aids and vehicles) of the so-called micro mobility, bicycles and cargo bikes of all types, velo mobiles, mobility aids such as wheelchairs and senior citizens' mobiles, electric light vehicles for transporting people and goods, and electric minicars. These are means of transport with a more human size in mind.

This spectrum of means of transport is currently virtually imperceptible as a cohesive segment. It is fragmented in terms of manufacturers, industry sectors, trade shows, trade magazines, distribution channels, sales outlets, and maintenance-, repair-, and aftermarket services. Vehicles and mobility aids in this segment can be found among sporting goods, toys and recreational products or motorized two-wheelers, in medical supply stores, sports markets, parent & child stores, bicycle and specialty bicycle stores, car dealerships, light vehicle/electric vehicle dealers, or industrial logistics shops.

We summarize this segment under the term "fine mobiles" and call mobility with these means of transport "fine mobility". The concept of fine mobility is aimed at selecting the smallest, lightest, finest option in the spectrum of all vehicles for the respective area of application. It thus is mobility that is both economical, and ecological. According to this, mobility needs and transport purposes are met with the

- smallest, lightest, most energy-saving and cost-effective as well as
- lowest emissions, most space-saving, quietest and most resource-efficient

Transportation option [11].

1.2 What is Fine Mobility Supposed to be Good for?

In order to replace a large proportion of today's car trips, fine mobility must not be limited to urban traffic, local and neighborhood mobility, or the so-called "last mile". If we want to move from "coarse" to fine mobility and if the turnaround in traffic is to be environmentally effective, there must be viable "fine mobility options" for distances of up to 100 km. After all, 75% of the annual mileage of passenger cars in Germany is accounted for by journeys of less than 100 km. [12].

Furthermore, vehicles of the fine mobility must take into account the main trip purposes of car trips: Just under half of car mileage is for commuting to work and for business activities. One quarter of the mileage is for leisure activities [12].

Driving routes for the aforementioned main travel purposes include the use of country roads and highways. To be allowed to operate on German autobahns, vehicles must have a maximum design speed of at least 60 km/h. The range of fine vehicles should therefore include those that can travel at speeds of at least 60 km/h.

Under the banner of fine mobility, means of transport and vehicles are to be defined as small, flexible, environmentally friendly, climate-friendly and moderately or expediently motorized, so that they can replace today's passenger cars to a significant extent. It is therefore not a matter of once again praising individual modes of transport, such as walking and cycling, or individual vehicle types, such as electric scooters, as alternatives to the car, but rather of taking a much more comprehensive approach that includes the aforementioned modes of transport: for almost everyone and for almost all travel or transport purposes in urban and regional transport, a fine alternative is being offered to the coarse mobility that prevails today. In order to make the variety of fine mobility comprehensible and to clearly distinguish it from the conventional, ever-growing passenger cars, a new classification is proposed for all road-based means of individual transport.

1.3   Existing Vehicle Categorizations

The vehicle segments defined by the European Commission and the German Federal Motor Transport Authority (KBA) with relevance for marketing and retail are listed in Table 1 with English translation. The segmentation does not include any vehicles below the smallest car and therefore no means of fine mobility. Rather, it shows by way of example how extensive and differentiated the European and German market for large passenger cars is

In contrast there is the EC classification, which is the basis for type approval in Germany and thus takes into account technical and functional characteristics. This comprises the following vehicle classes [15]:

- Category M: passenger transport, divided into classes M1, M2 and M3 according to the number of seats and the number of wheels.
- Category N: Carriage of goods, subdivided into classes N1, N2 and N3 according to permissible total mass and number of wheels.
- Category O: Trailers, subdivided into classes O1 to O4 according to maximum permissible mass.
- Category L: Light single and multi-track motor vehicles subdivided into classes L1e to L7e according to the number of wheels, maximum design speed, engine power, in some cases empty weight and maximum net power.

When looking at the subdivisions and characteristics of the classification, it is noticeable that only one class M1 is defined for conventional passenger cars for individual use - regardless of their size or dimensions, weight, engine power or maximum design speed. In contrast, light motor vehicles are differentiated and regulated according to a large number of (technical) features. Vehicles below the L class segment are not included in the EC classification [16].

It becomes apparent that the existing vehicle categorizations are not suitable for classifying all road-based means of transport according to their "fineness". They neither include all means of transport of fine mobility nor do they take into account the relevant characteristics for classification according to (urban) spatial and ecological effects. They are thus neither suitable as a basis for defining fine mobility nor for promoting it with regard to possible applications street design or in the form of monetary incentives for turning to humanly scaled means of transport.

Table 1. Vehicle segmentation of the European Commission and the German Federal Motor Transport Authority [13, 14].

European Commission	english translation	German Federal Motor Transport Authority	english translation
Kleinstwagen	mini	Minis (Kleinstwagen)	minis (smallest car)
Kleinwagen	small	Kleinwagen	small car
Mittelklasse	medium	Kompaktklasse	compact class
Obere Mittelklasse	large	Mittelklasse	middle class
Oberklasse	executive	Obere Mittelklasse	upper middle class
Luxusklasse	luxury	Oberklasse	upper class
Sportwagen	sports	Sportwagen	sports car
Multivan	multi-purpose	Minivan, Großraumvan	minivan, multi-purpose vehicle
Sport Utility Vehicle	sport utility (including off-road vehicles)	Sport Utility Vehicle	Sport Utility Vehicle
		Geländewagen	off-road vehicle

2 New Classification of Road-Bound Means of Transport

In order to define and promote fine mobility, a new classification of vehicles is necessary, on which

- technical regulations of infrastructure planning,
- road law and traffic regulations,
- the design of structural infrastructure,
- municipal traffic area designations as well as
- fiscal instruments such as road user or parking fees.

can be based.

According to this, the developed classification includes all road-based means of transport of individual passenger transport and light freight transport and not only motor vehicles in the current definition, and is oriented towards (urban) spatial and ecological compatibility. This means that not the technical-constructive characteristics of the registration law and the functional qualities for the occupants, but the qualities for the people in the roadside environment and the environment are considered. This includes effects on the following functions:

- Orientation in the roadside environment,
- Perception of safety in the roadside environment,
- Well-being and social interactions of people in roadside environment,
- Encounter traffic in narrow streets,
- Space requirements of moving equipment when stationary and in motion.

Accordingly, the aim of the classification is to distinguish between means of transport that enable mobility with a more human scale for more urban and quality of life as well as environmental and climate protection and vehicles that exceed the limits of (urban) spatial and ecological compatibility. For this purpose, the relevant characteristics have to be determined, which are suitable for the comparison and (classification) of all road-based means of transport with regard to the use cases already mentioned above and explained in more detail in Sect. 2.2.

2.1   (Urban) Spatial and Ecological Characteristics of Means of Transport

In a first step, any characteristics that could be relevant for the classification of means of transport under (urban) spatial and ecological aspects between "fine" and "coarse" and for different use cases of the new classification were compiled:

Size:
– Length
– Wide
– Floor space
– Height
– Space occupation (external volume)
– Turning circle

Weight:
– Empty weight
– Maximum permissible total weight
– Payload

Speed:
– maximum design speed

Weight and speed (in combination):
– kinetic energy ($E_{kin} = m * v^2$, with m = maximum permissible total weight and v = reference speeds to be set)

Environmental and resource impact:
– Material and energy consumption during production
– local emissions and energy consumption during operation
– Road and tire abrasion

For as large a number of reference vehicles as possible (selection of around 100 vehicle models as examples of vehicles and means of transport of their kind) in the broad spectrum of means of transport from unicycles to all-terrain vehicles, the corresponding characteristic values were determined for each of the above-mentioned features and

- where not available - qualified assumptions were made on the basis of comparison vehicles with similar technical and physical characteristics.

The list presents opportunities to rank the means of transport by different characteristics and test for similarities and differences of these characteristics of different vehicle types. It is evident, that there is a clear order of means of transport of all sizes according to their space occupation and related characteristics.

The analysis of the characteristic values by their respective ascending order shows a very strong, positive correlation of ,946 between the characteristic values of the space occupation and the maximum permissible total weight (gross vehicle weight) for the studied means of transport as seen in Table 2.

Table 2. Correlation between space occupation and gross vehicle weight (GVW)

Correlations			
		SPACE	GVW
SPACE	Pearson Correlation	1	.946**
	Sig. (2-tailed)		.000
	N	94	89
GVW	Pearson Correlation	.946**	1
	Sig. (2-tailed)	.000	
	N	89	90

**Correlation is significant at the 0.01 level (2-tailed).

Strong correlations between two or more features simplify the development of a new classification for means of transport, since it is possible to avoid looking at superfluous features that have little or no influence on the result.

2.2 Use Cases

In a second step, (urban) spatial and ecological use cases were defined for the formation and justification of a plausible, functional and as easily understandable as possible classification for means of transport. For each use case, the relevant characteristics of means of transport were recorded. The minimum consensus from the relevant characteristics of all use cases was then used to develop the calculation formula for the classification (see Table 4). The procedure ensures that each means of transport is assigned exclusively to one class and that each of the named use cases can refer to the same classification. Features that go beyond the minimum consensus and are relevant for individual use cases can optionally be added by the implementing actors for additional differentiation.

Table 3 shows that the minimum consensus of relevant characteristics across all the use cases considered are the characteristics of space occupation and turning circle. The two vehicle characteristics are able to reflect above all the (urban) spatial and, due to land use, also parts of the ecological compatibility of means of transport and thus the above-mentioned effects for people in the road space and for the environment. The developed

Table 3. Use cases and relevant vehicle characteristics.

Use case	Characteristics of the classification (minimal consensus from all application cases)	Features that can be added as additional regulation(s)	Exemplary implementation to promote fine mobility
Designation of staggered size parking stalls in public street space, parking areas, and parking garages or underground parking garages	Space occupation, turning circle	Width, length, drive type	Means of transport of fine mobility obtain parking areas in the immediate vicinity of public-intensive facilities. This allows more people to park closer to important facilities, and less land is sealed overall. If the type of drive is added as an additional characteristic, vehicles with combustion engines can be excluded, for example
Staggered pricing of parking spaces in public areas according to the use of urban space	Space occupation, turning circle	Drive type	Fine mobility vehicles may park free of charge. The parking of coarse vehicles is charged (in ascending order). This creates a monetary incentive to use fine(r) vehicles. If the type of drive is added as an additional characteristic, vehicles with combustion engines, for example, can be priced higher than electric vehicles
Differentiated ban on entering/passing through sensitive and special urban areas	Space occupation, turning circle	Width, local emissions, maximum permissible gross weight	Vehicles assigned to "coarse mobility" are not allowed to enter or pass through sensitive or special urban areas. This makes these urban areas less car-dependent, more traffic-calmed down and safer. If further characteristics are added, the selection of excluded means of transport can be further differentiated according to the type of sensitive urban area

(continued)

Table 3. (*continued*)

Use case	Characteristics of the classification (minimal consensus from all application cases)	Features that can be added as additional regulation(s)	Exemplary implementation to promote fine mobility
Differentiated fiscal treatment in the context of (city) tolling	Space occupation, turning circle	local emissions, weight	Vehicles assigned to "coarse mobility" are priced in ascending order within the framework of a (city) toll. This can create monetary incentives to use finer(r) vehicles. If local emissions are added as an additional feature, vehicles that are more harmful to the environment and health can be priced higher
Speed-based traffic areas (road space design in flowing traffic through speed limit per traffic area)	Space occupation, turning circle	Maximum design speed (necessary additional regulation) Maximum permissible total weight (possible additional regulation)	Vehicles are classified according to their class and maximum design speed on the various traffic areas with different speed limits (including roadway, bicycle path, sidewalk) to ensure compatibility between different means of transport on the same traffic area

classification thus particularly considers the spatially relevant factor of size (G) and is therefore called "G classification" in the following.

2.3 Selection of Characteristics for G Classification

The finding of strong correlations between several feature values and the minimum consensus with regard to the proposed use cases leads to concentrate the new classification on the most characteristic features for the street space, the space occupation and the turning circle. Other features, as far as necessary or useful for certain use cases, can be or have to be considered additionally (see Sect. 2.2).

Space occupation as a cuboid around the dimensions of a means of transport illustrates the occupied external volume and thus the space that cannot be used by people. Although calculated in a simplified manner and not corresponding to the actual external volume of the complexly designed vehicle bodies, the multiplication of the longest, widest and highest dimensions of a vehicle - i.e. the fiction of a tightly fitting cuboid around the vehicle - is plausible and useful as an indicator of space occupation. The actual occupation of space by a vehicle and the resulting strain on the physical and psychological environment follows the extreme values of length, width, and height. The areas and spaces within these extreme values are effectively lost to other road space use, particularly due to the need for provisioning.

The turning circle of means of transport, which is often given little attention, has an impact on the agility of vehicles and the design and layout of roads. The turning circle is included as a feature in the G classification because the maneuverability of a vehicle dictates the minimum amount of infrastructure required. The smaller the turning radius, the smaller the vehicle swept path, and the smaller sized the infrastructure and, in particular, parking maneuvering areas and curve radii at intersections and interchanges can be. In small-scale encounter traffic between vehicles among themselves or a moving vehicle and a human being, the turning radius represents a safety feature, since evasive maneuvers are relevant in avoiding accidents.

2.4 Class Differentiation, Thresholds and Calculation Formula

We propose to classify the entire range of movement means according to the seven classes that are familiar to everyone from the product world, such as the textile industry: XXS, XS, S, M, L, XL, XXL. The division into seven classes is sufficiently differentiated to reflect the variety of characteristic values of the selected features.

Threshold values for the classification of vehicles into the seven classes were defined for the characteristics of space occupation and turning circle (see Table 4). The lowest threshold value for the classification is based on human dimensions. In the case of space occupation, it is approximately one cubic meter, which is outlined in DIN 33402 as a person standing with arms outstretched upward and forward. In the case of the turning circle, we also orient ourselves to the meter, which is approximately the diameter of the movement area of a walking person.

The top threshold between the XL and XXL classes is based on upper limits of street space compatible dimensions. For the feature of space occupation, the combination of the dimensions of a parking space in vertical position without lateral boundaries with the average eye height of a German adult decides on the urban space incompatible dimension. Beyond this threshold, it can be interpreted that MIT vehicles take up additional space and the field of vision of pedestrians and cyclists is obstructed too much. For the turning circle feature, the outer diameter of the mini-roundabout, as defined by the German Road and Traffic Research Association is defined as the upper limit of urban compatibility. Turning circles of individual means of transport above this limit require additional space that is not available for the urban space for other use.

The thresholds between the respective lower and upper thresholds are linearly interpolated for the sake of comparability, under the condition that the distances between the class boundaries are identical.

The threshold values result in the classes XXS-XXL with class values between 1 (XXS) and 7 (XXL). The calculation rule for assigning a means of transport to a final G class with the class values in the calculation is as follows:

$$\text{G Classification} = \tag{1}$$

0.75 * the class value for the characteristic "space occupation"

+ 0.25 * the class value for the characteristic "turning circle",

Table 4. Derivation of thresholds.

Threshhold between classes	Characteristic	Criteria for thresholds
		Human measure
XXS and XS	Space occupation	The space taken by the 95th percentile of a German male corresponds to about one cubic meter at the largest possible extension according to DIN 33402 for the multiplication of length, width and height: 0.815 m as the reach of the arms forward (length) [17] * 0.555 m as width over the elbow [17] * 2.205 m as the reach of the arms upward (height) [17] = 0.988 cubic meter
XXS and XS	Turning circle	The square meter is the area required approximately by a walking person who can move freely in all directions
XS - XL XS - XL	Space occupation	Formation of thresholds by equal, numerical distances: 3800 cubic dimeters each Formation of thresholds by equal numerical distances: 240 cm each
		(urban) spatial and ecological incompatibility
XL and XXL	Space occupation	The multiplication of the footprint of a parking stall in vertical position with 12.5 m^2 as the upper limit of land use with 1.6 m as average eye height of a German adult results in 20 m^3
XXL	Turning circle	The mini-roundabout according to the Guidelines for the Design of Urban Roads of the German Road and Traffic Research Association offers with an outer diameter of at least 13 m an acceptable and usable guideline for what turning circle vehicles of the MIT should have at most [18]

where the final result of the G classification is between 1 (XXS) and 7 (XXL). Consequently, the threshold values between the g classes are at x.5.

The stronger weighting of the feature of space occupation is based on our assessment that the (urban) spatial and ecological impact of the vehicle's exterior volume is particularly high. The spatial dimension can also be understood figuratively in such a way that the length, width and height characteristics are each included in one characteristic and the proportions of means of transport are not restricted in each class, but the overall size is. Whenever necessary single characteristics of the means of transport can be restricted for use cases individually, but the basic g classification is not further influenced by this.

3   Definition of the Fine Mobility

Fine mobility is mobility with all the smallest possible, maneuverable, road-based means of transport that can be used by almost everyone to accomplish almost all transportation purposes.

By fine mobility we mean mobility with vehicles of classes XXS, XS and S. Fine mobility is thus everything that bears an "S": XXS, XS and S. It seems logical, plausible and in line with the classifications according to the individual characteristics to define class "M" as "medium". "Coarse" are thus vehicles of the L classes: L, XL, XXL.

As a rule of thumb (see also Fig. 2):

Fine = XXS, XS, S
Mean = M
coarse = L, XL, XXL

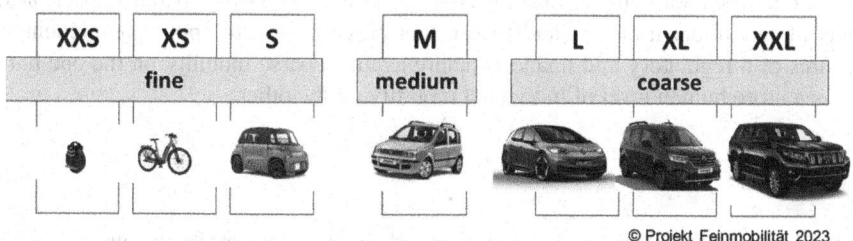

© Projekt Feinmobilität 2023

Fig. 2. Exemplary classification of movement means to the G classes (own representation).

Fine mobility vehicles can have up to four seats and a maximum design speed of up to 120 km/h. This is the case for some e-scooters, e-motorcycles, three- or four-wheeled light e-vehicles and minicars. Fine-mobility vehicles are thus basically suitable for distances of up to 100 km, are suitable for long-distance roads and for commuter, business and leisure trips, and could provide a relevant share of the mileage, as also shown in a DLR study on the potential of small and light electric vehicles [19].

A selected list of different means of transport for fine and coarse mobility is shown in Table 5.

The requirement to prefer the "finest possible" individual or transport vehicle is accompanied by the postulate of not buying and maintaining one vehicle for all purposes, the size of which is usually oriented to only a few trips a year (e.g., vacation trips, DIY store shopping). Instead, the variety of means of fine mobility opens up (initially theoretically) the possibility of always choosing the smallest, lowest-emission and most space-saving means of mobility for the specific purpose of travel. First of all, people must be made aware of the diversity of fine or finer means of transport and the relevance of vehicle sizes from an (urban) spatial and ecological point of view. The developed G-classification contributes to this. It can promote opportunities for equal use (and sharing) of finer-mobility vehicles and reduce the focus on pure electrification of any passenger car class. In the spirit of downsizing vessel sizes, fiscal preference to quantity and quality in the supply of fine-mobility vehicles should be given.

Table 5. Example list for means of transport of fine and coarse mobility.

G classes	Examples
XXS	Onewheels, Segways, Electric Wheelchairs
XS	Bikes, Pedelecs, Electric Scooters, small Cargo Bikes
S	Cargo Bikes, Velo Mobiles, Cabin Scooters, Mini Cars
M	Small City Cars
L	Compact Cars, Station Wagons
XL	Minivans, small SUVs, Crossovers
XXL	Vans, SUVs

The classification helps to use objective factors to assess the spatial impact of all means of individual transport and to use this classification to apply comprehensive measures of a regulatory and financial nature against coarse mobility on the one hand and for a more human level of individual mobility on the other.

References

1. Krail, M., et al.: Energie- und Treibhausgaswirkungen des automatisierten und vernetzten Fahrens im Straßenverkehr. Wissenschaftliche Beratung des BMVI zur Mobilitäts- und Kraftstoffstrategie. Fraunhofer-Institut für System- und Innovationsforschung (ISI) (Hrsg.), p. 156 (2019)
2. Winkle, T.: Sicherheitspotenzial automatisierter Fahrzeuge: Erkenntnisse aus der Unfallforschung. In: Maurer, M., Gerdes, JChristian, Lenz, B., Winner, H. (eds.) Autonomes Fahren, pp. 351–376. Springer, Heidelberg (2015). https://doi.org/10.1007/978-3-662-45854-9_17
3. Christ, J: Datenanalyse: Autos werdennicht erst seit dem SUV-Boom größer. https://www.rnd.de/wirtschaft/datenanalyse-autos-werden-nicht-erst-seit-dem-suv-boom-grosser-6GTM66RRNJEC7EYHR3FQS7Y24Y.html. Accessed 26 Feb 2023
4. The Urban Idea: EcoMobileum - Idee – Konzept – Entwicklung – Machbarkeit. https://cdn.website-editor.net/s/ec64f3a492c54135b5f9169a82bd6b17/files/uploaded/EcoMobileum_Idee%2520Konzept%2520Entwicklung%2520Machbarkeit_2020.pdf?Expires=1679602204&Signature=Ia63MnitJx93j2XOMYtZ-z~ptVe7BJXX8YUX-CPYjZNdRQ8-IhDE7L4AmuHlGF1vCl6rOLrEtcpeEAIOZNSNk9l6kSfvYPLUxKm725ABBKQnlG9QSFm9SsoXXCILatFZ4OZ3z6rrZCNma6K4zRAi9~k32QZIfYheiNBjrIlCk8vTPXj2x-jwiM0Fbox0PI6~LjbX5W~b1fEDp4-4hn3WIYUjCpfKBZiuQNIH85uVGemm31oONvylh2zS0LzU8K4gPhHQJTwRbvwEWySm6A0Bo4MtkBfEj5MPCID3O-EIpG8hw5htmRO~7Bmu2yIe~7lH74jS4mvfCamYVJrt1OByg__&Key-Pair-Id=K2NXBXLF010TJW. Accessed 12 Feb 2023
5. Statistisches Bundesamt: Pkw-Dichte in Deutschland in den vergangenen zehn Jahren um 12 % gestiegen. Pressemitteilung Nr. N 055 vom 11. September 2020. https://www.destatis.de/DE/Presse/Pressemitteilungen/2020/09/PD20_N055_461.html. Accessed 06 Feb 2023
6. Forschungsgesellschaft für Straßen- und Verkehrswesen, Arbeitsgruppe „Straßenentwurf", Richtlinien für Bemessungsfahrzeuge und Schleppkurven zur Überprüfung der Befahrbarkeit von Verkehrsflächen. RBSV. Ausgabe 2020, p. 5 (2020)

7. Road and Transportation Research Association, Working Group "Road Design", Guidelines for Design Vehicles and Drag Curves for Checking the Trafficability of Traffic Areas. RBSV. 2020 Edition, p. 5 (2020)

8. Schwarzer, C.M.: SUV statt Kleinwagen. In: ZEIT ONLINE, 05.01.22. https://www.zeit.de/mobilitaet/2021-12/elektromobilitaet-kleinwagen-suvs-modelle-falsche-anreize. Accessed 06 Feb 2023

9. BVA BikeMedia GmbH: Pedelec-Visionär Bio-Hybrid GmbH insolvent. https://radmarkt.de/vierraedriger-pedelec-visionaer-bio-hybrid-gmbh-insolvent/. Accessed 06 Feb 2023

10. Presse- und Informationsamt der Bundesregierung. https://www.bundesregierung.de/breg-de/themen/klimaschutz/eenergie-und-mobilitaet/faq-umweltbonus-1993830. Accessed 06 Feb 2023

11. The Urban Idea: Blatt Ökomobilität, Ideen zu Standards (2018)

12. infas, DLR, IVT und infas 360: Mobilität in Deutschland (im Auftrag des BMVI), pp. 71–72 (2018)

13. European Commission: Commission Regulation (EC) No 1400/2002 of 31 July 2002 on the application of Article 81(3) of the Treaty to categories of vertical agreements and concerted practices in the motor vehicle sector. Guidance Document (2002)

14. Kraftfahrt-Bundesamt: Segmente 2022. https://www.kba.de/DE/Statistik/Fahrzeuge/Bestand/Segmente/segmente_node.html. Accessed 06 Feb 2023

15. Kraftfahrt-Bundesamt: Verzeichnis zur Systematisierung von Kraftfahrzeugen und ihren Anhängern. Status: February 2022. Flensburg (2022)

16. European Union: REGULATION (EU) No. 168/2013 OF THE EUROPEAN PARLIAMENT AND OF THE COUNCIL of 15 January 2013 on the approval and market surveillance of two- or three-wheel and four-wheel vehicles, p. 94 (2013)

17. Deutsches Institut für Normung e.V. DIN 33402-2:2020-12. Ergonomie - Körpermaße des Menschen - Teil 2: Werte (2020)

18. Brost, M., Gebhardt, L., Ehrenberger, S., Dasgupta, I., Hahn, R., Seiffert, R.: The potential of light electric vehicles for climate. Protection through substitution for passenger car trips - Germany as a case study. Projectreport (2021)

19. Forschungsgesellschaft für Straßen- und Verkehrswesen: Richtlinien für die Anlage von Stadtstraßen RASt, FGSV-Nr.: 200, Ausgabe 2012, p. 115 (2006)

User Acceptance of Urban Air Mobility (UAM) for Passenger Transport: A Choice-Based Conjoint Study

Vivian Lotz[1]([⊠]) [iD], Ansgar Kirste[2] [iD], Chantal Lidynia[1] [iD], Eike Stumpf[2] [iD], and Martina Ziefle[1] [iD]

[1] Human-Computer Interaction Center, RWTH Aachen University, Campus-Boulevard 57, 52074 Aachen, Germany
{lotz,lidynia,ziefle}@comm.rwth-aachen.de
[2] Chair and Institute of Aerospace Systems, RWTH Aachen University, Wüllnerstraße 7, 52062 Aachen, Germany
{ansgar.kirste,eike.stumpf}@ilr.rwth-aachen.de

Abstract. Urban air mobility (UAM) has gained increased attention as a promising new solution to tackle issues such as traffic congestion. This study aims to assess users' perception and usage intentions and identify and weigh success-critical acceptance factors against each other. Further, user-related differences in the underlying decision patterns are analyzed. A mixed-methods approach was employed, including a prior interview study with laypeople and experts ($N = 16$), calculations of feasible prices via agent-based simulations, and the main survey study ($N = 135$) – a Choice-Based Conjoint (CBC) with the attributes: environmental impact, costs, placement of take-off and landing stations (vertiports), automation level, and onboard benefits. Results indicated that environmental impact, costs, and vertiport placement were the top three factors influencing user acceptance. The relative importance of these attributes differed, however, depending on the user group. Based on the results, three user segments were identified: automation-skeptical users, environmentally-conscious users, and cost-conscious users. The findings provide valuable insights for UAM providers and policymakers to understand the attitudes and needs of potential users and to design effective strategies for promoting the adoption of UAM.

Keywords: urban air mobility (UAM) · choice based conjoint · acceptance · air taxi

1 Introduction

Mobility is an important research area from several different perspectives. For one, there is the need to reach different places as a person. On the other, ever-increasing numbers of wares need to be transported to and from sometimes remote locations. What used to be provided via cars, delivery vehicles such as trucks, and public transit has gained negative attention on many bases. One major challenge of future mobility is avoiding unwanted pollution and emission of greenhouse gases [46]. While alternative fuels are being developed and researched (see, e.g., [34,42]), their active use is still off and therefore uncertain [37]. In addition, individual traffic, meaning the ownership of a car, especially if it is only used by a single person, is sought to be limited or at least decreased as, most of the time, this car is parked somewhere. Especially in cities and urban areas, space is a precious commodity, thus, sharing a vehicle with others might be a better solution [29]. However, car sharing and public transit cannot provide all mobility needs either. For one, pollution is still an issue. As is availability [4]. So, another attempt to reduce the number of cars and vehicles on the streets is to utilize a yet seldomly used area for traffic, namely airspace. Urban air mobility (UAM) could be a possible solution to reduce not only people but also wares transportation on streets [35,45]. This could also be another step toward the electrification of mobility. While other countries are already actively offering, for example, last-mile delivery via drones (e.g., [23]), this is only the first step. Passengers might also profit from short-distance traveling via so-called air taxis. While still in development, several start-ups have already begun to build working drones that are big enough to carry 1 to 6 passengers (e.g., [45]). Testing has also commenced [1]. However, even if such developments and tests are successful, their deployment is doubtful if people are unwilling to use them. Therefore, an early understanding of potential users' acceptance is important, and a first insight will be provided with this paper.

First, some background on the technological aspects of urban air mobility (UAM) is given. This will then be supplemented with insights into the acceptance of technologies, especially in the mobility concept. Based on this, the study design and empirical approach will be provided. The results will be detailed before they are discussed, and a final outlook will be given.

2   Background

This section will provide insight into the current state of research of urban air mobility by giving a short overview of current models in development for passenger air transportation. Afterward, a short synopsis of current acceptance research regarding drones and small aircraft, especially in urban areas, will be given.

2.1   State of the Art UAM

With the development of vertical take-off and landing (VTOL) vehicles, considering future UAM scenarios has gained importance. This potential of UAM systems is mainly based on shorter travel times compared to ground-based mobility, an on-demand service model, lower operating costs compared to complex helicopter systems, and electrically powered vehicles [26,36]. The latter is often defined as eVTOL (electric vertical take-off and landing) vehicles that can operate on vertiports near city centers or in rural areas [33]. In recent research, the overall UAM system is divided into aspects of eVTOL vehicles, infrastructure, social acceptance, certification, safety, and operations [17,18,36,47]. Regarding eVTOL vehicles, different technical configuration approaches are in focus to enable optimal fleet operations, namely wingless multicopters, lift and cruise configurations and tilting systems [3,33]. Firstly, multirotor configurations, such as *Volocopter 2X* and *E-Hang 184*, have a rather short range with mostly intracity implementation potentials due to the missing fixed wing and the resulting lack of aerodynamic lift in cruise flight and higher energy consumption in horizontal flight segments. Secondly, lift and cruise systems, such as *Wisk Cora* and *Boeing Passenger Air Vehicle*, are hybrid systems combining VTOL and longer range properties due to the additional fixed wing. With a travel speed of around 200 km/h, these configurations seem applicable in intracity and city-to-city transportation services, with a rather slow time to market. Thirdly, tilting concepts with tilting wings or rotors allowing VTOL capability are also applicable to both intracity and intercity services, as the cruise speed is relatively high. Disadvantages are the increased system complexity and the later time to market. Tilting air taxi configurations currently regarded in research are the *Airbus Vahana*, and the *Lilium Jet* [3,22,33]. Based on possible UAM scenarios, market potentials of up to 1% of demanded mobility were calculated using agent-based simulation, and fleet operational decisions such as charging approaches, station placement, and pricing were further analyzed [22,36].

2.2 Acceptance of UAM

Technology acceptance has been established as an important part of the future use of technologies [48]. While different technology acceptance models have been developed over the years, e.g., the Technology Acceptance Model (TAM) by Davies [14], or the Unified Theory of Acceptance and Use of Technology (UTAUT) by Venkatesh et al. [49], they have always had to be adapted to new technologies as those include aspects or intricacies that had not been present in the original acceptance models' technology [48]. However, the basic idea of central aspects such as *intention to use* or *perceived usefulness* can be translated well into new technologies and use cases.

In the context of mobility, especially with the addition of autonomous properties, factors influencing acceptance can stem from multiple sources. For one, there is trust [6], which goes hand in hand with safety [7]. This is also likely to hold true for air traffic as well [21,27].

Other aspects that might impact the acceptance of these small aircraft can be environmental factors. Here, not only gas emissions [13] but also noise emissions can lower the acceptance [17,41]. Perceived barriers to acceptance can also be found in the routes such aircraft are allowed to use and the locations they are allowed to land [24], as bystanders might perceive a loss of privacy or fear injury by a malfunctioning aircraft. Further, the willingness to share a flight, ticket pricing [44], and travel time in comparison to other modes of transportation affect the user acceptance [36,41,45]. For a more detailed overview of potential factors influencing the successful deployment of passenger drones, see, e.g., [1, 17,45].

While various factors affecting the acceptance of UAM have been identified by prior studies, little is known about which factors are ultimately decisive for (user) adoption when weighed against each other. Thus, the present study aimed to fill this gap by employing a holistic investigation of potential users' decisions for or against using urban air mobility and the underlying decision patterns. The research was guided by open-ended questions:

RQ1 Which acceptance attribute is most relevant for acceptance: costs, environmental impact, onboard benefits, automation level, or vertiport placement?
RQ2 Which attribute levels impact acceptance positively or negatively?
RQ3 Are there any differences in the decision patterns between different user profiles, and how do they differ?

3 Empirical Approach

In this section, the empirical approach is explained, starting with the overall research concept, followed by an explanation of the conjoint methodology. Subsequently, the selection of acceptance relevant attributes, levels, and the empirical study design are described. Finally, data preparation, analysis, and the sample are summarized.

3.1 Research Concept and Study Input Parameters

For this study, a three-step approach, combining a qualitative interview pre-study, an already conducted UAM cost optimization based on agent-based simulation [22], and the main study, an online survey – including a choice-based conjoint experiment, was used.

Simulations. Considering the UAM costs and scenario parameters, the results of Husemann et al. are used [22]. Here, the authors used the Multi-Agent Transport Simulation (MATSim) extended with UAM simulations (see Rothfeld et al. [36]) to determine the UAM demand in relation to both ground-based transportation and the time of day. As a simulation scenario, Husemann et al. considered the metropolitan region Rhine-Ruhr with a distribution of 100 vertiports. Further, input values in the base case were a horizontal cruise speed of 150 km/h for the air taxis, a hovering time of the vehicles for take-off and landing of 60 s, a minimum turnaround time of 3 min, a complete linkage of the UAM station network and a passenger capacity of up to 2 passengers per vehicle [22]. The simulated UAM trips were then applied in an optimization model using a total cost of ownership approach and other input data such as charging infrastructure, battery technology, and several cost-specific parameters. Including sensitivity studies for uncertainties in technological developments and UAM system design, resulting costs per passenger kilometer has been calculated to be around 1.50 euros [22].

Interview Pre-study. Guided interviews with laypeople ($N = 13$) and experts ($N = 3$) were conducted to identify UAM acceptance factors. The interview contained an introductory part about the reasons people use their current transport modes and a part about the requirements for air taxis (autonomous and piloted). For analysis, qualitative content analysis was used [25]. The identified topics and statements informed the development of the main studies' conjoint design.

3.2 Conjoint Analysis Method

A choice-based conjoint methodology was used to quantify and weigh the relevance of different decision factors against each other. Conjoint analysis is commonly used in technology acceptance research as it allows for examining individual preferences, decision patterns, trade-offs between decision-relevant attributes, and segmenting groups with distinct decision patterns [2]. Compared to traditional survey methods, choice-based conjoint (CBC) offers a more holistic view of complex decision patterns by mimicking real-world decision trade-offs. In CBC, participants are asked to make trade-off decisions between different product configurations, simulating realistic decision processes. This enables a calculation of how respondents' decisions are influenced by different product attributes and their corresponding levels. The choices are then decomposed into relative importance and part-worth utilities of attribute levels, providing insights into the overall impact of attributes on decisions and identifying tipping points of acceptance [30] [5,12]. For the present study, we included a dual response "none" query to assess the likelihood of participants opting for none of the offered product configurations [38].

3.3 Conjoint Attributes and Levels

The selection of relevant attributes was based on an initial literature review and a qualitative pre-study with laypeople ($N = 13$) and experts ($N = 3$). In total, five attributes (price, environmental impact, automation level, vertiport placement, and onboard benefits; cf. Table 1), were selected.

Table 1. Overview of attributes and levels for the CBC.

Attributes	Levels				
Price	56.25 € [-25 %]	67.50 € [- 10 %]	75 € [baseline]	82.50 € [+ 10 %]	93.50 € [+ 25 %]
Environmental impact	50 % less	identical	50 % more	100 % more	
Automation level	piloted	remote controlled	autonomous		
Vertiport placement	downtown	nearest airport	nearest train station	city centre	next to house or apartment
Onboard benefits	radio	wi-fi	comfortable seats	free carry on luggage	

Attribute levels were based on the interview insights and the results from the MATSim simulation with the aim to display realistic choices. In the following, the included levels are explained.

The Price per trip was included, as it was seen as a key requirement and exclusion criterion if too high [31] [44]. The base price was set at 1.50 euros per km, resulting in a ticket price of 75 €[22]. To limit complexity, the respondents only saw the total cost of the trip. The lowest price point was estimated, taking optimization potential into account. Based on the calculations, we included two low price points, which were 10 % and 25 % cheaper than the base price. Additionally, two price premiums (+ 10 % and + 25 % compared to the base price) were included. All included prices are listed in Table 1.

The Environmental Impact was included due to some concerns – voiced in the interviews and reviewed literature – regarding UAM's energy consumption and its impact upon the environment [13]. In light of recent energy crises and climate change, this is a highly relevant attribute when designing new transport solutions. However, operationalizing the impact on the environment is slightly tricky since respondents also indicated that they lack a feeling for judging energy consumption values. The matter becomes even more complex when, apart from the mere extent of power consumption, the power source (e.g., renewable vs. fossil fuels) is considered. For this study, environmental impact was operationalized as a rather abstract increase or decrease of the harmful effect on the environment. More precisely, the environmental impact represents a relative increase or decrease in the consumption of *non-regeneratively sourced* electricity compared to an average electric car.

The Automation Level was included since there were some reservations regarding the autonomous option in the interviews. For the CBC study, three automation levels were included: [1] a piloted option, meaning that a trained pilot would be on board; [2] a remote control option, meaning that the air taxi was remotely monitored by trained staff; and [3] a fully automated option where human supervision was no longer necessary and the air taxi was equipped to handle all situation by itself.

The Vertiport Placement as interviewees indicated that part appeal of air taxi services was that they might be closer, more easily reachable, and, thus, more flexible than conventional passenger flights. However, there might be concerns about noise [20,41] or visual pollution (visual density) [43] that could hinder acceptance when vertiports are too close to homes or city centers.

The Onboard Benefits were included to capture a whole assortment of comfort-related aspects mentioned in literature and interviews. For this study, we chose the most mentioned extras (comfortable seats, WI-FI, radio, and free carry-on luggage) and included them in the study design.

3.4 Online Survey Design

All used survey items were developed according to the interview insights and measured using 5-point likert scales. To avoid misunderstandings, the design was tested for comprehensibility and completion time by four pre-testers before distribution. In total, the survey consisted of four parts. First, an introductory part in which participants were informed about the study's purpose. It was stressed that, first and foremost, their personal opinion mattered. Further, it was explained that participation was [1] voluntary, [2] not compensated, and [3] could be canceled or interrupted at any time during the survey. Additionally, participants were informed that their data would be analyzed anonymously, i.e., none of their answers allowed conclusions about their person.

The second part was a warm-up in which respondents filled in their socio-demographics, such as age or gender. Additionally, we asked about mobility-related habits. Lastly, respondents stated their familiarity (i.e., perceived knowledge) with air taxis on a 5-point likert scale.

The main part contained the CBC experiment. To ensure every respondent had the same knowledge about the topic, information on air taxis and their use was given. Additionally, respondents received a detailed scenario description and descriptions of all attributes and attribute levels. Respondents were asked to imagine they would consider using a new air taxi service for a work or school commute (50 km), which would take 23 min and save about 22 min compared to commuting by car. Thereafter they were asked to choose the air taxi service configuration they would prefer out of a set of three. At the top of each choice task, an abbreviated version of the scenario was displayed. During the choice experiment, participants were offered an info button for the automation level, and benefits attribute levels to re-check their definitions. The choice tasks were a forced response with a dual-response none format. Thereby, we provided the respondents with the option to indicate if they would use their chosen alternative in real-life without limiting the precision of collected choice data [9]. With dual-response "none" the interpretation of utilities for all attributes apart from the "None of these" alternative remains unaffected. In Fig. 1, an example of a decision task is displayed.

Combining all possible profiles would have led to 1.200 (5 × 4 × 4 × 3 × 5) choice tasks. The conjoint analysis software offers a function to reduce the number of choice tasks. For this study, respondents completed 15 random and two fixed tasks (median efficiency = 93.37 %).

Lastly, we queried the respondents' willingness to use air taxi services (3 items, $\alpha = .72$, [13]), their current flight behavior (i.e., frequency, fear of high and flying), their automation concerns (4 items, $\alpha = .76$, adjusted from [10]), and their technology openness (7 items, $\alpha = .92$, based on [13,28]).

3.5   Data Preparation and Analysis

All constructs were checked for internal consistency (Cronbach's α), which was above the critical threshold of 0.7 for all used constructs. Additionally, analyses were performed to check for assumption violations. Participants with completion times under 10 min (below 35 % of the median) were removed. Based on the Hierarchical Bayesian (HB) analysis of the responses, a root-likelihood score (RLH) can be estimated for each participant. This fit statistic is higher the better the model fits the observed data, with the maximum being a value of 1. Random response behavior is generally poorly predictable and results in low RLH scores. Responses with an RLH threshold below 0.44 were removed, in line with the suggestions by Chrzan & Halversen (2020) [11].

Data were analyzed using descriptive (i.e., mean scores (M), Median (Mnd), standard deviations (SD), and interference statistics. The level of significance (p) was set at 0.05. For the analyses of the conjoint experiment, relative importance scores and part-worth utilities of each attribute level were calculated via HB using the Sawtooth Software [39]. The former indicates how important an attribute is for the overall decision in relation to every other included attribute. The part-worth utility scores hold information about the – positive or negative – contribution of each attribute level to the recorded decisions. Part-worth utilities cannot be compared between different attributes. To identify user segments, latent class analysis (LCA) was conducted. For the LCA analysis, information on the respondents' propensity to choose the "none" option was excluded from the calculations.

3.6 Sample Description

The survey was distributed in August 2022 online using snowball sampling and 135 complete responses were obtained. The sample is relatively young and highly educated compared to the German population. Most participants live in cities (71.8%), nearly all (94.1%) hold a driver's license, and most (80.7 %) also own a car. Additionally, just over half the sample own a public transport ticket (cf. Table 2).

The technology openness was, with $M = 3.79$ ($SD = 0.89$, min $= 1$, max $= 5$), higher than the center of the 5-point scale, indicating a rather technophile sam-

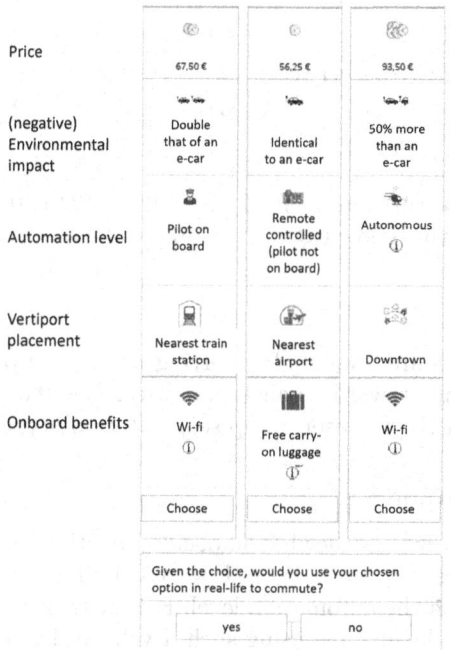

Fig. 1. Examplary choice task of the CBC experiment.

Table 2. Characteristics of the sample (N = 135).

Gender	Male	$n = 56$ (41.5 %)
	Female	$n = 79$ (58.5 %)
Age	M	31.96
	SD	13.01
Residential area	City center	$n = 60$ (44.4 %)
	Outskirts city	$n = 37$ (27.4 %)
	Suburban	$n = 24$ (17.8 %)
	Rural	$n = 14$ (10.4 %)
Occupation	Student	$n = 67$ (49.6 %)
	Vocational training	$n = 2$ (1.5 %)
	Employed	$n = 59$ (43.7 %)
	Retired	$n = 5$ (3.7 %)
	Seeking work	$n = 2$ (1.5 %)
Drivers license	Yes	$n = 127$ (94.1 %)
	No	$n = 8$ (5.9 %)
Car ownership	Yes	$n = 109$ (80.7 %)
	No	$n = 18$ (13.3 %)
Public transport ticket ownership	Yes	$n = 79$ (58.5 %)
	No	$n = 56$ (41.5 %)
Distance of regular commute	M	17.76 km
	SD	28.82
Frequency of flying (trips per year)	M	1.96
	SD	2.90

ple. Automation concerns were neutral ($M = 2.94$, $SD = 0.97$), and knowledge of air taxis were relatively low ($M = 2.19$, $SD = 0.90$).

4   Results

In this section, results are presented – starting with a section about the general attitude of participants toward an air taxi service, followed by the results of the conjoint analysis. Lastly, the user group segmentation is presented.

4.1   General Attitude

Overall, participants were somewhat undecided on the question if they would use air taxi services ($M = 3.05$, $SD = 1.15$). This willingness, however, changed depending on the automation level. For air taxis with a pilot on board, 72 % stated they could imagine using such a vehicle. For remote-controlled or autonomous air taxis, this share decreases to 57 % and 58 %, respectively (cf. Fig. 2).

4.2 Preferences for Air Taxi Services

On average, the environmental impact (30.04 %, $SD = 15.60$) was most important for participants' decisions, followed by the price (24.73 %, $SD = 13.12$), the
placement of vertiports (20.30 %, $SD = 8.79$), and the automation level
(16.82 %, $SD = 14.24$) (cf. Fig. 3). Least important were the offered onboard
benefits (8.11 %, $SD = 5.32$).

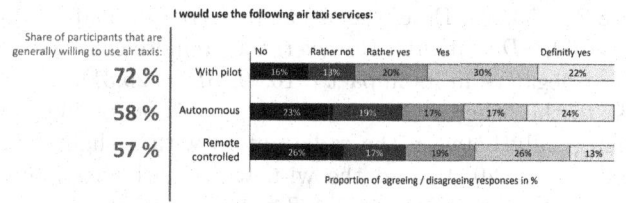

Fig. 2. Participants' willingness to use air taxi services with different levels of automation.

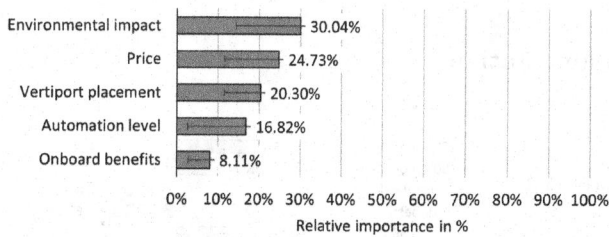

Fig. 3. Relative importance scores (with standard deviation) of the attributes in the
CBC study.

Additionally, part-worth utilities were examined (cf. Fig. 4). For the attribute
''environmental impact,'' the "50 % decrease" and the "identical" level were
acceptable, contributing both positively to the participants' decision (– 50+%:
+65.18, $SD = 49.47$; identical: +40.61, $SD = 27.20$). Overall, the results show
that the more the consumption of non-regenerative energy increases, the less
likely people are to choose the option. Consequently, an increase of 50 % (–
35.03, $SD = 27.48$) and 100 % (–70.76, $SD = 49.11$) both deterred respondents
from choosing the air taxi.

For the price the results indicate that the more affordable, the more positive the effect on the decision. The two cheapest price levels of 56.25 € (+52.38,
$SD = 43.67$) and 67.50 €(+25.47, $SD = 22.53$) had high positive contributions. In comparison, the 75 €price point only slightly positively impacted decisions (+1.37, $SD = 19.25$). Both price premiums showed large negative effects
(93.50 €: –59.84, $SD = 37.44$; 82.50 €: -19.38, $SD = 18.72$).

For the ''vertiport placement'' attribute, nearly all levels showed positive effects. The airport placement is the only exception (−58.12, SD = 39.77). Most positively contributed the "close to own home"-level (+29.61, SD = 20.91), followed by the placement at the closest train station (+14.52, SD = 21.21). The effects of the levels "downtown" (+8.58, SD = 16.92) and "city center"(+5.41, SD = 22.12) were comparably slight.

Within the attribute ''automation level'', both levels without a pilot on board had negative contributions, while the piloted option was deemed acceptable (+37.96, SD = 53.26). Here, the autonomous option had the highest negative impact (−21.00, SD = 35.51). However, the "remote controlled" option was only slightly less negative in its impact (−16.69, SD = 29.31).

In contrast to the other attributes, the utility scores of the benefit levels varied only marginally between the radio option with a slightly negative contribution (−9.2, SD = 20.89) and the wi-fi with the highest, but still small, positive contribution (4.31, SD = 16.80). The contributions of the carry-on luggage option (+3.25, SD = 16.76) and the comfortable seats (+1.64, SD = 16.10) were also slightly positive. After each choice task, we asked participants if they would use the selected option to commute. The results show a high positive utility for the ''none'' option, indicating that only an attractive combination of levels represents an air taxi service that would actually be used.

4.3   User Segmentation

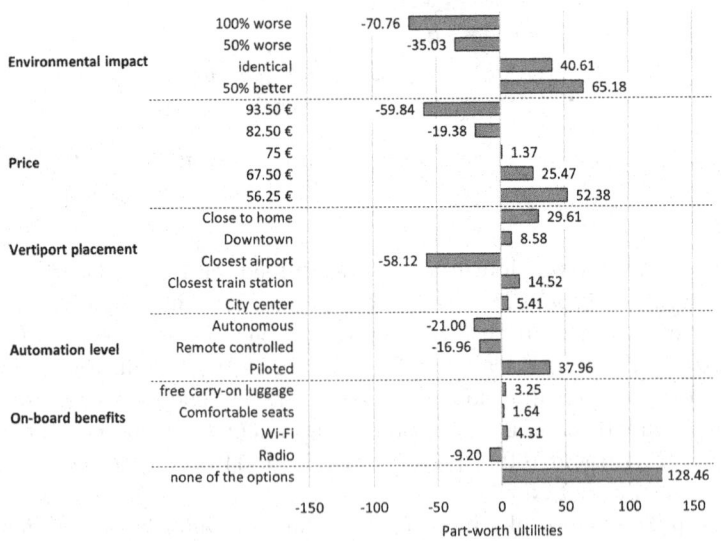

Fig. 4. Part-worth utility scores (zero-centered diffs) for each attribute level included in the CBC study. Note: Levels cannot be compared across attributes.

To identify clusters of respondents with differing preferences, latent class segmentation (LCA) was used [40]. In this analysis, utility scores for each found cluster and a probability score (i.e., the probability that this respondent belongs to the cluster) are calculated. To identify a suitable number of clusters, the criteria "percentage certainty", "consistent Akaike information criterion" (CAIC), and "relative Chi-square" were examined [40]. For the acquired data, a three-group solution showed the best fit.

Table 3. Overview of attributes and levels for the CBC.

Respondent characteristics	Group 1 "Automation sceptics" ($n = 22$)	Group 2 "Eco-conscious" ($n = 53$)	Group 3 "Price optimizers" ($n = 60$)	level of significance (p)	effect size
Age	$M = 35.68$, $SD = 13.17$	$M = 31.23$, $SD = 14.05$	$M = 31.23$, $SD = 11.95$	n.s	$\eta^2 = .02$
Gender	36.36 % male ($n = 8$)	56.60 % male ($n = 30$)	30.00 % male ($n = 18$)	$p \leq .05$	Cramér's V = .25
Technology openness (min = 1; max = 5)	$M = 3.64$, $SD = 0.75$	$M = 3.93$, $SD = 0.89$	$M = 3.73$, $SD = 0.93$	n.s	$\eta^2 = .02$
Automation concern (min = 1; max = 5)	$M = 3.83$, $SD = 0.82$	$M = 2.58$, $SD = 0.90$	$M = 2.93$, $SD = 0.88$	$p \leq .01$	$\eta^2 = .19$
Willingness to use air taxis (min = 1; max = 5)	$M = 2.36$, $SD = 0.1.03$	$M = 3.23$, $SD = 1.12$	$M = 3.14$, $SD = 1.14$	$p \leq .01$	$\eta^2 = .07$
perceived knowledge (min = 1; max = 5)	$Mdn = 2.00$, $SD = 0.87$	$Mdn = 3.00$, $SD = 0.95$	$Mdn = 2.00$, $SD = 0.83$	$p \leq .05$	Cramér's V = .26

To interpret the characteristics of the found clusters, demographics, and attitudinal characteristics were compared [2] via MANOVA and Chi²-tests. The results are listed in Table 3. For the MANOVA, Tukey Post Hoc tests were performed.

The one-way MANOVA showed a statistically significant difference between the groups F(8, 258) = 4.51, p ≤ .01, partial η^2 = .12, Wilk's λ = .77. Overall, differences were found for *gender, automation concerns, willingness to use,* and *perceived knowledge* (cf. Table 3). The share of male respondents was significantly lower in groups 1 (automation skeptics) and 3 (price optimizers). Further, group 2 (eco-conscious) showed a higher perceived knowledge. Automation concerns were significantly greater in group 1 (automation skeptics) than in the two remaining groups. Likewise, group 1's (automation skeptics) willingness to use air taxis was significantly lower than the willingness of groups 2 and 3. There were no significant differences regarding age or technology openness.

The relative attribute importance of each group was as follows (cf. Fig. 5): in **group 1**, the automation level was the most decisive decision criterion (55.18 %), followed by the vertiport placement (18.11 %), and the environmental impact (14.60 %). Offered benefits (5.75 %) and price (6.35 %) only played a minor role in group one's decision. For **group 2**, the most important factor was the environmental impact (54.59 %). Vertiport placement (15.70 %), price (16.07 %), and automation level (10.84 %) affected the decision less. Offered benefits only had a relative importance of 2.80 %. For the decisions of **group 3** respondents, the price was most decisive (43.43 %); but the vertiport placement also played a crucial role (27.28 %). Less influential were the environmental impact (16.82 %), the benefits (6.88 %), and the automation level (5.59 %).

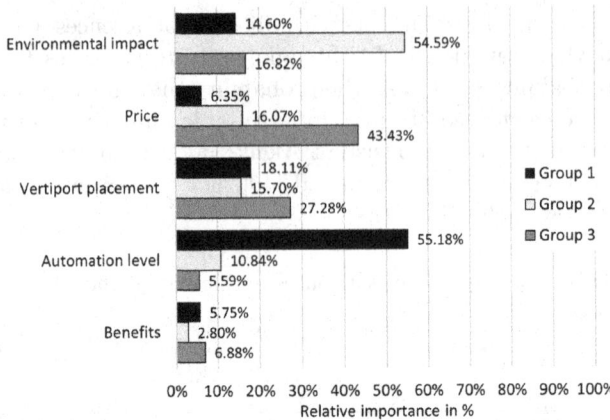

Fig. 5. Relative importance of the attributes for each identified preference group.

In Fig. 6, the average zero-centered diff part-worth utilities for each attribute level and all three groups are visualized. For the `environmental impact` attribute levels, the general preference tendencies are much the same across the three groups, with the "identical" and "50 % better" levels resulting in positive utilities. However, the range between the attribute levels is much more pronounced for group 2 than for both other groups. Especially for group 1, the "50 % worse" (−7.12) condition only had a slightly negative impact.

For group 3, the price level utility scores follow a linear line, following the principle of "the cheaper, the better". Here, acceptance tips into rejection after the 67.50 €price point. In contrast, for group 2, prices of 56.25 €(+24.49), 67.50 €(+17.15), and 75 €(+22.59) were all nearly equally (positively) affecting the decision. Lastly, for group 1, the price did not matter much, as indicated by the small range of utility scores. Still, lower prices were preferred slightly.

The preference pattern for the `vertiport placements` looks similar across groups. Notable is, however, that for group 1, a vertiport placement in the city center (−23.15) was not acceptable, while it was for groups 2 and 3 (group 2: +9.23; group 3: +8.36).

In group 1, the piloted option (+179.20) was highly preferred, and both automated options – remote controlled (−96.72) and fully autonomous (−82.49) – were equally highly rejected. For group 2, the preference pattern reflects the "the less human control, the worse" mindset and views the remote-controlled option (−0.72) rather neutrally while rejecting full automation (−26.74). Group 3 shows positive utility scores for both the piloted (+27.46) and the fully autonomous (+7.84) option but rejects the remote-controlled option (−26.74).

Differences between the preferences regarding the included `benefits` were observable for the "wi-fi" and the "comfortable seat" levels. Here, the wi-fi option was slightly negatively affecting the decision of group 2 (−2.57), while it had a positive impact on the decisions of group 1 (+10.15) and 3 (+8.29). Comfortable seats had a slight positive effect on the decisions of groups 2 (+4.93) and 3 (+8.20) and a slightly negative effect on group 1 (−5.41).

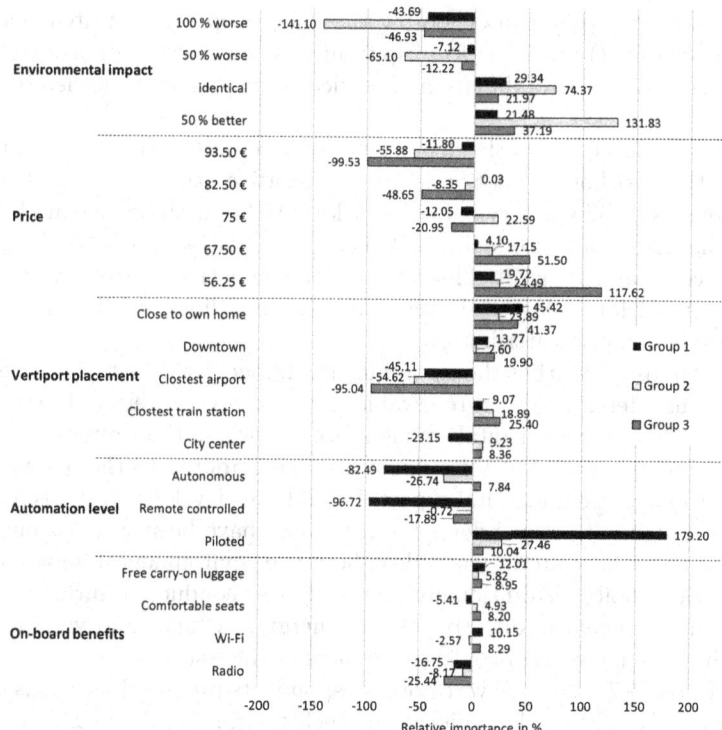

Fig. 6. Part-worth utility scores (zero-centered diffs) for each identified preference group.

5 Discussion

The present study aimed to investigate the user acceptance and acceptance influencing factors of air taxis for regular commute trips in Germany via a conjoint experiment approach. Before conducting the CBC, a literature review, a qualitative interview study, and a MATSim simulation were performed to ensure the relevance and accuracy of included attributes and levels. The obtained results are discussed in the following sections.

5.1 Insights on Decision Behaviors for Air Taxi Use

The present study assessed people's willingness to use UAM services. Results showed that the respondents are still undecided about using air taxis. While 72 % could imagine using air taxis if a pilot was accompanying the flight, this share dropped to only half for the remote and fully autonomous options, which still is a higher willingness than found in comparable recent studies [16,50]. Additionally, the high share of respondents choosing the "none" option indicated that most people are (as of yet) unwilling to swap from their current transportation choice to UAM. An attractive combination of service features may be needed to change this.

The results also confirm the dependence of acceptance on automation levels, as both options without a pilot onboard had negative utility scores and deterred people from using UAM services. The lacking trust in the automated options poses a challenge for introducing UAM, as staff wages substantially increase operating costs and make it difficult to offer competitive prices, which, however, are crucial for widespread acceptance as underlined by the high relative importance of the price attribute.

One key finding was that the environmental impact was most decisive for people's decisions. Here, little non-renewable energy use was preferred. Acceptance was tipped when environmental impacts became worse than alternative transportation choices. Several aspects might have contributed to the observed high importance of the environmental aspect. Possibly the level descriptions made the issue more tangible to respondents than it would have been in daily life, where people often only have an abstract idea about the environmental impact of their transportation choices. Additionally, the study was conducted during an energy crisis, when the media frequently covered energy shortages and volatile prices. This might have increased people's awareness of the issue.

Regarding the location of vertiports, respondents preferred locations close to their homes or easily reachable locations such as train stations. This finding is somewhat surprising as prior research suggests that concerns about noise and visual pollution caused by UAM are high [20,41,43] and that acceptance of UAM in people's immediate environment is limited [16]. This phenomenon is called NYMBY ("not in my backyard") and is typical for new and large-scale technologies deployed in public spaces [15,32].

One explanation that the NIMBY effect did not play a role in the current study might be that we looked at user acceptance rather than public acceptance. From a customer's point of view, travel time and effort are essential factors when choosing how to travel. Easily reachable locations are, thus, preferred. Especially since one of the perceived advantages of air taxis is their presumed flexibility and time savings, and far away or hard-to-reach locations might negate all expected advantages, as indicated by our pre-study. Lastly, the onboard benefits played only a neglectable role.

5.2   Explaining Divergent Decision Patterns

A high standard deviation could be observed for the level utility scores, which usually indicates diverging decision patterns due to user diversity factors. Indeed, an LCA revealed three preference subgroups differing in gender, automation concerns, willingness to use, and perceived knowledge: the automation skeptics, the eco-conscious, and the price optimizers. One of the major outcomes of the present study is that, even for a more or less homogeneous sample in terms of age and education, trade-off decisions are heterogeneous.

The first identified group, whose decisions were heavily influenced by the automation level, is not very likely to be at the forefront of adoption. Compared to the other user segments – even if they were less price sensitive – they were more unwilling to use it than the other groups. Their high automation concern and low perceived knowledge might explain this finding. Prior research shows the relationship between expertise and risk perception; i.e., higher domain-specific knowledge reduces risk perceptions [8].

In comparison, group 2 was more willing to use air taxis and also more willing to pay slightly higher prices if their environmental impact was better or at least identical to an electric car which makes them an attractive group for early adoption. Concerning the environmental impact, however, the question remains whether this is achievable by optimizing the aircraft or employing measures such as ride-pooling. Compared to the attributes mentioned above, the eco-conscious group placed little importance on the automation level. However, some preference for human-controlled UAM was observed. Remote-controlled aircraft were viewed neutrally, though full automation was rejected.

For the last group (prize optimizers), UAM has to be, above all, affordable and easily accessible. The automation level only played a minor role in their decision. Interestingly, the price optimizer group was the only group where full automation positively affected their task decision. In fact, the utility score of the fully autonomous air taxi was nearly as high as that of a piloted version which underlines how unconcerned this group is about flight automation. Interestingly, the remote-controlled option was rejected. At this point, it is unclear what might have contributed to this finding. Group 3 neither displayed a higher level of knowledge than the automation skeptic group nor were they more unconcerned about automation than the eco-conscious group. Their technology openness was also at the same level as for both other groups.

5.3   Limitations

Although this study provides valuable first insights into perceptions and decision patterns for air taxi use, some methodical limitations should be considered when interpreting the results.

Firstly, due to the early stages of UAM development, a scenario-based approach was adopted. The respondents lacked experience with the technology. Thus, like any social science study that examines attitudes and perceptions, the reported values do not allow an accurate prediction of consumer behavior – even if attitudes and behavior are, in most cases, closely linked, there still is a gap between them, especially for sustainability-related behavior (e.g., [19]).

Secondly, convenience sampling was used. The sample size was relatively small and homogeneous in terms of demographics. Still, even in this relatively homogenous sample, attitudinal and differences in the underlying preference patterns were observable. These might be even more pronounced in a larger and more diverse sample. Thus, further research should validate the findings with a larger census-representative sample.

Moreover, the chosen operationalization of the "environmental impact" attribute might need some adjustments to mimic a real-world scenario more closely, as directly showing a simplified score might have made the issue more salient and, thus, more decision-relevant. In a realistic situation, indications about ecological impacts are rarely tangible and almost always complex to asses. When considering the environmental impact of a transport choice prevailing mental models and consumer (mis-)conceptions play a crucial role. In other words: rather than the actual environmental impact, what people expect the environmental impact of air taxis is, might be more relevant for their transport decisions.

6   Conclusion – What Did We Learn

Overall, there was still a fair amount of skepticism, as indicated by the high share of respondents opting for the "none"-option. However, even if some skepticism about the use of air cabs is prevalent and the concept is still unfamiliar to most, most respondents were generally willing to use them if the price and environmental impact are right and vertiport locations are easily accessible. In other words: The new mobility option has to offer the user a tangible economic, time, or ecological advantage. In the present study particularly, the environmental impact played the most critical role in influencing usage intentions. This suggests that people increasingly consider the environmental impact of their transport mode choice. Recent developments, such as the uncertain energy supply in Germany and the increased salience of environmental issues, might fuel this finding. Especially for one group, the eco-conscious, the environmental factor plays a crucial role in their decision. Since this group is promising as an early adopter due to their general openness towards UAM and slightly larger price range acceptance, the environmental impact of new UAM technologies should be considered when designing such technologies. Furthermore, our results show that autonomy is still met with apprehension and skepticism by most user groups, especially so by automation skeptics, which after all, made up 16 % of this study's sample. Human-piloted air cabs are preferred overall. The observed apprehension most likely stems from missing trust in the system and has to be considered, especially during the early implementation phase.

Acknowledgements. This research project was funded by the Ministry of Culture and Science of the state of North Rhine-Westphalia with the funding code 327 321-8.03.07-127598 (Project ACCESS!). The authors would like to thank Linda Sinani research support. Further, we thank all study participants.

References

1. Al Haddad, C., Chaniotakis, E., Straubinger, A., Plötner, K., Antoniou, C.: Factors affecting the adoption and use of urban air mobility. Transport. Res. Part A: Policy Pract. **132**(March 2019), 696–712 (2020)
2. Arning, K.: Conjoint measurement. The International Encyclopedia of Communication Research Methods, pp. 1–10 (2017)
3. Bacchini, A., Cestino, E.: Electric VTOL configurations comparison. Aerospace **6**(3), 26 (2019)
4. Bansal, P., Kockelman, K.M., Singh, A.: Assessing public opinions of and interest in new vehicle technologies: an Austin perspective. Transport. Res. Part C: Emerg. Technol. **67**, 1–14 (2016)
5. Ben-Akiva, M., McFadden, D., Train, K., et al.: Foundations of stated preference elicitation: Consumer behavior and choice-based conjoint analysis. Found. Trends® Econometr. **10**(1–2), 1–144 (2019)
6. Biermann, H., Philipsen, R., Ziefle, M.: A matter of trust – identification and evaluation of user requirements and design concepts for a trust label in autonomous driving. In: Black, N.L., Neumann, W.P., Noy, I. (eds.) IEA 2021. LNNS, vol. 221, pp. 560–567. Springer, Cham (2021). https://doi.org/10.1007/978-3-030-74608-7_68
7. Biermann, H., Philipsen, R., Ziefle, M.: Crazy little thing called trust - user-specific attitudes and conditions to trust an on-demand autonomous shuttle service. In: Duffy, V.G., Landry, S.J., Lee, J.D., Stanton, N. (eds.) Human-Automation Interaction. Automation, Collaboration, & E-Services, vol. 11, pp. 235–252. Springer, Cham (2023). https://doi.org/10.1007/978-3-031-10784-9_14
8. Bostrom, A.: Risk perceptions: experts vs. lay people. Duke Envtl. L. & Pol'y F. **8**, 101 (1997)
9. Brazell, J.D., Diener, C.G., Karniouchina, E., Moore, W.L., Séverin, V., Uldry, P.F.: The no-choice option and dual response choice designs. Mark. Lett. **17**(4), 255–268 (2006)
10. Charness, N., Yoon, J.S., Souders, D., Stothart, C., Yehnert, C.: Predictors of attitudes toward autonomous vehicles: the roles of age, gender, prior knowledge, and personality. Front. Psychol. **9**, 2589 (2018)
11. Chrzan, K., Halversen, C.: Diagnostics for random respondents. In: Sawtooth Software European Conference (2020)
12. Chrzan, K., Orme, B.: An overview and comparison of design strategies for choice-based conjoint analysis. Sawtooth Softw. Res. Paper Ser. **98382**, 360 (2000)
13. Dannenberger, N., Schmid-Loertzer, V., Fischer, L., Schwarzbach, V., Kellermann, R., Biehle, T.: Verkehrslösung oder technikhype? ergebnisbericht zur einstellung der bürgerinnen und bürger gegenüber dem einsatz von lieferdrohnen und flugtaxis im städtischen luftraum in deutschland. Tech. rep., WiD, TU Berlin (06 2020)
14. Davis, F.: Perceived usefulness, perceived ease of use, and user acceptance of information technology. MIS Q. **13**(3), 319–340 (1989)
15. Devine-Wright, P.: Explaining "nimby" objections to a power line: the role of personal, place attachment and project-related factors. Environ. Behav. **45**(6), 761–781 (2013)
16. Eißfeldt, H., Biella, M.: The public acceptance of drones-challenges for advanced aerial mobility (aam). Transport. Res. Procedia **66**, 80–88 (2022)
17. European Union Aviation Safety Agency: study on the societal acceptance of urban air mobility in Europe. Tech. rep., European Union Aviation Safety Agency (2021). https://www.easa.europa.eu/en/downloads/127760/en

18. Garrow, L., German, B., Leonard, C.: Urban air mobility: a comprehensive review and comparative analysis with autonomous and electric ground transportation for informing future research. Transport. Res. Part C: Emerg. Technol. **132**, 103377 (2021)

19. Haider, S.W., Zhuang, G., Ali, S.: Identifying and bridging the attitude-behavior gap in sustainable transportation adoption. J. Ambient. Intell. Humaniz. Comput. **10**(9), 3723–3738 (2019). https://doi.org/10.1007/s12652-019-01405-z

20. Hansman, J., Vascik, P.: Operational aspects of aircraft based on-demand mobility. In: 2nd On-Demand Mobility and Emerging Aviation Technology Roadmapping Workshop, vol. 156 (2016)

21. Hoff, K.A., Bashir, M.: Trust in automation?: integrating empirical evidence on factors that influence trust. Hum. Factors **57**(3), 407–434 (2015)

22. Husemann, M., Kirste, A., Stumpf, E.: Analysis of cost-efficient urban air mobility systems: Optimization of operational and configurational fleet decisions. European Journal of Operational Research (2022)

23. Kellermann, R., Biehle, T., Fischer, L.: Drones for parcel and passenger transportation: a literature review. Transport. Res. Interdiscip. Perspect. **4**(January), 100088 (2020)

24. Salmon, P., Macquet, A.-C. (eds.): Advances in Human Factors in Sports and Outdoor Recreation. AISC, vol. 496. Springer, Cham (2017). https://doi.org/10.1007/978-3-319-41953-4

25. Mayring, P., Fenzl, T.: Qualitative Inhaltsanalyse. In: Handbuch Methoden der empirischen Sozialforschung, pp. 633–648. Springer, Wiesbaden (2019). https://doi.org/10.1007/978-3-658-21308-4_42

26. Melo, S., Cerdas, F., Barke, A., Thies, C., Spengler, T., Herrmann, C.: Life cycle engineering of future aircraft systems: The case of eVTOL vehicles. Procedia CIRP **90**, 297–302 (2020)

27. Mirnig, A.G., Trösterer, S., Meschtscherjakov, A., Gärtner, M., Tscheligi, M.: Trust in Automated Vehicles. I-Com **17**(1), 79–90 (2018)

28. Neyer, F., Felber, J., Gebhardt, C.: Kurzskala technikbereitschaft (tb, technology commitment). zusammenstellung sozialwissenschaftlicher items und skalen (zis) (2020). https://zis.gesis.org/skala/Neyer-Felber-Gebhardt-Kurzskala-Technikbereitschaft-(TB,-technology-commitment). Accessed 16 Aug 2022

29. Novikova, O.: The sharing economy and the future of personal mobility: new models based on car sharing. Telev. New Media **7**(8), 27–31 (2017)

30. Orme, B.: Interpreting the results of conjoint analysis. Getting Start. Conjoint Anal. Strateg. Product Design Pricing Res. **2**, 77–88 (2010)

31. Peeta, S., Paz, A., DeLaurentis, D.: Stated preference analysis of a new very light jet based on-demand air service. Transport. Res. Part A: Policy Pract. **42**(4), 629–645 (2008)

32. Pol, E., Di Masso, A., Castrechini, A., Bonet, M., Vidal, T.: Psychological parameters to understand and manage the nimby effect. Eur. Rev. Appl. Psychol. **56**(1), 43–51 (2006)

33. Rajendran, S., Srinivas, S.: Air taxi service for urban mobility: a critical review of recent developments, future challenges, and opportunities. Transport. Res. Part E: Logistics Transport. Rev. **143**(September), 102090 (2020)

34. Ramadhas, A.S. (ed.): Alternative Fuels for Transportation. CRC Press, Boca Raton (2011)

35. Reiche, C., McGillen, C., Siegel, J., Brody, F.: are we ready to weather Urban Air Mobility (UAM)? Integrated Communications, Navigation and Surveillance Conference, ICNS 2019-April, 2–8 (2019)

36. Rothfeld, R., Fu, M., Balac, M., Antoniou, C.: Potential urban air mobility travel time savings: An exploratory analysis of Munich, Paris, and San Francisco. Sustainability **13**(4), 297–302 (2021)
37. Salvi, B.L., Subramanian, K.A., Panwar, N.L.: Alternative fuels for transportation vehicles: a technical review. Renew. Sustain. Energy Rev. **25**, 404–419 (2013)
38. Sawtooth-Software: None option/dual-response none. https://legacy.sawtoothsoftware.com/help/lighthouse-studio/manual/hid_web_cbc_none.html. Accessed 31 Jan 2023
39. Sawtooth-Software: The CBC system for choice-based conjoint analysis - version 9.14.2 (2017). https://sawtoothsoftware.com/resources/technical-papers/cbc-technical-paper. Accessed 10 Jan 2023
40. Sawtooth-Software: the latent class. technical paper v4.8 (2021). https://sawtoothsoftware.com/resources/technical-papers/latent-class-technical-paper. Accessed 16 Jan 2023
41. Shaheen, S., Cohen, A., Farrar, E.: The potential societal barriers of urban air mobility (uam). Tech. rep, National Aeronautics and Space Administration (2018)
42. Stančin, H., Mikulčić, H., Wang, X., Duić, N.: A review on alternative fuels in future energy system. Renew. Sustain. Energy Rev. **128**, 109927 (2020)
43. Stolz, M., Laudien, T.: Assessing social acceptance of urban air mobility using virtual reality. In: 2022 IEEE/AIAA 41st Digital Avionics Systems Conference (DASC), pp. 1–9. IEEE (2022)
44. Straubinger, A., Kluge, U., Fu, M., Al Haddad, C., Ploetner, K.O., Antoniou, C.: Identifying demand and acceptance drivers for user friendly urban air mobility introduction. In: Müller, B., Meyer, G. (eds.) Towards User-Centric Transport in Europe 2. LNM, pp. 117–134. Springer, Cham (2020). https://doi.org/10.1007/978-3-030-38028-1_9
45. Straubinger, A., Rothfeld, R., Shamiyeh, M., Büchter, K.d.: An overview of current research and developments in urban air mobility - setting the scene for UAM introduction. J. Air Transp. Manage. **87**, 101852 (2020)
46. Straubinger, A., Verhoef, E.T., de Groot, H.L.: Going electric: environmental and welfare impacts of urban ground and air transport. Transp. Res. Part D: Transp. Environ. **102**, 103146 (2022)
47. Sun, X., Wandelt, S., Husemann, M., Stumpf, E.: Operational considerations regarding on-demand air mobility: a literature review and research challenges. J. Adv. Transport. **2021**, 3591034 (2021)
48. Turner, M., Kitchenham, B., Brereton, P., Charters, S., Budgen, D.: Does the technology acceptance model predict actual use? a systematic literature review. Inf. Softw. Technol. **52**(2), 463–479 (2010). https://ac.els-cdn.com/S0950584909002055/1-s2.0-S0950584909002055-main.pdf?_tid=spdf-a68c7a93-1965-43aa-ba13-c5d8ab7f5147&acdnat=1519223148_e293368d40117a6e6517b79da8dee0df
49. Venkatesh, V., Morris, M.G., Davis, G.B., Davis, F.D.: User acceptance of information technology: toward a unified view. MIS Q. **27**(3), 425–478 (2003)
50. Verbannt Unbemannte Luftfahrt (VUL): was denken die deutschen über unbemannte luftfahrt (2022). https://www.verband-unbemannte-luftfahrt.de/wp-content/uploads/2019/11/Akzeptanzumfrage.pdf. Accessed 26 Jan 2023

Research for Responsible Innovation: A Living-Lab Approach for Last-Mile Logistics Using a Self-Developed Autonomous Transport System

Nicola Marsden, Mihai Kocsis[✉], Nicole Dierolf, Claudia Herling, Jens Hujer,
and Raoul Zöllner

Heilbronn University of Applied Science, Max Planck Str. 39, 74081 Heilbronn, Germany
mihai.kocsis@hs-heilbronn.de

Abstract. This paper reports on research on responsible innovation for last-mile urban delivery. We designed, developed and tested a fleet of autonomous electric vehicles that operate in a new urban district developed as part of the federal horticulture exhibition 2019 in Heilbronn, Germany. Hereby not only the technical aspects play an important role but also the acceptance of the stakeholders. Thereby, the technical and technological innovation of the entire system is guided by the interactivity with the users. They are able to interact with the system by planning, modifying and monitoring their delivery requests in real time. All different usage scenarios together with the human machine interfaces were co-designed in a participatory way with users from the new neighborhood by an interdisciplinary team of engineers, logistics experts, behavioral scientists, designers, and developers. In this context, the research deals with both social and technical innovation. The ability to identify and describe processes that can be used as models for the implementation of such innovations and that can be reproduced in similar contexts is crucial to meet the challenges of digitalization and automation in mobility with regard to demographic change and the trend towards reurbanization.

Keywords: Co-creative participatory research · last mile logistics · autonomous driving

1 Introduction

Up to a third of urban inland traffic is commercial traffic, this means it serves for goods transportation [1]. In many cities, delivery vehicles parked in the street, while the driver is busy with the delivery, are now treated as a traffic violation. Also, the delivery vehicles are mostly equipped with internal combustion engines, which poses another problem, namely CO_2 emissions as well as noise pollution. One of the greatest potentials for improvement in urban logistics is therefore seen in the use of highly automated, small electric transportation vehicles [2, 3]. Cities in particular have become a place where the demands on logistics are characterized by a particular heterogeneity: the individualization of customer requests, online ordering options (with the expectation of short-term

H. Krömker (Ed.): HCII 2023, LNCS 14048, pp. 316–333, 2023.
https://doi.org/10.1007/978-3-031-35678-0_21

readiness for delivery in connection with a precisely defined delivery time), a changing society ("sustainable city"), increasing economic pressure on logistics service providers, in particular due to a lack of space and delays caused by traffic jams, but also increasingly due to limited staff availability, and political design wishes, e.g. to react to exceeding limit values for air pollutants and noise, meet in a small space. Urban logistics is therefore considered to be a particularly challenging future field [4].

Autonomous electric transport vehicles offer great potential for urban logistics of the last mile, but this is a solution that can only be successfully implemented if there is acceptance of the stakeholders, particularly the citizens populating the urban area, working or living there. Therefore, it is important to design and develop systems suited for their needs and wishes. Relevant stakeholders for the development of technological systems for last-mile delivery thus include residents, customers of stores, visitors of restaurants, citizens working in the urban area, traffic participants like pedestrians, drivers or cyclists, parcel delivery provider or courier express staff (CEP) – but also the municipality or the rescue services. One way to increase acceptance logistics with autonomous vehicles would be to ensure that the system offers features that might be laborious for conventional existing systems, but desired and useful.

The purpose of the research presented here was to identify possible features and develop a customized delivery system with autonomous vehicles rooted in the stakeholder's wishes. The result was a novel interactive system in which the user can specify the desired date, time, and delivery place, track the delivery in real time, as well as cancel and reschedule existing delivery requests. The system was implemented for four scenarios for parcel deliveries, neighbor to neighbor deliveries, return deliveries and waste disposal. The developed system and its acceptance were evaluated in a real-world use over a period of almost 6 months.

The outline of the paper is as follows: Sect. 2 presents a brief state of the art regarding living labs. Section 3 describes the living lab BUGA:log. In Sect. 4 and 5 the participatory research approach and resulting scenarios are presented. Section 6 offers a technical insight of the developed system based on the co-creative research. Section 7 deals with the evaluation of the system regarding user acceptance. Conclusion and further work are expounded in Sect. 8.

2 Living-Lab Research

Living-labs are a research infrastructure that can ensure "that science and innovation are carried out in the public interest by integrating methods to promote more democratic decision-making through greater involvement of major stakeholders who may be directly affected by the introduction of new technologies." [5]. Living labs are located at the interface between science and society and at the same time addresses the issue of sustainability. Living labs have emerged as an approach to support transitional processes at the interface of science and society [6].

The benefits of living labs are multi-faceted: they have the ability to drive societal change and transformation, serve as a platform for both scientific and societal learning, and generate transformation knowledge that can be applied [7]. Their research method employs transdisciplinary techniques with an eye towards scalability and transferability of findings. The research process is experimental, using real-world laboratories to perform transformation experiments, leading to the systematic evaluation of the entire project [8]. This experimental approach distinguishes itself from other experiments in several ways [9] and [10].

First of all, real-world labs are driven by hypotheses, however, these hypotheses may not always be clearly stated and new knowledge can be presented through various means, such as newspaper articles or art exhibitions [11]. The experimental aspect lies in the constructive use of uncertainties to generate new knowledge [12], with an emphasis on incorporating local knowledge and involving the community [13]. The research conducted in these labs is centered around the concept of sustainability and sustainable development, and researchers engage with values and normative ideas in a way that sets them apart from researchers in other disciplines [14].

Additionally, while someone must initiate the experiment and play a role in formulating hypotheses, designing the approach, evaluating results, and drawing conclusions, this person does not necessarily have to be a scientist. Non-scientists can be part of the team conducting the experiment and may even conduct experiments without the involvement of scientists. Practitioners and researchers collaborate to generate knowledge that is robust and anchored in a specific cultural, social, and geographical [15]. This democratizes the research process and acknowledges the impact of partial perspectives [16].

Finally, real-world experiments are not isolated from society; they take place in the real world, which means they have an immediate impact on society [17]. The "laboratization" of the world means that failures must be expected, as it is not possible to control all conditions, leading to an unpredictability of outcomes. This has significant ramifications for the social environment, requiring openness and a societal culture that is willing to accept failures. Society must be willing to embrace new practices and the experiment may generate resistance and opposition. Neither the interventions nor the outcomes can be fully controlled [18], so dealing with contingencies and being prepared for unexpected outcomes is a fundamental aspect of real-world labs [19]. This opens up opportunities for breaking past structures but also highlights the importance of considering consequences and responsibilities [9]. Thus, iterative and reflexive monitoring and evaluation must be a key component of real-world labs to facilitate continuous individual, organizational, and societal learning and support the broader systemic transition [8] and [10].

Real-world labs have evolved from their roots in technology-based innovation research and have expanded from a primarily technology-focused approach to encompass more socially-centered research [20]. Research on real-world labs is conducted in multiple fields, with a shift from a technology-focused perspective to a more generalized focus on innovation. Real-world labs play a role in transforming a production-based economy into an innovative service economy by involving users as co-creators in real-life environments [21]. A wider view of real-world labs goes beyond the development of specific products and adopts an action research approach that engages local people in an open, participatory process of improving their living spaces [20] and [22].

3   The BUGA: log Living Lab

The state of Baden-Württemberg (BaWü) has one of the largest collections of living labs in Europe funded by the state [7, 23] and [24] including seven living labs focusing on cities as experimentation areas for transformation [25]. BUGA:log is one of these living labs and was funded by the state of Baden-Württemberg in Germany between 2015 and 2019. It was coordinated by the University of Heilbronn and involved researchers from the fields of logistics, engineering, social sciences and HCI, who worked within the area of developing an automatic transport system for urban areas. Stakeholders from politics and business were also involved.

BUGA:log used the federal horticulture show 2019 (Bundesgartenschau 2019) in Heilbronn, Germany. The biennial federal expo covered both gardening and architecture and the grounds of the expo also included a new urban quarter, build on wasteland in the center of the city that had been fallow since the Second World War. This new residential area was part of the expo and designed to set a good example of sustainable urban development. In this context, the research project BUGA:log fits into an open and integrative environment to develop solutions for urban logistics in a responsible process.

Together with representative of the city of Heilbronn, representatives from the federal expo, other project partners from industry, residents and visitors of the expo, we developed and tested an autonomous transport system for the last mile of city logistics during the expo (April to October 2019). The aim of the research was to reduce traffic in urban districts and the transport demands of residents in the medium to long term, thereby making urban logistics more sustainable. Many of the supply processes that can be defined in the newly built "Neckarbogen", the residential area that was part of the expo, are exemplary for the logistical challenges already faced by existing downtown areas. These include supplying restaurants and stores as well as making deliveries to offices and households. In particular, supply processes on the last mile, i.e. delivery and collection in urban (residential) areas, remain a challenge regarding a more sustainable and thus more responsible urban development. The research format of a living lab integrates different perspectives regarding autonomous delivery in the development process. This way, we tried to achieve the greatest possible sensitivity and perception of different interests – including those by marginalized groups that are often overlooked - on this important subject.

We followed a living-lab approach to create an autonomous transport system [26]. A trans- and interdisciplinary team developed an electrically powered transport vehicle and a progressive web app, which we implemented and tested during the course of the expo. Our approach to the living lab comprised the four dimensions stakeholder/user space, methodological space, creative space, management space [25] and [27]. Research activities took place in four iterative phases. We used an iterative approach with four phases of co-creation, namely understand, co-design, implement, and evaluate (Fig. 1).

The development and construction of the "Neckarbogen" district at the expo presents typical logistical challenges that are comparable to processes that we already find in inner cities today. These include traditional supply processes such as the delivery of goods and waste disposal processes such as return deliveries, waste disposal. The use of autonomous vehicles within the scope of the project should provide insights into logistic processes, for example into the optimal handling of consignments in a district. Together

Fig. 1. Methodological space: Co-creation phases.

with relevant stakeholders we have identified, developed and implemented the logistical possibilities at the expo.

4   Methods Used in the Living-Lab Process

The BUGA:log living lab was designed as a participatory technology development process. The co-creation process of the autonomous electric delivery vehicles and the human-technology interaction was based on a broad bundle of quantitative and qualitative methods to ensure participation of the stakeholders. Some methods were used at the beginning of the project, some were used throughout the whole living-lab process, and some were used only during the six-month period during which the automatic transportation system was tested on the premises of the expo.

At the beginning of the three-year project, **exploratory interviews** were conducted with a variety of stakeholders. Parcel service providers, residents, city administration, pedestrians, etc. were interviewed in guide-based narrative interviews. The aim was to find out in an open and non-directive conversation which needs, ideas, concerns, etc. they had regarding urban logistics. The plan to incorporate autonomous transport vehicles into the expo to showcase a different type of urban logistics to be tested in a real environment was discussed with them. The audio recordings of the interviews were transcribed verbatim. The transcripts were subsequently analysed by means of a thematic analysis [28]. Based on this, items were developed for a quantitative survey of Heilbronn citizens and the results were fed into the co-creative process and the development of scenarios.

The following methods were used throughout the whole project spanning two iterations of the co-creation cycle from 2017 to 2019:

Third-space workshops [29] with citizens of the municipality as a space for mutual learning in which potential users became co-researchers and co-designers collaborating with the design team on equal terms. Between 2016 and 2019, eight third-space workshops were conducted with 4–12 participants. The content of the workshops was based on the four phases of the development process (understand – co-design – implement – evaluate) (Fig. 2).

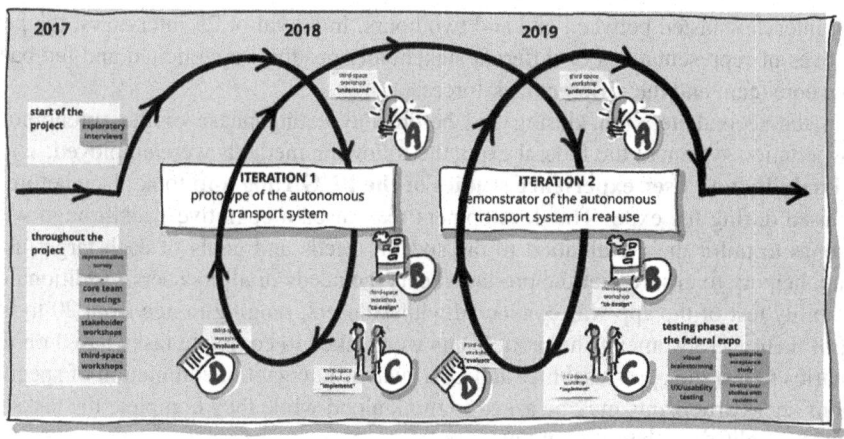

Fig. 2. Overview of the methods used in the two iterations.

Core team meetings took place once a month in 2017 and 2018, and weekly in 2019. The core team was an interdisciplinary team of researchers from engineering, logistics, behavioral sciences, design science, and software engineering. The design team consisted of 5 to 10 researchers, additionally, there were between 5 and 10 students involved working on their theses or as student assistants. There were approximately 40% women in the core team and among the student assistants. The core team was in charge of the coordination of the activities and the stakeholders. The core team not only co-designed the process but also initiated and managed the different task forces (development of software, hardware, exhibition, stakeholder involvement) and supplied relevant information to the different stakeholders and arenas [30].

Stakeholder workshops took place every three months. There was a fixed group of representatives of the municipality, the researchers, and the officials of the expo participating. Other stakeholders were invited and involved as needed throughout the living-lab process, e.g. the regulating traffic authority, representatives of urban planning, city development, traffic authorities, emergency services, politics, citizens' initiatives, the communication and marketing departments of the organizations involved, future inhabitants, exhibitors, event organizers, and business owners of the expo etc.

A **representative acceptance survey of the general population** was conducted using the "Heilbronn Barometer", a yearly panel of the citizens of the city of Heilbronn. Participants (938 respondents were surveyed in three consecutive years (2017: 500 participants, 2018: 477 participants, 2019: 938 participants). Participants were all over 18 years old, approximately half men and half women, 20–25% had a migration background. The participants were surveyed regarding their attitude towards autonomous transport vehicles in general, the attitude towards the use of autonomous vehicles in town, and the intention of use (in general and on the expo).

Qualitative interviews were conducted in every phase of the project. During selection of the interviewees, we took care to represent a varied cross-section of the population of Heilbronn in terms of geographical location, gender, age, migration background and different work and life phases. All interviews were conducted by phone or face-to-face.

Each interview lasted between one and two hours. In a total of 25 interviews, the perspectives of representatives of different stakeholder groups were elicited and fed back to the core team and the different task forces.

In the second iteration, during and before the testing phase of the autonomous transportation system at the federal expo, the following methods were employed:

Usability and user experience studies of the BUGA:log app took place before it was used during the expo. The development team used a cognitive walkthrough with personas to tailor their evaluation to the specific needs and goals of each target user group, helping to ensure that the product meets the needs of all its users. Additionally, a usability test of the app was conducted with 10 users, ranging in age from 20 to 68, half of them were women. The participants were asked to complete tasks based on the scenarios to test the app's usability and UX, such as navigation, completion of specific actions, etc. Participants may be asked to think aloud while they complete the tasks to provide insights into their thought process.

In-situ user studies with residents captured the context and the actual use of the BUGA:log transportation system during six-month period of the expo in 2019. A total of 20 households and 4 businesses in the Neckarbogen, the new quarter that had been built as part of the expo and that residents could move into starting in the summer of 2018, participated. They tested the different scenarios that had been developed for the co-creative process (some of the residents had participated in the third-space workshops from the beginning and had thus co-designed the scenarios for the BUGA:log autonomous transport system). Feedback from the residents was collected after each use of the BUGA:log system and in interviews at the end of the testing period.

Long-term visual brainstorming took place in the exhibition space onsite at the expo for the whole six months: The visitors of the BUGA:log showroom on the expo were encouraged to participate and brainstorm through several low-threshold methods: They received trigger cards to voice their opinions or ideas, these could either be handed in to the staff of the BUGA:log showroom, they could throw them in a mailbox or put them up on a wall that was specifically designed for this. They could participate in creating word clouds that were displayed in the showroom in real time. They could give input via tablets that were installed in the showroom. In addition, the staff of the show room took notes on relevant interactions with visitors.

An **onsite quantitative acceptance and acceptability study** [31] of visitors of the expo was conducted from April to October 2019 on the premises. The participants were either in the exhibition room featuring a model of the BUGA:log autonomous transportation vehicle or onsite the expo close by where the vehicle was driving. In total, 416 participants took part by filling out a digital questionnaire on tablet computers, six datasets had to be excluded during analysis, leaving a sample size of 409 participants (female 47,8%, male 47,8%, diverse 1,2%, not indicated 3,1%). All participants were older than 18 years old. Approximately 10% had a migration background.

5   Understanding and Co-design: Scenarios

In the first two phases of the first iteration, understanding and co-design, four scenarios were developed in the co-creation process: parcel deliveries, neighbor to neighbor deliveries, return deliveries, and waste disposal. These scenarios were further developed and fine-tuned in the first two phases of the second iteration and then implemented in real use at the expo.

Parcel delivery was the first scenario to come to life in the living lab: As soon as the parcel service delivers parcels to the parcel box integrated in the vehicle, residents can select the date, time and place for delivery of their parcel via the BUGA:log web app (Fig. 3 a). Similar to the parcel delivery scenario, we elaborated and implemented other scenarios, the neighborhood courier (neighbor to neighbor) and waste paper disposal for organizations. In this scenarios, residents can request a vehicle and access the parcel box in order, to send a delivery to another client, e.g. to send a birthday gift or a borrowed item (garden tool) to another recipient or neighbour (Fig. 3, b, c).

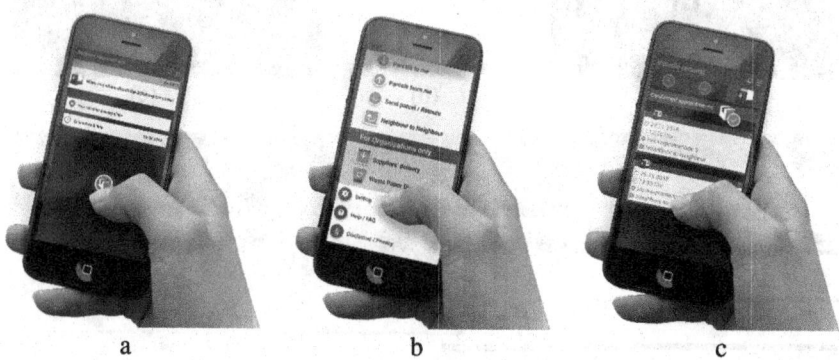

a   b   c

Fig. 3. Initial designs and GUIs of the client application: (a) selecting a delivery date and time; (b) overview with all possible scenarios; (c) two confirmed neighbor to neighbor deliveries.

In the participatory process with the relevant stakeholders searched for solutions the typical logistical challenges that were found in the Neckarbogen district and that are comparable to processes that we already find in inner cities today. These include traditional supply processes such as the delivery of goods and waste disposal processes such as return deliveries, waste disposal. The use of autonomous vehicles within the scope of the project should provide insights into logistic processes, for example into the optimal handling of consignments in a district. Together with relevant stakeholders (parcel service providers, residents, city administration, pedestrians, etc.) we have identified, developed and implemented the logistical possibilities at the expo.

The term "autonomous transport system" implies a system that acts independently and without human intervention. As a result, communication for users takes place via the desiged application, which allows the participants of the real experiment a flexible, independent and especially an intuitive handling of such a novel transport system. As soon as a parcel is delivered at the entrance of the expo site into the vehicle, the recipients receive a notification via the BUGA:log application and can the list with their open appointments (Fig. 4, a).

The residents must then click on the respective package to determine the desired delivery location and time of delivery (Fig. 4, b) and send the request. The system confirms the request if the vehicles is available and can make it in time (Fig. 4, c). If the vehicle cannot make it in time for the appointment, three alternative proposals are

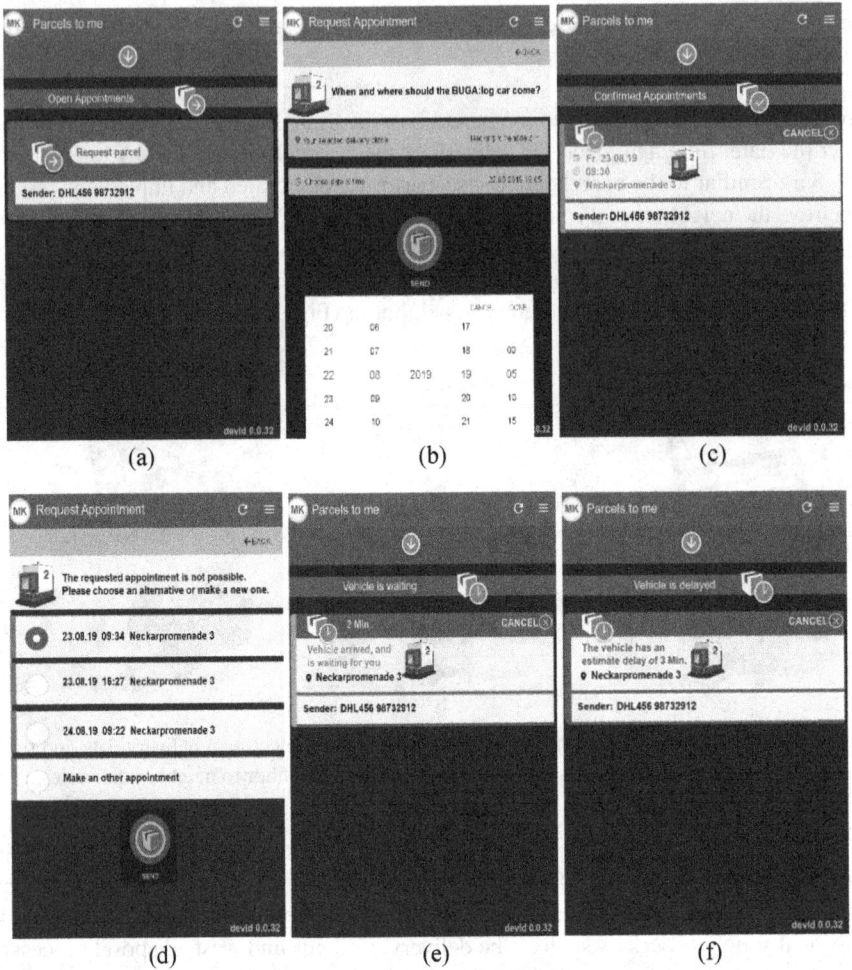

Fig. 4. User interfaces for request client application: (a) request list; (b) order interface; (c) confirmed order; (d) denied order with alternatives; (e) vehicle in waiting position notification; (f) vehicle is delayed notification.

suggested (Fig. 4, d). As soon as the vehicle arrives at the desired location of delivery, the users receive a message that the parcel is ready for collection (Fig. 4, e). In case of confirmed appointments, where the vehicle is delayed the client is also informed (Fig. 4, f). The parcel box, supplied by Erwin Renz Metallwarenfabrik GmbH & Co KG (RENZ), can be opened by entering a PIN as well as by using a previously assigned token (Fig. 5, f). The desire of the clients was also to be able to use the BUGA:log application on any device (smartphone, tablet, laptop, pc etc.). Therefor we decided to implement the BUGA:log app as a progressive web application.

Fig. 5. A resident identifying with her PIN on the screen of the parcel box.

6 Implementing the Scenarios: Vehicles and Mission Planning

Based on the co-creative collaboration the interactive planning and delivery system depicted in Fig. 6 was developed.

The system consists of the BUGA:log *client application* (progressive app) designed for the customer to make requests, receive status information and register to the system. The user data together with the request data are stored on a database connected via a *client & service management* server. The implemented system is designed to work for different type of services. Therefore, the database stores also for each service type a list of possible vehicles designed for the service type. Every request is forwarded, after a database update to the *planning & supervision of service execution* server. A request consists of an *ID*, the *client ID*, a *status* information, the list with *vehicle IDs* designed for the requested type of service, requested *date and time* for task execution, the *ID of the location* where the service has to be executed, a *comment* used for information update for the client and an *ID correlated to a previous request*, generated by the planning system in case a request could not be executed and alternative solutions were generated. The request date and time were added to the request definition as a result of the co-creative process. Also the ID correlated to a previous request is the result of this process and is needed to link a new request to a previous rejected request.

The package delivery scenario designed in the co-creative process was the most used one and generated the also the life cycle of a request. A request is introduced in the system either by the client itself via the BUGA:log app or by an auxiliary subsystem.

The *auxiliary subsystem* in the current work was the RENZ box (Fig. 7, a) mounted on the vehicle (Fig. 5). The interface of the RENZ system allows clients to log in via a token or a pin (Fig. 7, b) and choose between two options: introduce or collect packages (Fig. 7, c). When introducing a package into the box, the recipient out of a list (Fig. 7, d) and the size of box (Fig. 7, e) can be chosen. A box will open automatically (Fig. 7, g) and the client is guided to insert the package into the box (Fig. 7, f). After closing the

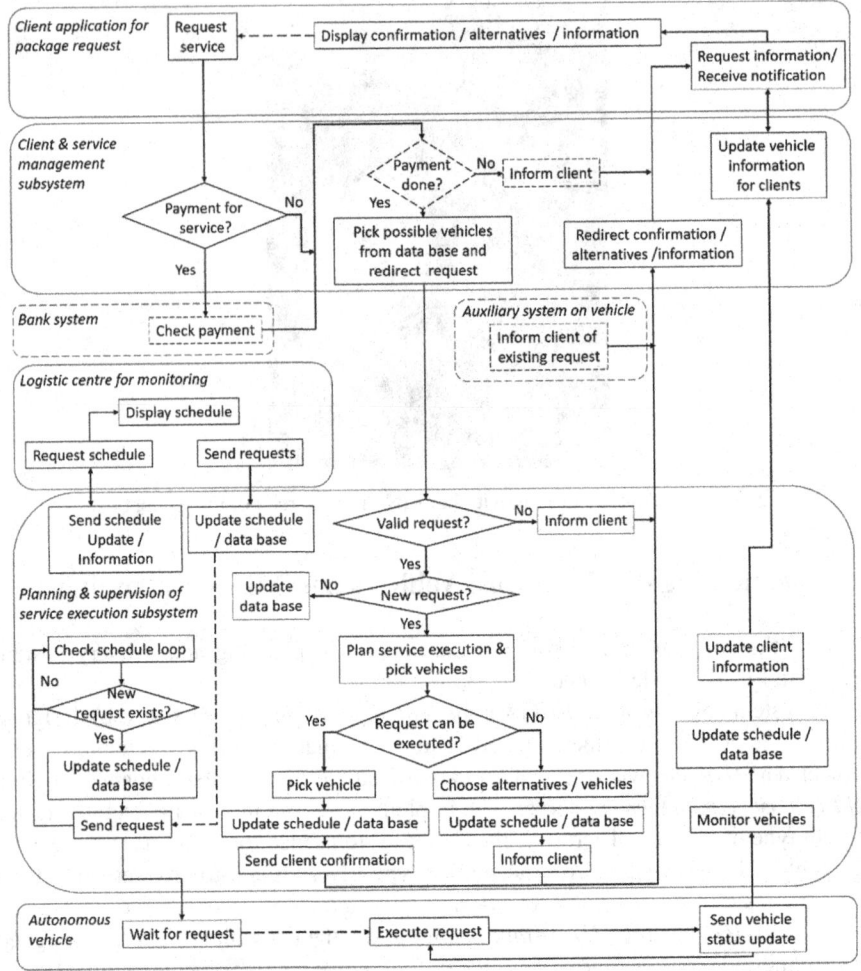

Fig. 6. Workflow of the implemented system.

opened box, the client is informed that the request has been successfully introduced into the system (Fig. 7 h). For the package pick up scenario the box(s) will open automatically after login and choosing the collect option (Fig. 7, c).

The status information of the request has the value STATION when a package is introduced into the box on the vehicle and PENDING after the user makes/answers a request using the client application.

The planning & supervision of service execution subsystem is the main component of the system and responsible for ad-hoc planning, direct communication and command of the vehicle fleet as well as for updating schedule data and sending out vehicle status information according to the current mission state. After receiving the request, a formal validity check is made. If the request is new a planning process starts. The planning algorithm was implemented according to the wishes desired in the participatory process.

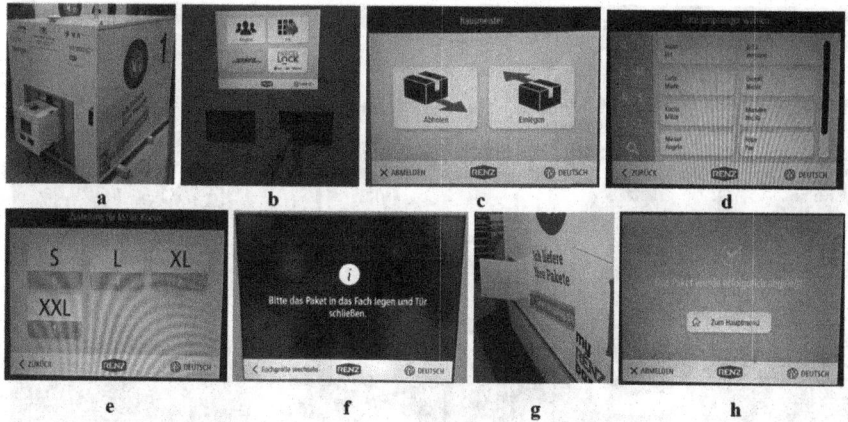

Fig. 7. User interfaces of the RENZ system: (a) RENZ box; (b) login interface; (c) option interface (pick and send package); (d) customer list; (e) box size options; (f) information interface regarding next user action; (g) opened box; (h) information interface regarding the request registration.

In case the request can be executed it's added to the data base and schedule of the vehicle and a confirmation to the client is send out (Fig. 4, c). The status of the request changes to ACCEPTED. Otherwise the planning process searches for three alternatives. The first alternative is the fastest delivery time after the requested one. The other two alternatives are the fastest delivery possibility in the following two operation slots. The status of the request is set to REJECTED and the client has one-minute response time to pick one of the proposed alternatives (Fig. 4, d), but also has the choice to cancel them and make a new request. A detailed technical description of the planning algorithm can be found in [32] and [33]. Planned requests can be canceled at any time by the client or by the system if for technical reasons they cannot be carried out. In this case, the status changes to: CANCELD_BY_CLIENT of CNACELD_BY_VEHICLE.

A separate process on the planning & supervision subsystem oversees the generated schedule of the autonomous vehicles and sends commands to execute the specific task at the scheduled time. The vehicles send information about their state in a 5 s cycle. Thus, it is possible to track the execution of the requests and inform the clients about the status of the request, e.g. in case of delays and when the vehicle is in waiting position (Fig. 4, e, f). Possible request status is DRIVING, DELAYED and WAITING. Also, clients can permanently view the location of the autonomous vehicles on a map. The entire mission planning and execution process is overseen by a human operator from a logistic center for monitoring (Fig. 8). The operator can view the schedule of each autonomous vehicle, their current state and location on a map, prioritize tasks and intervene by sending commands with a higher priority to the vehicle. The scope of the operator was to analyze the dealing with extraordinary situations e.g. blocked paths or technical difficulties. This is important in following research interest regarding the preparation in operating autonomous vehicles on level 4 of driving automation.

Since, at the beginning of the study, no autonomous vehicles for execution of logistics services existed, this had to be developed first. In order to not start this development from scratch existing vehicles were transformed. For the current study, three vehicles

Fig. 8. Logistic center for monitoring with human operator.

were transformed and homologated in order to operate in the pedestrian area of the living laboratory. The criteria that are taken into account when choosing the vehicle depend not only on the services that have to be performed and the operating environment, but also on a high acceptance for pedestrians. For acceptance of such vehicles in pedestrian areas, according to the clients wishes, low emissions and reduced noise pollution are essential. In order to meet these criteria, electric vehicles are clearly superior to those with internal combustion engines. First, they pollute less, and second, the transformation process, from a technical and homologation point of view, is easier. As the introduction of autonomous vehicles into pedestrian environments is still at the beginning, current acceptance by the population is difficult to assess. This criterion can be only assessed by tests of interaction of autonomous vehicles with pedestrians as traffic partners. For the selection and transformation of an existing vehicle into an autonomous vehicle, a design methodology was developed, which, in addition to meeting the functional requirements imposed, also aimed at developing modular architectures by implementing open-source subsystems, ensuring increased interchangeability and reduced costs [34]. The selected vehicle for transformation was an electric golf cart of type EZGO RXV (Fig. 9, a). Golf carts are being used for person/luggage transportation in densely populated environments, such as airports or railway stations. Moreover, in some Asian countries, this vehicle is often used for person transportation in urban areas. It is thus assumed that the use of this vehicle, transformed into an autonomous vehicle in pedestrian environments, will not be rejected by the public.

In order to homologate the vehicle after transformation a safety concept for vehicles operating in pedestrian areas, with the unique characteristic of an integrated bumper (Fig. 9, b), was developed [35]. The selection and integration of the box occurred after discussions with parcel delivery or courier express staff and manufacturers of parcel stations. The RENZ box fitted best the CEP demands. These kind of boxes are already in use, e.g. by DHL, and the process for a reliable delivery was already defined. Figure 9, c shows the first integration of this kind of boxes on the vehicle. Since the weight of the initial box was too high, and a homologation required a less weight, new lighter

Fig. 9. Vehicle development cycle: (a) chosen vehicle; (b) bumper integration; (c) parcel box integration; (d) final vehicle at charging station; (e) vehicle fleet.

boxes out of aluminium had to be manufactured. The final vehicle is depicted in Fig. 9, d, at a charging/waiting station and Fig. 9, e shows the entire fleet. For pedestrian not only the safety of the autonomous vehicles plays an important role but also a smooth and predictable movement. Therefore, navigation strategies were taken into account in the process of path and trajectory planning. This ensures generation of predictable trajectories for the traffic participants [33].

The co-creative collaboration with the stakeholders determined not only the services and the interaction concept used for development of the planning system, but also the design and the development of the autonomous vehicles together with the autonomous software stack development.

7 Evaluation: Results from Acceptance Research

Besides the functional evaluation of the implemented autonomous transportation system, the social, attitudinal, and intentional/motivational aspects of the innovation and the process of implementation were evaluated. The results of the acceptance research show a differentiated picture of high acceptance and comprehensive suggestions for improvement. They have been continuously incorporated into the further development of the hardware and software, the science communication and the trans-disciplinary process of the real laboratory as well as into the scientific community.

The analyses show that the assessments of the respondents who actually saw or used the autonomous transportation system at the expo were more positive than those of respondents to the representative panel. The annual panel also showed a positive change in attitude and acceptance among the population over the years.

In order to gain differentiated insights, migration background, gender, and age of the respondents were taken into account in the analysis. This showed, for example, that the assessment of autonomous vehicles is significantly more positive for women when they associate a concrete use case with it. The more positive assessment of the people surveyed at the expo compared to the representative panel of citizens is reflected in all areas. For example, there was a more positive assessment of the topic "Sharing is Caring": The people participating as residents or visitors at the expo appreciated the idea of sharing seldom-used everyday objects with the help of the autonomous delivery vehicles with other people in the neighborhood. In the iterations of the participatory process, a corresponding scenario ("neighborhood courier") had been implemented, i.e. the residents could send their neighbors things they wanted to borrow. This suggests that the living-lab approach might not only enable a development process, which is oriented towards the needs and interests of those affected, but it can also increase the acceptance of new developments, as the practical benefits can be experienced and are based on realistic usage scenarios.

The novel developed interactive planning method considered parameters that existing algorithms did not take into account. For example, the BUGA:log transport system, users could select the date and time of delivery, the response to requests happens right away and the users can change the request data anytime. Another benefit of this real-time interactivity is the availability of customers at the requested destination for delivery. In the case of deliveries made by domain companies (DHL, UPS, etc.), it often happens that customers are not available at the time of delivery, and they are usually notified to collect the package from a post office. With these benefits, the system acceptance for using autonomous vehicles for delivery services increased by 15% (from 76% to 91%) [36]. This increase comes at a cost of performance loss regarding the planning time needed to execute the deliveries. Our evaluation shows that the loss in terms of planned/executed time is 36.39% [33]. This loss is "compensated" at quality level from the customer point of view. Also 97% of the deliveries were made on time or with a delay of less than 10 min [33], which is in the participants' threshold of acceptability.

8   Conclusion and Further Work

BUGA:log was one of seven real laboratories supported in the German state of Baden-Württemberg. As transdisciplinary research, the project was a co-creative collaboration between researchers and stakeholders from the municipality, the stakeholders from the expo, industry, and citizens. With the new residential quarter on the grounds of the expo and the six-month event of the federal horticultural and architectural show as a living lab, an automated transport system in a public urban space was developed and evaluated. We followed a participatory research approach of co-creative phases in two iterations. Citizens and potential users were involved from the very beginning and throughout the entire human-centered design process.

The living lab BUGA:log showed the necessity of a participative design process as a prerequisite for a well-designed system and the acceptance of logistics innovations in urban space. In this context, the living lab dealt with both social and technical innovation. The ability to identify and describe processes that can be used as models for the implementation of such innovations and that can be reproduced in similar contexts is crucial to meet the challenges of digitalization and automation in mobility with regard to demographic change and the trend towards reurbanization.

Moving forward, the research results can be reused for other applications customized to new and today's demands. Since the pandemic, the growing competition from online retail poses major challenges for inner cities. A large, still underutilized potential lies in the cooperation between actors in the city center and research institutions. Structures such as real-world and innovation laboratories offer a special opportunity to jointly develop innovative approaches for the attractiveness of inner cities and stationary trade and to test them in an application-oriented environment. This significantly strengthens the cooperation between trade and science and provides targeted digitalization and innovation impulses in small and medium-sized trading companies. The developed fleet of autonomous vehicles together with the management system for interactive mission/service planning has since been fulfilling other transportation tasks in the city center of Heilbronn within the project Urban Innovation Hub [37]. With the Urban Innovation Hub, an inner-city center of attraction and exchange for different customer and target groups is being created in Heilbronn. It offers demonstrators, exhibits, and theme worlds, as well as events, lectures, and workshops to continuously engage the different target groups with attractive contributions – and continues the living lab research on co-creating innovative technologies.

References

1. Arndt, W.-H.: Aktuelle Entwicklungen und Konzepte im urbanen Lieferverkehr in Wulf-Holger Arndt und Tobias Klein (Hg.): Lieferkonzepte in Quartieren - die letzte Meile nachhaltig gestalten. Lösungen mit Lastenrädern, City-Cruisern und Mikro-Hubs, pp. 5–9 (2018)
2. DHL, Robotics in Logistics A DPDHL perspective on implications and use case for logistics industry. https://www.dhl.com/content/dam/downloads/g0/about_us/logistics_insights/dhl_trendreport_robotics.pdf. Accessed 02 Feb 2023
3. DPD, AutonomesFahren in der Paketzustellung, DPD und Aachener Automobilexperten präsentieren The-sen und Szenarien zur autonomen Paketzustellung. https://www.fka.de/images/pressemitteilun-gen/2017/2017-02-06-17fka0002-DPD-Presseinformation-Autonomes-Fahren_Thesenpapier.pdf. Accessed 02 Feb 2023
4. Bauer, M., Bienzeisler, B., Bernecker, T., Knoll-Mridha, S.: Anforderungen und Chancenfür Wirtschaftsverkehre in der Stadt mit automatisiert fahrenden E-Fahrzeugen, Stuttgart/Heilbronn (2018)
5. Jirotka, M., Grimpe, B., Stahl, B., Eden, G., Hartswood, M.: Responsible research and innovation in the digital age. Commun. ACM 60(5) (2017)
6. Trencher, G., Bai, X., Evans, J., McCormick, K., Yarime, M.: University partnerships for co-designing and co-producing urban sustainability. Glob. Environ. Chang. 28, 153–165 (2014)
7. Schäpke, N., et al.: Reallabore im Kontext transformativer Forschung: Ansatzpunkte zur Konzeption und Einbettung in den internationalen Forschungsstand. IETSR Discussion papers in Transdisciplinary Sustainability Research 1 (2017). http://hdl.handle.net/10419/168596. Accessed 02 Feb 2023
8. Luederitz, C., et al.: Learning through evaluation–a tentative evaluative scheme for sustainability transition experiments. J. Clean. Prod. 169, 61–76 (2017)
9. Weiland, S., Bleicher, A., Polzin, C., Rauschmayer, F., Rode, J.: The nature of experiments for sustainability transformations: a search for common ground. J. Clean. Prod. 169, 30–38 (2017)

10. Bos, J.J., Brown, R.R., Farrelly, M.A.: A design framework for creating social learning situations. Glob. Environ. Chang. **23**, 398–412 (2013)
11. Mohr, H., Landau, F.: Interventionen als kreative Praxisform: Die Suche nach Neuheit als gesellschaftliches Phänomen. In: Reinermann, J.-L., Behr, F. (eds.) Die Experimentalstadt, pp. 59–76. Springer, Wiesbaden (2017). https://doi.org/10.1007/978-3-658-14981-9_4
12. Groß, M.: Experimentelle Kultur und die Governance des Nichtwissens. In: Reinermann, J.-L., Behr, F. (eds.) Die Experimentalstadt, pp. 21–40. Springer, Wiesbaden (2017). https://doi.org/10.1007/978-3-658-14981-9_2
13. Dezuanni, M., Foth, M., Mallan, K., Hughes, H.: Digital Participation Through Social Living Labs: Valuing Local Knowledge, Enhancing Engagement. Chandos Publishing (2017)
14. Lang, D.J., et al.: Transdisciplinary research in sustainability science: practice, principles, and challenges. Sustain. Sci. **7**, 25–43 (2012)
15. Karvonen, A., Heur, B.: Urban laboratories: experiments in reworking cities. Int. J. Urban Reg. Res. **38**, 379–392 (2014)
16. Hillgren, P.-A.: Democratizing the city. Democratic configurations and imagination. Scandinavian J. Inf. Syst. **29**, 93–98 (2017)
17. Guggenheim, M.: Laboratizing and de-laboratizing the world: changing sociological concepts for places of knowledge production. Hist. Hum. Sci. **25**, 99–118 (2012)
18. Caniglia, G., et al.: Experiments and evidence in sustainability science: a typology. J. Clean. Prod. **169**, 39–47 (2017)
19. Abson, D.J., et al.: Leverage points for sustainability transformation. Ambio **46**(1), 30–39 (2016). https://doi.org/10.1007/s13280-016-0800-y
20. Franz, Y.: Designing social living labs in urban research. Info **17**(4), 53–66 (2015)
21. Følstad, A.: Living labs for innovation and development of information and communication technology: a literature review (2008)
22. Bødker, S., Kyng M.: Participatory design that matters – facing the big issues. ACM Trans. Comput.-Hum. Interact. (TOCHI) **25**, 4:1–4:31 (2018)
23. Ministerium fürWissenschaft, Forschung und Kunst Baden-Württemberg: Förderlinien "Reallabore" und "Reallabore Stadt" (2018)
24. Defila, R., Di Giulio, A.: Transdisziplinär und transformativ forschen. Eine Methodensammlung. Springer, Wiesbaden (2018). https://doi.org/10.1007/978-3-658-21530-9
25. Schäpke, N., et al.: Urban BaWü-labs: challenges and solutions when expanding the real-world lab infrastructure. GAIA-Ecol. Perspect. Sci. Soc. **26**(4), 366–368 (2017)
26. Marsden, N., Bernecker, T., Zöllner, R., Sußmann, N., Kapser, S.: BUGA:log – a real-world laboratory approach to designing an automated transport system for goods in urban areas. In: IEEE International Conference on Engineering, Technology and Innovation (ICE/ITMC) (2018)
27. Ogonowski, C., Jakobi, T., Müller, C., Hess, J.: PRAXLABS: a sustainable framework for user-centered ICT development Cultivating research experiences from Living Labs in the home. In: Wulf, V, Pipek, V., Randall, D., Rohde, M., Schmidt, K., Stevens, G. (eds.). Socio Informatics – A Practice-based Perspective on the Design and Use of IT Artefacts. Oxford University Press, New York (2018)
28. Braun, V., Clarke, V.: Using thematic analysis in psychology. Qual. Res. Psychol. **3**, 77–101 (2006)
29. Mucha, H., et al.: Collaborative speculations on future themes for participatory design in germany. i-com **21**(2), 283–298 (2022). https://doi.org/10.1515/icom-2021-0030. Accessed 09 Feb 2023
30. Nevens, F., Frantzeskaki, N., Gorissen, L., Loorbach, D.: Urban transition labs: co-creating transformative action for sustainable cities. J. Clean. Prod. **50**, 111–122 (2013)
31. Pröbster, M., Marsden, N., (submitted): The Social Perception of Autonomous Delivery Vehicles Based on the Stereotype Content Model (2023)

32. Kocsis, M., Winckler, J., Sußmann, N., Zöllner, R., Interactive mission planning system of an autonomous vehicle fleet that executes services. In: 23rd International Conference on Intelligent Transportation Systems (ITSC), pp. 1–6 (2020)
33. Kocsis, M., Zöllner, R., Mogan, G.: Interactive system for package delivery in pedestrian areas using a self-developed fleet of autonomous vehicles. J. Electron. **11**, 748 (2022)
34. Kocsis, M., Schultz, A., Zöllner, R., Mogan, G.: A method for transforming electric vehicles to become autonomous vehicles. In: CONAT International Congress of Automotive and Transport Engineering, pp. 752–761 (2016)
35. Kocsis, M., Sußmann, N., Buyer, J., Zöllner, R.: Safety concept for autonomous vehicles that operate in pedestrian areas. In: IEEE/SICE International Symposium on System Integration, Taipei, Taiwan (2017)
36. Allhoff, J., Dierolf, N., Marsden, N.: Forschung zur Akzeptanz logistischer Nahversorgung im urbanen Raum mit automatisierten Transporteinheiten (Research Report). In: Heilbronner Institut Für Angewandte Marktforschung; H-InfaM: Heilbronn, Germany; Appendix IV (2020)
37. Urban Innovation Hub. https://www.hs-heilbronn.de/de/uih. Accessed 02 Feb 2023

Challenges of Operators for Autonomous Shuttles

Cindy Mayas$^{(\boxtimes)}$ (ID), Tobias Steinert, Heidi Krömker (ID), Fabia Kohlhoff,
and Matthias Hirth (ID)

Technische Universität Ilmenau, Ilmenau, Germany
{cindy.mayas,tobias.steinert,heidi.kroemker,fabia.kohlhoff,
matthias.hirth}@tu-ilmenau.de

Abstract. Autonomous shuttles can extend the flexibility of micro-mobility to small groups, families, and persons with mobility limitations and impairments. When integrating autonomous shuttles into public transportation systems, the challenge is to shift the tasks of the driving personnel to the operators in the traffic control center. From there, the autonomous shuttles are centrally monitored and controlled. At the same time, the service quality for the passengers should be fully maintained. This leads to the research question, "Which new tasks arise for operators in the traffic control center in the dispositive control of autonomous shuttles?" A task analysis of the previous tasks of the driving personnel was conducted to answer the research question. The analysis consists of a method mix of inductive and deductive methods to compensate for the disadvantages of both. The result is a systematic of tasks that have to be shifted to the operators in the traffic control center. Furthermore, the technical potentials to support these tasks by assistive systems are described. The results mean a human-centered design of autonomous shuttles for passengers and a basis of task design for operators in the traffic control center for public transport companies.

Keywords: Autonomous Shuttles · Operator · Driving Personnel · Traffic Control Center · Task Analysis

1 Introduction

The autonomous driving technology is still in development with the ultimate goal that vehicles move safely in traffic without driving personnel. This level of autonomy corresponds to level 4, "High Driving Automation," and level 5, "Full Driving Automation," according to the SAE International categorization, shown in Table 1. In this state of development, the system takes over all aspects of the dynamic driving task in all scenarios that can be handled by human drivers [3]. According to recent studies, autonomous shuttles are already being used on over 300 test routes in more than 30 countries worldwide [1].

Autonomous shuttles complement the public transportation system and can be used in both scheduled and on-demand services. The shuttles can increase

H. Krömker (Ed.): HCII 2023, LNCS 14048, pp. 334–346, 2023.
https://doi.org/10.1007/978-3-031-35678-0_22

Table 1. SAE Levels of Autonomous Driving [3].

SAE Level	SAE Name	Definition
0	No Driving Automation	The performance by the driver of the entire dynamic driving task, even when enhanced by active safety systems
1	Driver Assistance	The sustained and operational design domain-specific execution by a driving automation system of either the lateral or the longitudinal vehicle motion control subtask of the dynamic driving task (but not both simultaneously) with the expectation that the driver performs the remainder of the dynamic driving task
2	Partial Driving Automation	The sustained and operational design domain-specific execution by a driving automation system of both the lateral and longitudinal vehicle motion control subtasks of the dynamic driving task with the expectation that the driver completes the object and event detection and response subtask and supervises the driving automation system
3	Conditional Driving Automation	The sustained and operational design domain-specific performance by an automated driving system of the entire dynamic driving task with the expectation that the dynamic driving task fallback-ready user is receptive to automated driving system-issued requests to intervene, as well as to dynamic driving task performance relevant system failures in other vehicle systems, and will respond appropriately
4	High Driving Automation	The sustained and operational design domain-specific performance by an automated driving system of the entire dynamic driving task and dynamic driving task fallback without any expectation that a user will respond to a request to intervene
5	Full Driving Automation	The sustained and unconditional (i.e., not operational design domain specific) performance by an automated driving system of the entire dynamic driving task and dynamic driving task fallback without any expectation that a user will respond to a request to intervene

the variety of mobility offers and make an essential contribution to improving the connection of the rural population to towns and cities [8]. The aim is to make public transport services more flexible and open new routes. In this way, public transport can attract new target groups who have previously preferred private cars.

The acceptance of autonomous shuttles is a crucial factor for this resource-efficient mobility. However, this acceptance depends not only on the passengers but also on the possibility of integrating autonomous shuttles into the central traffic control center of public transport companies.

The challenges for driving personnel has to be addressed to integrate autonomous shuttles into the regular public transportation system. To this end, the tasks previously performed by the driving personnel must be transferred to the operators in the traffic control center. The driving personnel consists, e.g., of a person who drives the vehicle or is carried in a vehicle as part of his duties to be available for driving if necessary. Until now, the driving personnel has prepared the trip by, e.g., checking the vehicle's safety. During the trip, the driving personnel has to respond to stop requests, inform passengers, and assist them with boarding and alighting. However, the driving personnel was also responsible for ensuring that rules of conduct were followed in the vehicle. For example, to ensure that no passengers were being inconvenienced. In emergencies, the driving personnel was responsible for providing appropriate assistance to passengers. Driving personnel also provides passenger and fare information. From the passenger's point of view, certain services were vital, such as recovering lost property.

Table 2 lists the typical tasks of the driving personnel and divides them into three phases: trip preparation, trip, and post-processing. It is not intended to be exhaustive, because it is a first approximation.

The operators of a traffic control center were previously responsible for dispositive traffic control. The tasks of the operators in the traffic control center are now enriched. They are to take over the tasks of the driving personnel in the vehicle, i.e., they are connected to the vehicle via sensors and video and have to perform tasks previously done on-site. Therefore, autonomous shuttles have a strong impact on the operator's tasks. To optimize the user-oriented design of autonomous shuttles, the perspective of the driving personnel, the operators and the passengers is also highly relevant. The paper, therefore, contributes to the systematization of the tasks of the driving personnel, which must be shifted to the operators in the traffic control center.

These tasks have not been adequately described so far, but their description is relevant for successfully integrating autonomous shuttles into public transport. The roles and tasks have been empirically determined and systematized for selected scenarios in extensive studies in test fields and regular operations. This paper aims to answer the research question:

Which new tasks arise for operators in the traffic control center in the dispositive control of autonomous shuttles?

The study results are the basis for designing generic technical information interfaces between the traffic control center and the autonomous shuttles.

2   Background and Related Work

The scientific reappraisal of integrating autonomous shuttles into traffic occurs against the background of entirely different disciplines, such as computer science, human-machine interaction, shuttle technology, safety engineering, or even traffic science, in widely scattered publications of varying depths. The activities on autonomous driving [13], as a young research field, are loosely assigned, e.g., on different levels [4] as shown in Fig. 1. The research in the presented project is in the area of human factors. Human factors influence not only the design of technology but also ecological, legal, and social issues, as illustrated in Fig. 1. In particular, social acceptance is closely related to legal regulations and economic issues, such as the frequency of use of autonomous shuttles.

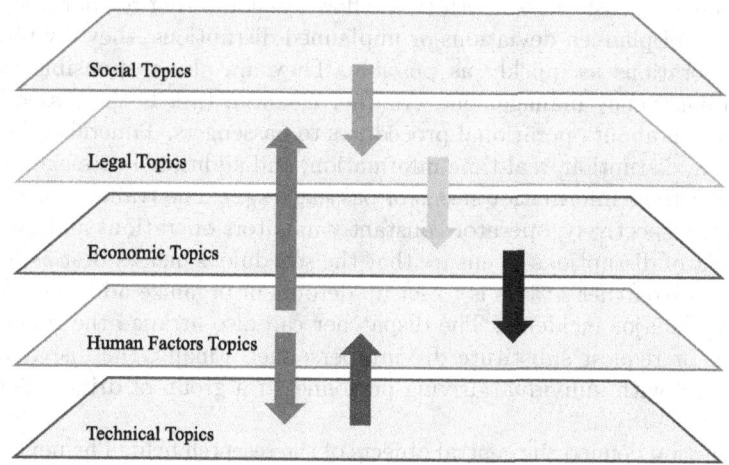

Fig. 1. Different Topic Areas of Autonomous Driving According to VDI [4]

It is first essential to analyze the existing definitions of the tasks of driving personnel and operators, respectively, dispatchers in the traffic control center, to answer the research question raised in Sect. 1. For Europe, these tasks are mainly laid down in country-specific regulations.

2.1   Driving Personnel

The European Regulation No. 561/2006 defines the driving personnel as "any person who drives the vehicle, even for a short period of time, or who is carried in a vehicle as part of his duties in order to be available for driving if necessary" [5]. In addition to driving, driving personnel performs many other activities before, during, and after the public transport service. For example, before the passenger transport, the driving personnel must check the operational and roadworthy condition of the vehicle, start the vehicle systems, and, if necessary, disconnect the vehicle from the loading system. During passenger transportation, the driving personnel takes over the control of the vehicle and the concentrated perception of the road traffic. The driving personnel has the primary responsibility for the safe transportation of passengers. However, driving personnel is also responsible for providing passenger and fare information. The personnel is also responsible for delivering boarding assistance and communicating with passengers and the control center in an emergency or accident [6]. Table 2 lists typical tasks of the driving personnel and divides them into the phases of preparation, passenger travel, and post-processing.

2.2   Traffic Control Center and Operator

The staff of a traffic control center is responsible for traffic control. The staff is responsible for ensuring smooth traffic flow by monitoring regular operations. In the event of planned deviations or unplanned disruptions, they try to restore smooth operations as quickly as possible. They are also responsible for incident and emergency management. Another essential task is the dissemination of information about operational procedures to passengers. Timetable data, fare information, disruption, real-time information, and additional messages are provided to meet the information needs of passengers [2]. The traffic control center dispatcher, respectively, operator constantly monitors operations and intervenes in the event of disruptions to ensure that the schedule is met as best as possible. Part of the dispatcher's tasks is to set up detours or organize additional trips in the event of major incidents. The dispatcher can also arrange the replacement of vehicles or request substitute driving personnel. Finally, the dispatcher can communicate with individual driving personnel or a group of driving personnel [12].

This section defined the central objects of the research field. The next chapter describes how the driving personnel's tasks are methodically recorded to shift them to the operators in the traffic control center.

3   Methodology

The mobility domain has a heterogeneous nature and autonomous shuttles are mainly used in test fields. Therefore, there are no established processes to integrate autonomous shuttles into public transportation system.

In order to describe this heterogeneous field a structured analysis of the transformation process is required. Analyzing the transformation of tasks for operators from manual to autonomous driven shuttles requires a combination of established inductive and deductive methods to compensate for the disadvantages of both [9].

An explorative study of the application field was conducted with the qualitative method of interviews. A qualitative content analysis of the interview protocols reveals very detailed information on a few aspects of the operator's task in autonomous shuttles. Therefore, the results are extended by the structural

Table 2. Task of Driving Personnel in Non-autonomous Vehicles.

Task	Description
Phase: Trip Preparation	
Checking the vehicle	Inspection of operational and traffic safety, vehicle condition [6]; The vehicle is to be inspected for defects, damage and completeness of equipment [6]
Activating	Start of systems, separation of vehicle/power supply (E-buses) [6]
Signing on	Logon with driver ID or circulation ID/retrieval of schedule data [6]
Phase: Trip	
Driving	The driving personnel must concentrate on driving the vehicle and road traffic during the journey. [6]
Arriving at a stop	If necessary, check whether passengers are standing at the stop or want to leave the vehicle at the stop (stop request)
Ticketing	Issuing and checking tickets [6]
Informing passengers	Answering questions from passengers, e.g., about the route, stops, etc
Supporting to get on/off	If necessary, assistance with boarding and alighting (folding out the ramp, etc.) [6]
Departing from a stop	Checking the door area, the vehicle surroundings and the route when leaving the stop; departing from the stop according to the timetable [6]
Monitoring behavioral codes	The driving personnel pays attention to the observance of rules of conduct in the vehicle
Making emergency calls	Direct communication with passengers; Immediate report to the control center [6]; coordination of measures and further procedure [6]
Helping passengers	Care of helpless or injured persons inside and outside the vehicle until they are taken over by the rescue service[?]
Emergency stopping	Checking the condition of the vehicle, the emergency situation, etc. Execution of the emergency stop as well as evacuation of the vehicle [6] if necessary fire fighting [6]
Offering services	Securing of lost property, control of the vehicle condition, if necessary removal of smaller impurities
Phase: Post-Processing	
Parking the vehicle	Approach of the specified parking place
Signing off	If necessary, connecting the vehicle with a loading device
Checking the vehicle	Checking the condition of the vehicle (cleanliness, lost property, remaining passengers), if necessary technical check and forwarding error messages or problem reports

and contextual information from established usage scenarios in public transport [14]. In contrast to the structured approach of a hierarchical task analysis [7], the synthesis of the applied method mix allows a faster derivation of the tasks and a better use of the results in unknown application areas.

The analyzed data for the qualitative, inductive study consists of 15 expert interviews [17] with stakeholders of pilot projects with autonomous shuttles in German public transport. The interviews are conducted by telephone from May to August 2022 and reveal changes in typical tasks and touchpoints of operators in the handling of autonomous shuttles. The interview guideline includes

- open-ended questions about personal opinions on and experiences with autonomous driving,
- detailed questions about the experienced changes in operators' tasks and workloads with autonomous shuttles,
- descriptive questions about the perception of user-oriented quality factors.

The deductive analysis extends the data from the current pilot projects with the typical system usage of public transport. Therefore, scenarios [16], which describe typical usage situations from the passengers' perspective, are adapted to autonomous shuttles.

Scenario "Standard". This scenario describes a typical journey, which runs as planned and expected. Nevertheless, the passengers need clear information about travel times and stops, which the system usually provides automatically.

Scenario "Service". This scenario represents service cases, such as requests for additional passenger information, requests for cleanliness, or lost property. Usually, service requests can be solved directly by the driving personnel, or the personnel forwards them to the traffic control center.

Scenario "Event". This scenario represents both planned and unplanned, positive and negative events that influence the journey's course and the passengers' feelings. In this situation, dispatching actions need to be taken, and passengers need to be informed about the impact on their journey and possible trip alternatives.

Scenario "Emergency". This scenario represents emergencies on the shuttle, such as an accident or fire, and emergencies within the shuttle, such as medical emergencies or crime. Emergencies require a quick analysis of the situation and a decision about the required interventions. Clear instructions must be given to the passengers, and the success of interventions must be monitored.

The concluding description of the tasks follows informal documentation, including typical usage contexts [10]. The tasks represent a first structured approach that does not aim at a complete description.

4 Results

As a first step, the transformation of tasks for autonomous driving is described based on the classical tasks described in Table 2. Second, opportunities for assistive technologies for operators of autonomous shuttles are derived to reduce their workload. Finally, potential integration strategies for different mobility systems with different complexity of transport and traffic systems are presented.

4.1 Tasks for Operators of Autonomous Shuttles

A task distribution exists between humans and machines, in which specific tasks taken over by humans can be systematized. As a prerequisite for the description of the operator's tasks, the following assumptions about the task execution of autonomous shuttles are considered:

- driving tasks on defined routes with traffic signs and crossroads,
- arriving at, waiting at, and departing from stations with defined waiting times,
- automated door controlling and automated boarding aids.

Table 3 shows the shift of handling tasks in real-world situations to monitoring and decision tasks for technical systems via digitally perceived information, which might be incomplete, inconsistent, or even inaccurate. This situation causes higher uncertainties of effects and consequences of decisions and also increases the mental workload of operators.

4.2 Assisting Technologies for Operators of Autonomous Shuttles

Assistive technologies are required to reduce the estimated workload for monitoring and decision-making. According to the identified scenarios for deriving the tasks from Sect. 3, the following recommendations are developed for the operators of autonomous shuttles.

Scenario "Standard". It is recommended that the monitoring tasks of operators for multiple autonomous shuttles are supported by an intelligent support system bundling a large amount of sensor data to a focused status information of each shuttle and even a holistic status for a mobility area.

Scenario "Service". Additional services have high relevance for the user experience of passengers in the service-oriented mobility sector, e.g., for feeling comfortable and safe. It is recommended to transfer existing services, e.g., for accessibility aid, lost properties, or cleanliness, to autonomous shuttles. For this reason, additional digital communications requests or channels for services have to be introduced which do not overlay safety-critical and operation-related information.

Table 3. Tasks of Driving Personnel in Autonomous Shuttles.

Task	Description
Phase: Trip Preparation	
Checking the vehicle	Checking shuttle functions and additional status information on infrastructure and route
Activating	Starting systems, Initiating data transfer
Signing on	Assigning trip information to shuttle
Phase: Trip	
Driving	Monitoring of shuttle status and operations
Arriving at a stop	Monitoring of waiting and boarding persons
Ticketing	No task
Informing passengers	Answering passenger requests via additional communication system
Supporting to get on/off	No task
Departing from a stop	Monitoring of waiting and alighting persons
Monitoring behavioral codes	Monitoring passenger behavior in case of abnormality
Making emergency calls	Communicating directly with passengers and traffic control center, coordinating of measures and further procedure
Helping passengers	Taking decision about the appropriate measures for taking care of helpless or injured persons inside and outside the shuttle
Emergency stopping	Taking decision about appropriate measures for emergency situation, if necessary initiating emergency stops and passenger evacuation
Offering services	Taking decision about the appropriate measures for service calls, e.g. securing of lost property
Phase: Post-Processing	
Parking the vehicle	Assigning parking information to shuttle
Signing off	No task
Checking the vehicle	Checking shuttle status and taking decision about the appropriate measures, if necessary technical checking and forwarding of error messages or problem reports

Scenario "Event". It is recommended that standardized information and measures are implemented, between which the operator can easily choose in case of an event, such as delays, route changes, or cancellations. A decision support system might facilitate these decisions and tasks for informing passengers and adapting system information.

Scenario "Emergency". In an emergency, bi-directional flexible communication between operators and passengers is required. On the one hand, the operator needs additional video data to analyze the situation, and on the other hand, synchronous voice communication can help to make decisions and take action more quickly. In addition, voice communication can directly instruct the passenger to take the necessary action, ask for more details, or calm the situation. When transmitting video data, it is necessary to consider not only the privacy of the passengers but also the emotional health of the operators, who must not be exposed to cruel scenes.

4.3 Integration Strategies

Another result of the investigation of the test fields is that in some countries, there is a three-phase transition, as shown in Fig. 2, from the driver-controlled bus to the autonomous shuttle. Specifically, this means that the autonomous shuttles are accompanied by driving personnel in the first introduction phase. The manufacturers of autonomous shuttles do not foresee this phase, i.e., there is no workplace in the autonomous shuttle for the driving personnel, as shown in Fig. 2a. In the second phase, the operator is already located in the traffic control center and monitors the trip of one or more autonomous shuttles from a special workstation, as shown in Fig. 2b. Only in the third phase are the operator's tasks completely integrated into the control center, as shown in Fig. 2c. Next to the monitoring, communicating, and decision-making tasks, the operators may conduct other tasks in the traffic control center or may be supported by other operators of the control center in the case of emergencies. The integration in the traffic control center also enables a further specialization and distribution of tasks for autonomous shuttles, e.g., in a specialized communicating operator.

In practice, there are also mixed forms of these roles, which communication interfaces must support and consider in the long run. The identified roles illustrate an integration strategy for autonomous shuttles in the public transportation system. The generic results for the operators' tasks can also be transferred to future strategies for controlling autonomous shuttles by passengers in private transport.

5 Discussion

The research question stated at the beginning could be answered completely:

> Which new tasks arise for operators in the traffic control center in the dispositive control of autonomous shuttles?

A generic classification of driving personnel tasks could be extracted, ranging from trip preparation to trip and post-processing as shown in Table 3.

The literature review shows no comprehensive studies yet on developing and evaluating the human-machine interface for autonomous shuttles in a human-centered way. This gap in the literature is unexpected, as it is crucial to the success of the technology. Therefore, existing definitions from technology-oriented standards had to be used to systematize the research field.

(a) *Operator on Board*: One or more safety drivers accompany the trip as operators in the shuttle. A connection to the control center is provided on demand via additional communication devices and a manufacturer-specific interface.

(b) *Operator on Standby*: A teleoperator accompanies the trip of one or more shuttles from a separate, specially equipped workstation outside the shuttle. The required data connections from the shuttle are proprietary provided by the manufacturer.

(c) *Operator on System*: A control center operator monitors the movement of multiple shuttles from a workstation in the control center, which may be connected to other control center tasks. Real-time data and voice connections to and from the shuttle are provided via a standardized interface.

Fig. 2. Integration Process for Operators of Autonomous Shuttles.

The young research field was explored using qualitative social research methods. The interview results of the 15 experts were subjected to qualitative content analysis and generalized according to defined rules. Even if the generalization of statements is criticized from the point of view of the theory of science [11,15], Nevertheless, qualitative content analysis is currently one of the few methods that allow access to new fields of research and hypothesis generation. Examples for hypothesis may include:

- Operators have an increased mental load due to the increased amount of information and increased stress due to the large number of decisions that would have to be made.
- Assistive systems compensate the reduced opportunities for intervention.

Methodological limitations also included limited access to test sites and the fact that autonomous shuttles are still very little used in everyday life. However, the results provide a basis for a more detailed elaboration of the usability and user experience factors in further research projects.

6   Conclusion

This paper provides a basis for structuring the research field of operator tasks for autonomous shuttles. The presented tasks, recommendations for assistive systems, and integration strategies reduce the complexity of the challenges and point to future research fields. It is necessary to link the worldwide research results of the test fields of autonomous shuttles in public transport more closely to optimize the operator's workplace regarding usage-oriented qualities and standardize the interface between autonomous shuttles and the traffic control center.

The paper provides principles for assistive systems for operators in the context of autonomous driving. In future research, usage-oriented methods, such as user journeys and experience maps, can be used to refine the requirements of the operators. Further, disciples can conclude the organizational development of the traffic control center regarding economic and technical aspects. At least, the results provide information for staff development to train future operators of autonomous shuttles.

Acknowledgements. Part of this work was funded by the German Federal Ministry for Digital and Transport (BMDV) grant number 45AVF3004G within the project OeV-LeitmotiF-KI.

References

1. Automatisierte Shuttles: Ein Markt-Überblick. https://www.rms-consult.de/news/automatisierte-shuttles-ein-markt-ueberblick/. Accessed 15 Feb 2023
2. VDV-Recommendation No. 731 "Operational Requirements for Public Transport Control Centers". Recommendation, Association of German Transport Companies (VDV), Köln, July 2015
3. Ground Vehicle Standard (J3016_201806) Taxonomy and Definitions for Terms Related to Driving Automation Systems for On-Road Motor Vehicles. Recommended practice, SAE International, May 2018
4. VDI-Handlungsempfehlung Automatisiertes und autonomes Fahren. Recommendation, The Association of German Engineers (VDI), December 2019
5. DIN 5566-1 Railway Vehicles - Driver Cabs - Part 1: General Requirements. Standard, German Institute for Standardization (DIN), May 2020
6. VDV-Recommendation No. 709 "Service Instruction for the Driving Service with Buses". Recommendation, Association of German Transport Companies (VDV), Köln, April 2022
7. Annett, J.: Hierarchical Task Analysis. The Handbook of Task Analysis for Human-Computer Interaction (2004)
8. Canzler, W.: Voraussetzung für einen Wirksamen Klimaschutz: Die Verkehrswende in den Städten. In: Lozán, J.L., Breckle, S.W., Graßl, H., Kuttler, W., Matzarakis, A. (eds.) Warnsignal Klima: Die Städte, pp. 286–292. Verlag Wissenschaftliche Auswertungen in Kooperation mit GEO Magazin-Hamburg, November 2019
9. Diaper, D.: Understanding Task Analysis for Human-Computer Interaction. In: The Handbook of Task Analysis for Human-Computer Interaction, pp. 5–47 (2004)
10. Ergonomics of Human-System Interaction - Part 11: Usability: Definitions and Concepts. Standard, International Organization for Standardization, Geneva, CH, March 2018
11. Lincoln, Y., Guba, E.: Naturalistic Inquiry. SAGE Publications, Inc. (1985)
12. Maurer, M., Gerdes, J.C., Lenz, B., Winner, H.: IT-Systeme für Verkehrsunternehmen: Informationstechnik im öffentlichen Personenverkehr. dpunkt.verlag GmbH (2011)

13. Maurer, M., Gerdes, J.C., Lenz, B., Winner, H. (eds.): Autonomes Fahren: technische, rechtliche und gesellschaftliche Aspekte. Springer, Heidelberg (2015). https://doi.org/10.1007/978-3-662-45854-9

14. Mayas, C., Hörold, S., Krömker, H.: Meeting the challenges of individual passenger information with personas. In: Advances in Human Aspects of Road and Rail Transportation, pp. 822–831 (2012)

15. Popper, K.: Logik der Forschung: Zur Erkenntnistheorie der modernen Naturwissenschaft. Springer, Vienna (1935)

16. Rosson, M.B., Carroll, J.M.: Scenario-based usability engineering. In: Proceedings of the 4th Conference on Designing Interactive Systems: Processes, Practices, Methods, and Techniques. DIS '02, p. 413. Association for Computing Machinery, New York, NY, USA (2002). https://doi.org/10.1145/778712.778776

17. Wilson, C.: Chapter 2 - Semi-structured interviews. In: Wilson, C. (ed.) Interview Techniques for UX Practitioners, pp. 23–41. Morgan Kaufmann, Boston (2014). https://doi.org/10.1016/B978-0-12-410393-1.00002-8

Investigating the Impact of Anthropomorphic Framing and Product Value on User Acceptance of Delivery Robots

Eileen Roesler[1]([envelope]) [ID], Johannes Pickl[1], and Felix W. Siebert[2] [ID]

[1] TU Berlin, Berlin, Germany
eileen.roesler@tu-berlin.de, Johannes.pickl@p3-group.com
[2] DTU Kopenhagen, Kongens Lyngby, Denmark
felix@dtu.dk

Abstract. Delivery robots can contribute to efficient transportation and handover of goods in urban areas. To realize the full potential of this technology, users need to accept these novel systems. One possibility to increase acceptance is adapting the design of robots. The design of delivery robots, which are meant to function as tools but also interact closely with humans during transportation and delivery, presents both challenges and opportunities for increasing user acceptance. Specifically, anthropomorphic framing, or ascribing human-like characteristics to the robot, may influence acceptance. In addition, the type of goods being transported by the robot may also affect user acceptance. We used a video-based online experiment, to investigate how anthropomorphic framing and product value affect the individual and general acceptance and intention to use robots for transporting goods. In addition, we operationalized the perceived value of the robot's service through the willingness to pay for this delivery service. The data of 189 participants were retrieved in our between-subjects online study. The study revealed no differences in general acceptance, intention to use, and willingness to pay for the service robot. However, anthropomorphic framing and the prize of the transported goods mattered for the individual acceptance of the delivery robots. In particular, anthropomorphically framed robots and robots transporting inexpensive goods were accepted significantly more. As the services of transporting inexpensive products were accepted more, the successful implementation and actual usage could lead to an extension of the acceptance of more expensive goods transportation. This result could be the basis for a gradual market entry strategy that starts with low-cost product transportation like food delivery.

Keywords: Mobile Robots · Anthropomorphism · Human-Robot Interaction

© The Author(s), under exclusive license to Springer Nature Switzerland AG 2023
H. Krömker (Ed.): HCII 2023, LNCS 14048, pp. 347–357, 2023.
https://doi.org/10.1007/978-3-031-35678-0_23

1   Introduction

The ongoing Covid-19 pandemic has led to increased measures for reducing close contact and maintaining distance. As a result, contactless home delivery has become a crucial tool for preventing the spread of infection [1]. In addition to reducing the risk of infection, robots also have other benefits in delivery services, such as being available at all times. This has led to the deployment of mobile robots for deliveries by various companies in many cities across the United States and Europe [6,16]. However, the implementation of these new technologies in everyday human life poses new challenges for the collaboration and coexistence of humans and robots [27]. From a psychological perspective, it is particularly interesting that people do not just treat these robots like technical machines but interact with them in unexpected ways. As one example, food delivery robots that were launched at an American university were embraced by students (e.g., dressed up for holiday, or coined as "marshmallows") [12].

It is not unusual for people to assign human-like characteristics to service robots. For example, in the early research on vacuum cleaning robots, one interviewed family stated, that they named their vacuum cleaning robot Manuel because it has a personality [8]. This early observation of anthropomorphizing robots can still be observed today, as a quick google search reveals inspiration for robot naming such as Clean Elisabeth or Cleanardo Di Caprio. What we see well illustrated in these examples is the human tendency to anthropomorphize non-living objects like robots [29]. This tendency is the basis for the effectiveness of anthropomorphic robot design [7], as purposely designed appearance, communication, movement, or context of robots can further increase the ascription of human-like features [20]. In the examples above, people applied linguistic framing via names and personality ascription bottom-up. However, an anthropomorphic linguistic framing can also be used to appeal top-down to the mental models of the human that is interacting with the robot [13].

Interestingly, the effect of such anthropomorphic framing in the service area, although anecdotally reported quite often, is not well understood [22]. Especially for appearance, it is evident that the desirability of anthropomorphism is dependent on the application domain and the specific task of the robot [9,14,22]. Whereas anthropomorphism seems to be desirable in the social domain and for tasks with higher sociability, a reciprocal preference can be found in the industrial domain and for tasks requiring lower task sociability [9,24]. Considering that service robots are supposed to fulfill their task as a tool, but can still benefit from social aspects when interacting with humans, anthropomorphism is both a challenge and an opportunity [22]. It is therefore not surprising that the few existing studies on anthropomorphism in the service domain do not provide a comprehensive picture of the desirability and effectiveness of anthropomorphism [14,24]. Nevertheless, the lack of generalizability is also related to the lack of research especially in the service domain and in regard to anthropomorphic framing [22], which thus underlines the importance of the current study.

In line with the results concerning anthropomorphic appearance, earlier studies on anthropomorphic framing in social human-robot interaction (HRI) showed mainly positive effects [2,17], whereas the effects in task-related settings were mixed [13,19]. Moreover, most studies focused on the effects on empathy [2,17,19]. Whereas empathy seems to be especially important in social HRI, attitudes like acceptance are assumed to be important in more task-related settings, as acceptance determines the intention to use and actual usage of service robots [15]. Due to the task-relatedness of delivery robots, we hypothesized that a technically framed robot is accepted more than an anthropomorphically framed one. In addition, we assumed that the intention to use and to pay for a technically framed service robot is higher compared to an anthropomorphically framed one.

However, not only the framing but also the transported content of a delivery robot is important to answer the question of acceptance. Delivery robots can carry many kinds of goods, from low-cost food items to expensive high-tech [10,11]. One of the few studies that investigated the role of the transported good in the usage of follow-me robots showed that people were less stressed and more comfortable with the robot if it transported a low-value item [26]. In line with this study, we hypothesized that transporting low-priced goods leads to a higher acceptance as well as intention to use and pay for the robot compared to transporting high-priced goods. In addition, research in the industrial domain showed that anthropomorphism can reduce the perceived reliability of robots [18,25]. Therefore, we had the hypotheses that, as product value increases, the decline in acceptance, intention to use and pay for the robot is more pronounced for anthropomorphically framed delivery robots than for technically framed delivery robots. All proposed hypotheses were investigated in a video-based experiment, which was conducted online due to the Covid-19 pandemic [4]. Each participant was introduced to a commercially available Starship delivery robot via a video and self-recorded audio description, as well as subtitles. The robot was framed either anthropomorphically or technically, and transported either a cheaper or more expensive product. After the exposure to the video clip, we asked participants how much they would accept and intent to use the robot for the respective scenario, as well as how much they would be willing to pay for the service.

2   Methods

This research complied with the tenets of the Declaration of Helsinki, and the experiment was controlled by a checklist of the local ethics committee. The checklist as well as the preregistration can be obtained via the Open Science Framework (https://osf.io/6sje2/).

2.1   Participants

We conducted an a priori power analysis to determine our sample size using $G*Power$ [3]. The calculation was based on an ANOVA without repeated measures with an alpha level of .05 and a medium effect size of $f = .25$, resulting in a targeted sample size of $N = 179$. We recruited $N = 204$ participants, predominantly from a local university participant pool. However, 15 participants needed to be excluded from further analysis as their audio output was not working. After exclusion, the remaining data set resulted in $N = 189$ participants. The study sample had an average age of $M = 24.16$ ($SD = 5.13$) with the majority being female (141 female, 46 male, 2 non-binary).

2.2 Task and Materials

The introduction was the same for all conditions and stated the relevance of service robots for the future as following: *Robots will play an increasingly important role for humans in the future as they are increasingly used in areas close to everyday life, for example as delivery robots. The task of a delivery robot is to deliver packages. You will see a video of such a robot on the following page. This is followed by further instructions.* The following video (i.e., introducing the robot and framing) and vignettes (i.e., describing the product value) were different for the respective conditions.

Video. The original video we used was uploaded under the Creative Commons (CC) license [21]. It shows a delivery robot from the company Starship Technologies (Fig. 1). The video was edited by separating and deleting the audio track from the video. Then the video was cut into a 25-second clip which included scenes showing the robot in motion. The video was dubbed with self-recorded audio (and subtitles) to frame the scenario that followed. One audio commentary framed the robot anthropomorphically and the other framed the robot technically. The audio of the anthropomorphic and technical framing conditions were described as follows:

Anthropomorphic: *This is Karl. Karl is a modern delivery robot that transports packages through the city. To do this, he knows the road traffic regulations and consistently follows them. He can communicate with people through natural language and perceive his environment to deliver packages to their destination. Karl works during the week from 8 a.m. to 4 p.m. However, he needs a break after 4 h to recover from work. Karl is free on weekends.*

Technical: *In the video, you can see the modern X30 series delivery robot transporting packages through the city. Part of the programming is consistent compliance with road traffic regulations. X30 is equipped with a display and various types of sensors to deliver packages to the destination. X30 is in service Monday through Friday from 8 a.m. to 4 p.m. and requires a charging break after 4 h to recharge the batteries. On weekends, the robot is switched off.*

Vignettes. The vignettes were mainly used to manipulate the value of the transported product. However, the vignettes also contained aspects of anthropomorphism as the robot was further referred to as *Karl* or *X30*. They describe the delivery robot's procedure for delivering goods. In the low-priced conditions, a low-cost novel technical electronic product is supplied, while under the high-priced conditions, the financial value of the electronic product was higher. In particular, participants were asked to imagine buying a product that is delivered via the robot. In the high-priced conditions, they were asked to imagine that they would buy a new, high-priced technical electronic product, e.g., a smartphone or a laptop, which cost over 1000 Euro. In the low-priced conditions, they

Fig. 1. One scene of the video used in this study without subtitles.

were asked to imagine the purchase of a low-priced technical electronic product, for example, headphones or a power bank, which cost less than 50 Euro.

Dependent Measures. Acceptance was measured via two single items. First, the general societal acceptance of the robot was assessed via the item *How much do you accept the robot as part of the society?*. Second, the individual acceptance was examined via the item *How much do you accept that the robot delivers the ordered technical device to you?*. Both items were rated on a scale from 0–100%. Intention to use the robot (i.e., *How often would you choose the robot?*) was examined on a 5-point Likert scale (i.e., from *never* to *always*). In addition, the willingness to pay for the service was assessed via the item *How much money would you spend for this service [in Euro]?*. Participants could enter an amount with two decimal places as an answer to this question.

We included two control variables. First, the affinity towards technology was assessed via two subscales (i.e., cognitive and affective) on a six-point Likert scale [5]. Second, we asked participants to fill in a 5-item short version of the Interindividual Differences in Anthropomorphism Questionnaire, which had to be rated on an eleven-point Likert Scale [29]. Both control measures were used to check for possible confounding effects with regard to participants' attitudes toward technology. To test whether the manipulation of anthropomorphism was successful we incorporated a questionnaire with ten items that addressed aspects of anthropomorphic context (e.g., the character, task, and preferences of the robot). All items were rated on a 0–100% human-likeness scale.

Procedure. The study was conducted as an online survey using the platform *SoSci Survey*. Participants completed the study on their private computers without any oversight by an experimenter. All participants were randomly assigned to one of the four conditions. First, participants were requested to activate their sound output and a corresponding question checked whether this was successful. Subsequently, participants received corresponding written and video-based instructions including the framing of the robot and the price of the transported product. After the respective introduction of the robot via text and video, participants had to answer the questionnaire consisting of the dependent variables, followed by control and sociodemographic variables. The entire procedure lasted around 20 min, and participants received course credit after they finished the survey.

Design. The study consisted of a 2 × 2 between-subjects design with the two between-factors robot framing (anthropomorphic vs. technical) and product value (low vs. high).

2.3   Results

Control Variables. First, the variables regarding the individual attitude concerning affinity for technology and the tendency to anthropomorphize were analyzed between the four conditions using one-way ANOVAs. The analyses revealed no significant differences between the four groups in the affinity to technology ($F(3,185) = 0.66$, $p = .575$), as well as no difference in the tendency to anthropomorphize ($F(3,185) = 0.83$, $p = .481$).

Manipulation Check. To check whether the intended experimental anthropomorphic framing was successful the context anthropomorphism was compared between the anthropomorphic and technical conditions via an independent t-test. The analysis revealed that the anthropomorphic framing ($M = 36.63$; $SD = 15.51$) was perceived as significantly more anthropomorphic than the technical one ($M = 31.61$; $SD = 12.72$) ($t(187) = 2.43$, $p = .016$).

Dependent Variables. Acceptance, intention to use, and willingness to pay were analyzed by 2 (robot framing) x 2 (product value) between-subjects ANOVAs.

The analysis of general acceptance revealed neither a significant main effect of framing ($F(1,185) = 2.12$, $p = .147$) nor of product value ($F(1,185) = 0.09$, $p = .760$). Moreover, no significant interaction effect was found ($F(1,185) = 1.94$, $p = .165$).

The means and standard errors for individual acceptance are displayed in Fig. 2. The analysis of individual acceptance revealed a significant main effect of framing; $F(1,185) = 4.18$, $p = .042$. Surprisingly, the anthropomorphically framed robot ($M = 79.66$; $SD = 26.42$) was accepted significantly more for the delivery of products than the technically framed one ($M = 71.53$; $SD = 31.31$). In addition, the analysis revealed a significant main effect of product value; $F(1,185) = 10.56$, $p = .001$. As expected, the delivery of low-priced products ($M = 82.10$; $SD = 24.59$) was accepted significantly more than the delivery of high-priced products ($M = 68.84$; $SD = 32.04$). Moreover, the analysis showed no significant interaction effect of robot framing and product value; $F(1,185) = 0.93$, $p = .337$.

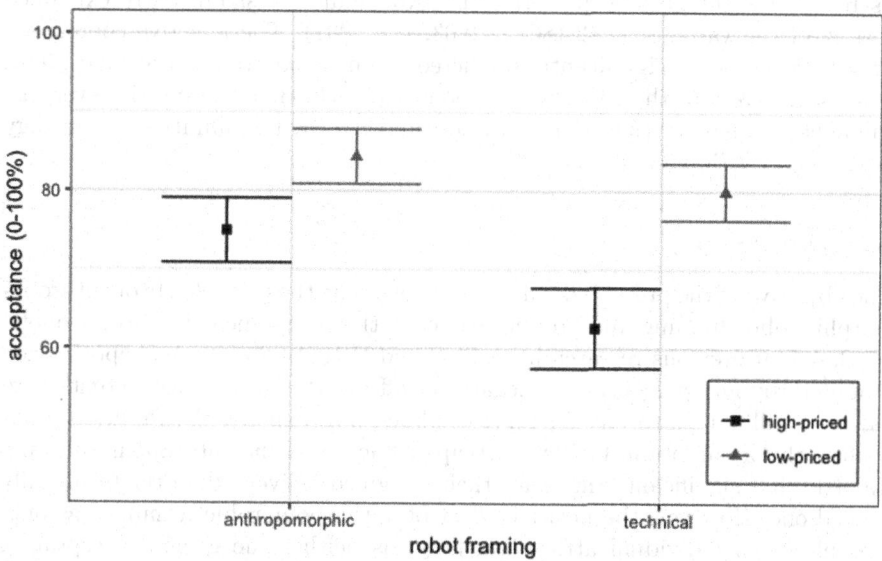

Fig. 2. Participant's individual acceptance for robot framing and product value conditions.

The analysis of intention to use the robot showed neither a significant main effect of framing ($F(1,185) = 2.37$, $p = .125$), or product value ($F(1,185) = 1.85$, $p = .175$), nor a significant interaction effect ($F(1,185) = 0.04$, $p = .837$). Furthermore, no significant main effects of robot framing ($F(1,185) = 0.12$, $p = .727$) and product value ($F(1,185) = 2.58$, $p = .110$) were revealed for the analysis of willingness to pay for the service. In addition, no significant interaction effect was found for the interaction effect ($F(1,185) = 0.70$, $p = .403$).

Exploratory Analysis. Even though there were no differences in affinity for technology between the groups, it can still be assumed that interindividual differences in this variable might be relevant for the acceptance of novel technologies like delivery robots. To investigate this assumption, multiple linear regressions were conducted. The results of the regression for general acceptance indicated that the two predictors affective and cognitive component of affinity for technology significantly explained 5.91 % of the variance ($F(2,186) = 6.90$, $p = .001$). Interestingly, only the affective component ($\beta = 6.15, p = .049$), but not the cognitive component ($\beta = 3.71, p = .113$) significantly predicted the general acceptance of the robot. In line with these results, the regression for individual acceptance showed that the two predictors affective and cognitive component of affinity for technology significantly explained 5.43 % of the variance ($F(2,186) = 6.39$, $p = .002$). Again, only the affective component ($\beta = 8.92, p = .003$), but not the cognitive component ($\beta = 0.12, p = .958$) significantly predicted the acceptance of the robot. For the intention to use the robot, the analysis revealed a comparable pattern. The two predictors significantly explained 8.41 % of the variance ($F(2,186) = 9.63$, $p < .001$). The affective component ($\beta = 0.35, p < .001$) significantly predicted the intention to use the robot. This was not the case for the cognitive component ($\beta = 0.03, p = .668$). However, the willingness to pay for the service was not predicted by the affinity of technology ($F(2,186) = 0.75$, $p = .472$).

2.4 Discussion

The objective of the presented study was to examine the joint effects of anthropomorphic robot framing and product value for the acceptance of delivery robots.

Based on previous research in task-related HRI [19,25], it was hypothesized that anthropomorphic framing negatively affects the acceptance, intention to use, and willingness to pay for service robots. Surprisingly, the results did not confirm this assumption. In fact, participants accepted the anthropomorphically framed robot significantly more for their imagined delivery than the technically framed one. However, the positive effect of anthropomorphic framing was only present on an individual attitudinal level, as neither the general acceptance, the intention to use the robot nor the willingness to pay for the service was significantly related to the anthropomorphic framing. As intention to use and willingness to pay were operated as hypothetical behavioral indicators, the lack of evidence may be explained by the online nature of the study. Past research already illustrated that depictions of robots are a suitable approach to investigating perceptions and attitudes. However, behavior and even possibly behavioral indicators seem to be less influenced by anthropomorphism in depicted compared to embodied HRI [23,28]. Given these findings, it is important to note that the online format of our study may have limitations and could impact the validity of our results. Even though the video-based approach represents the possible interaction more dynamically than a static depiction, the experience remains limited. In terms of future research, it would be useful to extend the current findings by examining the acceptance and actual behavior towards differently framed robots in real-life delivery scenarios. Nonetheless, our findings further highlight that the effectiveness of anthropomorphism is depending on the interaction context [9,22,24]. Our results represent the first indication that anthropomorphic framing can facilitate the acceptance of service robots.

Besides anthropomorphism also the price of the transported product influenced individual acceptance. The result that the delivery of less valuable products is accepted more is consistent with earlier research [26]. On a practical side, this is good news for the acceptance of already implemented delivery robots, as they are currently used predominantly for (low-priced) food deliveries [12]. However, the incorporation of these robots to more diverse and maybe also more expensive product ranges [11], poses challenges to the acceptance of delivery robots.

A possible starting point for enhancing the acceptance of robots might be addressing more hedonic aspects of service robots. Instead of purely focusing on performance-based robot attributes, which might address the cognitive level, advertisement and exposure can be also used to facilitate the fascination for robots. On the one hand, this assumption is based on the finding that anthropomorphic framing leads to higher acceptance compared to purely technical framing. On the other hand, the affective component of affinity for technology predicted the individual and general acceptance as well as the intention to use the delivery robot. Having in mind that the affective component of affinity for technology includes fascination of and desire to own technologies [5], this might be a reasonable strategy for the market introduction of delivery robots.

The present study aimed to enhance our understanding of user acceptance of delivery robots. Despite the limitations of using a depicted exposure to the robot and not being able to assess actual behavior, the study yielded several key findings. Firstly, it was found that anthropomorphic framing can be used to increase the acceptance of delivery robots. Secondly, the acceptance of delivery robots was found to be higher for low-priced products compared to high-priced products. Thirdly, the affective component of affinity towards technology was identified as a crucial predictor for acceptance and intention to use delivery robots.

References

1. Abrar, M.M., Islam, R., Shanto, M.A.H.: An autonomous delivery robot to prevent the spread of coronavirus in product delivery system. In: 2020 11th IEEE Annual Ubiquitous Computing, Electronics and Mobile Communication Conference (UEMCON), pp. 0461–0466 (2020). https://doi.org/10.1109/UEMCON51285.2020.9298108
2. Darling, K., Nandy, P., Breazeal, C.: Empathic concern and the effect of stories in human-robot interaction. In: 2015 24th IEEE International Symposium on Robot and Human Interactive Communication (RO-MAN), pp. 770–775. IEEE (2015)
3. Faul, F., Erdfelder, E., Lang, A.G., Buchner, A.: G*power 3: a flexible statistical power analysis program for the social, behavioral, and biomedical sciences. Behav. Res. Methods **39**(2), 175–191 (2007). https://doi.org/10.3758/BF03193146
4. Feil-Seifer, D., Haring, K.S., Rossi, S., Wagner, A.R., Williams, T.: Where to next? The impact of Covid-19 on human-robot interaction research (2020)
5. Feuerberg, B.V., Bahner, J.E., Manzey, D.: Interindividuelle unterschiede im umgang mit automation-entwicklung eines fragebogens zur erfassung des

complacency-potentials. In: In L. Urbas and C. Steffens (Hrsg.), Zustandserkennung und Systemgestaltung. 6. Berliner Werkstatt Mensch-Maschine-Systeme. (Reihe 22, Fortschrittberichte VDI, Nr. 22) (S. 199–202). Düsseldorf: VDI (2005)

6. Figliozzi, M., Jennings, D.: Autonomous delivery robots and their potential impacts on urban freight energy consumption and emissions. Transport. Res. Procedia **46**, 21–28 (2020)

7. Fischer, K.: Tracking anthropomorphizing behavior in human-robot interaction. J. Hum.-Robot Interact. **11**(1) (2022). https://doi.org/10.1145/3442677

8. Forlizzi, J., DiSalvo, C.: Service robots in the domestic environment: a study of the Roomba vacuum in the home. In: Proceedings of the 1st ACM SIGCHI/SIGART Conference on Human-Robot Interaction, pp. 258–265 (2006)

9. Goetz, J., Kiesler, S., Powers, A.: Matching robot appearance and behavior to tasks to improve human-robot cooperation. In: The 12th IEEE International Workshop on Robot and Human Interactive Communication, Proceedings. ROMAN 2003, pp. 55–60. IEEE (2003)

10. Hwang, J., Choe, J.Y.J.: Exploring perceived risk in building successful drone food delivery services. Int. J. Contemp. Hosp. Manag. (2019)

11. Jung, S., Kim, H.: Analysis of amazon prime air UAV delivery service. J. Knowl. Inf. Technol. Syst. **12**(2), 253–266 (2017)

12. Kim, M.J., Kohn, S., Shaw, T.: Does long-term exposure to robots affect mind perception? An exploratory study. In: Proceedings of the Human Factors and Ergonomics Society Annual Meeting, vol. 64, pp. 1820–1824. SAGE Publications Sage CA: Los Angeles, CA (2020)

13. Kopp, T., Baumgartner, M., Kinkel, S.: How linguistic framing affects factory workers' initial trust in collaborative robots: the interplay between anthropomorphism and technological replacement. Int. J. Hum Comput Stud. **158**, 102730 (2022)

14. Li, D., Rau, P.L., Li, Y.: A cross-cultural study: effect of robot appearance and task. Int. J. Soc. Robot. **2**(2), 175–186 (2010)

15. Lin, H., Chi, O.H., Gursoy, D.: Antecedents of customers' acceptance of artificially intelligent robotic device use in hospitality services. J. Hosp. Market. Manag. **29**(5), 530–549 (2020)

16. Mims, C.: The scramble for delivery robots is on and startups can barely keep up. Wall Street J. (2020)

17. Nijssen, S.R., Müller, B.C., Baaren, R.B.v., Paulus, M.: Saving the robot or the human? Robots who feel deserve moral care. Soc. Cogn. **37**(1), 41–S2 (2019)

18. Onnasch, L., Hildebrandt, C.L.: Impact of anthropomorphic robot design on trust and attention in industrial human-robot interaction. ACM Trans. Hum.-Robot Interact. (THRI) **11**(1), 1–24 (2021)

19. Onnasch, L., Roesler, E.: Anthropomorphizing robots: the effect of framing in human-robot collaboration. In: Proceedings of the Human Factors and Ergonomics Society Annual Meeting, vol. 63, pp. 1311–1315. SAGE Publications Sage CA: Los Angeles, CA (2019)

20. Onnasch, L., Roesler, E.: A taxonomy to structure and analyze human-robot interaction. Int. J. Soc. Robot. **13**(4), 833–849 (2021)

21. Richards, G.: Starship Campus Delivery Service with Robots (2019). https://www.youtube.com/watch?v=Ftc0AVQEF6s

22. Roesler, E., Manzey, D., Onnasch, L.: A meta-analysis on the effectiveness of anthropomorphism in human-robot interaction. Sci. Robot. **6**(58), eabj5425 (2021)

23. Roesler, E., Manzey, D., Onnasch, L.: Embodiment matters in social HRI research: effectiveness of anthropomorphism on subjective and objective outcomes. ACM Trans. Hum.-Robot Interact. (2022)

24. Roesler, E., Naendrup-Poell, L., Manzey, D., Onnasch, L.: Why context matters: the influence of application domain on preferred degree of anthropomorphism and gender attribution in human-robot interaction. Int. J. Soc. Robot. 1–12 (2022)

25. Roesler, E., Onnasch, L., Majer, J.I.: The effect of anthropomorphism and failure comprehensibility on human-robot trust. In: Proceedings of the Human Factors and Ergonomics Society Annual Meeting, vol. 64, pp. 107–111. SAGE Publications Sage CA: Los Angeles, CA (2020)

26. Shanee, H.S., Dror, K., Tal, O.G., Yael, E.: The influence of following angle on performance metrics of a human-following robot. In: 2016 25th IEEE International Symposium on Robot and Human Interactive Communication (RO-MAN), pp. 593–598. IEEE (2016)

27. Sheridan, T.B.: Human-robot interaction: status and challenges. Hum. Factors **58**(4), 525–532 (2016)

28. Wainer, J., Feil-Seifer, D.J., Shell, D.A., Mataric, M.J.: The role of physical embodiment in human-robot interaction. In: ROMAN 2006-The 15th IEEE International Symposium on Robot and Human Interactive Communication, pp. 117–122. IEEE (2006)

29. Waytz, A., Cacioppo, J., Epley, N.: Who sees human? The stability and importance of individual differences in anthropomorphism. Perspect. Psychol. Sci. **5**(3), 219–232 (2010)

Let's Use VR! A Focus Group Study on Challenges and Opportunities for Citizen Participation in Traffic Planning

Marc Schwarzkopf[✉], André Dettmann, Adelina Heinz, Melissa Miethe, Holger Hoffmann, and Angelika C. Bullinger

Chemnitz University of Technology, 09126 Chemnitz, SA, Germany
marc.schwarzkopf@mb.tu-chemnitz.de
https://www.tu-chemnitz.de/mb/ArbeitsWiss/

Abstract. Processes in urban planning are predominantly based on laws and regulations rather than user experience, potentially leading to low acceptance and usage of completed infrastructure projects. An evaluation of the completed project is usually not within the project's scope, making it difficult to identify and address any issues in the process or design of completed measures – and hence device learnings for future projects. In this paper, we describe the current shift in local urban planners' objectives and processes to be able to include citizens into the planning process and evaluate the results of infrastructure projects to learn for future endeavors. To foster a deeper understanding for their work we conducted a structured workshop with urban planners ($N = 7$) with a focus on facilitation of citizen engagement and evaluation of infrastructure measures. Using the specific scenario of a cycling infrastructure project, we identified challenges, opportunities and requirements to support the planning process using digital tools and evaluated a virtual reality (VR) application as a potential digital support method. Our results indicate that urban planners are eager to increase citizen participation, but are limited by regulations and resources. They see digital tools as a way to increase engagement and present planning content in a more accessible manner.

Keywords: VR · urban planning · traffic planning · citizen participation · requirement engineering · focus group study · workshop

1 Background

Urban development and inner-city mobility are undergoing changes driven by social and technological trends such as demographic shifts [1] and digitalization [2]. For the future (re)design of cities, this means adjustments in urban development and inner-city mobility [3, 4]. Addionally, there is also a need to adjust mobility behavior of citizens. This has been the focus of many research projects and has been influenced through gamification [5], financial incentives and political marketing measures to increase the use of alternative modes of transportation like walking, cycling, and public transportation [6, 7].

H. Krömker (Ed.): HCII 2023, LNCS 14048, pp. 358–371, 2023.
https://doi.org/10.1007/978-3-031-35678-0_24

Measures to increase the use of alternative modes of transportation can be divided into two categories: soft and hard policies [8]. Soft policies focus on improving access to information, motivation, and communication, while hard policies involve physical changes to infrastructure, increased costs for car use, or control of road space. Research shows that soft policies have a short-term impact on behavior, while hard policies have a higher potential for long-term change [7, 9]. Studies have found that well-designed bike lanes separated from roads can increase cycling trips by 55% and reduce car commuting [10]. Factors such as safety, comfort, and overall satisfaction with the infrastructure are important indicators of usage [10]. A positive relationship has been found between satisfaction with the availability of bike parking spaces, and actual bike usage [11].

Evaluating infrastructure and its impact on safety, comfort, and satisfaction is often done through on-site inspections of roads, traffic intersections, and surrounding areas. This process is resource-intensive and limited by factors such as the time of day, weather, season, traffic conditions, and availability of participants for interviews. With these limitations, only a snapshot of the current state of the infrastructure can be captured. This is also true for pre-assessment or longitudinal studies, which are particularly challenging due to the complexity, resource requirements, and duration of implementation of new interventions [12].

Therefore, from the perspective of urban planners, problems for pedestrians, cyclists or people with impairments regarding safety, comfort and satisfaction of existing infrastructure implementations are rarely identified and corresponding recommendations for action are not derived for future projects. For example, infrastructure measures for cyclists in Germany are only planned according to recommendations (ERA 10) [13], not according to concrete specifications. As a result, infrastructure measures for the same planning challenge vary from city to city and rarely consider factors such as perceived safety, comfort experience, and, building on this, satisfaction.

Another approach besides the direct assessment of the infrastructure measure is citizen participation. Citizens in Germany are, in theory, able to directly participate in planning processes and give feedback on infrastructure solutions. The same problems arise here as with the evaluation approach, since citizens are in reality rarely or not at all involved in the planning processes [14]. In addition to regulations and scarce governmental resources, the reason could be a lack of low-threshold support methods for participation processes for planners. This further limits the ability to identify problems from the user's perspective during the planning phase.

City infrastructure projects in Germany are conducted using a Design-Bid-Build (DBB) approach, which is often used as a standard for infrastructure projects around the world, e.g. the United States [15]. DBB projects are conducted along a linear process of one task following the completion of a previous task without overlap (refer to Table 1). For Germany, this process is well documented in a directive of the Federal Government, the "Honorarordnung für Architekten und Ingenieure (HOAI)" [16]. In the HOAI the tasks to be completed are defined as well as the billable fees for the individual tasks as a percentage of the overall cost for different types of projects.

360   M. Schwarzkopf et al.

Table 1. HOAI fee percentage distribution over project phases in different project types [15]

	Project Phase	Buildings (§ 34)		Traffic Installations (§ 47)	
Design	1 – Basic Requirement Elicitation	2%	52%	2%	70%
	2 – Preliminary Planning	7%		20%	
	3 – Conceptual Design	15%		25%	
	4 – Permit Application	3%		8%	
	5 – Execution Planning	25%		15%	
Bid	6 – Bidding Preparation	10%	14%	10%	14%
	7 – Bidding Coordination	4%		4%	
Build	8 – Construction Supervision	32%	34%	15%	16%
	9 – Acceptance Process	2%		1%	

In general, project phases 1 through 5 can be attributed to the design stage of the project, where basic requirements are collected, designs are drawn up and building permits are requested. Phases 6 through 9 are less creative and more of a procedural nature, dealing with the bidding process as well as supervision of the building process and acceptance of the structure.

One of the major shortcomings of this process felt by urban planners is the late involvement of different target groups in the process. Current practice based on the legal directive is to only involve citicens in phase 5, when all object planning and designs have already been finalized, a construction permit has been applied for and depending on the project type more than 55% of the labor-cost have been spent. This limits the involvement of citicens as the everyday users of public buildings and traffic installations mostly to a passive role of being informed about pre-fixed facts. Other officials, like city councilpersons approving or disapproving the project as well as employees of the building authorities are only involved at marginally earlier stages where again most creative work in phases 2 and 3 have been completed. As a result, the Design-Bid-Build process can turn into what practitioners sometimes refer to as Design-Bid-Redesign-Rebid loop [17], where prior research has shown the detrimental effect of change orders on project cost and project schedule in many different areas, including public transportation infrastructure [18].

Another drawback of DBB processes is the lack of resources for evaluating the outcome. There is no mechanism in place to ensure the final implementation is accepted by its users and to understand why it may be rejected. This makes it harder to identify and fix mistakes, and may lead to the continuation of faulty processes, or implementations. To overcome this, involving potential users in the design stage (phases 1 through 3) is essential to produce a user-friendly result. However, resources in these early stages are limited, so the participation methods must be cost-effective, easy to use for both planners and users, and presented in a way that appeals to the target group. A possible solution to the problem could be the use of intuitive, low-threshold technologies, such as VR applications. These can, for example, help users to evaluate infrastructure measures in an effective, easy and cost-effective manner [19].

To solve the issues of inadequate citizen involvement in planning, inadequate evaluation of infrastructure projects, and insufficient resource allocation, we conducted a study to gain more detailed insights into the work practices of urban planners and citizen participation experts. The main objective was to identify opportunities for digital tools to help planners engage citizens in the planning process. This includes promoting early citizen participation in infrastructure planning and evaluating completed projects from the user's perspective. We hence conducted a requirements analysis through a structured workshop to gather insights on the functionalities and requirements of digital support. Finally, we will present a discussion on the methods for involving citizens and evaluating infrastructure to assist urban planners and authorities.

2 Method

The requirements analysis was carried out in the form of a workshop based on a structured guideline. Also, a demonstration of an VR application was included.

2.1 Participants

$N = 7$ experts in urban planning and citizen participation were involved in the study. The subjects were directly acquired from the city administration of Chemnitz. Participants had at least five years of working experience in the field of urban planning e. g. traffic infrastructure & leisure facilities ($n = 6$) and/or citizen participation ($n = 7$). The respondents rated their experience in conducting citizen participation processes as high. According to the ATI technology affinity scale [20], four respondents are not technology affine, two respondents have a low technology affinity and one respondent has a medium technology affinity. One participant had minor experience in the field of virtual reality, all other participants had none. The experts took part in the study during their work hours and received no further compensation for their participation.

2.2 Workshop Objectives

The workshop focused on two main points: getting to know the everyday work of urban planners and participation experts with the focus on citizen participation and finding possible points of contact in this everyday work for digital support methods to derive requirements for such tools. A specific goal of this workshop was also to find out whether an existing prototype of a VR application is considered suitable for the application area of citizen participation and which functions such a tool would have to contain from the perspective of urban planners and participation specialists.

2.3 Material and Procedure

One week before the workshop started, a short questionnaire was sent to the participants by mail. This questionnaire contained questions about the work activity and whether it included planning processes and citizen participation. Next, the participants were asked to rate their experience in conducting citizen participation and about their affinity for technology and their previous experience with VR technology.

The two interviewers conducted the workshop based on a structured guide. This guide was divided into six phases (refer to Table 2) and was derived from a requirements analysis based on Zimmermann et al. [21]. This requirement analysis was chosen as it focuses on the work processes of a specified target group and obtains insights for the creation of a particular system/method through a structured workshop with the target group. The authors initially examine the daily work routine and specific work procedures to generate suggestions for the future system.

The phases were proposed by Zimmermann et al. [21] and adapted to the recommendations by Renner and Jacob [22] by a focus group of $N = 4$ experts from the field of urban planning and user-centred design. The guide was created with the goal of gaining insight into the everyday work of urban planners and those responsible for citizen participation, including approach to project planning, regulations, experiences in the implementation of planning and participation processes, the evaluation of these processes and the identification of potential starting points for digital support methods. Details are displayed in Table 2.

Table 2. Summary of the contents of the guide

Phase and objective	Content & Questions
1 Introduction Warm-up and get to know	· Introduction of the participants · Privacy and consent · Presentation of the schedule
2 Citizen participation in Urban Planning Insights into the daily process of citizen participation	· How are planning activities carried out? · How is the term "successful" participation defined? · What is important for planners in citizen participation? · What factors do urban planners have influence on? · What is important for citizens in public participation from the perspective of planners? · Which tools are used for participation processes? · From the perspective of urban planners, what are the problems with citizen participation?
3 Evaluation Gaining insights into the evaluation of infrastructure measures.	· How are planning activities evaluated? · How are finished projects evaluated? · What competencies for conducting an evaluation are available? · Would urban planners like to see more evaluations of completed measures?

(continued)

Table 2. (*continued*)

Phase and objective	Content & Questions
4 Example digital support method Start a creative process to find possible starting points for a support tool.	· Demonstration of a VR Tool which was designed to evaluate the subjective safety of infrastructure measures. · Could such a tool be used in participation processes?
5 Requirement analysis: digital support tool Identification of the needs for support tools	· Which funcionalities should be included in a digital support tool for citizen participation? · Where would these tools be applicable? · How could the tool be integrated into the daily work of planners?
6 Conclusion Clarify open questions, close workshop	· Summary of the workshop findings · Farewell to the participants

The implementation of all phases followed the same scheme with the exception of phase 4. Here, a technology demonstration of a potential support method for participation processes took place. A VR application was presented via a VR headset (HTC Vive). The purpose of the application is to evaluate the subjective safety of cycling infrastructure [19]. In the application, 360° images of various cycle paths from different safety categories are shown and subjects are asked to evaluate perceived safety via a mixed method approach (refer to Fig. 1). The basic functionalities of the applications can also compare or visualise other infrastructure measures and projects, digital surveys can be implemented as well. The application requirements were extracted from the literature [19]. It will be presented to the target group of urban planners and participation experts for the first time during the workshop. The purpose of this technology demonstration was to start a creative process to identify possible links and functionalities for digital support methods.

The workshop took place over a period of three hours. There was a break after 80 minutes. At the end of phase 3, two experts had to leave the workshop due to work. In addition to the questions displayed in Table 2 after each phase the interviewers summarized the answers and asked the questions "did we miss anything" and "is there anything you want to add"? The auditory content of the workshop was recorded with three tablets (Samsung Galaxy). There was no video recording. The transcription was carried out by student assistants, the evaluation was based on qualitative content analysis according to Mayring [23].

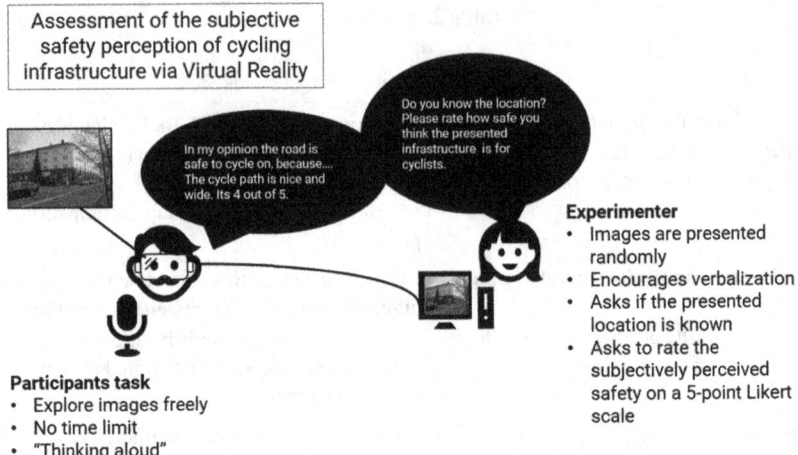

Fig. 1. Methodological structure of the VR-Demonstration [18]

3   Results

To present the workshop results in a structured manner, only responses that have received consensus from all participants will be included in this section. The answers are categorized based on the phases (excluding phases 1 and 7, as they are the introductory and farewell phases and contain no valuable information for this paper). The presentation will emphasize the key areas of interest:

1. Gaining an understanding of the daily work of urban planners and participation specialists, with a focus on citizen participation,
2. identifying potential points of interaction and requirements for (digital) support tools and
3. and the potential applicability of a VR application for the field of citizen participation.

The evaluation is based on the questions listed in Table 2 under "Content and Questions". The responses provided here are summarised extracts.

3.1   Phase 2 - Citizen Participation

How are planning activities carried out? The planning process follows the DBB procedure of the HOAI (refer to Table 1). The length of the planning phases may vary based on the complexity of the project, and multiple offices may be involved. Citizen participation can occur at any stage, but is sometimes mandatory in stage 5 (depending on the project type). The aim is to initiate participation as early as possible, though this is primarily dependent on available resources. In some cases, planning activities may be outsourced to external planning offices, who may take over the complete planning process, including citizen participation, depending on the budget. City-employed planners serve a supervisory role in these scenarios.

How is the term "successful participation" defined?. Successful participation is defined by reaching as many citizens as possible, especially the target audience (e.g. direct residents), and selecting an appropriate format (analog, digital or hybrid). It is also crucial to clearly communicate the purpose of the participation, whether it involves actual participation or simply providing information. This sets the foundation for a mutual understanding between citizens and authorities, reducing misunderstandings and promoting calm, informative participation. Ideally, participation should begin in the early planning stages (HOAI phase 2 or 3, refer to Table 1), and as many citizen concerns as possible should be incorporated into the plans.

What is important for planners in citizen participation?. The priority is to ensure accessibility for all citizens and maximize their involvement. It has been observed that presentation of planning projects can be overly complicated, with too much technical information (such as planning data and maps). It is crucial to simplify the content and make it easily understandable, without the need for extensive explanation.

What factors do urban planners have an influence on?. Citizen participation is purely optional in Germany and only partially required in phase 5 of the HOAI (refer to Table 1). The planning and execution of participatory measures, as well as the selection of formats, are entirely the responsibility of the planners, given that each project is considered unique. The limiting factors for such measures include resources, time, and knowledge, including moderation, digitalization strategies, and ease of understanding.

What is important for citizens in public participation from the perspective of planners?. It is crucial to keep citizens informed, provide updates on project progress, and communicate any changes or cancellations with clear reasoning. Transparency in communication is essential and requires addressing the reasons behind decisions, even if they are complex. Effective communication must also be tailored to the target audience, considering factors such as age, digital access, and proximity, and must be presented in an easily understandable format. Participation should be time efficient, as it is carried out in the citizens' free time and should use motivating formats (e.g. in connection with district festivals). Above all, citizens must feel that their opinions are valued and heard.

Which tools are used for participation processes?. The tools used are oriented towards the planning project. Common methods include mailers, physical events like temporary venues (for around a week), citizen interviews, on-site inspections, and digital surveys using government-provided tools. Traditional methods like newspaper ads and posters are still widely used. The use of innovative technologies is limited and mostly restricted to private planning offices.

From the perspective of urban planners, what are the problems with citizen participation?. The major issue is that not all demographic groups are adequately engaged. Typically, participation formats attract individuals from the age group 60 and above, while young people and those with a migration background are seldom represented. To enhance the sense of ownership and empowerment among citizens, it is necessary to increase the direct involvement of residents and encourage discussion of their proposals. These problems are also due to the limited digitalization of participation processes. There is a need for more resources and planning to establish a stronger digital infrastructure, including the use of online surveys, the ability to propose initiatives, and the integration of digital technologies.

3.2 Phase 3 – Evaluation

How are participation processes evaluated?. A structured evaluation of participation processes is not carried out. Personal impressions and reactions in the form of letters, telephone calls and e-mails are used as a basis for deciding whether the respective participation process is considered a perceived success.

How are finished projects evaluated?. Completed infrastructure measures are not evaluated in a structured manner, as no resource allocation is provided for this in the process (refer to Table 1). Whether completed infrastructure measures are "successful," i.e., used in the intended sense, is determined in rare cases via on-site inspections. In addition, in individual cases urban planners carry out subjective observations in their free time. Increasing use is being made of a digital government survey tool, which is still only used in specific cases. Personal opinions and feedback in the form of letters, phone calls, and emails are also used as a form of evaluation.

What competencies for conducting an evaluation are available?. The participants competencies with regard to conducting and evaluating an evaluation are described as fundamental. The urban planners and citizen participation experts who were interviewed have the capability to create basic questionnaires (mainly consisting of quantitative, yes/no answers), but for more in-depth evaluations and analyses, expert support is necessary.

Would urban planners like to see more evaluations of completed measures?. Subjective observations or binary responses were considered enough for evaluation, but it was expressed that there was a desire to gain deeper insight into citizens' behavior, such as the reasons behind their usage, non-usage, or altered usage of implemented measures.

3.3 Phase 4 - Example Digital Support Method

Could such a tool be used in participation processes?. The tool is envisioned to be useful in engaging the public in urban planning processes. Urban planners and citizen participation experts believe that it can present best practices in infrastructure solutions and create a sense of inclusion for participants. This would also be ideal for visualizing various planning options and comprehending the reasoning behind decisions made. The tool is also seen as appealing and motivating, particularly for younger people, and as a way to easily visualize accessibility issues and consider the perspectives of children.

The participants agreed that it can also evaluate implemented measures effectively and efficiently and help to persuade political decision-makers to support planned implementations.

3.4 Phase 5 – Requirement Analysis: Digital Support Tool

Which funcionalities should be included in a digital support tool for citizen participation?. A user-friendly digital tool is required that both urban planners and citizens can understand easily. Urban planners should be able to easily add and modify content (e. g. surveys). The tool should provide fun and encourage participation. Desired features include

- conducting and evaluating surveys
- visualizing existing models and planning views in an easy way
- considering accessibility aspects (different visual filters, perspective of wheelchair users, etc.)
- comparing different planning models
- displaying videos or allow movements in the model.

Where would these tools be applicable?. Possible areas of application would be:

- Preparation of citizen surveys
- Support for on-site events
- Simplified visualisation of complex planning data
- Sensitisation to the needs of children, people with disabilities and the elderly
- Support with data collection and evaluation
- Sharing ideas within the planning office
- Convincing decision-makers

How could the tool be integrated into the daily work of planners?. The tool must be user-friendly for both urban planners and participation specialists, requiring no external assistance. The interface should be intuitive to use, with a low learning curve. Ideally, the tool should offer customizable content (e.g. surveys) to simplify its use.

4   Discussion

Urban planners and experts in citizen participation highlight the lack of resources for participation and evaluation in their daily work. This limits the scope for citizen involvement and the ability to assess infrastructure measures from a user-centric perspective. Although workshop participants perceive this as an issue, only the limited participation was seen as a problem. A thorough evaluation of completed infrastructure measures, such as perceived safety, was considered desirable, but is regarded as beyond the available resources according to the planners, also the complexity of potential survey formats was seen as a crucial factor.

The participants also agreed that partizipation processes should take place as early as possible, ideally during HOAI phase 2 (refer to Table 1), and it is also their endeavour to initiate these processes as early as possible. However, the available methods of citizen participation were often criticized as inadequate, as the content is often too complex for citizens (e.g. planning maps), the approach is mostly analogue and not very targeted, and the methods currently used are not very attractive to all social groups.

Unsurprisingly, the expected characteristics for potential (digital) support methods in citizen participation and user-centered evaluation of infrastructure measures are simplicity, efficiency, and appeal. The two emerging key issues are how urban planners and specialists in citizen participation can be assisted in

a) a way that enhances the output of participation and evaluation processes, which

b) does not consume excessive time and financial resources.

Both issues are not easy to address, as they require a more complex analysis of typical participation formats, appropriate methods as well as a dedicated investigation of the potential of user-centered infrastructure evaluation. However, first proposals can be defined by extracting the answers from the workshop, which reduces the scope of future investigations.

First of all, potential solutions to both questions would need to build on existing processes and digital infrastructure. Major invasive interventions in both structures would possibly require a reorganization of everyday work and existing processes and would therefore have little chance of success.

According to the participants, to address issue a), the scope of participation must be expanded and directly impacted citizens should be involved in a targeted manner, using digital, analog, and hybrid channels. Another issue was the lack of consistent communication, with gaps often occurring between information, participation, and implementation processes or inadequate notification of project stops or termination, leading to a decrease in citizens' motivation for participating. Thus, continuous communication should be maintained, combining past information and current updates, for example, through a simple blog site or similar digital medium, though it may not be accessible for older people. On-site events were also considered as challenging. Firstly, because not all age and social groups are represented at this type of event and the content presented is often complex and not very attractive. Accordingly, such events would have to be designed more attractively for a broad target group. For example, by using innovative technologies, haptic models or mock-ups to make changes comprehensible in a low-threshold way. Furthermore, presented content would have to be edited to make it easily accessible as well. For both proposals, however, there is a need for best practice solutions that can be easily used in a slightly modified way in different participation formats in order to consider the requirements of issue b). In order to enable user-centered evaluation of infrastructure measures, it must be considered that the method knowledge of city planners and experts in the field of citizen participation is often rudimentary. Therefore, it also requires the use of adaptable methods, such as questionnaires or questions, as well as a mostly automated evaluation of these methods. Those evaluation formats may be also combined with the innovative technology used at on-site events. In line with planners' requirements, the long-term perspective for planning projects would be for citizens to shape participation in part by themselves and to discuss proposals with the city or among themselves independently of already planned projects. This may be achieved by providing a platform where citizens can be informed about individual projects and also have the ability to share their suggestions within feasibility limits. For designing support methods, its also important to consider the needs of people with disabilities. These methods should provide accessible features for all stakeholders in public life, e. g. including wheelchair users and children, and account for any potential visual limitations.

To address question b), it's necessary to implement prototypes that meet the requirements from question a). These implementations should ideally be intuitive systems or existing systems/methods that have been adapted or expanded to minimize learning efforts, considering time and financial aspects. Ideally, planners should have access to a toolkit for participation and evaluation that includes easily adaptable templates or digital applications. We have tried to outline what such a system might look like through a technology demonstration and to gather additional requirements.

In the technology demonstration, we showed the VR application presented in Sect. 2.3 to the urban planners and participation experts and asked for an assessment regarding its applicability. The experts were convinced that such an application could make participation measures more motivating for citizens and better conveyed the planning project. The demonstration featured only 360° photos of various infrastructure measures, but this basic implementation was deemed helpful for communicating potential plans, comparing different options, understanding physical constraints, and preparing information for decision makers. The technology was also considered a foundation for evaluations. The workshop participants also appreciated the ability to implement digital questionnaires and directly record citizen feedback. Both urban planners and participation experts rated the tool and its potential highly, seeing it as a valuable support in future participation initiatives. Accordingly, such a tool can be a part of the proposed toolbox for participations and evaluations if it is adapted to the mentioned requirements of urban planners and participation experts.

4.1 Limitations

The basis for the workshop and requirements analysis was a guideline created by 4 urban planning and user-centered design experts, so not all relevant aspects may have been included. The sample of 7 experts interviewed during the workshop is small and may not be representative, as participation was voluntary, participants may have a heightened interest in the topics discussed. Accordingly, real requirements could differ. Another important point is that the survey was only conducted in one city, Chemnitz, and according to the participants, this city has implemented little digitization in the area of citizen participation so far. Other German cities are already significantly advanced in this area and might have different requirements. Only answers agreed upon by all participants were used, potentially leading to distorted results, but this was mitigated by the absence of conflicting statements, only abstentions due to non-existing experience arose. During the workshop, various disciplines also came together, from public transportation planning, road planning, to child and youth participation. Therefore, a dedicated study of the requirements of the individual areas would make sense. In addition, the presentation of the VR app in phase 4 may have influenced all subsequent statements by the participants.

4.2 Future Work

The results of this paper provide a foundation for enhancing user-centered participation processes, meaning they consider both, the perspective of urban planners and citizens alike. Therefore, participation processes need to be further investigated in order to select appropriate methods and to adapt these processes to the reported requirements. For this purpose, analog (mock-ups, miniature models, classical surveys, etc.) as well as digital methods (VR application, citizen platforms, etc.) will be investigated and adapted or developed in the following studies and tested directly in the field. The goal is to develop a form of participation that is jointly designed by public authorities and citizens and that gives both actors the feeling of communicating requirements and needs in a way that is appropriate for the target group.

Funding Reference. This research work was supported by the German Federal Ministry of Education and Research (BMBF) (project "NUMIC 2.0 – Neues urbanes Mobilitätsbewusstsein in Chemnitz 2.0 (FKZ 01UR2204B)") as part of the funding program "Teilprojekt B: Erforschung und Implementierung partizipativer Umsetzungsstrategien und Unterstützungsformate für Bürgerbeteiligungsprozesse in der Stadt- und Verkehrsplanung". The funder had no influence on the study design, the collection, analysis and interpretation of the data, the writing of the report or the submission of the article.

References

1. Buffel, T., Phillipson, C., Scharf, T.: Ageing in urban environments: developing 'age-friendly' cities. Crit. Soc. Policy **32**(4), 597–617 (2012). https://doi.org/10.1177/0261018311430457

2. Kramers, A., Höjer, M., Lövehagen, N., Wangel, J.: Smart sustainable cities – exploring ICT solutions for reduced energy use in cities. Environ. Model. Softw. 52–62 (2014)

3. Burns, L.: A vision of our transport future. Nature **497**, 181–182 (2013). https://doi.org/10.1038/497181a

4. Loorbach, D., Shiroyama, H.: The challenge of sustainable urban development and transforming cities. In: Loorbach, D., Wittmayer, J.M., Shiroyama, H., Fujino, J., Mizuguchi, S. (eds.) Governance of Urban Sustainability Transitions. TPUST, pp. 3–12. Springer, Tokyo (2016). https://doi.org/10.1007/978-4-431-55426-4_1

5. Torres-Toukoumidis, A., Vintimilla-Leon, D., De-Santis, A., López-López, P.C.: Gamification in ecology-oriented mobile applications—typologies and purposes. Societies **12**, 1-12 (2022). https://doi.org/10.3390/soc12020040032

6. Thøgersen, J.: Promoting public transport as a subscription service: effects of a free month travel card. Transp. Policy **16**(6), 335–343 (2009)

7. Bamberg, S., Schmidt, P.: Theory-driven subgroup-specific evaluation of an intervention to reduce private car use. J. Appl. Soc. Psychol. **31**(6), 1300–1329 (2001)

8. Gärling, T., Bamberg, S., Friman, M., Fujii, S., Richter, J.: Implementation of soft transport policy measures to reduce private car use in urban areas (2009)

9. Wardman, M., Tight, M., Page, M.: Factors influencing the propensity to cycle to work. Transport. Res. Part A: Policy Pract. **41**(4), 339–350 (2007)

10. Springer, S., Kreußlein, M., Krems, J.F.: Shedding light on the dark-field of cyclists' safety critical events: a feasibility study in Germany. In: Proceedings of 9th International Cycling Safety Conference ICSC 2021. Lund, Sweden (2021)

11. Martens, K.: Promoting bike-and-ride: the Dutch experience. Transport. Res. Part A: Policy Pract. **41**(4), 326–338 (2007). https://doi.org/10.1016/j.tra.2006.09.010

12. Oliveira, V., Pinho, P.: Evaluation in urban planning: advances and prospects. J. Plan. Lit. **24**(4), 343–361 (2010)

13. Alrutz, D.: Die Empfehlungen für Radverkehrsanlagen (ERA) (2011)

14. Swapan, M.S.H.: Who participates and who doesn't? Adapting community participation model for developing countries. Cities **53**, 70–77 (2016)

15. Tran, D.Q., Diraviam, G., Minchin, R.E., Jr.: Performance of highway design-bid-build and design-build projects by work types. J. Constr. Eng. Manag. **144**(2), 04017112 (2018)

16. Korbion, H., Mantscheff, J., Vygen, K.: Honorarordnung für Architekten und Ingenieure (HOAI) (2021)

17. Levy, S.M.: Construction Process Planning and Management: An Owner's Guide to Successful Projects. Butterworth-Heinemann (2009)
18. Shrestha, P.P., Maharjan, R.: Effect of change orders on cost and schedule for small low-bid highway contracts. J. Leg. Aff. Disput. Resolut. Eng. Constr. 11(4), 04519025 (2019)
19. Schwarzkopf, M., Dettmann, A., Trezl, J., Bullinger, A.C.: Do you bike virtually safe? An explorative VR study assessing the safety of bicycle infrastructure. In: de Waard, D., et al. (eds.), Proceedings of the Human Factors and Ergonomics Society Europe Chapter 2022 Annual Conference: Enhancing Safety Critical Performance, pp. 101–110. Human Factors and Ergonomics Society. https://www.hfes-europe.org/largefiles/proceedingshfeseurope 2022.pdf
20. Franke, T., Attig, C., Wessel, D.: A personal resource for technology interaction: development and validation of the affinity for technology interaction (ATI) scale. Int. J. Hum.-Comput. Interact. 35(6), 456–467 (2019)
21. Zimmermann, J., Konrad, S., Nerdinger, F.W.: Bedarfs- und Anforderungsanalysezur Entwicklung einer internetbasierten Kommunikationsplattform zur Unterstützung des Forschungstransfers. Universität Rostock (2009). https://doi.org/10.18453/ROSDOK_ID0 0002241
22. Renner, K.H., Jacob, N.C.: Konzeption und Erstellung eines Interviewleitfadens. In: Renner, K.H., Jacob, N.C. (eds.) Das Interview. Basiswissen Psychologie, pp. 47–64. Springer, Heidelberg (2020). https://doi.org/10.1007/978-3-662-60441-0_4
23. Mayring, P., Fenzl, T.: Qualitative inhaltsanalyse. In: Handbuch Methoden der empirischen s, pp. 633–648. Springer VS, Wiesbaden (2019). https://doi.org/10.1007/978-3-531-18939-0_38

Sustainable Mobility

Effects of Visual and Cognitive Load on User Interface of Electric Vehicle - Using Eye Tracking Predictive Technology

Gan Huang[✉] and Yumiao Chen

School of Art Design and Media, East China University of Science and Technology, Shanghai, China

1026344208@qq.com, cym@ecust.edu.com

Abstract. Purpose - With the increasing integration of car functions, as well as the increasing operation and information on the central control screen, we explored how to improve the user interface, in order to reduce cognitive load and improve reading efficiency. **Methodology** - This paper applied the neural network-based eye-tracking prediction model to analyze the eye-tracking data of mainstream smart electric vehicle center control screens. Through analyzing and discussing the attention map, clarity map, regions of interest, etc., we assess the usability of user interface and propose design guidelines. **Conclusion** - In a landscape central control screen, dock bar is more visually significant on the left side. The layout should avoid scattering, the shape of the function card should avoid using long stripes, and the information should not be too concentrated. Important information should be designed with high contrast and distinctive colors, and filled types icons should be used. Important text should be succinct, enlarged, bolded, and not be too dense. Concentrated text is more likely to attract users' attention, but it will also cause higher cognitive load.

Keywords: vehicle user interface · eye-tracking predictive technology · electric car · visual effects

1 Introduction

Automobiles are gradually developing towards intelligence, and the functions of human machine interface(HMI) are continuously integrated, with more and more information and complex interface relationships. The car's center console has also transformed from physical buttons such as keys, knobs and switches to virtual touch screen buttons.

Lacking the tactile feel of physical buttons, drivers are forced to look at the screen during operation, which affects driver performance, increases cognitive and workload, and can even lead to driver distraction, causing road safety problems [1]. According to statistics, 87.5% of the major traffic accidents in China are caused by drivers' own factors, of which 14%–33% are caused by distracted driving [2]. In order to reduce the driver's operational difficulty, improve operational efficiency, and enhance the driver's driving comfort experience, a more applicable human-machine interface for smart electric vehicles needs to be designed.

© The Author(s), under exclusive license to Springer Nature Switzerland AG 2023
H. Krömker (Ed.): HCII 2023, LNCS 14048, pp. 375–384, 2023.
https://doi.org/10.1007/978-3-031-35678-0_25

2   Literature Review

In a study of the information structure of automotive interfaces, Engström J et al. [1] systematically investigated the effects of visual and cognitive demand on driving performance and driver state, founding that visual load lead to reduced speed and increased lane keeping variation. Castellano et al. [3] studied the visualization of massive data structures in a limited screen space in order to improve user cognitive efficiency. Schmidt et al. [4] studied the automotive interactive interface and proposed the correlation between the design of automotive interface and driver distraction combined to safe driving.

Sun B. et al. [5] studied the automotive human machine interface hierarchy, identified elements such as color, shape, text, layout, navigation, and transition, proposed the principles of hierarchical design. Through eye-tracking technology, Ren H. et al. [6] explored the influence of dock bar's position and interface layout on driver, founding that signal-column display is better than double-column display in cognitive efficiency, and the dock bar bottom-left layout is more efficient than the left layout. Sun B. et al. [7] analyzed the color elements of the automotive human machine interface and derived the most comfortable color scheme through eye-tracking experiments. Jin X. [8] et. Evaluated the size of button in interface through user testing, and concluded that the proposed size of center touch screen button is 9–15 mm, which can be adapted to different sizes after formula conversion. Xi J. [9] derived usability design guidelines for in-car interface icons through eye-tracking analysis and satisfaction questionnaires. Tan H. [10] analyzed the multi-screen interaction experience of car navigation and derived the design guidance of multi-screen interface. Guilei S. et. [11] and Ping Z. et al. [12] provide a reasonable layout and color scheme for the car dashboard through eye-tracking analysis.

As the journal reviewed, in terms of research subjects, intelligent networked vehicles are the hot spot that scholars have focused on in recent years. The studies for the dashboard are more than the center control screen. It's because the dashboard has been applied for a long time, while the center control screen has just emerged in recent years.In terms of research content, many scholars have conducted subjective and objective studies on various layers of HMI, including color, text, buttons, icons, etc. In terms of research methods, eye-tracking experiments are still the main means of HMI research.

In this paper, for the center control screen of smart electric vehicles, by using an eye-tracking prediction model based on neural networks, we obtain the eye-tracking data of the first 3–5 s when users view the interface of the center control screen, evaluate the usability of the interface layout, and propose design guidelines for the center control screen to improve the cognitive and operational efficiency of drivers, reduce the distraction while driving, and optimize the interaction experience of the interface.

3   Method

The eyes are a window into user's behavior and assessment of the subconscious mind. Eye tracking experiments can capture user's gaze behavior; which usually takes several days and requires expensive eye-tracking instruments. With the development of artificial intelligence technology, some scholars have trained machine learning models to predict users' eye tracking data by artificial neural network (ANN), which is fast and reliable, does not require any instrumentation, and can obtain eye tracking analysis data in seconds by uploading images. In the field of cognitive psychology in China, scholars have already applied it to the eye-tracking analysis of text reading [13]. Eye-tracking prediction models have been commercialized and have been applied in software engineering, medical, and gaming fields [14]. Relevant companies providing eye-tracking prediction services include EyeQuant (UK), Neurons (Denmark), etc.

This eye-tracking prediction model works by analyzing nearly one million data from hundreds of participants to train ANN eye-tracking prediction models for quantifying people's gaze patterns and eye movements, such as sweeping (fast tracking), gazing (steady gaze), and smooth pursuit (following a moving object), to detect which design features (luminance, architecture, saturation, structural, edge density. etc.) attract attention in the first 3–5 s of attention. The accuracy of this prediction model can reach more than 90%.

Eye tracking prediction models generate attention maps, clarity maps and regions of interest. It can help UX designers, product managers, and marketers to quickly understand how users react when they see an interface. It can help us make data-driven decisions and optimize the interface to increase conversion rates and improve user experience. This study used EyeQuant's eye-tracking prediction model to obtain eye-tracking data from the center control screen of smart vehicles.

4 Material

4.1 Subject

At present, the brands that produce electric vehicles are traditional brands such as Volkswagen, Mercedes-Benz, BMW and Audi, as well as new brands such as Tesla, NIO, Xpeng and Li. Due to its purely electric feature and "future" positioning, electric vehicles tend to use more and bigger screens than traditional gas cars, and the operation button are also highly integrated in the central control screen. In recent years, Tesla and other new brands have become the choice of more and more young people, but there is little research on them. In this paper, the central control screen of Tesla, NIO, Xpeng, and Li, which are the mainstream electric vehicle brands in China, are selected for study. In order to get close to the car scene in most cases, the navigation state is selected as the test interface. The selected vehicle models are the latest models in 2022, and the interface information of the center control screen is summarized in Table 1.

Table 1. Interface Information.

NO.	Model	Orientation	Size	Dock position	Function card
A	Tesla Model Y	Landscape	1920 * 1200	Bottom	Switch display
B	Xpeng P7	Landscape	2400 * 1200	Left	Fixed left
C	NIO ES7	Portrait	1600 * 1400	Bottom	Fixed bottom
D	Li ONE	Landscape	2608 * 720	Lower left	Switch display

4.2   Interface Analysis

Tesla. The interface of Tesla Model Y is selected as a study case, and the interface is shown in Fig. 1. As Tesla canceled the dashboard, it's center screen is displayed in partitions, with the left 1/3 area as the car information page, the right 2/3 area as the car system page. The navigation is used as the main page, and the dock bar is located at the bottom.

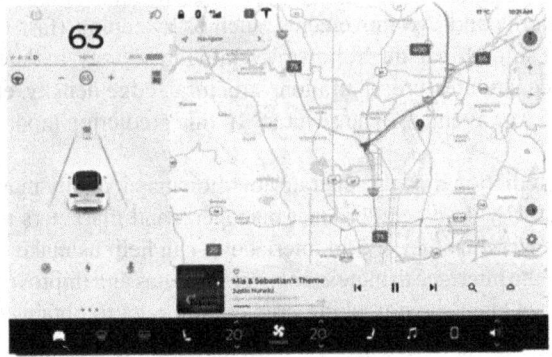

Fig. 1. The navigation page of Tesla model Y center screen.

Xpeng. The Xpeng P7's dashboard screen and the center screen are designed in a crossover style, and the center screen is shown in Fig. 2. The Xpeng P7's center screen interface is divided into three main parts: the dock bar, the function cards, and the content display page from left to right. The content will be displayed as a pop-up window on top of the map when tapped.

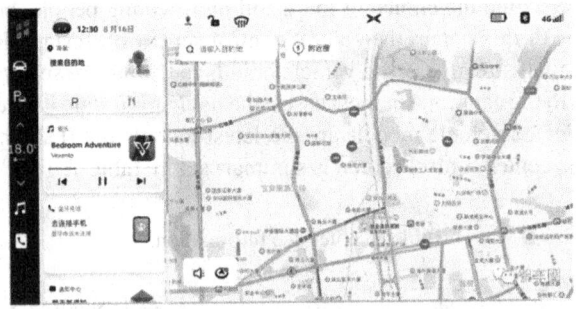

Fig. 2. The navigation page of Xpeng P7 center screen.

NIO. The interface of the NIO ES7 is displayed on a vertical screen, as shown in Fig. 3. The interface uses navigation as the main page, with two cards hovering above. Near the driver's side is the music card, while the right side is the rest of function cards. The dock bar is located at the bottom of the interface.

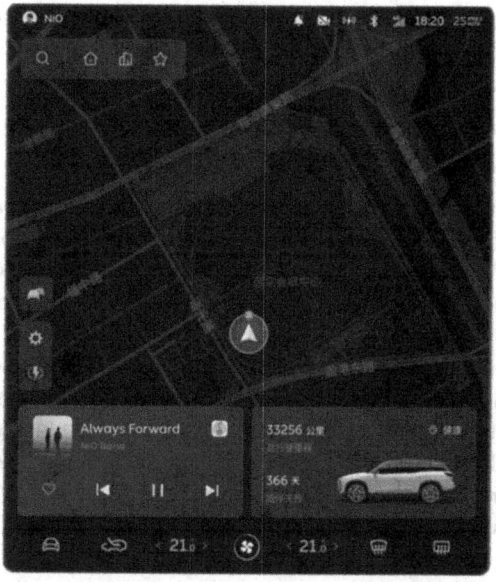

Fig. 3. The navigation page of NIO ES7 center screen.

Li. The main page of Li ONE adopts a card-like function cards design (Fig. 4). The left side of the central control screen displays common functions such as time, music, voice assistant. There are five shortcut keys in the bottom. The right side is the card area of each common function. The cards are arranged horizontally, and display switched.

Fig. 4. The navigation page of Li ONE center screen.

5   Result and Analysis

Input the pictures of A. Tesla, B. Xpeng, C. NIO and D. Li into the eye tracking prediction model, and the attention map, hot spots, clarity score and clarity map are calculated by model. After dividing the area, the regions of interest can be obtained, which showing the visual significance difference of each region. The analysis of these eye-tracking data can helping analyze the influence of interface layout, color, shape and text on user's cognitive load and recognition efficiency.

5.1 Attention Map and Hot Spots - Recognition Efficiency

Analysis of the attention map can provides an objective view of the elements that users are most likely to notice. If users can't find what they're looking for in a few seconds, taking their eyes off the road for long periods of time can create a host of dangers. The attention map and hot spots shown what the user sees in the first few seconds (Fig. 5, Fig. 6).

The analysis results are as follows: **(1)** In the horizontal center screen, the dock bar is easier to see on the left side than on the bottom side. **(2)** The eye-saccade distance of A in L-shaped layout is significantly larger than B in left-right layout, which will result in higher cognitive load and require more attention. **(3)** A square dock bar is easier to be noticed than a long dock bar. **(4)** High contrast, complex images, large bold text and focused information are easier to be seen by users.

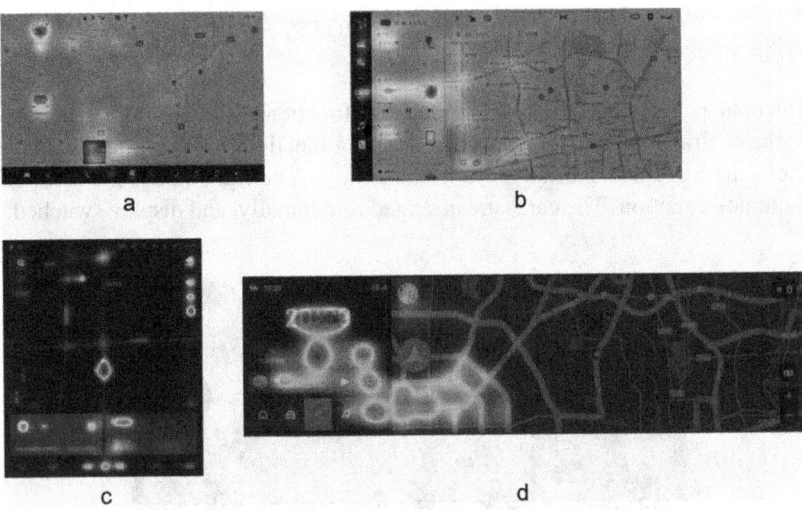

Fig. 5. Attention map.

5.2 Clarity Score - Cognitive Load

The clarity score measures the degree of clutter or clarity of the interface design. A lower score indicates more clutter and a higher score indicates more clarity, and the results of the analysis are shown in Table 2. In the clarity map (Fig. 7), the areas highlighted in red caused a higher cognitive load than the green areas.

(1) B has the lowest clarity, due to it's amount of dense text and on the function cards. The text of the C's map is less, so the definition is higher. **(2)** The filled type icon

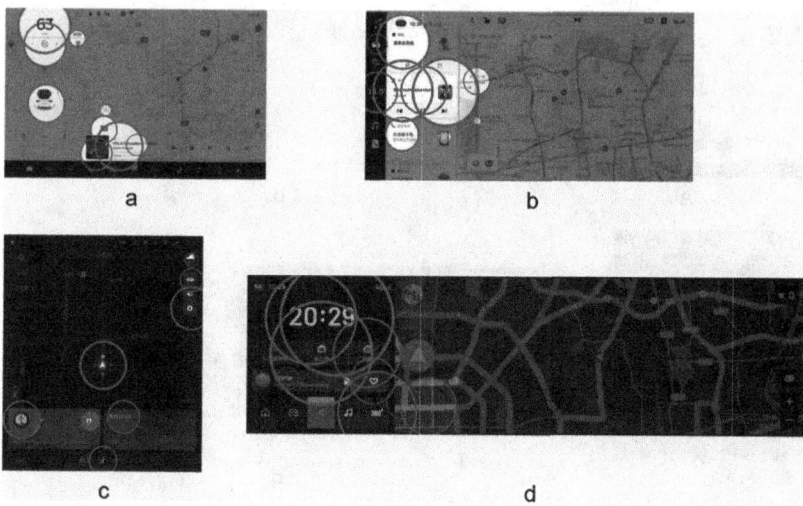

Fig. 6. Hot spots.

is clearer and has a lower cognitive load. **(3)** Compared the function cards of A and C, it can be found that the color of dense text information such as song name and singer is more red, and it can be inferred that dense text will causes higher cognitive load.

However, the different sizes and appearance of maps will affect the clarity score, which is the limitation of this paper. In future research, more variables can be controlled to obtain more objective data.

Table 2. Clarity score comparison.

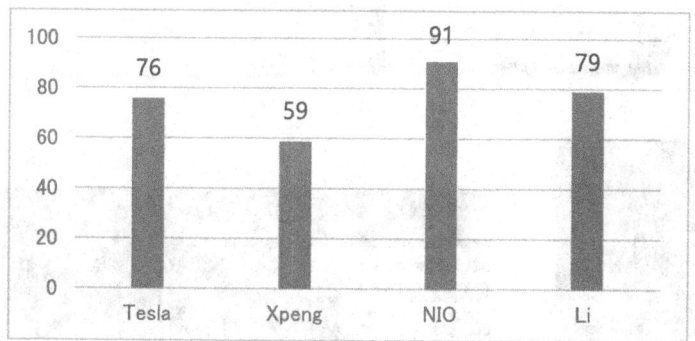

5.3 Analysis of Regions of Interest - Division is More Significant

The regions of interest measures the average significance of pixels within the selected region and compares it to the average significance of all areas on the entire page, which

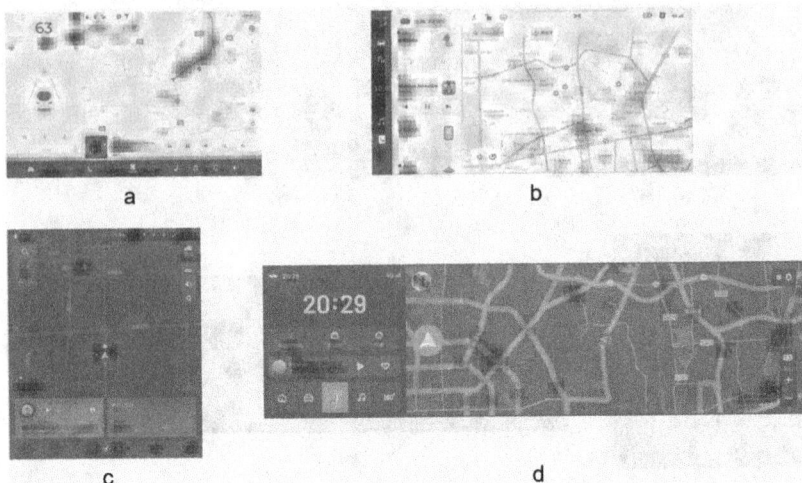

Fig. 7. Clarity map.

shows the visual significance of the region compared to the entire interface (Fig. 8). And Table 3 shows the scores of different regions in percentage. We can find out by comparing the score of the dock bar that B and D performs better with their dock bar positions on the left. It can be tentatively concluded that the more it tends to be square, the higher the visual significance.

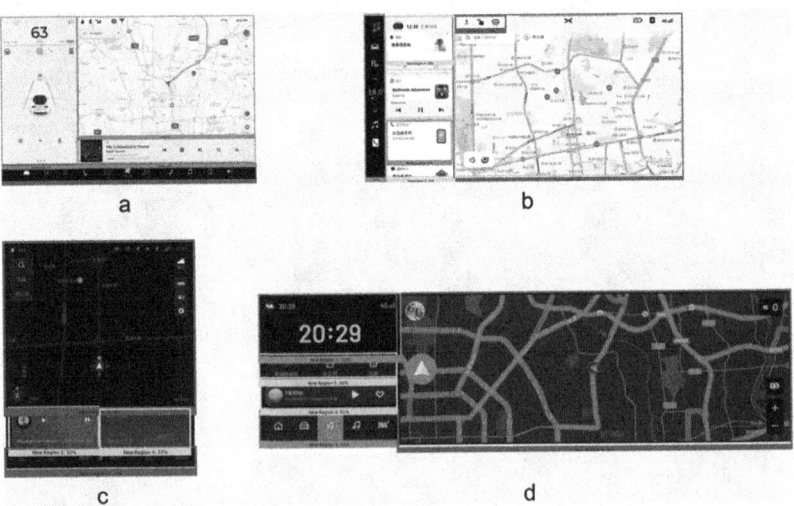

Fig. 8. Regions of interest

Table 3. Regions of interest comparison.

NO.	Model	Navigation page	Dock bar	Function card	Status bar
A	Tesla Model Y	0	−21	14	−
B	Xpeng P7	−9	12	29	−19
C	NIO ES7	−6	−6	34	−
D	Li ONE	−9	15	51	12

5.4 Discussion

Through the analysis and inference of attention map, clarity score, clarity map and regions of interest, the following design guidelines can be initially derived. It can provide a reference basis for the design of the smart electric vehicle's center screen interface, reduce the cognitive load of users in driving, improve recognition efficiency and operation efficiency, improve user experience and driving safety.

(1) **In landscape screen interface, the dock bar should be laid out on the left side.** According to the analysis results of attention map and area of interest score, the dock bar on the left side has higher visual prominence than on the bottom side, which is easier to be noticed by users and thus can reduce the time spent by users in cognition.

(2) **The layout of each area should avoid scattering, and the shape of the area should avoid using long stripes.** According to the attention map, it is inferred that a square dock bar is easier to be noticed than a long dock bar, and a centralized layout is more efficient than a scattered layout. Therefore, common functions should be distributed as centrally as possible. However, it should be noted that the information in the area should not be distributed too centrally in order to ensure clarity.

(3) **Important information should be designed with high contrast and distinctive colors, and icons should be chosen as filled type.** In the interface design, graphic logo design with intuitive image is needed to effectively convey information to users and avoid users spending too much attention on text recognition.

(4) **Important text should be succinct, enlarged, bolded, and not too dense.** Concentrated text is easier to attract users' attention, but it will also cause higher cognitive load. Therefore, the layout of the text should be reasonably arranged according to the importance of the text.

5.5 Limitation

In this paper, a comparative study of the center screen interfaces of four models is conducted and the general design direction is analyzed. However, the size of the four center screen interfaces, as well as the position of the center screen in the car, are different. In the actual driving process, these factors will affect the user's viewing effect. In addition, Tesla integrates the dashboard and the center screen together, which is quite different from the others. The NIO has a HUD in addition to the dashboard, which may not require looking at the center screen during actual driving like the Tesla does.

Overall, this paper uses an eye-tracking prediction model to make a general comparison of the center screens of four mainstream electric vehicles models, which is a relatively new attempt. In future research, the design of centre screen interface can be optimized for the results of this eye-tracking analysis, and eye-tracking analysis can be used again to make comparative comparisons.

6   Conclusion

Aiming at the current situation of highly integrated, complex functions and increasing information of the center control screen on smart electric cars, we analyzed the eye-tracking data of the screen interfaces through eye-tracking prediction model, which based on neural networks. Through in-depth analysis of data such as attention map, hot spots, clarity score, clarity map, and regions of interest etc., discovered the problems in the existing interface and proposed the relevant principles of interface design, which helps to reduce the cognitive load of users during the driving process and improve the recognition efficiency, user experience and driving safety.

References

1. Engström, J., Johansson, E., Östlund, J.: Effects of visual and cognitive load in real and simulated motorway driving. Transport. Res. F: Traffic Psychol. Behav. **8**(2), 97–120 (2005)
2. Qin, J., Liu, Z., Liu, J.: On the road traffic safety risk based on accident tree. Value Eng. **16**, 192–195 (2015). (in Chinese)
3. Castellano, G., Cimino, M.G.C.A., Fanelli, A.M., et al.: A multiagent system for enabling collaborative situation awareness via position-based stigmergy and neuro-fuzzy learning. Neurocomputing **135**(13), 86–97 (2014)
4. Schmidt, A., Spiessl, W., Kern, D.: Driving automotive user interface research. IEEE Pervasive Comput. **9**(1), 85–88 (2009)
5. Sun, B., Yang, J., Sun, Y.: Research on interface hierarchy design for human vehicle interaction. J. Mach. Des. **36**(02), 121–125 (2019). (in Chinese)
6. Ren, H., Tan, Y.: Watching behavior analysis of vehicle touch screen based on eye movement experiment. Packag. Eng. **41**(20), 97–101 (2020). (in Chinese)
7. Sun, B., Yang, J., Sun, Y., Yang, H., Li, S.: Color design of human vehicle interaction based on eye-tracking experiment. Packag. Eng. **40**(02), 23–30 (2019). (in Chinese)
8. Jin, X., Li, L., Yang, Y., Fu, M., Li, Y., You, F.: Touch key of in-vehicle display and control screen based on vehicle HMI evaluation. Packag. Eng. **42**(18), 151–158 (2021). (in Chinese)
9. Xi, J.: The Usability Design Research of Interface Icons in In-Vehicle Infotainment System. Master, Jiangsu University (2017). (in Chinese)
10. Tan, H., Xu, S.: Multi-screen interactive experience of car navigation based on vehicular CPS. Packag. Eng. **38**(20), 17–22 (2017). (in Chinese)
11. Guilei, S., Qin, L., Yanhua, M., Linghua, R., Peiwen, F.: Design of car dashboard based on eye movement analysis. Packag. Eng. **41**(02), 148–153+160 (2020). (in Chinese)
12. Ping, Z., Yizhi, X., Yanyi, X.: Design of intelligent car dashboard based visual color bias. Design **35**(08), 123–127 (2022). (in Chinese)
13. Xiaoming, W., Xinbo, Z.: Eye movement prediction of individuals while reading based on deep neural networks. J. Tsinghua Univ. (Sci. Technol.) **59**(06), 468–475. (2019). (in Chinese)
14. Akshay, S., Megha, Y.J., Shetty, C.B.: Machine learning algorithm to identify eye movement metrics using raw eye tracking data. In: 2020 Third International Conference on Smart Systems and Inventive Technology (ICSSIT), pp. 949–955 (2020)

Recommendation of Sustainable Route Optimization for Travel and Tourism

Raja Kiruthika, Le Yiping$^{(\boxtimes)}$, Tipporn Laohakangvalvit$^{(\boxtimes)}$, Peeraya Sripian$^{(\boxtimes)}$, and Midori Sugaya$^{(\boxtimes)}$

College of Engineering, Shibaura Institute of Technology, 3-7-5, Toyosu, Koto-Ku, Tokyo 135-8548, Japan
{am20007,leyp,tipporn,peeraya,doly}@shibaura-it.ac.jp

Abstract. In our study, we aim to provide a sustainable route for tourism by considering the carbon emission by different transportation methods in an attempt to create a sustainable travel itinerary for each user. A model for route optimization between selected travel spots using an actual dataset of popular tourist spots in Tokyo, the road network, and available train connections is proposed. We try different vehicle routing algorithms such as Dijkstra's algorithm and genetic algorithm. This is followed by creating clusters for close points and providing a walking/bicycle route option. We test our model by comparing it with Google Maps. The result of this paper shows the relationship between time and carbon emission, and the improvement in terms of carbon emission between different routes.

Keywords: Route Optimization · Genetic Algorithm · CO2 emissions · Travel and Tourism

1 Introduction

The mobility and transportation industries have significantly improved as a result of global development and technological innovation. At the 26th Conference of the Parties of the United Nations Framework Convention on Climate Change in Glasgow in 2021, the currently growing issue of climate crisis was discussed. It was noted that the tourism industry among others is to play a crucial role in achieving net zero emissions [1].

When considering factors such as population growth, rising incomes, and increased access to various modes of transportation such as cars, trains, and airplanes, it is clear that transportation demand is going to experience significant growth in the upcoming years. According to the International Energy Agency's Energy Technology Perspectives report, the global demand for transportation, measured by passenger kilometers, is expected to double. Additionally, car ownership rates are predicted to rise by 60% and the demand for passenger and freight aviation is estimated to triple by the year 2070 [2]. This substantial increase in transportation demand will inevitably lead to a substantial rise in transport-related emissions.

H. Krömker (Ed.): HCII 2023, LNCS 14048, pp. 385–396, 2023.
https://doi.org/10.1007/978-3-031-35678-0_26

In recent years however the scale of traffic growth has been reduced as a result of policy initiatives, especially in urban centers. However, congestion and transport carbon dioxide (CO_2) emissions continue to rise in various areas around the world. Transportation modes such as walking, cycling, and bus use are usually static at best and often still in long-term decline.

The growing concern over climate change and the implementation of policies aimed at reducing carbon emissions have made it necessary for the transportation industry, including the tourism sector, to consider the impact of their operations on the environment. The low-carbon concept, which promotes low energy consumption, low pollution, and low emissions, has gained prominence. The transportation industry consumes a significant amount of energy and is a significant contributor to carbon emissions. As such, the tourism sector must take into account the costs associated with new environmental policies as well as fuel consumption costs when developing operations strategies.

In this context, there is a need for tourism route optimization that considers CO_2 emissions. By optimizing the routes taken by tourists, it may be possible to reduce energy consumption, lower carbon emissions, and reduce the costs associated with these emissions. Additionally, by reducing the environmental impact of tourism, the industry can demonstrate its commitment to sustainability and help to mitigate the effects of climate change.

Google Maps is a popular site that is used by travellers and tourists for planning their travel, generating travel routes, and coming up with a travel itinerary. It provides users with route recommendations from their selected origin spot to their destination. If users wish to add other destinations, they can search and add locations manually to their route. Users can create a travel itinerary of upto 10 spots manually and this generates a route for them. However, they do not get route recommendations to more than one destination at a time from Google Maps. The users must instead organize their itinerary in a suitable way that allows them to visit as many places as possible. Therefore, there is a need to optimize these results to provide users with a route recommendation that allows them to visit as many places as possible without the limit of 10 spots.

Our research goal is to provide the best route recommendation for tourists and travellers once they select their destination spots. In addition, we propose to provide sustainable route recommendations by combing transportation modes such as walking and cycling. To achieve this goal, we consider travel locations in Tokyo that tourists would be interested to visit. In order to provide an optimized route recommendation, we look into typical methods to solve vehicle routing problems such as the Dijkstra's algorithm and Genetic Algorithm. Once an optimized route is generated, different modes of transportation are considered and a travel route recommendation considering the combination of different transportation modes is provided.

2 Literature Review

The vehicle routing problem (VRP) is a generic name given to a set of problems in which the VRP aims to form a route with the lowest cost to serve all customers. The VRP is a major problem in distribution and logistics and was first described in 1959, by Dantizg and Ramser, and was called The Truck Dispatch Problem [3].

Zhang et al. studied the VRP with the consideration of fuel consumption and carbon emission, specifically incorporating the idea of fuel, carbon emission, and vehicle usage costs into the existing VRP [4]. Similarly, Pu et al. considered a multi-depot VRP with minimized logistic, fuel, and carbon emission costs [5]. Palmer established an integrated routing and carbon dioxide emission model for vehicle routes and highlighted the role of speed in reducing CO_2 emission [6]. Xiao et al. proposed a hybrid optimization model that combines a genetic algorithm to study time-dependent VRP to create vehicle scheduling plans with lower CO_2 emissions and fuel consumption [7].

In recent years, multimodal transportation has been recognized as an effective solution for being able to reduce carbon emissions. However, the idea of multimodal transportation is still in its infancy and has not been widely accepted by the public [8]. Xu et al. discussed how this concept can effectively solve travel demand by proposing a personalized multimodal travel service based on the concept of Smart Product Service System (SPSS).

Recommendation systems have been widely used in our daily life these days from movie recommendations on Netflix to music recommendations on Spotify. These recommendation systems can also be applied in the tourism and travel industry to provide users with travel route recommendation. Gunawan et al. carried out research on creating a recommendation system for tourist attraction places in Indonesian cities, specifically Malang where they give the shortest path for selected spots [9].

Based on the literature review, we propose to implement a route recommendation for travelers by looking into VRP and optimizing it considering CO_2 emissions and multimodal transport.

3 Route Optimization

Several studies have been carried out on finding routes using various algorithms, such as the Dijkstra's algorithm, the Floyd-Warshall algorithm, the Johnson's algorithm, the genetic algorithm, etc. Dijkstra algorithm and genetic algorithm are commonly used for vehicle routing problems [10]. Bagheri et al. compared Dijkstra's and Genetic algorithm in their paper and found that the genetic algorithm finds the shortest path faster than the Dijkstra's algorithm [11]. However, this is used for a different purpose and different dataset. In our research, we look into the Dijkstra's algorithm and genetic algorithm for developing our model using our dataset. Both algorithms are briefly described in the following subsections.

3.1 Dijkstra's Algorithm

Dijkstra's algorithm is an algorithm commonly used for finding the shortest path from one node to another based on a weighted graph. It returns the shortest path tree, containing the shortest path from a starting vertex to each other vertex, but not necessarily the shortest route that visits all other vertices. It is commonly used in research related to VRP and even recommendation systems. The algorithm has been used to plan the optimal path for an automated guided vehicle that was eventually able to minimize energy consumption and decrement of operation time [12].

However, our problem is not in finding the shortest way between two points, but in making a route between all the optimal points. Once we have the optimal route, we can use Dijkstra's algorithm to find the shortest path between each point in the route.

3.2 Genetic Algorithm

A genetic algorithm is a type of search strategy that takes inspiration from Charles Darwin's theory of evolution. The algorithm was first proposed by J.H. Holland, which mimics the process of natural selection where the fittest individuals are selected for breeding to produce the next generation [13]. Chromosomes hold an individual's genetic information and consist of genes that determine their traits. These traits are passed down to offspring through breeding. Individuals with advantageous genes will survive and pass on their traits to future generations, improving the overall traits of the population. Genetic algorithms apply this concept by combining existing solutions to find the optimal answer. Like other optimization algorithms, a genetic algorithm begins by defining the optimization variables, the cost function, and the cost It ends like other optimization algorithms too, by testing for convergence.

Pseudocode for Genetic Algorithm. The Code Below Explains the Pseudocode for the Genetic Algorithm [11].

```
Repeat
   Select two routes, route 1 and route 2 from R
   Crossover the route 1 * route 2 to C1, C2
   For each child path C do
      Mutate C with a small probability
      Replace one solution path in R with C
   End of for
Until the stopping criterion is satisfied
```

Genetic Representation. The Cost Matrix $W = [W_{ij}]$ specifies the costs or the weights which represent the cost of transmitting a packet on the link (i, j) [11]. The cost matrix can be defined as shown in the Eq. (1).

$$W_{ij} = N (if\ exists\ the\ link\ from\ node\ i\ to\ j\ is\ positive)$$
$$W_{ij} = 0 (otherwise) \tag{1}$$

In a genetic algorithm, all the results or solutions are stored as an individual chromosome. An example of a routing path encoded from a starting point to a goal is shown in Figs. 1 and 2.

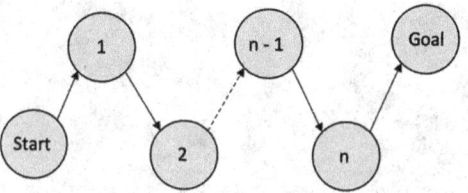

Fig. 1. Example of our routing path [11]

Fig. 2. Example of how a route is stored as a chromosome [11]

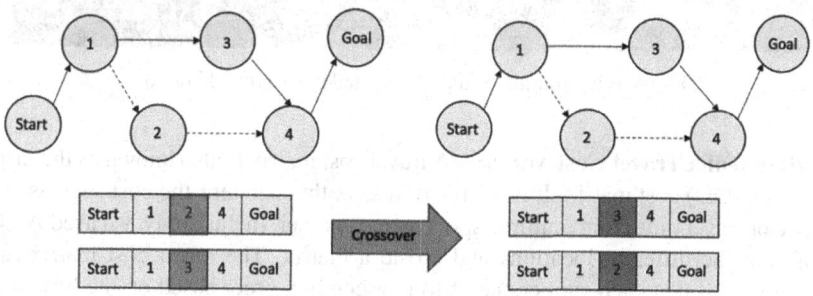

Fig. 3. Example of the crossover operation [11]

Figure 3 shows an example of the crossover operation. Crossover examines the current results by partially exchanging two routes to find better ones. This is carried out and is followed by mutations which randomly flip some of the paths in the route until the stopping criterion is met.

4 Methodology

This section presents our methodology for data collection and model construction to solve the VRP and recommend a suitable route path. The details are described next.

4.1 Data Collection

Travel Spots. For our travel spots, we randomly selected popular tourist locations in the area of Tokyo, Japan as shown in Fig. 4. These locations were used as the destination points in the model to be constructed in the next step. The origin and final spot is set as spot (1) which is Roppongi Hills. A total of 13 selected spots are (1) Roppongi Hills, (2) Hachiko Statue, (3) Mansei Bashi, (4) Yebisu Beer Brewery, (5) Asakusa Shrine, (6) Imperial Palace, (7) Tsukiji Outer Market, (8) Tokyo Station, (9) Akhibara, (10) Meiji Jingu Shrine, (11) Kanda Myoji Shrine, (12) Tokyo Skytree, and (13) Tokyo Tower.

Fig. 4. Distribution of the 13 selected spots around Tokyo.

Calculation of a Travel Cost Matrix. A travel cost matrix is also known as the origin-destination (OD) cost matrix. It is a table or matrix that contains the cost such as travel distance or travel time from multiple spots to one another. In this study, we used ArcGIS to map out the different locations and create a matrix. The travel cost matrix ranks the destinations that each spot connects to in ascending order based on the lowest cost required to travel between the two spots. Similarly, we also considered the train routes by creating a dataset of existing train stations and their transit points, with each node considering the distance between each station. In ArcGIS, we can make use of parameters such as travel mode that considers transportation modes such as driving a car/truck or walking.

4.2   Route Optimization

Model Construction using Dijkstra's and Genetic Algorithms. Using the collected travel spots and the calculated travel cost matrix, we implemented the model as discussed in the previous section using Dijkstra's and genetic algorithms.

For the Dijkstra's algorithm, travel spot 1 (Roppongi Hills) was selected to be our initial and destination nodes. The cost between each node was obtained from the OD cost matrix.

For the genetic algorithm, each of the tourist spots (1 to 13) represents the gene as explained in the previous section. The chromosomes have many possible solutions which are randomly composed of the genes (i.e., tourist spots), one of which includes our optimal path. These chromosomes were randomly initialized. Crossover and mutation probabilities were set as 0.6 and 0.2, respectively. The genetic operation was repeated until the best path was found.

Model Comparison. As a result, the models constructed by Dijkstra's and genetic algorithms gave the same route since the cost matrix was the same. Figure 5 shows the result of the best route obtained, which represents the route as follows: (1) Roppongi Hills → (4) Yebisu Beer Brewery → (2) Hachiko Statue → (10) Meiji Jingu Mae → (6) Imperial Palace → (3) Mansei Bashi → (11) Kanda Myoji Shrine → (9) Akihabara → (5) Asakusa Shrine → (12) Tokyo Skytree → (8) Tokyo Station → (7) Tsukiji Outer Market → (13) Tokyo Tower → (1) Roppongi. The starting point and end goal is set to be the same in this case as Roppongi hills.

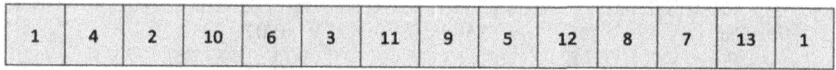

| 1 | 4 | 2 | 10 | 6 | 3 | 11 | 9 | 5 | 12 | 8 | 7 | 13 | 1 |

Fig. 5. Model result for the best route from the selected 13 travel spots.

4.3 Sustainable Route Recommendation

Overview. During traveling, the generation of a huge amount of CO_2 is due to transportation. To minimize the environmental impact of traveling, we try to find solutions that can evaluate the amount of CO_2 emission. Literature review suggests that considering multimodal transportation is a key factor to satisfy the increasing travel demand while also considering sustainability. It is said that the most efficient ways of travel are via walking, bicycle, or train. By using a bike instead of a car, CO_2 emission can be reduced by 75%, while taking the train for a medium-length trip can reduce CO_2 emission by 80% [14]. We thereby considered different transportation modes and kept in mind the CO_2 emission generated by each method of route, then used genetic algorithm to optimize the route. According to Navitime, the carbon footprint for trains is 20g/km. Table 1 shows the transportation modes and their related information.

Clustering. To incorporate sustainable approach into our research, we applied clustering method to the list of travel spots based on distance and transportation modes such as using walk or bicycle rental services instead. The k-means clustering algorithm, which is a method of vector quantization, originally came from signal processing, was used in this step. It is used to partition n observations into k clusters in which the observations belong to the cluster with the nearest mean (i.e., cluster center). A cluster center represents the cluster.

In our research, we have selected 13 spots hence the number of observations (n) is 13. In order to choose the appropriate number of clusters, there exist various methods such as the elbow method and the silhouette method. We did a preliminary trial using the number of clusters (k) ranging from 2 to 12. We use the inbuilt feature in ArcGIS to determine the optimal number of clusters. This is done using the Davies-Bouldin index if k value isn't specified, which optimizes it based on the similarities within a cluster and the differences between them. For the given data we use 2 and 4 clusters.

4.4 Model Verification

Google maps is a popular site for generating routes and travel plans among tourists after they decide on what places they want to visit. We use Google Maps to compare our

Table 1. Transportation modes and their related information.

Transportation Mode	Average Speed (km/hr)	CO2 emission (kg/km)
Walk	4	0.012
Bicycle	15	0.003
Motorcycle	27	0.03
Car	35	0.07
Bus	28	0.05
Train	36	0.02

model and test it. It tends to provide the user with a route from origin to destination. If the tourists want to add other locations, they can add them one by one with a maximum limit of 10 locations. However, it does not consider the best route plan and one would have to organize the itinerary. Using the results obtained for different transportation methods, we compare it with our models optimized result.

5   Results

Google Maps is used to obtain the general travel route. It provides a travel route with a distance of more than 50 km for just 9 spots as shown in Table 2. After applying the genetic algorithm to the same travel spots, we obtained a total distance of 45.6 km. Figure 6 shows the optimized route for the selected 13 spots. By applying the sustainable approach, although the time increased for the total travel, the CO_2 emission was reduced significantly depending on the modes of transportation.

Tables 2 and 3 show travel distances, times, and CO_2 emissions as the results obtained from the optimized routes by different transportation modes based on Google Maps and our proposed method, respectively. However, since Google Maps has a limit of generating routes only for 10 locations, we generated the route for spots 1 to 9 and back to spot 1. Also, Google Maps does not have the option for providing a train route for multiple locations. So, this value was not included in our result.

We generated sustainable route recommendation by combining different transportation modes. To do this, we selected transportation modes that yielded the least CO_2 emissions. In this order, the transportation modes such as car were removed, and we focused only on train, walking, and cycle. We employed the clustering method as described in Sect. 4 to generate the optimized routes while considering the distance and transportation modes. Figure 7 illustrates the generated routes using 2 and 4 clusters. We then generated the routes using two combination of transportation modes: (1) train + walk, and (2) train + cycle as seen in Tables 4 and 5.

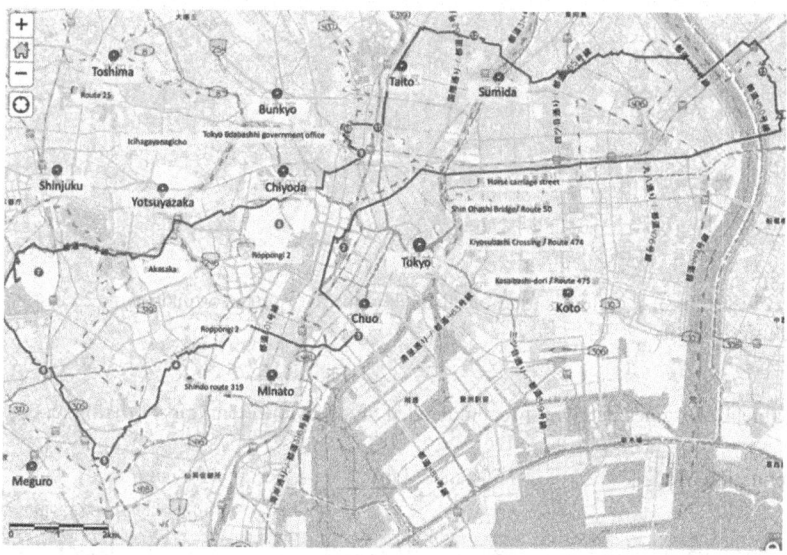

Fig. 6. A map that shows the optimized travel route for 13 randomly chosen popular tourist spots in Tokyo.

Table 2. Travel details for the optimized route of 9 spots using Google Maps.

Transportation Mode	Distance (km)	Time (hr:mins)	CO2 emission (kg/km)
Walking	55.8+	12:10	0.67+
Cycling	60.6+	4:31	0.18+
Car	71+	2:24	4.24+

Table 3. Travel details for the optimized route of 13 spots using our proposed method.

Transportation Mode	Distance (km)	Time (hr:mins)	CO2 emission (kg/km)
Walking	45.6	9:48	0.55
Cycling	45.6	4:45	0.24
Car	49	1:53	3.43
Train	61	3:49	1.22

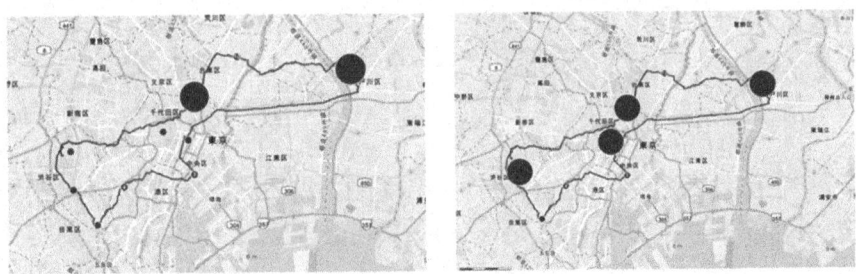

Fig. 7. Clustering of locations based on distance with the routes formed using 2 clusters (left) and 4 clusters (right).

Table 4. Distance details for recommended routes by different transportation modes.

Transportation Mode	Distance (km)			
	Train	Walk	Cycle	Total
Train + Walk (2 clusters)	57	4	-	61
Train + Cycle (2 clusters)	57	–	4	61
Train + Walk (4 clusters)	51	7	–	58
Train + Cycle (4 clusters)	51	–	7	58

Table 5. Carbon emission details for recommended routes by different transportation modes.

Transportation Mode	CO_2 emission (kg/km)			
	Train	Walk	Cycle	Total
Train + Walk (2 clusters)	1.14	0.05	–	1.19
Train + Cycle (2 clusters)	1.14	–	0.01	1.15
Train + Walk (4 clusters)	1.02	0.08	–	1.10
Train + Cycle (4 clusters)	1.02	–	0.02	1.04

6   Discussion

The comparison results between our proposed method using the model constructed by genetic algorithm and the Google Maps for optimizing travel routes are presented. Google Maps provided a travel route with a walking distance of more than 55.8 km for just 9 spots (Table 2), whereas our proposed method was able to generate a route that visits all 13 spots within a walking distance of 45.6 km (Table 3) which is more efficient reducing the distance by 9.4 km.

The model was also tested for walking, cycling, and train routes, which yielded the distances of 45.6 km, 45.6 km, and 61 km, respectively. The model created two sustainable routes by forming clusters of points close together and using the combination of walking, train, and cycle. For the transportation method of train + walk the 2-cluster hybrid route had a distance of 61 km (Table 4) and a CO_2 emission of 1.19 kg/km (Table 5), while the 4-cluster hybrid route had a distance of 58 km (Table 4) and a CO_2 emission of 1.10 kg/km (Table 5). Similarly for train + cycle, the 2-cluster hybrid route had a distance of 61 km (Table 4) and a CO_2 emission of 1.04 kg/km (Table 5), while the 4-cluster hybrid route had a distance of 58 km (Table 4) and a CO_2 emission of 1.10 kg/km (Table 5). In both hybrid routes, we see that the routes significantly reduced the total CO_2 emissions and distance when compared to google maps. In addition, the 4-cluster route of train + cycle reduced CO_2 emissions and distance.

The comparison results between our proposed method and Google Maps show that the optimized route created by our model constructed by genetic algorithm significantly lowered the distance compared to Google Maps result. The optimization not only reduces the distance but also leads to a decrease in CO_2 emissions.

The results of the recommended route after optimization show us that considering sustainable modes of transportation while generating routes can help reduce CO_2 emissions in the travel and tourism industry.

7 Conclusion

This research focuses on the recommendation of travel routes that provide an efficient way of travel in term of distance and CO_2 emission. Our goal was to create a sustainable route recommendation for travelers using a dataset of actual spots in Tokyo. We explored different route optimization algorithms and literature reviews and found the genetic algorithm to work the best. To improve the existing route recommendation, we formed clusters of nearby spots and recommended using sustainable transportation modes such as cycling or walking around to get to these locations. As a result, our proposed method for route recommendation suggests a better route than Google Maps in term of distance and CO_2 emission which contributes to a sustainability in travel and tourism industry.

For our future work, we will compare more route optimization algorithms and consider different methods for implementing a sustainable approach other than clustering. Furthermore, we will continue to create a travel recommendation system which also considers user preferences for generating the route.

References

1. Higham, J.P., Font, X., Wu, J.: Code red for sustainable tourism. J. Sustain. Tour. **30**(1), 1–13 (2021)
2. Our World in Data: Cars, planes, trains: where do CO_2 emissions from transport come from?.https://ourworldindata.org/co2-emissions-from-transport. Accessed 10 Feb 2020
3. Dantzig, G., Ramser, J.: The truck dispatching problem. Manag. Sci. **6**, 80–91 (1959)
4. Zhang, J., Zhao, Y., Xue, W., Li, J.: Vehicle routing problem with fuel consumption and carbon emission. Int. J. Prod. Econ. **170**, 234–242 (2015)
5. Pu, X., Lu, X., Han, G.: An improved optimization algorithm for a multi-depot vehicle routing problem considering carbon emissions. Environ. Sci. Pollut. Res. **29**(36), 54940–54955 (2022). https://doi.org/10.1007/s11356-022-19370-0

6. Palmer, A.C.: The development of an integrated routing and carbon dioxide emissions model for goods vehicles (2007). https://dspace.lib.cranfield.ac.uk/handle/1826/2547

7. Xiao, Y., Konak, A.: A genetic algorithm with exact dynamic programming for the green vehicle routing & scheduling problem. J. Clean. Prod. **167**, 1450–1463 (2017)

8. Xu, G., Zhang, R., Xu, S.X., Kou, X., Qiu, X.: Personalized multimodal travel service design for sustainable intercity transport. J. Clean. Prod. **308**, 0959–6526 (2021)

9. Gunawan, E.P., Tho, C.: Development of an application for tourism route recommendations with the Dijkstra algorithm. In: International Conference on Information Management and Technology, pp. 343–347. Jakarta, Indonesia (2021)

10. Yuan, C., Uehara, M.: Improvement of multi-purpose travel route recommendation system based on genetic algorithm. In: 7th International Symposium on Computing and Networking Workshops, pp. 305–308. Nagasaki, Japan (2019)

11. Bagheri, A., Akbarzadeh, T., Mohammad, R., Saraee, M.: Finding the shortest path with learning algorithms. Int. J. Artif. Intell. **01**, 86–95 (2008)

12. Kim, S., Jin, H., Seo, M., Har, D.: Optimal path planning of automated guided vehicle using Dijkstra algorithm under dynamic conditions. In: 7th International Conference on Robot Intelligence Technology and Applications, pp. 231–236. Daejeon, South Korea (2019)

13. Holland, J.H.: Adaptation in Natural and Artificial Systems. University of Michigan Press, Ann Arbor (1975)

14. Our World in Data: Which form of transport has the smallest carbon footprint? https://ourworldindata.org/travel-carbon-footprint. Accessed 10 Feb 2020

Platform Design for a Minimum Viable Academic Mobility as a Service Ecosystem

Lisa Kraus[✉] [iD] and Heike Proff[iD]

Chair of General Business Administration and International Automotive Management,
University of Duisburg-Essen, Duisburg, Germany
{lisa.kraus,heike.proff}@uni-due.de

Abstract. Mobility as a Service (MaaS) has been hailed as a new ecosystem that is transforming transportation systems and individual mobility behavior. However, due to new phenomena such as Covid-19, adoption in the mobility industry has stagnated, and not all ecosystems, i.e., novel partnership networks, could scale. The most important component of both ecosystems in general and MaaS is the platform that combines all other building blocks and enables exchanges between the supply and demand sides. This research therefore aims to design the platform of a minimum viable MaaS ecosystem using academic MaaS as an example, with a qualitative content analysis of supply-side interviews and demand-side focus group discussions and workshops. The focus lies on an academic offer, as MaaS is typically targeted at young individuals and persons with high education. The results show that the standardization of programming interfaces and the provision of user data for all participating institutions are key drivers.

Keywords: MaaS · Platform · Student Mobility · Germany

1 Introduction

Transportation is being transformed by digitalization [1]. In the form of information and communication technologies (ICT), it enables the provision of new mobility services via a platform [2]. Thus, ICT facilitates the entry of new service providers into the mobility market [3] and is changing the entire market in the direction of ecosystems [4].

Against the background of this change, Mobility as a Service (MaaS) has been discussed in academia and industry in recent years. First proposed in 2014 by Hietanen [5], MaaS is viewed as an ecosystem, a novel partner network with an overarching value proposition. The ecosystem perspective of MaaS has also been formulated in academia [6]. This is partly reflected in the most commonly used definitions: according to Hensher [7], MaaS consists of the definitional elements of budgets, bundles and brokers (plaform ecosystem orchestrators). Kamargianni and Matyas [8] defined MaaS as a "a user-centric, intelligent mobility distribution model in which all mobility service

Supported by the Ministry of Transport of the State of North Rhine-Westphalia [grant number 2020 18 111].

H. Krömker (Ed.): HCII 2023, LNCS 14048, pp. 397–407, 2023.
https://doi.org/10.1007/978-3-031-35678-0_27

providers' offerings are aggregated by a sole mobility provider, the MaaS provider, and supplied to users through a single digital platform" (p. 4). Jittrapirom et al. [9] compiled the key characteristics of MaaS, focusing on multiple players, demand-responsiveness, and one platform, among others. According to the analysis of Jovic and Baron [10], MaaS is classified by integration, individualization, collaborative consumption, and a platform. All these definitions have in common user-centricity (i.e. human interaction) and the presence of a digital platform, which requires a focus on human-computer interaction. MaaS was hit hard by Covid-19 [11], so many MaaS offers did not survive the pilot phase. A thorough analysis of how to design the platform of a minimum viable MaaS ecosystem is therefore essential for future success.

One user group of particular interest, due to their specific mobility behaviour [12] are university members, consisting of young students and academics [13] as the former are less interested in owning a car [14] and are more open to new sharing technologies [15]. To date, research on MaaS for students has focused only on the potential integration of carpooling in Brasil [16].

Therefore, a research gap becomes evident in terms of a) exploring platform design as an enabler of the MaaS ecosystem and b) focusing on an academic MaaS offer. This research gap is addressed by this contribution: The MaaS platform as a building block for designing the minimum viable ecosystem is explained in terms of strategy science and applied to an exemplary MaaS offer.

The contribution is organized as follows: Sect. 2 provides a background on the platform in a minimum viable ecosystem. The qualitative content analysis as methodology is described in Sect. 3. The results are presented and discussed in Sect. 4 before the contribution finishes with a conclusion and outlook in Sect. 5.

2 Background: Platform in a Minimum Viable Ecosystem

Ecosystems, a new form of value network in strategy science, are an emerging topic in academia and industry [17]. The concept of ecosystems was transferred from biology to business by Moore [18]. More recently, a unified understanding of ecosystems has emerged in strategy science: ecosystems are now understood as partnership networks in which an overarching value proposition (joint customer solution) is tailored by firms' individual value streams [e.g. 19, 20]. Jacobides et al. [17] defined ecosystems as "a set of actors with varying degrees of multilateral, non-generic complementarities that are not fully hierarchically controlled" (p. 2264). An emphasis is placed on modularization, and non-generic complementarities from complementarity resources must be present in both consumption and production. Adner [21] defined ecosystems as structure as "the alignment structure of the multilateral set of partners that need to interact in order for a focal value proposition to materialize" (p. 42). The importance of structural ecosystems is increasing exponentially in the literature [22]. Since not all ecosystems survive, it is important to create a minimum viable ecosystem [23], i.e., an initial liveable version of the ecosystem, and thus to think in advance about the precise design of focal building blocks [23–25]:

1. Focal value proposition;
2. Operating model;

3. Value drivers;
4. Governance mechanisms; and
5. Platform (see Fig. 1).

All five building blocks serve to create joint value and to capture it by the ecosystem players. The starting point for the design of the minimum viable ecosystem is the **focal value proposition** based on customer information [25] according to the service-dominant logic [e.g. 26], which states that individualization and integration in particular are part of a focal value proposition [27]. The customer learns the value proposition through the platform.

To deliver the proposed value to the customer, the required value elements are arranged in a particular way, dictating the **operating model** [24], with financial flows and relationships between multilaterally aligned partners [18, 23] and the platform at the center of the operating model [22, 28].

The **platform** is an intermediary for the direct exchange of information and resources to achieve platform effects through **value drivers** [23, 29]. Such value drivers are complementarities in production as supply-side economies of scope and complementarities in consumption as demand-side economies of scope [17]. In addition, modularization through standard interfaces [30], network effects as demand-side economies of scale [31] and price effects through cross-subsidization in multi-sided markets [32] are value drivers. A specific value driver for digital platforms is data-based learning enabled by artificial intelligence [33].

The interdependencies in multilateral ecosystem relationships are beyond the control of a single actor [21]. Therefore, **governance** mechanisms [17, 34] are needed to regulate platform access and control [21]. If the minimum viable ecosystem has demonstrated that the design phase has been successful, it enters the alignment phase; otherwhise, redesign is required.

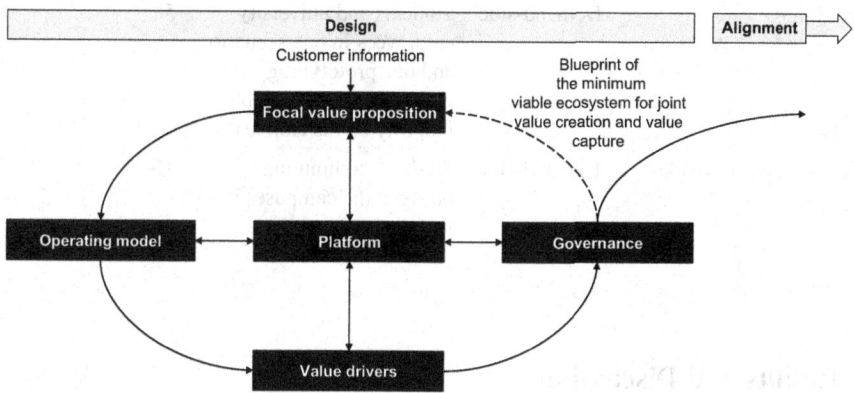

Fig. 1. Building blocks of the blueprint of a minimum viable ecosystem [Based on 20, 23–25]

3   Methodology: Qualitative Content Analysis

Information about the platform of a MaaS ecosystem and the categories of building blocks that influence or are influenced by the platform were obtained through qualitative research to identify how and why things happen [35]. The MaxQDA software was used for coding following the qualitative data analysis model by Kuckartz [36]. Strategies suggested by Merriam [35] to promote reliability and validity were used, such as triangulation using multiple sources, variation in sample selection, and peer review [37].

Since fully functional MaaS ecosystems are not yet available [38], it was not possible to evaluate a real-world application. Therefore, the study hypothesizes how an academic MaaS ecosystem should be built [39], using information from individuals and institutions with initial experience with MaaS applications. Therefore, the data collection was conducted as part of the research project "InnaMoRuhr" (Integrated Sustainable Mobility for the University Alliance Ruhr), which investigates the mobility of members of the three universities University of Duisburg-Essen, Ruhr-University Bochum and the Technical University of Dortmund in the metropolitan Ruhr region, Germany [40]. The qualitative data collected is composed of the sources shown in Table 1.

Table 1. Data sources for the qualitative analysis

Data source	Platform side	Participants	No. of participants
Semi-structured interviews	Supply-side	Bike-sharing (BS), car sharing (CS) and e-scooter sharing (ES) providers, public transport association (PTAs) and one public transport operator (PTO)	7
Workshops	Demand-side	Students and university employees in one scenario and one prototyping workshop on sustainable mobility on the campus	58
Focus group discussions	Demand-side	Students commuting between the campuses in two discussions	16

4   Results and Discussion

4.1   Platform

The platform for interaction was mentioned by many participants. In particular, the exchange of information between providers was mentioned: *"They get reports from us and can see where the bikes are, how many rentals have been made at which stations, and then they can also see exactly where the customers who register as [PTO] customers are traveling"* (BS provider). Information provision should be *"non-discriminatory"* so that all partners are equally visible (PTA1). Visibility was named as a benefit of participating in the platform (PTA1). This digital visibility was also considered by the scenario workshop participants. The importance of the software to the overall solution was also pointed out (CS provider). The shared information should be made available to the customer also (PTA2).

It was mentioned that resource sharing via a platform was already taking place to provide a complementary product between BS and public transportation, but without further specifying (BS provider). Scenario workshop participants said that *"collaboration with stakeholders"* in *"fleet allocation"* was a prerequisite. In this sense, collaboration leads to the ability to *"ultimately evaluate when is the station empty and when do we need to fill it up, when are too many lanes coming, when are too few coming"*. A link between software and available hardware is essential, as one focus group participant reckoned, *"because the main problem, I think, is the availability of the transportation modes"*. The data generated by a MaaS app should be used for resource planning by the sharing providers: *"The goal is to study where the main commuting flows are, which we can then individually incorporate into our network planning, station planning, and also the further expansion of the system, especially at the university locations"* (BS provider). Among the other mentions, the technical architecture of the platform was cited as the biggest issue (PTA1, scenario workshop participant), especially with regard to recent updates (CS provider).

4.2 Other Building Block Categories Interacting with the Platform

With regard to the focal value proposition, the platform appeared primarily in the context of the integration value proposition of multimodal mobility apps (BS provider). In this context, different levels of integration were mentioned (ES provider, see [41]) and the possibility of *"[integration at a] collection point [...], even if the app then might not have all depth integrated, that this way you have the ideal solution: one account for everything"* (BS provider). It is necessary for the integration value of the services, formulated as *"reservation, commissioning, payment, schedule information and route planning are handled via apps and smartphones"* (scenario workshop participant). Additional considerations for universities only could be the integration of students' semester and weekly schedules into route planning, as well as integration with existing university apps that also provide other information, such as about the libraries or the canteens and how to get there, as well as general routing on university campuses, as expressed by the focus group participants. The need to have multiple apps ("multi-homing") for the above steps should be avoided as this reduces the user experience for the customer (PTA1, ES provider, BS provider). This cannot be clearly confirmed in the literature [42]. However, the PTA would consider a necessary app change between the steps as competition and not complementarity. Subsequently, market research should be conducted among the university members themselves to determine if multi-homing is acceptable when full platform integration is not technically feasible. The BS provider expressed concerns about feasibility because deep integrations present technical difficulties: Integrating payments and invoices into an app is difficult, as are tax regulations. Partial integration already exist for other platforms. A filter should be implemented into the platform to increase individualization (prototyping workshop participant). Also, the app should be cleanly designed (ES provider). *"Usability plays a strong role in how well I can use it [...]"* (prototyping workshop participant). In terms of platform integration, only a few MaaS platforms can be classified as highly integrated so far [41].

Regarding the operating model, different platform orchestrators were mentioned as ecosystem leaders, e.g. the PTA would take on this role (PTA1), which was also requested by the participants of the prototyping workshop. The orchestrator should control the data and access to APIs (Application Programming Interfaces). PTA1 stated that a subscription for all modes is not desirable because it would reduce revenue and/or remain mostly with the PTO. This is inconsistent with the focal value proposition of an integrated offer. The CS provider stated that platform costs need to be redistributed across all partners. According to the ES provider, a small contribution or percentage could be paid if the orchestrator produces valuable data for the businesses. The value should be equitably redistributed based on the percentage of modes used (ES provider).

In terms of value drivers, modularization in joint value creation was stated as open sources (BS provider) for open data to provide APIs. The provider also mentioned the value added by this adaptability to other platforms. API integration was also named by the ES provider in relation to Google Maps, stating that *"such a Google Maps integration is a no-brainer, you just need to take some time for such a mini project"*. Modularization thus lowers transaction costs. PTA1 also mentioned their open API: *"This is our daily business, we have an open interface. There is both an interface that points directly to our information and to an open data portal, our information is made available again via a different format"*. Furthermore, PTA1 said that integrating other services into their own platform is preferred because of the speed, because *"the added value at this point would simply be if you had a system that already has several other partners integrated into it, that if you only have to connect that one system, it's significantly less effort than if you only have to connect each one individually"*. Private mobility service providers are less likely to use standards, which increases costs for PTOs: *"They say "here this is my interface" and then, the respective transport company has to pay for the effort, so to speak, to integrate that again"* (PTA2). In this context, the CS provider mentioned the need for willingness to cooperate: *"There is always a lack of interest from several sides to merge the individual software"*. It is important to *"break new ground"* (CS provider). According to PTA3, standardization is a key issue in digitalization. The need for open APIs was also expressed in the prototyping workshop. A modular physical change of transportation modes is already given by certain cooperation models: *"We are also together at some stations. At mobile stations, so there is a logical connection to public transport, which is often also a partner of ours [...]. And then there's also car sharing, which is also together with the [PTO] and then that's a complete package. And that's another place where it's worth sharing data."* (BS provider). Complementarity in production is high between PTO resources and cooperating sharing services, as noted by the PTO. PTA1 mentioned that the added value is only given *"if the additional traffic fills gaps that currently exist"*, contrary to the opinion of the BS provider who also emphasizes complexity. Complementarity in consumption is already partially implemented in other apps, as PTA1 noted: *"With the new developments that we have, we are now deviating from our old way, towards an information system where basically pure public transport information is offered and other transportation modes are added only at the request of the user.*" Hence, it is the user who actively expresses the desire to use certain other transportation modes. This was also stated by the ES provider as a prerequisite for increasing the integration value for the user. Co-creation is hence a prerequisite for individualization. The ES provider suggested an in-app customer survey to decide whether routing features should be integrated based on the potential benefit to the customer. Co-creation was also desired by the prototyping workshop participants: "*We imagined the app like Google Maps, only better. [...]*"*I only want to use mobility XY*"." This aspect also shows a link between the individualization value and co-creation, as complementarity in consumption drives value co-creation. Data-based learning was also mentioned by one participant of the second focus group as a nice-to-have. The need for open standards for MaaS to thrive has also been widely discussed in the literature [43]. In this regard, Germany is a progressive country in terms of API standardization, especially through initiatives such as the Mobility Data Space [44].

Governance was mentioned only in terms of the legal requirements that must be met. The influence of external governance is obvious: "*But there are now, for example, also system architectures that MaaS players could/should follow [...], the described standards for open mobility platforms*" (PTA2). Open data legislation must therefore be demanded at the state and federal level (PTA2): "*Especially when it comes to mobility and mobility data, a huge mountain of legal regulations has accumulated in recent years*" (PTA2). PTA3 corrobated this information: "*Legal requirements in the EU requirements are then implemented in national laws [...]. There has been a lot of movement in this direction, in the direction of the platform economy*".

The BS provider mentioned that the participation should lead to new knowledge and that all participants have to share their data. Open data as modularization thus interacts with knowledge-sharing routines. PTA1 saw a conflict here: "*It's simply because sharing service providers often want the customer data themselves, while the transport companies want to see the customer data.*" PTA1 would define success if "*it moves our systems forward*". Non-disclosure agreements by the firms to safe company data must also be adhered to (ES provider) while transparency is also required for customers (PTA1). Data compliance is an important aspect, especially due to customer privacy concerns [45]. Data governance is therefore important regarding data ownership [43]. All platform requirements of the ecosystem building blocks are summarized in Table 2.

Table 2. Platform design requirements for an academic MaaS ecosystem

Building block	Definition of platform requirements
Platform	Establishment of supply-side customer information exchange
	Digital visibility of suppliers to users
	Transparent information provision for customers about data shared
	Data analysis of commuter flows

(continued)

Table 2. (*continued*)

Building block	Definition of platform requirements
	Matching physical transportation modes with data
	Fleet allocation through cooperation with stakeholders
	Updated technical platform architecture
Focal value proposition	Integration via a collection point to have one account for everything
	Integration of university-specific information
	Campus routing
	Absence of multi-homing for specific customer groups
	Filter implementation in mobility service selection for route planning
	High usability
Operating model	Equal platform cost and proportional value distribution among supply-side
Value drivers	Open APIs of supply-side partners
	Standardization of supply-side platform architectures
	Machine learning algorithm implementation to propose personalized routes
Governance	Adherence to EU- and country-wide regulations on mobility data

5   Conclusion and Outlook

In summary, the most important aspect of platform design is the conflict of access to customer data. All supply-side partners want access to the data to optimize their individual operating models, e.g. regarding fleet allocation and utilization rates. However, if all partners have the data, none can have a competitive advantage. Coopetition is hence visible. One enabler for integration is the standardization of APIs, which has not yet been implemented by all micromobility providers. Therefore, binding legal requirements are necessary to encourage the development of MaaS apps. Level 3-integration is necessary to differentiate from existing MaaS apps.

Regarding the academic MaaS offer, the study showed that no outstanding differences are needed. What is needed is a geographical focus on the university campuses and the mobility services offered there. These modifications do not necessarily require platform adaptation.

The platform is the prerequisite for providing the MaaS service and distributing the value generated. Minimum viable mobility ecosystems often destroy value instead of creating it [46]. With regard to MaaS, no concept has been profitable to date, which is why most pilot projects have failed to survive [6]. Hence, network effects for scaling in mobility are difficult to achieve, as hardware availability is a limiting factor, especially due to fixed-step costs. As a result, additional novel revenue streams must be provided through the platform.

This is one topic for future research: These revenue streams should have a value proposition tailored to the students' and university employees' needs. In addition, the technical implementation of the platform requirements found in this study should be further explored. The potential of artificial intelligence regarding the MaaS offer and business process optimization should be analyzed in detail due to the critical role of data in the ecosystem. Complementing this, the economic feasibility of MaaS should also be investigated in terms of formulating the platform business model for the MaaS orchestrator. This study is the first to categorically examine the platform requirements for a minimum viable MaaS ecosystem and provides a good basis for doing so.

References

1. Stopka, U., Pessier, R., Günther, C.: Mobility as a service (MaaS) based on intermodal electronic platforms in public transport. In: Kurosu, M. (ed.) HCI 2018. LNCS, vol. 10902, pp. 419–439. Springer, Cham (2018). https://doi.org/10.1007/978-3-319-91244-8_34
2. Covarrubias, A.V.: When disruptors converge: the last automobile revolution. Int. J. Automot. Technol. Manage. 18, 81–104 (2018). https://doi.org/10.1504/IJATM.2018.092184
3. Stopka, U.: Multimodal mobility packages - concepts and methodological design approaches. In: Krömker, H. (ed.) HCII 2020. LNCS, vol. 12213, pp. 318–339. Springer, Cham (2020). https://doi.org/10.1007/978-3-030-50537-0_24
4. Adner, R., Lieberman, M.: Disruption through complements. Strategy Sci. 6, 91–109 (2021). https://doi.org/10.1287/stsc.2021.0125
5. Hietanen, S.: 'Mobility as a Service' – the new transport model? Eurotransport, vol. 12 (2014)
6. Hensher, D.A., Ho, C.Q., Mulley, C., Nelson, J.D., Smith, G., Wong, Y.Z.: Understanding Mobility as a Service (MaaS). Elsevier, Amsterdam (2020)
7. Hensher, D.A.: Future bus transport contracts under a mobility as a service (MaaS) regime in the digital age: are they likely to change? Transp. Res. Part A: Policy Practice 98, 86–96 (2017). https://doi.org/10.1016/j.tra.2017.02.006
8. Kamargianni, M., Matyas, M.: The Business Ecosystem of Mobility-as-a-Service (2017). https://www.google.com/url?sa=t&rct=j&q=&esrc=s&source=web&cd=&cad=rja&uact= 8&ved=2ahUKEwjiIIrSs9_3AhW1SPEDHW3HDIkQFnoECAgQAQ&url=https%3A% 2F%2Fdiscovery.ucl.ac.uk%2F10037890%2F1%2Fa2135d_445259f704474f0f8116ccb6 25bdf7f8.pdf&usg=AOvVaw3mgv8k8aMTRCkerg6vH-JU
9. Jittrapirom, P., Caiati, V., Feneri, A.-M., Ebrahimigharehbaghi, S., Alonso-González, M.J., Narayan, J.: Mobility as a service: a critical review of definitions, assessments of schemes, and key challenges. Urban Planning 2, 13–25 (2017)
10. Jovic, J., Baron, P.: Implications for improving attitudes and the usage intention of Mobility-as-a-Service - Defining core characteristics and conducting focus group discussions. ICoMaaS 2019 - Proceedings, pp. 182–197 (2019)
11. Das, S., Boruah, A., Banerjee, A., Raoniar, R., Nama, S., Maurya, A.K.: Impact of COVID-19: a radical modal shift from public to private transport mode. Transp. Policy 109, 1–11 (2021). https://doi.org/10.1016/j.tranpol.2021.05.005
12. Bonham, J., Koth, B.: Universities and the cycling culture. Transp. Res. Part D: Transp. Environ. 15, 94–102 (2010). https://doi.org/10.1016/j.trd.2009.09.006
13. Hafezi, M.H., Daisy, N.S., Liu, L., Millward, H.: Modelling transport-related pollution emissions for the synthetic baseline population of a large Canadian university. Int. J. Urban Sci. 23, 519–533 (2019). https://doi.org/10.1080/12265934.2019.1571432
14. Proff, H.: Multinationale Automobilunternehmen in Zeiten des Umbruchs. Springer Fachmedien Wiesbaden, Wiesbaden (2019)

15. Nash, S., Mitra, R.: University students' transportation patterns, and the role of neighbourhood types and attitudes. J. Transp. Geogr. **76**, 200–211 (2019). https://doi.org/10.1016/j.jtrangeo.2019.03.013

16. Gandia, R.M., Antonialli, F., Oliveira, J.R., Sugano, J.Y., Nicolaï, I., Cardoso Oliveira, I.R.: Willingness to use maas in a developing country. Int. J. TDI, vol. 5, 57–68 (2021). doi: https://doi.org/10.2495/TDI-V5-N1-57-68

17. Jacobides, M.G., Cennamo, C., Gawer, A.: Towards a theory of ecosystems. Strat. Mgmt. J. **39**, 2255–2276 (2018)

18. Moore, J.F.: Predators and Prey: A New Ecology of Competition. Harv. Bus. Rev. May-June, 75–86 (1993)

19. Leclercq, T., Hammedi, W., Poncin, I.: Ten years of value cocreation: an integrative review. Recherche et Applications en Marketing (English Edition) **31**, 26–60 (2016). https://doi.org/10.1177/2051570716650172

20. Dattée, B., Alexy, O., Autio, E.: Maneuvering in poor visibility: how firms play the ecosystem game when uncertainty is high. Acad. Mgmt. J. **61**, 466–498 (2018). https://doi.org/10.5465/amj.2015.0869

21. Adner, R.: Ecosystem as structure. J. Manag. **43**, 39–58 (2017)

22. Kapoor, R.: Ecosystems: broadening the locus of value creation. J. Organ. Des. **7**(1), 1–16 (2018). https://doi.org/10.1186/s41469-018-0035-4

23. Jaspers, D., Knobbe, F., Proff, H., Salmen, S., Schmid-Szybisty, G.: Prototypisches Ecosystem für die induktive Taxi-Ladung. In: Proff, H. (ed.) Induktive Taxiladung für den öffentlichen Raum, pp. 181–162. SpringerGabler, Wiesbaden (2023)

24. Lewrick, M., Link, P., Leifer, L.: The Design Thinking Playbook: Mindful Digital Transformation of Teams, Products, Services, Businesses and Ecosystems. Wiley, Hoboken (2018)

25. Adner, R.: Winning the right game. How to disrupt, defend, and deliver in a changing world. The MIT Press, Cambridge, Massachusetts (2021)

26. Lusch, R.F., Vargo, S.L.: Service-dominant logic: reactions, reflections and refinements. Marketing theory, vol. 6 (2006)

27. Kraus, L., Proff, H., Giesing, C.: Composition of a Mobility as a Service offer for university students based on willingness to pay and its determinants. Int. J. Automot. Technol. Manage. **23**, 1–30 (2023, in press). https://doi.org/10.1504/IJATM.2023.10053474

28. Rochet, J.-C., Tirole, J.: Platform competition in two-sided markets. J. Eur. Econ. Assoc. **1**, 990–1029 (2003)

29. Hagiu, A., Rothman, S.: Network effects aren't enough. Harv. Bus. Rev. **94**, 17 (2016)

30. Hagiu, A., Wright, J.: Multi-sided platforms. Int. J. Ind. Organ. **43**, 162–174 (2015). https://doi.org/10.1016/j.ijindorg.2015.03.003

31. Katz, M.L., Shapiro, C.: Network externalities, competition, and compatibility. Am. Econ. Rev. **75**, 424–440 (1985)

32. Eisenmann, T., Parker, G., van Alstyne, M.W.: Strategies for Two-Sided Markets. Harv. Bus. Rev. October, 1–11 (2006)

33. Iansiti, M., Lakhani, K.R.: Competing in the age of AI. Strategy and leadership when algorithms and networks run the world. Harvard Business Review Press, Boston, MA (2020)

34. Cusumano, M.A., Gawer, A., Yoffie, D.B.: Can self-regulation save digital platforms? Ind. Corp. Chang. **30**, 1259–1285 (2021). https://doi.org/10.1093/icc/dtab052

35. Merriam, S.B.: Qualitative research. A guide to design and implementation. Jossey-Bass, San Francisco (2009)

36. Kuckartz, U.: Qualitative text analysis: a systematic approach. In: Kaiser, G., Presmeg, N. (eds.) Compendium for Early Career Researchers in Mathematics Education. IM, pp. 181–197. Springer, Cham (2019). https://doi.org/10.1007/978-3-030-15636-7_8

37. Campbell, J.L., Quincy, C., Osserman, J., Pedersen, O.K.: Coding In-depth Semistructured Interviews. Sociological Methods Res. **42**, 294–320 (2013). https://doi.org/10.1177/004912 4113500475

38. Karlsson, I., et al.: Development and implementation of mobility-as-a-service – a qualitative study of barriers and enabling factors. Transp. Res. Part A: Policy Practice **131**, 283–295 (2020). https://doi.org/10.1016/j.tra.2019.09.028

39. Talmar, M., Walrave, B., Podoynitsyna, K.S., Holmström, J., Romme, A.G.L.: Mapping, analyzing and designing innovation ecosystems: the Ecosystem Pie Model. Long Range Plan. **53**, 101850 (2020). https://doi.org/10.1016/j.lrp.2018.09.002

40. Handte, M., et al.: Visualizing urban mobility options for InnaMoRuhr. In: Proff, H. (ed.) Transforming Mobility - What Next? Technische und betriebswirtschaftliche Aspekte. Research, pp. 645–658. Springer Gabler, Wiesbaden, Germany (2022). https://doi.org/10. 1007/978-3-658-36430-4_37

41. Sochor, J., Arby, H., Karlsson, I.M., Sarasini, S.: A topological approach to Mobility as a Service: A proposed tool for understanding requirements and effects, and for aiding the integration of societal goals. Res. Transp. Bus. Manag. **27**, 3–14 (2018). https://doi.org/10. 1016/j.rtbm.2018.12.003

42. Park, J.H., Kang, Y.J.: Evaluation index for sporty engine sound reflecting evaluators' tastes, developed using K-means cluster analysis. Int. J. Automot. Technol. **21**(6), 1379–1389 (2020). https://doi.org/10.1007/s12239-020-0130-8

43. MaaS Alliance: Data makes MaaS happen. MaaS Alliance Vision Paper on Data (2018). https://maas-alliance.eu/wp-content/uploads/sites/7/2018/11/Data-MaaS-FINAL-after-ple nary-1.pdf

44. Tagesspiegel Background: EU: Das steht 2023 auf der Mobility Agenda (2018). https://www. google.com/url?sa=t&rct=j&q=&esrc=s&source=newssearch&cd=&cad=rja&uact=8& ved=2ahUKEwiT7avj7eT8AhXzX_EDHSmUAZQQxfQBKAB6BAgaEAE&url=https% 3A%2F%2Fbackground.tagesspiegel.de%2Fnewsletter%2F3j13rXmKbvMJsQ4NDbP aud&usg=AOvVaw2XoCd8_5QqcNwvae3rvt8k

45. Cottrill, C.D.: MaaS surveillance: Privacy considerations in mobility as a service. Transp. Res. Part A: Policy Practice **131**, 50–57 (2020). https://doi.org/10.1016/j.tra.2019.09.026

46. Schulz, T., Zimmermann, S., Böhm, M., Gewald, H., Krcmar, H.: Value co-creation and co-destruction in service eosystems: the case of the reach now app. Technol. Forecast. Soc. Chang. **170**, 120926 (2021). https://doi.org/10.1016/0040-1625(88)90012-1

Research on the Design Framework of Bike-Sharing App Based on the Theory of Perceived Affordances

Miao Liu and Yufeng Wu[✉]

East China University of Science and Technology, Shanghai 200237, People's Republic of China
756554566@qq.com

Abstract. With the continuous development of bike-sharing apps, the competition has become more and more fierce, and the products tend to be homogenized. The study also aims to build a framework of "demand-perceived-performance-function" in bike-sharing apps, and to point out the core dimensions of user experience design in bike-sharing apps. The study helps to improve the exploration of perceptual schematic theory in the field of information interaction and shared mobility, and provides a reference basis for the current mainstream user experience design of bike-sharing apps.

Keywords: Perceived Affordances · User Experience Design · Shared Bicycle · Content Analysis

1 Introduction

Shared bicycle is a new product in the era of sharing economy, which mainly refers to the provision of bicycle sharing and rental services in cities and campuses and other places, mainly to solve the problem of last-mile travel. At present, the domestic vehicle search is carried out through the shared bicycle app, and the bicycle is unlocked by one key using intelligent methods such as sweeping code, and the emergence of shared bicycle app has provided great convenience and security for the general public in their daily travel. According to the 50th Statistical Report on Internet Development in China, as of June 2022, the number of Internet users using cell phones to access the Internet in China reached 1.051 billion [1], of which 300 million are bike-sharing users, and this number is still growing, and bike-sharing apps have become an indispensable digital application in people's daily travel. However, there are more than 100 bike-sharing apps on the mainstream mobile application market, and the homogeneity among these apps is serious [2], which has intensified the market competition to a certain extent. To increase user satisfaction and thus product competitiveness, it is necessary to conduct user experience evaluation for bike-sharing apps.

Research on bicycle sharing at home and abroad is mainly focused on operation and management, and research on user experience is mostly focused on the analysis of users' overall experience of bicycle sharing process, and lacks research focusing on

H. Krömker (Ed.): HCII 2023, LNCS 14048, pp. 408–427, 2023.
https://doi.org/10.1007/978-3-031-35678-0_28

software interaction. The interaction design theory is especially important to improve the experience of bicycle sharing APP because they involve multiple stakeholders, multiple scenarios and multi-dimensional interactions. Therefore, this paper introduces the theory of perceived affordances into the study of user experience of bike-sharing APP. It was first proposed by ecologist Gibson J in 1979 to discuss a set of attributes in the environment that can be perceived and used by actors [3]. Norman then introduced the theory of sensibility to the field of industrial design, explaining its connotation from the perspective of user experience, emphasizing that perceived sensibility is the subjective feeling and evaluation of users on the sensibility of product design, and designers should pay more attention to the cognitive experience of users [4]. With the expansion of theories, perceived affordances has been gradually enriched to the fields of human-computer interaction and information behavior practice. Therefore, there is a close relationship between perceptual sensibility and user experience, and the introduction of the theoretical framework of perceived affordances can, to a certain extent, guide designers to understand users' experience requirements and continuously optimize the user experience of products. Based on this, this paper introduces the version iteration data and the latest version function data of three mainstream bike-sharing apps (Hello Bike, Meituan Bike and Qingju), and forms the conceptual framework of "demand-perceptual performance-function" of bike-sharing apps based on content analysis, so as to provide reference for system developers in version iteration. This will provide reference for system developers in version iteration.

2 Literature Review

2.1 User Experience Design of Bicycle Sharing App

The concept of user experience first emerged in the field of human-computer interaction in the 1940s and is based on usability and user-centered design. User experience includes the whole process of user interaction with a product throughout its life cycle. Enhancing user experience can greatly increase the utility and satisfaction of users when interacting with products [5].

At present, there are few research on the user experience design of shared bicycle APP at home and abroad, and most of the literature focuses on the overall user experience of the shared bicycle use process, which includes the hardware end and software end, as well as the operation and management level. For example, foreign scholars SUN C et al. [6] proposed the method of shared bicycle operation mode, in which users select the riding mode through APP and scan and unlock the bicycle through QR code and settle the fees, which ensures the efficiency of travel time and improves the user riding experience. XIE Y [7] proposed a bicycle sharing parking lock system that connects and provides information through a smart lock system, a personal APP and a cloud platform to ensure the parking location of dockless shared bicycles, avoiding the problem of random parking and difficult management of shared bicycles. TANG Y [8] proposed a method to prevent excessive gathering of shared bicycles for parking, through a shared bicycle APP, encouraging users to park in the collection area for a short time and short distance to dismantle shared bicycles, and this method can effectively reduce the operating costs of shared bicycle enterprises and avoid collective over-parking. HE F et al. [9] studied a shared bicycle parking device for autopilot, interconnecting the vehicle intelligent management device, autopilot device, and cell phone APP, thus providing a device that can determine whether the vehicle is parked in a legal area.

Domestic scholars Zhao Li et al. [10] analyzed the operation status of shared bicycles and focused on user experience. By comparing Mo-bike, OFO, and government public bicycles, they proposed that the reasons for poor user experience include vehicle damage problems, GPS finding vehicle problems, and refunding deposit process problems, and put forward relevant suggestions to improve user experience. Jiang Yujie et al. [11] analyzed the user satisfaction of bicycle sharing users and showed that user experience, security and facility level positively affect user satisfaction, thus suggesting that bicycle sharing enterprises should conduct regular promotions for users, establish regulations for protecting users' personal information, reasonably place bicycles with the help of big data technology, and provide real-time and accurate information about bicycles at each parking site. Based on the perspective of user experience, Zhu Yong [12] constructed an evaluation index system of natural and social factors, used quantitative research methods to study the influence of each factor on the number of users, and developed a bike-sharing pricing strategy. Yuehui Wang [13] obtained users' evaluation of bicycle sharing experience quality through text mining technology and constructed a quality concept model for measurement and evaluation, and concluded that experience consists of three parts: emotional experience, word-of-mouth experience, and utility value, and guided management practice.

In summary, the research on user experience of bike-sharing APP at home and abroad mainly focuses on exploring new operation and management modes and technical support to solve the pain points of the enterprise and user side, but lacks a systematic theoretical guiding framework to improve the interaction experience of the software side, therefore, the introduction of the perceptual schematic framework can provide a reference basis for this purpose.

2.2 Research on the Theory and Application of Perceived Affordances

The concept of Affordance was first proposed by Gibson [3], an ecological psychologist, to refer to the substances provided and supplied to animals by a specific environment. Later, Norman [4] introduced the theory of Affordance to the field of industrial design, emphasizing that it is better to discuss real oscillatory properties in everyday interaction design than to focus on perceptual oscillatory properties, i.e., designers should pay more attention to the user's perceptual experience. Foreign scholars such as Boy [14] introduced the Perceived Affordances into information visualization design, proposed design principles and verified its rationality by using experiments. Domestic scholars such as Zhao Yuxiang [15] introduced the framework of perceptual performance to the user experience evaluation of mobile music APP, and obtained the user experience rating ranking of four mainstream music APPs by constructing user evaluation indexes and using AHP-entropy weighting method to analyze the index evaluation, which provides reference for future related research. Based on the previous research, Zou Zhenbo [5] and other scholars introduced the perceptual schematic framework into the user experience design research of mobile map APP, coded the version iteration data of three mainstream mobile map APPs in China by using content analysis method, constructed the perceptual schematic framework of mobile map APP, and obtained the core design dimensions of mobile map APP, which provides reference for its future user experience design.

In summary, this study introduces the theory of Perceived Affordances from the perspective of user needs to the study of bike-sharing APP framework, and constructs the theoretical framework of "needs-perceptual performance-function" for bike-sharing APP.

3 Study Design

3.1 Research Methodology

The version iteration data of the bike-sharing app can trace the problems and improvement directions that product operators and designers pay attention to in terms of product functions, processes and experiences, and the interpretation of the version iteration data from the users' perspective can reflect the actual experience and real feelings of users on the bike-sharing app to a certain extent [15]. In this study, three users with rich experience in using bike-sharing apps were invited as coders to analyze the version iteration data of bike-sharing apps using content analysis method and analyze the connection between the functional design of bike-sharing apps and users' needs using the theory of perceived schematic performance.

This study is divided into three stages, namely, open coding, spindle coding and theoretical mapping. To ensure the rationality of the first two stages of coding, the reliability of the coding needs to be tested. In this paper, Cohen'sKappa coefficient was selected as a reliability indicator to test the coding results. Cohen'sKappa coefficient is a statistical measure of the coder's qualitative labeling and categorization of content in terms of fit [16]. Specifically in this study, Cohen'sKappa coefficient reflects the degree of agreement between two coders in classifying N concepts into C category categories, as shown in Eq. 1.

$$K = \frac{Pr(a) - Pr(e)}{1 - Pr(e)} \tag{1}$$

where $Pr(a)$ is the relative percentage of observed agreement between the two coders, as shown in Eq. 2.

$$Pr(a) = \sum_i \frac{a_{ii}}{N} \tag{2}$$

where a_{ii} is the number of coding results where both coders agree and N denotes the total number of concepts.

$$Pr(e) = \sum_i \frac{n_{.i}}{N} \times \frac{n_{i.}}{N} \tag{3}$$

$Pr(e)$ is the percentage of consistency expected by the two coders, as shown in Eq. 3.

where $n_{.i}$ and $n_{i.}$. Denote the total number of rows and columns corresponding to the ith category in the coding result matrix, i.e., the number of concepts for the two coders in the ith category, respectively. If the coders' coding results are perfectly consistent,

K = 1; if the coders have no consistent coding results except for the expected chance consistency, K = 0. Usually, the coding results have good reliability when the value of K is greater than 0.75. If the reliability of the coding results did not reach 0.75, recoding was required.

3.2 Data Acquisition and Processing

As of January 2023, the top three apps in the travel bike-sharing App category on the iOS platform and the travel bike-sharing App category on the Xiaomi App Store, based on user rating ranking, installation count ranking and overall ranking, are Hello, Meituan and Drip Qingju, with ratings exceeding 4.5 points and a total number of installations exceeding 4.5 billion. Thus, it seems that Hello, Meituan and Qingju have good representativeness and typicality, but since the bike-sharing business of Meituan and Hello are conducted as sub-modules of the software, picking the updated iteration data about bike-sharing in them as a reference, this paper selects the version data of the above three bike-sharing apps as a data source for research. This paper uses Octopus data crawler to crawl the version data from January 2018 to January 2023, and saves the data in an Excel spreadsheet, including version number, update time and update log, etc. Sample

Table 1. Sample version iteration data crawled

Shared Bicycle App	Version number	Version update date	Example of update log
Ha Luo	6.32.1	2019/06/16	Hello Bike] June 18 bike riding task online, daily ride to win cash rewards, more ride more earn - June 17 national free ride public welfare day, "Yi ride wake up hearing impaired children small ears"
			[Hello bicycle] bonus car function is online, 100 yuan wool, welcome to grip
			The new version of the home page of Hello Windy Ride is more convenient to book a windy ride -Automated orders for owners are fully online -Friday windy day, passengers get \$12 off and owners get a big subsidy -Invite friends to send cash, more invitations, more money
			Share games to earn energy and help you adopt a bike soon
			Hello Station] More interactive forms, voting function is online!

(continued)

Table 1. (*continued*)

Shared Bicycle App	Version number	Version update date	Example of update log
MeiTuan	12.6.406	2019/12/16	Must experience] 1. Meituan sweep code free deposit to use the car, support to unlock all the orange and yellow models of Mobike bicycle 2. Search "hot spring hotel", receive 50 yuan red packet, dating, walking children, friends get together must sleep in a good hotel are here!
DidiQingju	3.7.8	2023/01/03	No deposit at 0 threshold, easy to scan the code, instant departure This issue optimizes some of the content features and provides a better riding experience, so come and explore the new city!

crawl results are shown in Table 1. After crawling the data, the data content is cleaned, excluding meaningless special symbols and other irrelevant descriptive information, and only retaining information about new function launch The cleaned data results are summarized in Table 2. At the same time, the latest versions of the three mobile map software were selected for installation on the smartphone side, and each function was manually analyzed and recorded according to the software function structure, and incorporated into the processed version iteration data to finally form the data set of this study.

Table 2. Results of cleaned version iteration data

Bicycle Sharing App Platform	Version data time span	Number of crawled versions	Valid data bars
Ha Luo	2018/01/07–2023/01/07	405	121
MeiTuan	2018/01/07–2023/01/07	371	42
DidiQingju	2018/01/07–2023/01/07	238	38

4   Data Analysis

To ensure the reliability of the coding, the coders were first screened and trained, and three coders were finally identified to form the coding team for this study. All three coders had more than 5 years of experience in using bike-sharing apps and had used more than three bike-sharing software, and had a deep understanding of the bike-sharing app in this study. In the first stage of open coding, two members of the coding team collaborated to refine and code the functional concepts in the data and summarize the application categories based on their own knowledge and experience of using the product; in the second stage, the coders used the conceptual framework based on perceived performance [15] as the main category template for this stage of coding, and categorized and sorted out the application categories; in the third stage, the coders drew on a number of psychological theories. In the third stage, the study will interpret the user's needs when adopting and using the product with the help of some psychological theories, and map them with the main category of perceptual indicativity, and finally construct the "need-perceptual indicativity-function" framework of the bicycle sharing app.

4.1   Open Coding Based on Version Iteration Data

1) Coding process. The purpose of this phase of open coding is to comprehensively sort out and summarize the functions in the popular bike-sharing app today. Based on careful study and understanding of the dataset, members of the coding team collaborated on coding naming the data bars as the unit of analysis. To ensure that all concepts have clear meanings and are mutually exclusive, the coding team changed or eliminated functional concepts with similar meanings in the coding results, and functional concepts that are exactly the same or essentially the same were merged to finally obtain all functional concepts. Due to space limitation, only some of the coding results are listed in Table 3. Then, the three coders collaborated to further refine all functional concepts into categories. After the initial determination of the application categories, two coders independently coded the correspondence and performed a reliability test. If the confidence level was met, the application categories were clear and covered all functional concepts and could be collated as the final result; if the confidence level was not met, the third coder initiated a discussion to revise the concepts of the application categories and then recoded the correspondence.

2) Coding results. In this stage, the three coders obtained a total of 57 functional concepts, numbered from a1 to a57. All functional concepts were then organized and summarized to form application categories. After several adjustments and calculations, the kappa coefficients of each application category in the final coding results were all greater than 0.75, indicating that the coders basically agreed among themselves and the coding results had a high degree of reliability. A total of 20 application categories (A1 to A20) were coded, as shown in Table 4.

4.2   Spindle Coding Based on the Theory of Perceptual Oscillatory Properties

1) Coding process. The purpose of this phase of coding is to further deduce and generalize the results of the previous phase of coding and to link them with the perceptual schematic theory in user experience design. Specifically, this phase of coding is based on the theoretical framework of perceptual schematics [15], and identifies perceptual physical schematics, perceptual cognitive schematics, perceptual affective schematics, perceptual control schematics, and perceptual participation schematics as the master category concepts for this phase of coding. Then, after ensuring that

the coders learned and understood the connotations of each master category concept, this study asked the coders to organize the application categories from the previous round of coding results and to correspond to the master category concepts. However, it is worth noting that application and perceptual energetics are not inherently many-to-one relationships, and an application can have multiple perceptual energetics in different contexts [17]. In this study, for example, the community space in the bike-sharing app is designed for users to facilitate their interaction and connection with the product and other users to a certain extent, which also increases the emotional value of using the product; at the same time, the community space design of the product also allows them to experience the connection and communication from the product designer and the society, which results in positive emotions. Thus, the community space in the bike-sharing app can be classified as either perceived control or perceived emotion. However, considering the practicality of the study and the simplicity of the results, the coders were asked to code the scope of the app to correspond to the core of its perceived omnipotence.

2) Coding results. After defining the concept of application categories and understanding the connotations of the main categories of perceptual representativeness, the three coders organized 20 application categories such as usage pattern management into the five main categories of perceptual representativeness. In the reliability test of the coding results, the Kappa coefficients of each main category were all greater than 0.75, indicating that the consistency among the coders was satisfactory and the reliability of the coding results was high. The specific application category connotations and categorization results are shown in Table

Table 3. Results of functional concept coding (partial)

Functional concept	Functional concepts to be incorporated	initial data
a12 car return process optimization	Car return process upgrade It's easier to find a return point	Hello bike: Did you ever hear the first blessing of 2018? this time, I also brought you the New Year new experience return process more smooth, experience new upgrade Qingju Bike: Optimize the experience of finding the return point in cycling, make it more convenient to find the return point, optimize the positioning and refresh logic, and make it more accurate positioning

(continued)

Table 3. (*continued*)

Functional concept	Functional concepts to be incorporated	initial data
a8 trip is safe	Cycling security Safety warning	Qingju Bike: Add a new "emergency contact" function to enhance the security of cycling. After adding an emergency contact, the contact can call the street Rabbit customer service to confirm your travel information; in an emergency, the customer service will contact the contact Hello bike: new safety warning in the trip, pay close attention to the safety of the trip, and make timely reminder of abnormal situations;
a9 zero threshold of deposit free	Heavy welfare No pressure from now on	"Hello bike: the national deposit free", from now on no "bet" force ~ this is a small ha silently hold back for a long time the big move! The weather is warm, the sun is just right, quickly ride on our small car to the place you want to go, see the people you want to see! Qingju bike: 0 threshold free deposit, easy to scan the code, immediately start! This issue will optimize some functions to provide a better cycling experience. Come to explore the novelty of the city Meituan bike: Meituan scan code free deposit car, no matter Mobike Meituan, orange and yellow or bike moped, with Meituan, can drive!
a4 UI interface optimization	Parking point display optimization Optimize the ride-ui interface	Meituan bike: optimize the display strategy of parking points and no-parking areas, and make the car more convenient! Qingju Bike: Optimize the UI interface in cycling, improve the experience of finding and returning points, optimize the return process experience, and reduce the probability of users touching the order by mistake

Table 4. Phase I open coding results

Application category (Kappa coefficient)	Functional concept
A1 Location reservation function (0.903)	A 1 query the vehicle in the area, A 2 reserve the vehicle, A 3 locate any position…
A2 Bluetooth Lock Lock Service (0.876)	A 4 Bluetooth unlock, A 5 scan code for unlocking, A 6 remote unlock, A 23 scan code for car, A 22 number for unlock…
A3 Online Payment Service (0.931)	A 27 wallet, A 29 billing rules, A 33 free secret payment, A 35 ride before pay, A 42 invoice…
A4 Intelligent Recommended Parking Solution (0.847)	A 9 red envelope car, a12 application parking, a43 find parking…
A5 Message Push Service (0.915)	A 32 special purchase card, A 40 cal consumption…
A6 precise positioning (0.923)	A 8 search, A 9 red envelope, A 24 refresh current position…
A7 Site Navigation (0.875)	A 43 look for parking spots, a24 refresh the current position, A 3 locate any position…
A8 Vehicle status function (0.922)	A 44 vehicle failure, A 46 vehicle remaining power, A 45 vehicle normal A 45 vehicle normal…
A9 Temporary Parking (0.826)	A 12 apply for parking spot, a24 refresh current location…
View of vehicle / parking area near A10 (0.935)	A 31 map car cars, a57 find parking…
A11 Fun Small Features (1.000)	A 14 carbon traveler, A 18 sign-in courtesy, A 36 red envelopes…
A12 fault repair report (0.824)	A 10 customer service Center, A 11 fault fault report, A 13 fault violation report…
A13 deposit mechanism (0.951)	A 20 learn cycling rules, A 21 car rules…
A14 intelligent billing (0.922)	A 19 buy a monthly card, a37 bike card…
A15 itinerary sharing (0.962)	A 25 cumulative cycling, A 26 usage time, A 28 travel route, A 38 travel record…
A16 Multiple interaction pathways (1.000)	A 49 gesture operation return, A 55 voice reminder, A 56 parking record location…
A17 Current Affairs-Based Services (0.978)	A 15 Today today, A 16 outbreak map, A 17 Mission Center…
A18 Content Creation (0.921)	A 47 cycling route, A 48 Riding Life, A 50 activity square…
A19 puts out its own views (0.933)	A 51 Travel pk, A 52 Challenge, A 53 Cycling today…

(continued)

Table 4. (*continued*)

Application category (Kappa coefficient)	Functional concept
A20 Member rights system (0.918)	A 7 low carbon members, a30 personal information collection list, A 32 special purchase card, A 34 cycling rights...

Table 5. Second stage spindle coding results

Main category (Kappa coefficient)	Application category	Application category connotation
Perceived physical performance (0.927)	A16 Multiple interaction pathways	Provide a variety of human-computer interaction channels, to mobilize people's vision, listening, smell
	A4 intelligent recommended parking solution	Recommend the parking place and scheme for users to facilitate their timely parking processing
	A2 Bluetooth lock unlocking service	Provide users to open the Bluetooth after opening the service, improve the fluency of use
	The A13 deposit mechanism	It provides users to provide credit free service before use to facilitate subsequent use
	A12 fault repair report	After the user uses the damaged vehicle, it is convenient for follow-up maintenance and improve the user experience
Perceived cognitive performance (0.921)	A17 services based on current affairs	Provide special periods such as holidays, epidemic periods and other services to timely meet the needs of users
	A3 online payment service	After the user's use, it is convenient to settle the online fees

(*continued*)

Table 5. (*continued*)

Main category (Kappa coefficient)	Application category	Application category connotation
	A14 intelligent billing	According to the billing principle, the user should be informed before the use to facilitate the subsequent billing
	A1 positioning reservation function	According to the vehicles needed by users, accurate search and reservation are provided and convenient for users to use
	A6 Accurate positioning	Provide services according to the location of the user, convenient for users to understand their location, surrounding and other relevant information
	Vehicle / parking area near A10 View	Provide nearby vehicle search and parking area search services for users to use and return
Perceived emotional performance (0.899)	The A5 message push service	Provide the push of key information in special periods such as product and service upgrade, special holidays and epidemic situations to meet users' needs for information
	A11 fun small function	Game elements are added to the design to improve the fun of the platform and stimulate users' desire to use it
	A8 vehicle status function	The vehicle condition will be displayed to the user to facilitate the user to use the fault-free vehicle and improve the use experience
	A15 itinerary sharing	Provide users to share the trip before or after the ride to improve the emotional communication between users

(*continued*)

Table 5. (*continued*)

Main category (Kappa coefficient)	Application category	Application category connotation
Perceptual control performance (0.932)	A20 member rights and interests system	Provide members with high-quality rights and interests system services to enhance the experience of users
	A7 Site navigation	Provide the navigation and positioning of the target site to facilitate users to choose their routes
	A9 temporary parking	Provide optional temporary parking service to meet the multi-dimensional needs of users
Perceived engagement (0.813) performance (0.813)	A18 content creation	Users can create a variety of content related to their travel, such as DIY routes, picking cycling races, etc
	A19 gives its opinion	Provide an evaluation and feedback approach that enables users to express their views on the content and functional design of shared bikes

4.3   Theoretical Mapping of User Needs and Perceived Affordances

1) Mapping process. This phase attempts to link the five dimensions of perceived performance in the bike-sharing app to user needs and explore the specific user needs that each perceived performance can satisfy. Studying the association between user needs and perceived performance can effectively reveal how the iterative design and adjustment of the platform dynamically dovetails with and meets user needs [18]. In this study, it is concluded that a single dimension of perceptual performance can satisfy different needs in different contexts, and a single need can be satisfied by multiple dimensions of perceptual performance. For example, the multiple interaction paths in the bike-sharing app in the area of perceived physical performance greatly cater to users' habits in different contexts and give them greater autonomy to choose, meeting their needs for ease of use and usability, autonomy and self-reliance. For example, the fun features in the bike-sharing app in the area of perceived emotional performance allow users to donate through carbon energy collection on their rides, thus helping others and society; the content creation service in the area of perceived participation performance allows users to improve their riding routes in their leisure time, so that others can get accurate and detailed information.

In view of this, the coders, after considering the application categories and functional concepts of the bike-sharing app, refer to the previous categorization of user motivations for ICT use [19], and organize the formation of user needs categories, and map them to the main category of perceptual indicativeness, and the results are shown in Table 6.

Table 6. Perceived Affordances-Theoretical mapping of user needs

Perceived representability	User requirements (in part)
Perceived physical performance	Autonomy and self, ease of use, and availability
Perceived cognitive performance	Autonomy and self, and information acquisition
Perceived emotional performance	Feelings and emotions, altruism and reciprocity, entertainment, integrity, ability and achievement
Perceptual control performance	Autonomy and self, ease of use and availability, and security
Perceived engagement performance	Relevance, leadership and following ability, altruism and reciprocity

2) Conceptual framework construction. In this study, the 57 functional concepts in the bike-sharing app were first divided into 20 application categories, and the 20 application categories were mapped to 5 main categories of perceptual performance, then 11 demand categories based on psychology and other related theories were mapped to 5 perceived affordances, and finally the conceptual framework of "demand-perceptual performance-function" in the user experience design of the bike-sharing app was constructed, as shown in Fig. 1.

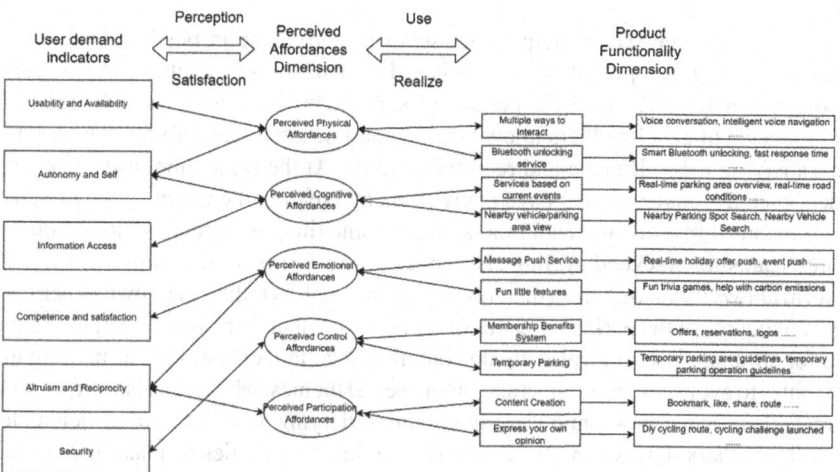

Fig. 1. The conceptual framework of "demand-perception-performance-function" in the user experience design of bike-sharing app

This framework considers that the five perceptual signatures of physical, cognitive, emotional, control and participation are richly reflected in the bike-sharing context, linking user needs and product functions in a multi-dimensional way. In the bike-sharing context, users further experience the product by perceiving its performance, while the use of bike-sharing products also greatly enhances users' perceptual performance, thus satisfying their diverse needs. For example, in terms of perceived control of performance, users can independently choose or change their riding routes according to their actual conditions and personal preferences, and they can also use special voice packages or theme interfaces according to their preferences. These personalized management functions of membership rights greatly satisfy users' needs for independent choice and self-expression, and they can also receive relevant messages to participate in activities at the first time after they are issued. The platform optimizes the experience before and after the operation of temporary parking, and streamlines the common functions such as code unlocking and number unlocking into one-key operation, which improves the efficiency of the users, and the planning and navigation of the area in advance for temporary parking. Meet the user requirements for ease of use and usability of the product. As shown in Fig. 2.

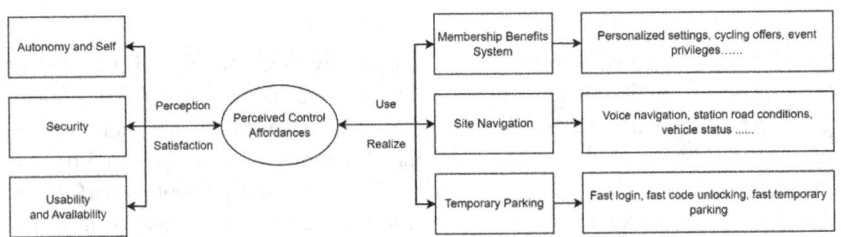

Fig. 2. Application system of Perceived Control Affordances in bike-sharing app

In terms of perceived participation, users can participate in posting diy map ride routes, improving the optimal ride routes based on actual road conditions, making short videos, personalized voices and other services, in the process, users not only get a little extra money, but also let others know accurate and detailed real information about the ride, satisfying their altruistic and reciprocal needs. At the same time, users can act as leaders such as "captains" and have many peers by creating their own cycling teams to satisfy the need to unite and lead others; at the same time, they can also join their own favored teams for weekend cycling challenges to satisfy the need to follow others. The platform allows users to express their own opinions, such as telling their own experiences in the comments of the parking area for other users to "plant" or "avoid", satisfying their need for altruism. Users can also create content to a certain extent, such as making their own routes to avoid epidemic areas and other special themes, which can also attract others to create their own routes while recording them, satisfying users' need to associate, lead and follow others. Users can also establish or join communities to make friends with similar interests and enhance their connection with others. This is shown in Fig. 3.

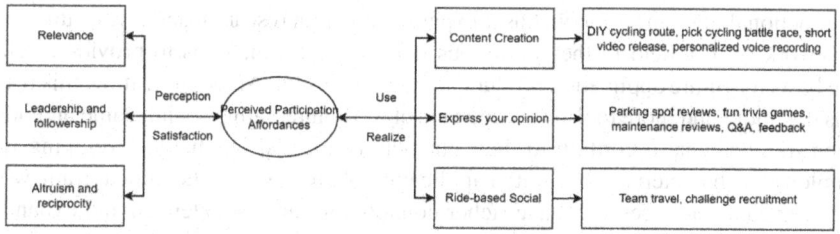

Fig. 3. Application system for Perceived Participation Affordances in bike-sharing apps

5   Exploring the Conceptual Model of Perceived Affordances in Bike-Sharing Usage Scenarios

5.1   Theoretical Contributions

This paper refers to the NAF framework proposed by Karahanna et al. and uses the five specific dimensions (physical, cognitive, emotional, control, and involvement) distilled from existing research on perceptual omnipotence theory as the intermediary bridge between user needs and product functions to build the concept of "needs-perceptual omnipotence-function" for bike-sharing apps. The framework of "demand-perceptual-energy-function" for bike-sharing apps is developed to remedy the lack of interaction design theory in the field of bike-sharing apps. This study emphasizes the mediation of perceptual performance between the demand domain and the function domain, and suggests that users try to use the application function of the bike-sharing app after perceiving the five perceptual performance dimensions of the app; the application function of the app realizes the five perceptual performance dimensions, feeds back and satisfies specific user needs, and motivates users to use the product. At the same time, this paper supports Norman's idea that "the product designed by the designer should be mapped to the user's psychological needs [20]" and proposes the psychological state of the user when interacting with the bike-sharing app from the perspective of needs, which provides a new perspective for the study of user information behavior in the field of bike-sharing. It provides a new perspective for the study of user information behavior in the bike-sharing field. In the subsequent research, researchers can use the conceptual framework proposed in this paper to further explore the deep-seated reasons for the connection between user needs and product functions in bike-sharing apps, and to evaluate the user experience design of bike-sharing apps in the market. Meanwhile, considering the homogeneity and universality of users' psychological needs, the framework can be transferred to other application contexts, and to a certain extent, it can provide reference for the research of user experience design framework of other similar application products.

In addition, this study introduces the perceptual schematic theory into the user experience design of bicycle sharing app for the first time, which enriches the contextual exploration of the perceptual schematic theory and verifies the rationality and transferability of the existing perceptual schematic theoretical framework. In the subsequent research, researchers in related industries can refer to the technical route proposed in this paper and use the theoretical framework of perceptual energetics to bridge the demand domain and functional domain to conduct user experience design research, and form a universal framework in ICT field in the continuous iterative research, so as to provide a reference basis for future application product interaction design. At the same time, this paper finds that more and more tool-based apps are cleverly integrating social, game and other hedonistic elements according to their characteristics, and gradually developing into complex and characteristic integrated application platforms. The theoretical framework of perceived affordances will have richer connotations and be extended and expanded in the subsequent contextualization exploration, so as to better guide interaction design and user experience practice.

5.2 Practice Inspiration

This study comprehensively and systematically compares the functions in the mainstream bike-sharing apps through preliminary data capture, and uses perceived affordances as a theoretical guide to link product functions with user needs and provide ideas for subsequent bike-sharing app design. Among them, in terms of perceived physical performance, the platform should allow users to freely switch the usage mode according to the usage context and interact with the platform through visual, auditory and tactile senses. For example, day and night mode can be switched in case of low light, and voice interaction can be used in case of concentration during cycling, etc., to meet users' needs for autonomy and requirements for product ease of use.

In terms of Perceived Cognitive Affordances, the platform should provide detailed and timely travel information so that users can flexibly choose their travel routes; provide information about nearby vehicles according to users' locations so that users can understand the situation of the surrounding bicycle sharing; organize and classify information on the platform and provide multiple information search channels so that users can quickly access information; provide detailed guidance to help users learn how to use the product; and record users' usage behavior so that they can trace their usage history. It also provides detailed usage guidance to help users learn how to use the product; records users' usage behavior to facilitate users' retrospection of usage history. By influencing the user's mind, the platform can meet the user's needs for information access and autonomy and improve user satisfaction. In terms of perceived emotional indicativeness, platforms should use value-added services such as generating carbon energy for users at the end of shared rides to improve users' satisfaction at the end of the ride and enhance their action efficiency; reduce users' use costs and improve their stickiness by introducing preferential activities; establish a membership rights system to consolidate users' stickiness and enjoy more personalized services; add gamification design elements to stimulate users' We have also introduced public welfare activities and warm-hearted reminders to build a good social image of the platform, and finally satisfy users' needs of feeling a sense of achievement, altruism and reciprocity, and entertainment.

In terms of Perceived Control Affordances, users can independently choose or change their riding routes according to their actual situation and personal preferences, and can also use special voice packages or theme interfaces according to their preferences. These personalized management functions of membership rights and benefits greatly satisfy users' needs for independent choice and self-expression and enhance their sense of security; common functions such as code unlocking and number unlocking are streamlined into one-key The operation improves the efficiency of users' use, and the planning and navigation of the area will be carried out in advance when parking temporarily, which meets users' requirements for product ease of use and usability. Finally, users can participate in publishing routes for DIY map rides, perfecting the optimal route for rides based on actual road conditions, making short videos, personalized voice and other services. This process not only allows users to gain a little extra money, but also allows others to learn accurate and detailed real information about the ride, satisfying their needs for altruism and reciprocity. Develop social functions to establish connections between users to enhance user stickiness; allow users to express their opinions on parking locations, platforms, etc. to better solve the cold start problem of the platform's community function and meet users' altruistic and reciprocal needs. At the same time, in order to avoid homogenization of products, bike-sharing app designers can also develop functions based on the requirements and perceived performance dimensions proposed in this framework to further improve user satisfaction.

In addition, this study sorts out some application categories with their corresponding perceived demonstrative dimensions and user needs, which can be used as a reference for other similar application product designs. Designers should be sensitive to user needs and public opinion trends, update or eliminate outdated and underutilized features, and design features that meet specific user needs in a timely manner, such as modules on how to ride safely during an epidemic that meet impromptu and emergency user needs, which will greatly enhance user satisfaction and improve user stickiness.

6   Concluding Remarks

In this paper, we analyze the existing functions and version iterations of three mainstream bike-sharing apps in China with the help of the existing theoretical framework of perceptual performance to address the limitations of interaction design in the field of bike-sharing apps. After three stages of research, including open coding, spindle coding and theoretical mapping, this paper finally constructs a "demand-perceptual-performance-function" framework for the bike-sharing app industry. The insights of the research are of reference value for the contextualized application of the theoretical framework of perceptual energetics and the user experience design of bike-sharing apps. Future research can be conducted in four aspects.

First, the version iteration data does not directly represent the user needs, in order to explore the real user needs and pain points, questionnaire research and interviews can be conducted in the form of refining the mapping relationship between user needs and perceived performance. Second. From the direction of historical data version update, the current version update can be compared with the historical one to find out the change direction of update content, which can reflect the change of user's demand to some extent. Thirdly, because the use of bicycle sharing app requires users' daily practice, the perceptual performance dimension in this context can be further considered to be divided by combining time attributes and socio-cultural elements, and finally, a multi-dimensional hierarchical analysis of the functions of the bicycle sharing app can be conducted by drawing on the AHP hierarchical analysis, and on this basis, the corresponding perceptual performance can be further associated, and the performance can be refined from the perspective of interaction design. In the end, the multi-dimensional hierarchical analysis of bike-sharing app functions can be based on the AHP hierarchy analysis.

References

1. The 50th Statistical Report on the Development of China's Internet. Inf. Syst. Eng. **2022**(10), 4–5 (2022)
2. Miao, W., Li, L.: A survey on the language use of shared bike apps and applets in Guangdong, Hong Kong, Macao and the Greater Bay Area. Language Life Paper - Report on the State of Language Life in Guangdong, Hong Kong, Macao and the Greater Bay Area **2021**, 249–261 (2021). https://doi.org/10.26914/c.cnkihy.2021.037043
3. Gibson, J.: The Ecological Approach to Visual Perception, pp. 292–293. Houghton Mifflin, Boston (1979)
4. Norman, D.: Affordance, conventions, and design. Interactions **6**, 38–42 (1999)
5. Zou, Z., Yong, Y., Zhao, Y., Xue, X.: A conceptual framework for mobile map app user experience design based on perceptual performance theory. Library Forum, pp. 1–12
6. Sun, C.: Method for sharing operation mode of traveling application shared bicycle, involves utilizing application by user for selecting riding mode and buyers, and performing expense settlement operation by user after ending riding process
7. Xie, Y.: Shared bicycle parking locking system, has front lock head hinged with front lock body, and personal APP which provides prompt information such as front lock of shared bicycle does not match position of front wheel
8. Tang, Y., Tang, F.: Method for preventing bicycle sharing from over-aggregation parking, involves realizing moving service function, and installing bicycle APP to encourage user to remove shared bicycle at short time and distance in parking area
9. He, F., Yao, H., Ye, X.: Shared bicycle parking device for automatic driving, has vehicle intelligent management control device bidirectionally connected with automatic driving device through communication, and server unit fed back to mobile phone APP unit
10. Zhao, L., Guo, X.L.: Research on user experience of shared bicycles in China--a survey based on users of Mobike, ofo and government public bicycles. Times Bus. Econ. **2018**(12), 17–18 (2018). https://doi.org/10.19463/j.cnki.sdjm.2018.12.005
11. Jiang, Y.J., Zhang, B.: Research on the analysis of user satisfaction of shared bicycle use and improvement strategies. China Bicycle **03**, 78–83 (2020)
12. Zhu, Y., Pan, J.: An empirical study on pricing strategy of shared bicycle based on user experience perspective. J. Chongqing Jiaotong Univ. (Social Science Edition) **20**(02), 36–42 (2020)
13. Wang, Y., Wang, X., Tang, S., Wu, S.: Measurement and empirical study on the quality of travel experience of shared bicycle users. China Soft Sci. **2020**(S1), 133–146 (2020)
14. Boy, J., Eveillard, L., Detienne, F., et al.: Suggested interactivity: seeking perceived affordances for information visualization. IEEE Trans. Visual Comput. Graphics **22**(1), 639–648 (2015)
15. Xue, X., Zhao, Y.: A study on user experience evaluation of mobile music app based on the theoretical framework of perceived affordances. Library Intell. Knowl. **2020**(06), 88–100+156 (2020). https://doi.org/10.13366/j.dik.2020.06.088
16. Smeeton, N.C.: Early history of the kappa statistic. Biometrics **41**, 795 (1985)

17. Markus, M.L., Silver, M.S.: A foundation for the study of IT effects: a new look at DeSanctis and Poole's concepts of structural features and spirit. J. Assoc. Inf. Syst. **9**(10), 5 (2008)
18. Zhao, Y., Zhang, Y., Tang, J., et al.: Affordances for information practices: theorizing engagement among people, technology, and sociocultural environments. J. Documentation **77**(1), 229–250 (2021)
19. Zhang, P.: Motivational affordances: Reasons for ICT design and use. Commun. ACM **51**(11), 145–147 (2008)
20. Norman, D.A.: The psychology of everyday things. Basic books (1988)

Owners (& Frequent Users) of E-Scooters – Who Are They?

Tibor Petzoldt[1] , Madlen Ringhand[1] , Juliane Anke[1]([⊠]) , and Tina Gehlert[2]

[1] Technische Universität Dresden, 01062 Dresden, Germany
tibor.petzoldt@tu-dresden.de
[2] German Insurers Accident Research, Wilhelmstraße 43-43G, 10117 Berlin, Germany

Abstract. While shared e-scooters are still highly prevalent in many urban areas, we also see an increasing share of privately owned e-scooters on our roads. However, so far, not much is known about the users of such privately owned e-scooters. As there is reason to suspect that those who ride their own e-scooter differ considerably from occasional users of rental vehicles, the question is what the characteristics of those riders are and how their use of vehicles can be described. To address these questions, we analysed data from an online survey (99 usable data sets) and several focus group discussions (20 participants) with frequent users and owners of e-scooters. Results show that these frequent riders have a high degree of identification with the vehicle, and often use it as their main means of transport. The majority indicated that they used their e-scooter daily or almost daily, with commuting as their most frequent trip purpose. They are rather safety conscious, as is highlighted by their use of protective gear and the installation of additional safety equipment on their vehicles. The results indicate that a specific focus on frequent users is warranted, as they appear to be quite distinct from the user population that occasionally uses shared services. With the number (and share) of privately owned e-scooters continuously increasing, targeted approaches that facilitate their safe and efficient participation in traffic will be required.

Keywords: Micromobility · User characteristics · Usage behaviour

1 Introduction

Although e-scooters have been available only for a few years, they have seen massive growth in usage since their introduction. Initially, this growth was mainly driven by shared e-scooters. The relative novelty of these vehicles, coupled with the shared mobility aspect, made them attractive mostly to occasional users, with infrequent use (once a month or less) being the rule rather the exception (e.g., Siebert et al. 2021). In line with this, the majority of trips taken by e-scooter were reported to be related to leisure activities (e.g., Deutscher Verkehrssicherheitsrat e.V. 2020; Portland Bureau of Transportation 2019; Ringhand et al. 2021).

However, while shared e-scooters are still highly prevalent in many urban areas, we now see an increasing share of privately owned e-scooters on our roads (e.g., Haworth

H. Krömker (Ed.): HCII 2023, LNCS 14048, pp. 428–437, 2023.
https://doi.org/10.1007/978-3-031-35678-0_29

et al. 2021). The fact that users decided to buy their own vehicles might be seen as an indication that the sharing services, for whatever reason, did not fulfil their mobility needs. However, little is known about who these individuals are, and what differentiates them from the occasional users of shared e-scooters. In more general terms, the question is what the characteristics of frequent e-scooter users are, and how their use of these vehicles can be described.

To address these questions, we analysed data collected in a project that looked into the safety of e-scooters and how it could potentially be improved (Anke et al. 2022). Within this project, we conducted an online survey and several focus group discussions with frequent users and owners of e-scooters. Among the information collected in these studies were the characteristics of frequent users (e.g., age, gender, degree of identification with the e-scooter as a mode of transport) as well as aspects of their e-scooter usage behaviour (e.g., frequency of use, trip purposes, travelled distances, use of protective gear).

2 Method

2.1 Online Survey

Structure and Content. The online survey consisted of three parts. Part 1 included questions about the usage background and riding experience. The second part (not relevant for the analyses presented in this paper) addressed safety relevant aspects of e-scooter use, e.g. experiences with critical situations. The survey concluded with part 3 on socio-demographic information.

Implementation. The survey was created using Sosci Survey (Leiner and Leiner 2021) and was freely accessible from 03.09.2021 to 30.09.2021. In the run-up to the publication of the survey, a pre-test was conducted with five people, the survey was revised according to the comments and underwent a final technical function test.

In order to reach as many frequent e-scooter riders as possible, the online survey was promoted via groups on the topic of "e-scooters" via blogs and social media channels. In order to increase the response rate, it was pointed out in the cover letter (mail contact with potential participants) and on the welcome page (page 1 of the survey) that participation could contribute to gaining knowledge about the still new means of transport. In addition, participants could win one of five 20€ vouchers. They also got the opportunity to compare their answers with those of the previous participants in a short evaluation of two items at the end of the survey.

Data Preparation. At the end of the data collection period, 154 individuals had taken part in the survey. However, not all of them had completed the questionnaire. Participants that had not completed the last relevant page were excluded from further analysis. In addition, some response sets had a high proportion of missing values (participants skipping questions). These were excluded as well. If participants were unreasonably fast in completing the survey, we critically examined their responses for any obvious irregularities, and, if necessary, excluded them. We also, unfortunately, had to exclude two data sets in which participants stated that they used a monowheel or an electric skateboard instead of an e-scooter. At the end, a sample of 99 respondents remained for analysis.

2.2 Online Focus Groups

Structure and Content. Complementary to the survey, semi-structured script-based focus group interviews were conducted with frequent and experienced e-scooter riders. Similar to the survey, safety relevant aspects of e-scooter use, which are not subject to the analyses presented in this paper, were one of the main topics that were discussed. However, the focus groups also covered general aspects of usage behaviour (how often do riders use the e-scooter, for what types of trips, do they use protective gear etc.), as well as participant demographics. A script was developed that contained all the relevant topics as well as some guiding questions for these topics, in order keep the different focus groups as structured as possible.

Implementation. A first version of the focus group script was pre-tested on a single focus group with participants that had cycling and e-scooter riding experience. Based on the gathered insights, adjustments were made to the script and the schedule. At the same time, the focus group moderators were trained on the contents and structure of the script. They received information on how to deal with possible conflicts, and got some additional communication tips.

For the acquisition of participants, about 40 people were contacted who had given their consent to this in the online survey. Further acquisition took place via social media, calls or advertisements on eBay Classifieds, in e-scooter forums and a newsletter.

Due to the restrictions resulting from the COVID-19 pandemic, we opted for an online implementation. Six online focus groups were conducted in November 2021 using the Zoom video meeting service. During the discussions, two moderators guided the participants through the different parts of the script, with short breaks in between. At the end, the participants were asked for their feedback and information was given about the allowance (30€). The whole process took between two and two and a half hours.

Data Preparation. The focus group discussions were recorded as video/audio files and then transcribed. The participants' statements were then coded using the MAXQDA software (VERBI Software, 1989–2021) with the help of a coding scheme, which was based on the contents defined in the script, and which was successively extended into subcategories. Central to the analyses presented in this paper were mostly information gathered during the introductory stage, including the participants' age and gender, the background for their e-scooter use, the trips they used it for, etc.

3 Results

3.1 Sample/General User Characteristics

The final data set of the online survey included a total of 99 e-scooter users who use an e-scooter at least once a month. The characteristics of the sample are presented in detail in Table 1. The mean age of the participants was 37.8 years ($SD = 12.1$). The high proportion of male users (91%) is striking. While the vast majority of frequent users owned the vehicle they were riding, interestingly, one participant reported to use a company vehicle, which is certainly still a novel concept when it comes to e-scooters.

Also worth highlighting is the fact that, while rental vehicles are available primarily in larger urban areas, frequent users / private owners can also be found in smaller towns and rural areas.

Table 1. Sociodemographic characteristics of survey sample.

	%	N
gender	**100**	**80**
female	6.3	5
male	91.3	73
diverse	2.5	2
age	**100**	**91**
<18 years	2.2	2
19–30 years	29.7	27
31–40 years	30.8	28
41–50 years	20.9	19
>50 years	16.5	15
place of residence	**100**	**96**
rural area (small town < 5,000 inhabitants)	10.4	10
small town (5,000 to 20,000 inhabitants)	19.8	19
medium-sized town (20,000 to 100,000 inhabitants)	22.9	22
large city (100,000 to 500,000 inhabitants)	25.0	24
small metropolis (500,000 to 1 million inhabitants)	8.3	8
large metropolis (over 1 million inhabitants)	13.5	13
e-scooter ownership	**100**	**99**
yes	97.0	96
no, use of company vehicle	1.0	1
no, use of rental vehicle	2.0	2

The characteristics of the focus group sample largely reflect those of the survey sample. Seventeen (85%) participants were male, three (15%) female, with a mean age of 36.7 years ($SD = 9.5$). Of the 20 participants, 17 owned an e-scooter, the other three used rental vehicles.

3.2 Use of/Identification with Different Modes of Transport

When asked about their main means of transport, about 41% of the participants in the online survey indicated that the e-scooter was indeed the most important mode for them (Fig. 1). For one third of the participants, it was the car. When asked with what mode of transport they would identify the most, about one quarter of the survey participants indicated that they indeed felt like primarily being an e-scooter user, while about one third considered themselves car drivers (Fig. 2). When, in a separate question, asked specifically to what degree the considered themselves being an e-scooter rider, more than three quarters (78%) of the respondents indicated the highest possible level of identification (5 on a scale from 1 to 5).

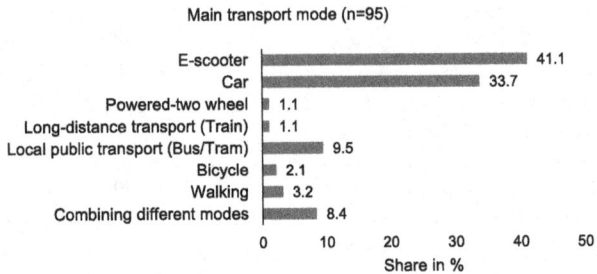

Fig. 1. Relative frequencies of responses to question "What mode of transport do you mainly use?".

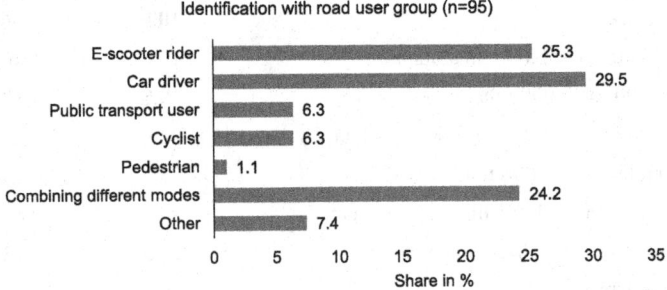

Fig. 2. Relative frequencies of responses to question "With what road user group do you identify the most?".

3.3   Travel Purposes, Frequency of Use and Distances

The majority of participants in the online survey indicated that they used their e-scooter daily or almost daily. For more than half of them, their most frequent trip by e-scooter was longer than 5 km. The purpose of this most frequent trip with the e-scooter was the journey to work or school for nearly half of all respondents (Fig. 3). At the same time, riding the e-scooter for purposes of leisure (either with a specific destination, or just for the joy of riding as such) was the most typical / frequent trip purpose for nearly one third of the participants.

These results are in line with the findings from the focus groups. There, too, participants reported that the main purpose of use was commuting to work, also for long distances and in combination with long-distance and regional transport. Leisure related

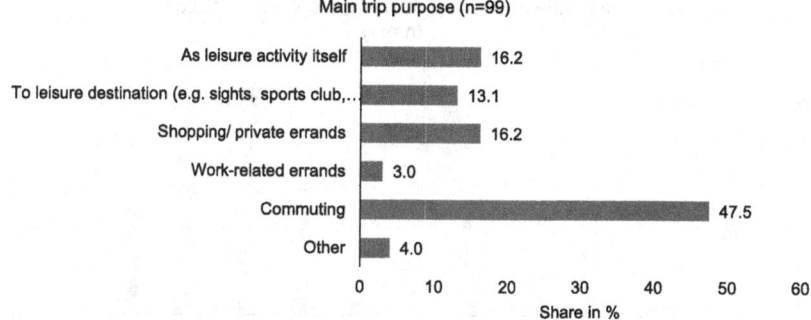

Fig. 3. Relative frequencies of responses to question "For what purpose do you ride e-scooters most often?".

trips did play a certain role as well. The following are sample quotes from the focus group discussions:

> *"I use the e-scooter mainly to - once a) overcome the last mile, which now relates to the [name of a city] transport provider [...]. Secondly, of course, the commute to work or the trip to sports, or actually to go shopping, that also regularly. I actually ride almost every day."*

(male private vehicle user, 39 years).

> *"... That only since April this year I have been using the thing then also for commuting to work. [...] I travel by train [...]. The big advantage of the scooter for me is that it folds up and fits under the seat, unlike a bicycle. In [city 1] it's another 10-12 kilometres to my workplace. That's the commuting distance. So [city 2] from the flat to the main station, and then [city 1] main station office and return. That's the most common or almost daily route. Otherwise, well, sometimes in my free time I go somewhere in [city 2] and fool around in the city, or sometimes I go out to my parents' house, who also live in the countryside, where the bus goes three times a day. The scooter is also very useful there and of course also for shopping in the city, just put a bigger backpack on and you've almost done your weekly shopping."*

(male private vehicle user, 39 years).

3.4 Perceived Safety and Use of Protective Equipment

The results of the online survey show that the frequent e-scooter riders feel relatively safe in traffic when they ride e-scooters (Fig. 4).

About half (50.5%) of the surveyed riders stated that they use a helmet and/or other protective equipment when riding an e-scooter (36.4% helmet only, 12.1% helmet and additional equipment, 2.0% equipment without helmet). In addition, the respondents had the opportunity to provide more detailed information on other protective equipment they used. They mentioned:

- gloves (6x),

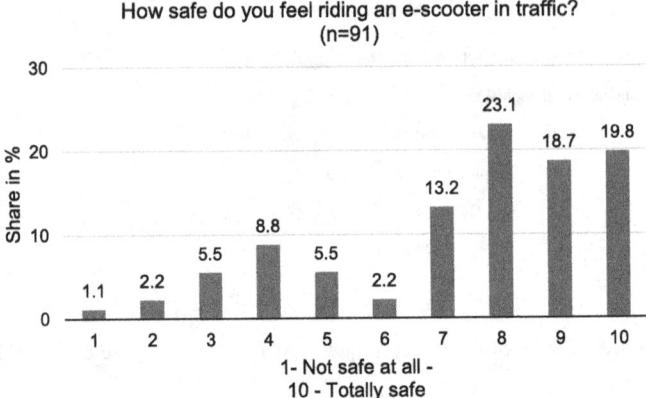

Fig. 4. Relative frequencies of responses to question "How safe do you feel riding an e-scooter in traffic?".

- additional light (2x),
- protectors (1x),
- signal clothing (1x),
- long clothing (1x) and
- reflectors (1x).

In the online focus group, eight out of twenty participants stated that they use a helmet, including one of the rental vehicle users. Other protective equipment reported in the focus groups was similar to that reported in the online survey:

- gloves (3x),
- safety boots (2x),
- day-glow jacket (2x),
- reflectors (3x),
- long clothing (1x),
- additional light (1x) and
- coloured brake cables (1x).

During the discussions, it became clear that the aspect of visibility played a prominent role, as the following quotes illustrate:

"And I got myself reflective silicone cables for the wiring so that I can be seen better from the side."

(male private vehicle user, 35 years).

"What also comes to my mind is what I use to travel in the dark season that we are starting to have now: I usually, actually always have a backpack on my back. And I have a red light on the back of the bag on the side, just on the left side. A flashing light. Because I think that the low red light of the scooters is not seen in traffic anyway. So that makes no sense at all. That's why I have another one a bit higher up and hope that I'll be seen a bit better that way."

(male private vehicle user, 37 years).

3.5 Tuning of the E-Scooter

About two-thirds of the surveyed e-scooter owners can imagine tuning their vehicle (27.4%), or have already done that (33.7%). This high proportion in favour of (illegal) modifications is also reflected in the statements collected during the focus groups. Interestingly, (increasing) speed was not the sole focus of such modifications. The adjustments mentioned in the focus groups are presented below in paraphrased form:

- tuning for speed (9x),
- retrofitting of turn signals (or in one case turn signals had just been ordered) (4x),
- additional mirror on handlebars (3x), or even second mirror (1x),
- change of light settings for continuous lighting (2x),
- handlebar extension in the sense of an extension to mount indicators, mobile phone holder, light etc., to be seen better and to have a better grip (1x),
- other tyres mounted, in the sense of winter tyres (1x),
- setting the "international mode", which enables a cruise control function and thus also contributes to better handling of the right-turn situation (1x).

4 Discussion

Aim of the analysis presented in this paper was to describe frequent users and owners of e-scooters, and identify differences to as well as commonalities with occasional users of rental vehicles. Results show that our participating riders had a high degree of identification with the vehicle, and often used it as their main means of transport. The majority indicated that they used their e-scooter daily or almost daily, with the commute their most frequent trip purpose.

When comparing our survey participants' demographics to those of an earlier project (Ringhand et al. 2021), in which we observed and interviewed e-scooter users on Berlin roads, some differences become apparent. While less than one in ten (9%) of the participating frequent users were female, we observed a ratio of about one out of four (24%) on the street. Age-wise, our frequent users were slightly older (38 years) than those interviewed in the field (30 years). Given that 94% of the vehicles observed in Berlin were rentals, it is reasonable to assume that the differences in demographics would somewhat reflect general differences in user characteristics between frequent riders (primarily of their own vehicles) and occasional users (of rental vehicles).

Corresponding differences were also found with regard to other aspects of vehicle usage. Among the respondents in the field study, around 27% were using the e-scooter for the first time, while only 16% indicated to use an e-scooter on a (nearly) daily basis. Not surprisingly, usage was much more frequent among our sample of frequent e-scooter riders. More importantly, the majority (90%) of the on-street respondents indicated that their e-scooter ride was leisure related, in line with previous findings on users of rental vehicles (Orozco-Fontalvo et al., 2022). This trip purpose only played a secondary role for our frequent riders, while the commute was clearly the most relevant use case.

Among the findings that stuck out was the fact that a considerable share of participants lived in smaller towns or even villages. Accordingly, relevant usage contexts might not solely be urban or suburban areas, which are typically served by providers of shared vehicles. Considerations regarding the integration of e-scooters (and other forms of micromobility) into the traffic and transport system should therefore not only focus on those larger cities that currently struggle to accommodate the large fleets of rental vehicles. In fact, privately owned e-scooters might actually be considered a reasonable approach to tackle the first / last mile issue in less urbanised areas that cannot reasonably be addressed by rental vehicles.

It is important to note that frequent riders and private owners seem to be very safety conscious. Not only do they use protective gear to a much larger extent than the average user (Huemer et al. 2022). They also actively modify their vehicles by adding equipment that is supposed to increase the vehicles' safety. This approach might, to some degree, be linked to trip purposes and usage contexts (longer commutes, in parts potentially on rural roads), potentially resulting in a higher exposure to safety critical events. In principle, however, this finding indicates that many negative preconceptions regarding e-scooters that can be found should not be generalised to all users. Indeed, during the focus group discussions, frequent users repeatedly tried to emphasise the gap between themselves (described as serious, experienced and responsible users) and the others (described as those who don't know what they are doing and have no respect for the rules).

While our dataset was comparatively small and somewhat limited, our results indicate that a specific focus on frequent users seems warranted, especially given their usage patterns that seem to deviate substantially from those of occasional users. It becomes clear that frequent users and private owners of e-scooters are a different kind than the user population that we saw initially for shared services. With the number (and share) of privately owned e-scooters continuously increasing, targeted approaches that facilitate their safe and efficient participation in traffic will be required.

References

Anke, J., Ringhand, M., Petzoldt, T., Gehlert, T.: Präventionsmaßnahmen für E-Scooter-Nutzer:innen. Forschungsbericht Nr. 87 (2022). Unfallforschung der Versicherer. Gesamtverband der Deutschen Versicherungswirtschaft e.V., Berlin. https://www.udv.de/resource/blob/113224/957b47e9a6ee47bdc794223867f864d8/87-praeventionsmassnahmen-fuer-e-scooter-nutzer-innen-data.pdf

Deutscher Verkehrssicherheitsrat e.V.: Roll ohne Risiko (2020). https://www.dvr.de/praevention/kampagnen/roll-ohne-risiko

Haworth, N., Schramm, A., Twisk, D.: Changes in shared and private e-scooter use in Brisbane, Australia and their safety implications. Accid. Anal. Prev. **163**, 106451 (2021). https://doi.org/10.1016/j.aap.2021.106451

Huemer, A.K., Banach, E., Bolten, N., Helweg, S., Koch, A., Martin, T.: Secondary task engagement, risk-taking, and safety-related equipment use in German bicycle and e-scooter riders – an observation. Accid. Anal. Prev. **172**, 106685 (2022). https://doi.org/10.1016/j.aap.2022.106685

Leiner, D. & Leiner, S. (2021). SoSci Survey (Program version 3.2.33) [Computer Software]: SoSci Survey GmbH. https://www.soscisurvey.de/de/index

Orozco-Fontalvo, M., Llerena, L., Cantillo, V.: Dockless electric scooters: a review of a growing micromobility mode. Int. J. Sustain. Transp. 1–17 (2022). https://doi.org/10.1080/15568318.2022.2044097

Portland Bureau of Transportation. 2018 E-Scooter Findings Report (Portland Bureau of Transportation, Ed.). Portland Bureau of Transportation (2019). https://www.portlandoregon.gov/transportation/article/709719

Ringhand, M., Anke, J., Petzoldt, T., Gehlert, T.: Verkehrssicherheit von E-Scootern. Forschungs-bericht Nr. 75. Unfallforschung der Versicherer. Gesamtverband der Deutschen Ver-sicherungswirtschaft e.V., Berlin (2021). https://www.udv.de/resource/blob/79908/1d2bc0eee dae8b30ff521bec9b708115/75-verkehrssicherheit-von-e-scootern-download-data.pdf https:// www.udv.de/resource/blob/79908/1d2bc0eeedae8b30ff521bec9b708115/75-verkehrssicherh eit-von-e-scootern-download-data.pdf
Siebert, F.W., Ringhand, M., Englert, F., Hoffknecht, M., Edwards, T., Rötting, M.: Braking bad – ergonomic design and implications for the safe use of shared E-scooters. Saf. Sci. **140**(6), 105294 (2021). https://doi.org/10.1016/j.ssci.2021.105294
VERBI Software. (1989–2021). MAXQDA. Qualitative data analysis software [Computer Software]. Berlin: Consult. Sozialforschung GmbH

Riding e-Scooters Day and Night – Observation of User Characteristics, Risky Behavior, and Rule Violations

Madlen Ringhand[1]([✉]) [ID], Juliane Anke[1] [ID], Tibor Petzoldt[1] [ID], and Tina Gehlert[2]

[1] Technische Universität Dresden, 01062 Dresden, Germany
madlen.ringhand@tu-dresden.de
[2] German Insurers Accident Research, Wilhelmstraße 43-43G, 10117 Berlin, Germany

Abstract. Crash statistics and hospital data show that injured e-scooter riders arrive at hospitals often at night and on weekends. Subsequently, the crash risk at night is higher compared to the daytime. A possible explanation might be increased rule violations, safety-critical behaviors, and changes in the user group at night compared to daytime. Therefore, we aimed to conduct an observational study analyzing the interrelationships of risky behaviors, rule violations, and user characteristics of e-scooter riders in two German cities. A total of 732 observations were analyzed with Chi-Squared tests and Generalized Estimating Equations. The results show increased rates of tandem riding at night compared to the daytime and increased rule violations of adolescents compared to older e-scooter riders regardless of the time of day. Rates of helmet use, wrong-way riding, headphone use, smartphone use, and luggage transport were comparable for daytime and night observations. The results suggest that educational campaigns should focus on tandem riding, especially targeting the user group of teenage riders. This study brings e-scooter riding at night into the light and emphasizes riders' nightly behaviors for policymakers and traffic safety.

Keywords: Micromobility · Traffic safety · Night riding

1 Introduction

Riding e-scooters in Germany started in 2019, followed by a rush of shared e-scooters in the cities with a predominantly male and young user group (Haworth et al. 2021b; Haworth and Schramm 2019; Huemer et al. 2022). Especially in the early stages, e-scooter riding at night was associated in the popular press with people being drunk, riding e-scooters in groups, and severe crashes (Noack 2019; Tapper 2019). Confirming these reports, crash statistics and hospital data show that injured e-scooter riders arrive at hospitals more often at night and on weekends compared to data on cyclists (Kleinertz et al. 2021; Stigson et al. 2021). In addition, Shah and Cherry (2022) showed a higher crash risk at night compared to daytime when putting the number of car-e-scooter crashes into perspective on the number of trips made.

© The Author(s), under exclusive license to Springer Nature Switzerland AG 2023
H. Krömker (Ed.): HCII 2023, LNCS 14048, pp. 438–449, 2023.
https://doi.org/10.1007/978-3-031-35678-0_30

A possible explanation for increased crashes and injuries at night might be increased rule violations, safety-critical behaviors, and changes in the user group compared to daytime. To raise riders' awareness and reduce crash risk at night, policymakers and traffic safety specialists would benefit from findings about the changes from day to night regarding the behavior and characteristics of e-scooter riders. With this, they could address e-scooter riders at risk more specifically and aim to reduce critical behaviors that might occur more often at night than in the daytime.

Demographic characteristics, safety-critical behavior, and rule violations of e-scooter riders were the focus of international scientific research in recent years without particular differentiation between day and night. Evidence shows that e-scooter riders are primarily young and male (Curl and Fitt 2020; Haworth et al. 2021b; Laa and Leth 2020; Orozco-Fontalvo et al. 2022; Portland Bureau of Transportation 2020; Siebert et al. 2021). With this, e-scooter riders might be more prone to risky riding than cyclists because they are inexperienced in riding and can be affected by a self-enhancement bias known for this age group (Harré et al. 2005; Sibley and Harré 2009). By overestimating their abilities, young e-scooter riders might perform safety-critical behavior like tandem riding (two people on one e-scooter), riding against the direction of travel, and using the wrong infrastructure (i.e., footpaths, depending on national regulations). A first hint that safety-critical behavior is enhanced for younger e-scooter riders compared to older riders was shown by the observations of Huemer et al. (2022) for a sample consisting of e-scooter riders and cyclists. Still, a replication of this finding is missing.

For riding e-scooters at night, the rate of young people using e-scooters at night might be increased compared to the daytime if they head to bars, clubs, or pubs. In addition, these young users could be strongly affected by riding with their peers. Research on car drivers shows that adolescents' risky behavior increases with peers' being passengers (Leadbeater et al. 2008) and by having risky friends or peer pressure (Simons-Morton et al. 2012). Regarding e-scooter riding, the question arises whether adolescents travel primarily in groups with peers and whether this is associated with increased safety-critical behavior, especially at night.

An observational study could gather robust data on riding in groups and the safety-critical behaviors of e-scooter riders both day and night. Previous research used camera-based methods or human observers to study the demographic characteristics and rule violations of e-scooter riders (Arellano and Fang 2019; Haworth et al., 2021a, 2021b; Haworth and Schramm 2019; Huemer et al. 2022; Siebert et al. 2021). With camera-based methods, riding in groups cannot be captured well because of too-small image detail, especially when aiming to rate the togetherness of individuals at greater distances. In addition, at night, the image of camera-based observations loses sharpness and coloration, making the categorization of gender and age more difficult. A human observer, however, can study an e-scooter rider over an extended period and effortlessly recognize a person's social affiliation with others regardless of the time of day. Since there have been no previous studies on the nighttime use of e-scooters or riding in groups, this study aims to fill this gap through observations with human observers, also covering the behaviors of young e-scooter riders.

To sum up, we aimed to analyze the following research questions with an observational study:

- Which safety-critical behavior of e-scooter riders can be observed at night compared to daytime?
- Are there differences in the demographic profile of e-scooter riders between day and night?
- Does the percentage of riding in groups increase at night compared to the daytime?
- Is riding in groups or age correlated with increased safety-critical behavior at night compared to daytime?

2   Methods

2.1   Observation Plan and Sites

To compare the user characteristics, risky behavior, and rule violations of e-scooter riders between day and night, we chose four sites in two German cities (Dresden and Berlin) that guaranteed high e-scooter usage during the daytime. In addition, these sites were located close to pubs, restaurants, and clubs, which enabled nighttime observations. Each site had two clearly defined observation areas (two sides of the road) with an expanse of around 70 m without junctions or intersections. An exception to this arrangement is site two because both a main street (with two observation areas), and a side street adjacent to it were observed there. Every observation area had one observer. The data collection was made in two weeks, in August and September 2020. The nighttime observations were conducted on Friday and Saturday evenings (9 pm to 0.30 am). We added the afternoon of the same day for daytime comparisons (2 pm to 6.30 pm) and another afternoon on a weekday (2 pm to 6.30 pm). Indicators for the chosen times were the evaluations of Tack et al. (2019) on the usage of rental e-scooters. Table 1 gives an overview of the observation plan.

Table 1. Observation plan for the four sites.

	Site 1	Site 2	Site 3	Site 4
City	Berlin	Berlin	Dresden	Dresden
Location	Warschauer Brücke	Unter den Linden	Albertplatz	Kulturpalast
Day (Tuesday/Wednesday) - observation times	Tuesday: 2 pm–4 pm / 4.30 pm–6.30 pm	Wednesday: 2 pm–4 pm / 4.30 pm–6.30 pm	Tuesday: 2 pm–4 pm / 4.30 pm–6.30 pm	Wednesday: 2 pm–4 pm / 4.30 pm–6.30 pm
Day (Friday / Saturday)–observation times	Friday: 2 pm–4 pm / 4.30 pm–6.30 pm	Saturday: 2 pm–4 pm / 4.30 pm–6.30 pm	Friday: 2 pm–4 pm / 4.30 pm–6.30 pm	Saturday: 2 pm–4 pm / 4.30 pm–6.30 pm
Evening / night (Friday / Saturday)–observation times	Friday: 9 pm–11 pm / 11.30 pm–0.30 am	Saturday: 9–11 pm	Friday: 9 pm–11 pm / 11.30 pm–0.30 am	Saturday: 9–11 pm

2.2 Observation Categories

The following variables were observed: age, gender, group size, vehicle type (rental, owner), tandem riding (two persons on one e-scooter), sidewalk riding, helmet use, used road infrastructure, riding against the direction of traffic, headphone use, handheld smartphone use as well as luggage on/at the e-scooter.

The following definitions were made:

- The category group size reflects the number of e-scooters within one group, i.e., the number of vehicles.
- A group is defined as a socially cohesive group of people.
- Within a group of two or more e-scooters, the person driving in front is observed for all other categories.
- For tandem riding, the person steering is observed for all other categories.
- Age categories were the following: children (<14 years), adolescents (14–20 years), young adults (20–40 years), middle-aged adults (41–65 years), and pensioners (>65 years)

Observers were trained in observational areas and categories before starting the official data collection. For observations, they used a tablet-based observation tool (Vollrath 2019). Inter-rater reliability was excellent for observations of group size, helmet use, tandem riding, smartphone use, and gender (each $\kappa = 1.00$), almost perfect for headphone use and type of vehicle ($\kappa = 0.88$), substantial for luggage on/at the e-scooter ($\kappa = 0.75$), and moderate for age ($\kappa = 0.43$) (Landis and Koch 1977).

2.3 Data Analysis

We collected the data for all observed variables separated into two measures of daytime and one of nighttime. The total number of observations was $N = 732$. The influence of the observation time on distributions of demographic characteristics, group size, and safety-critical behavior was tested using Pearson Chi-squared tests. We performed post hoc tests by comparing significance levels of adjusted standardized residuals with Bonferroni correction (Beasley and Schumacker 1995). The influence of age and group size on safety-critical behavior at night was tested for the variables wrong-way riding, sidewalk riding, and tandem riding with Generalized Estimating Equations with a binary logic link function. The predictors were observation time (day vs. night), observed group size (riding alone vs. riding in groups), age (adolescents vs. older), and their two-way interactions.

3 Results

First, we report the results on the differences between daytime and nighttime e-scooter riding on observed gender, age, group size, and vehicle type. Table 2 shows the descriptive numbers.

Around three-quarters of all observed e-scooter riders were male, and one-quarter were female, regardless of the time of day, $\chi 2 (2) = 1.21, p = .559$, Cramer's $V = .040$.

Table 2. Demographic characteristics of e-scooter riders, group size, and vehicle type depending on the time of observation (N = 732). Exception of sample sizes due to impossible classification for vehicle type: n = 182 for the day [Tuesday/Wednesday] and n = 200 for evening /night.

Variable		Day Tuesday/Wednesday		Day Friday/Saturday		Night Friday / Saturday	
		n	%	n	%	n	%
gender	male	137	74.9	270	77.5	148	73.6
	female	46	25.1	78	22.5	53	26.4
age in years	children	5	2.7	4	1.2	1	0.5
	adolescents	37	20.2	48	13.8	41	20.6
	young adults	116	63.4	241	69.2	134	67.3
	middle-aged adults	25	13.7	55	15.9	23	11.6
	pensioners		–		–		–
group size	1 e-scooter	105	57.4	128	36.8	80	39.8
	≥2 e-scooter	78	42.6	220	63.2	121	60.2
vehicle	rental	167	91.8	330	94.8	196	98.0
	privately owned	15	8.2	18	5.2	4	2.0

The number of teenage e-scooter riders was slightly higher on Friday/Saturday nights than on the same days in the daytime, but without a statistically significant difference $\chi2 (4) = 6.85, p = .144$, *Cramer's V* = .069.

The observations of people riding e-scooters in groups with peers significantly changed when testing for the effect of observation time, $\chi^2 (4) = 29.41, p < .001$, *Cramer's V* = .142. Post hoc tests showed that in the daytime in the middle of the week, the relative number of people riding alone was higher than expected ($p < .001$). In contrast, riding alone in the daytime of a Friday or Saturday day was lower than expected ($p = .017$). Riding in groups with peers was observed more often on Fridays/Saturdays (60% and 63%) than during the week (43%). An assumed increase in riding in groups at night hours cannot be confirmed.

The observation time significantly influenced how many rental or privately owned e-scooters were observed, $\chi2 (2) = 7.73, p = .021$, *Cramer's V* = .103. At night the percentage of rental e-scooters was higher than for daytime observations. With the rather small global effect, the post hoc tests with Bonferroni correction showed no significant differences ($p = .128$).

Next, the results on safety-critical behavior are reported. Table 3 shows the descriptive statistics of helmet use, wrong-way riding, sidewalk riding, tandem riding, headphone use, smartphone use, and luggage transport, depending on the observation time. As can be seen, the reported rates of safety-critical behaviors varied strongly. Overall, helmet and smartphone use were observed only in a few cases. For both, no influence of the observation time is given, helmet use: $\chi2$ (2) = 1.62, p = .446, *Cramer's V* = .047; smartphone use: $\chi2$ (2) = 3.50, p = .173, *Cramer's V* = .069. Similarly, for the transport of luggage at/with the e-scooter, no influence of observation time on observed rates is given, $\chi2$ (2) = 4.43, p = .109, *Cramer's V* = .078.

For both wrong-way riding and sidewalk riding, we see similar patterns: The observed amounts are high in the daytime during the week, then become lower for the daytime of Friday/Saturday, and finally increase again for the night observations. Wrong-way riding was observed at night for every one in ten and sidewalk riding every three in ten. For both variables, a significant influence of observation time is given: wrong-way riding: $\chi2$ (2) = 7.50, p = .024, *Cramer's V* = .101; sidewalk riding: $\chi2$ (2) = 21.04, p < .001, *Cramer's V* = .170. For wrong-way riding with a relatively small global effect, the post hoc tests showed no significant differences. There are tendencies that observations of wrong-way riding were higher than expected in the daytime during the week (p = .104) and lower than expected in the daytime on Fridays/Saturdays (p = .088). For sidewalk riding, post hoc tests showed that sidewalk riding was higher than expected for daytime during the week (p = .010) and lower than expected for daytimes of Fridays/Saturdays (p < .001).

Tandem riding (two persons on one e-scooter) was increased for the nighttime observations compared to daytime observations regardless of the day of the week, with every one in ten observed e-scooter riders. An overall significant effect of observation time was found, χ^2 (2) = 10.40, p = .006, *Cramer's V* = .119. Post hoc tests showed that observations of tandem riding at night were significantly higher than expected (p = .002). These results indicate more rule violations of tandem riding for night observations.

Lastly, the percentage of headphone use was observed more often during the daytime in the week compared to Fridays/Saturdays regardless of daytime or night. The influence is significant, χ^2 (2) = 9.97, p = .007, *Cramer's V* = .117. Post hoc tests showed that the observed headphone use for daytime during the week was higher than expected (p = .015).

To quantify the effects on wrong-way, sidewalk, and tandem riding, we tested the influence of age, group size, observation time, and their 2-way interactions with Generalized Estimating Equations. Table 4 shows the variables and levels being used in the model. Table 5 shows the results of the models. For *wrong-way riding*, we see a significant influence of age, indicating that adolescents ride more often against the direction of travel than older users of e-scooters, regardless of the time of day. There was no effect on wrong-way riding for group size, observation time, and two-way interactions. The model for *sidewalk riding* showed no significant influences at all, implying similar rates of users riding the e-scooter illegally on a sidewalk regardless of observation time, group size, and age. For *tandem riding*, several significant influences were found. Results show that tandem riding is increased for observations at night compared to daytime. One can also see a significant interaction between observation time and group size. It shows that tandem riding is less likely to occur when riding with two or more e-scooters at night than in other combinations. For a better understanding, Fig. 1 illustrates this interaction with descriptive numbers showing that the observed tandem riding rate is particularly enhanced for two people sharing one e-scooter at night. Finally, the model showed a

Table 3. Safety-critical behavior depending on the time of observation (N = 732).

	Day – working days[1] (n = 183)		Day - Friday/Saturday (n = 348)		Night - Friday / Saturday (n = 201)	
	n	%	n	%	n	%
helmet use	2	1.10	6	1.70	1	0.50
smartphone use	2	1.10	6	1.70	0	0.00
wrong-way riding	26	14.20	24	6.90	21	10.40
sidewalk riding	60	32.80	58	16.70	59	29.40
tandem riding	9	4.90	12	3.40	20	10.00
headphone use	26	14.20	27	7.80	11	5.50
luggage transport	17	9.30	26	7.50	8	4.00

Table 4. Descriptive statistics of categories used for Generalized Estimating Equations, N = 720 (without children).

Variable		n
Wrong-way riding	no	653
	yes	67
Sidewalk riding	no	550
	yes	170
Tandem riding	no	680
	yes	40
Observation time	night	198
	day	522
Group size	$\geq$ 2 e-scooter	412
	1 e-scooter	308
Age	adolescents (15–20 years)	126
	older than 21 years	594

significant influence of age on tandem riding. The result indicates that adolescents ride more often together on one e-scooter than older users do, regardless of the time of day.

---

[1] Except for Friday.

Table 5. Generalized estimating equation model results of predicting wrong-way riding, sidewalk riding, and tandem riding vs. showing no wrong-way riding*, no sidewalk riding*, and no tandem riding*. Bold highlighted results are significant at $\alpha < .05$. * Referenced category.

	Estimate	SE	Wald	P
Wrong-way riding				
Intercept	−2.26	0.22	107.14	<.001
night vs. day*	−0.33	0.36	0.82	.365
≥ 2 e-scooter vs. 1 e-scooter*	−0.47	0.41	1.33	.248
adolescent vs. older*	**1.33**	**0.54**	**6.07**	**.014**
night & ≥ 2 e-scooter vs. other combinations*	0.78	0.47	2.72	.099
night & adolescent vs. other combinations*	−0.41	0.52	0.63	.426
≥ 2 e-scooter & adolescent vs. other combinations*	−0.47	0.61	0.60	.441
QIC	445.87			
Sidewalk riding				
Intercept	−1.87	0.22	72.38	<.001
night vs. day*	0.13	0.34	0.14	.711
≥ 2 e-scooter vs. 1 e-scooter*	0.11	0.26	0.18	.674
adolescent vs. older*	0.80	0.51	2.46	.117
night & ≥ 2 e-scooter vs. other combinations*	0.18	0.46	0.16	.690
night & adolescent vs. other combinations*	0.22	0.44	0.26	.612
≥ 2 e-scooter & adolescent vs. other combinations*	−0.47	0.65	0.54	.464
QIC	824.98			
Tandem riding				
Intercept	−3.23	0.37	74.76	<.001
night vs. day*	**1.84**	**0.55**	**11.29**	**<.001**
≥ 2 e-scooter vs. 1 e-scooter*	−0.22	0.52	0.18	.668
adolescent vs. older*	**1.54**	**0.72**	**4.52**	**.034**
night & ≥ 2 e-scooter vs. other combinations*	**−3.08**	**1.22**	**6.35**	**.012**
night & adolescent vs. other combinations*	−0.70	0.82	0.73	.394
≥ 2 e-scooter & adolescent vs. other combinations*	−2.26	1.29	3.09	.079
QIC	273.37			

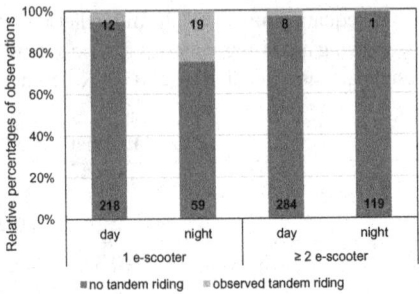

Fig. 1. Observed rates of tandem riding depending on observation time and group size.

4   Discussion

We conducted an observational study in two German cities to analyze possible differences between daytime and night for e-scooter riders concerning safety-critical behavior and the demographic profile of users.

Most interestingly, we found an increase in tandem riding at night compared to the daytime. Moreover, tandem riding at night was more likely to be shown for e-scooter riders using one e-scooter compared to those riding in groups with two or more e-scooters. This finding implies that tandem riding at night is a cause of concern, especially for pairs renting only one e-scooter. Reasons for increased tandem riding at night could be reduced availability of e-scooters, increased disinhibition by alcohol or drug abuse, shorter trips, peer dynamics, and pricing. It would be necessary to analyze in more detail what factors lead to renting only one e-scooter at night and sharing it with someone, for instance, with interviews in the nighttime. From a traffic safety perspective, some might say that tandem riding at night is a reason to prohibit the usage of e-scooters for specific hours. For instance, with the rental ban on shared e-scooter at night in Atlanta (after four deaths of e-scooter), Anderson et al. (2021) showed a reduction in e-scooter riders arriving at the hospital compared to times without the rental ban. However, tandem riding might also be reduced by less drastic traffic measures like price offers for pairs or promotional campaigns in the e-scooter apps. In addition, our results show that safety-critical behavior and rule violations for all other variables did not change when comparing day and night observations (helmet use, wrong-way riding, headphone use, smartphone use, and luggage transport). This implies that e-scooter riding at night is not automatically related to irregular behavior.

For adolescents (approx. 14–20 years old), we found an increased rate of wrong-way riding and tandem riding compared to older categorized e-scooter riders. With this, younger users appear to be more prone to these safety-critical behaviors, which is in line with the findings of Huemer et al. (2022). The result indicates that the self-enhancement bias, as observed in young car drivers (Harré et al. 2005; Sibley and Harré 2009), might also be relevant in e-scooter riding. Inexperience with e-scooters and traffic rules, in general, could also contribute to the irregular behavior of young riders (Petzoldt et al. 2021). Indeed, more research is needed with the latest numbers to confirm these assumptions. No significant differences between day and night were found for the demographic profile of e-scooter riders. The observed distributions of gender and age correspond to previous findings for daytime observations (Haworth and Schramm 2019; Huemer et al. 2022), with three-quarters being male and a main age range between 20 and 40 years.

A further relevant finding relates to the differences between both daytime observations. During the week, the rate of headphone use, wrong-way riding, sidewalk riding, and people riding with one e-scooter was higher than for precisely the same (day) times' observations on Fridays and Saturdays. Conversely, we observed lower group riding rates during the week than on Fridays and Saturdays. Bringing these results together might indicate a change in the trip purposes. For work days, e-scooters could be mainly used for commuting trips (using only one e-scooter, using headphones). In contrast, at weekends, people might use e-scooters for leisure activities with accompanying peers. This interpretation of data matches the observation of more privately-owned vehicles than shared ones during the week, which might also be commute-related. Such differences in trip purposes and maybe also user characteristics (except for age and gender) could also be a reason for increased crash rates of e-scooter riders on weekends. Further research should focus on rule violations in light of differentiations between rides of commuters and leisure activities, similar to Huemer (2018) with cyclists.

This research provided valuable results on safety-critical behaviors and demographic characteristics of e-scooter riders at night. However, some limitations must be acknowledged. A shortcoming of our observational study is the fact that the human observer could only categorize one person at the same time. This means that within a group of several people, the other ones might be under-represented in our data. For this reason, we suggest further studies aiming to analyze riding in groups by combining human and video observation. With the help of time-synchronized video data, it would be possible to encode the other e-scooter riders. Another limitation is the small number of sites that were covered. We know that the locations used for observations can have their peculiarities, and data might be biased. For this reason, further research is needed to validate our findings.

5 Conclusion

The observational study being presented in this article analyzed safety-critical behavior, rule violations, and user characteristics of e-scooter riders during daytime and night. The results indicate a particular relevance of tandem riding at night and increased rule violations of adolescents regardless of the time of day. The results suggest that traffic safety education should focus on tandem riding, especially targeting the user group of teenage riders. With this study, we bring e-scooter riding at night into the light and emphasize rider's nightly behaviors for policymakers.

Acknowledgments. The research presented in this paper was funded by German Insurers Accident Research (UDV).

References

Anderson, B., Rupp, J.D., Moran, T.P., Hudak, L.A., Wu, D.T.: The effect of nighttime rental restrictions on e-scooter injuries at a large urban tertiary care center. Int. J. Environ. Res. Pub. Health **18**(19). https://doi.org/10.3390/ijerph181910281

448   M. Ringhand et al.

Arellano, J.F., Fang, K.: Sunday drivers, or too fast and too furious? Analyz-ing speed and rider behaviour of e-scooter riders in San Jose, California (2nd Highest Scoring Mas-ters/Undergraduate Abstract Award Sponsored by HNTB Corporation - Great Lakes Region). J. Transp. Health **14**, 100725 (2019). https://doi.org/10.1016/j.jth.2019.100725

Curl, A., Fitt, H.: Same same, but different? Cycling and e-scootering in a rapidly changing urban transport landscape. NZ Geogr. **76**(3), 194–206 (2020). https://doi.org/10.1111/nzg.12271

Harré, N., Foster, S., O'neill, M.: Self-enhancement, crash-risk optimism and the impact of safety advertisements on young drivers. Br. J. Psychol. (London, England: 1953), 96(Pt 2), 215–230 (2005). https://doi.org/10.1348/000712605X36019

Haworth, N., Schramm, A.: Illegal and risky riding of electric scooters in Bris-bane. Med. J. Australia. Advance online publication (2019). https://doi.org/10.5694/mja2.50275

Haworth, N., Schramm, A., Twisk, D.: Changes in shared and private e-scooter use in Brisbane, Australia and their safety implications. Accid. Anal. Prevent. **163**, 106451 (2021). https://doi.org/10.1016/j.aap.2021.106451

Haworth, N., Schramm, A., Twisk, D.: Comparing the risky behaviours of shared and private e-scooter and bicycle riders in downtown Brisbane, Australia. Accid. Anal. Prevent. **152**, 105981 (2021). https://doi.org/10.1016/j.aap.2021.105981

Huemer, A.K.: Motivating and deterring factors for two common traffic-rule viola-tions of cyclists in Germany. Transp. Res. Part F: Traffic Psychol. Beh. **54**, 223–235 (2018). https://doi.org/10.1016/j.trf.2018.02.012

Huemer, A.K., Banach, E., Bolten, N., Helweg, S., Koch, A., Martin, T.: Second-ary task engage-ment, risk-taking, and safety-related equipment use in German bicycle and e-scooter riders - An observation. Accid. Anal. Prevent. **172**, 106685 (2022). https://doi.org/10.1016/j.aap.2022.106685

Kleinertz, H., Ntalos, D., Hennes, F., Nüchtern, J.V., Frosch, K.-H., Thiesen, D.M.: Accident mechanisms and injury patterns in e-scooter users. Deutsches Arzteblatt Int. **118**(8), 117–121 (2021). https://doi.org/10.3238/arztebl.m2021.0019

Laa, B., Leth, U.: Survey of E-scooter users in Vienna: Who they are and how they ride. J. Transp. Geogr. **89**, 102874 (2020). https://doi.org/10.1016/j.jtrangeo.2020.102874

Landis, J.R., Koch, G.G.: The measurement of observer agreement for categorical data. Biometrics **33**(1), 159 (1977). https://doi.org/10.2307/2529310

Leadbeater, B.J., Foran, K., Grove-White, A.: How much can you drink before driving? The influence of riding with impaired adults and peers on the driving behaviors of urban and rural youth. Addiction (Abingdon, England) **103**(4), 629–637 (2008). https://doi.org/10.1111/j.1360-0443.2008.02139.x

Noack, R.: Electric scooters have arrived in Europe — and a lot of people there hate them too. The Washington Post, 9 July 2019. https://www.washingtonpost.com/world/2019/07/09/electric-scooters-have-arrived-europe-lot-people-hate-them-too/

Orozco-Fontalvo, M., Llerena, L., Cantillo, V.: Dockless electric scooters: a re-view of a growing micromobility mode. Int. J. Sustain. Transp. 1–17 (2022). https://doi.org/10.1080/15568318.2022.2044097

Petzoldt, T., Ringhand, M., Anke, J., Schekatz, N.: Do German (Non)users of e-scooters know the rules (and Do They Agree with Them)? In: Krömker, H. (ed.) HCII 2021. LNCS, vol. 12791, pp. 425–435. Springer, Cham (2021). https://doi.org/10.1007/978-3-030-78358-7_29

Portland Bureau of Transportation: 2019 E-Scooter Findings Report. Portland Bu-reau of Trans-portation (2020). https://www.portland.gov/sites/default/files/2020-09/pbot_escooter_report_final.pdf

Shah, N.R., Cherry, C.R.: Riding an e-scooter at nighttime is more dangerous than at daytime. In: Contributions to the 10th International Cycling Safety Conference 2022 (ICSC2022), pp. 60–62. Technische Universität Dresden (2022). https://doi.org/10.25368/2022.436

Sibley, C.G., Harré, N.: The impact of different styles of traffic safety advertise-ment on young drivers' explicit and implicit self-enhancement biases. Transp. Res. Part F: Traffic Psychol. Beh. **12**(2), 159–167 (2009). https://doi.org/10.1016/j.trf.2008.11.001

Siebert, F.W., Ringhand, M., Englert, F., Hoffknecht, M., Edwards, T., Rötting, M.: Braking bad – ergonomic design and implications for the safe use of shared e-scooters. Saf. Sci. **140**(6), 105294 (2021). https://doi.org/10.1016/j.ssci.2021.105294

Simons-Morton, B.G., et al.: Peer influence predicts speeding prevalence among teenage drivers. J. Safety Res. **43**(5–6), 397–403 (2012). https://doi.org/10.1016/j.jsr.2012.10.002

Stigson, H., Malakuti, I., Klingegård, M.: Electric scooters accidents: analyses of two Swedish accident data sets. Accid. Anal. Prevent. **163**, 106466 (2021). https://doi.org/10.1016/j.aap.2021.106466

Tack, A., Klein, A., Bock, B.: E-Scooter in Deutschland: Ein datenbasierter Debattenbeitrag. civity Management Consultants GmbH & Co. KG. http://scooters.civity.de

Tapper, J.: Invasion of the electric scooter: can our cities cope? The Guardi-an, 15 July 2019. https://www.theguardian.com/cities/2019/jul/15/invasion-electric-scooter-backlash

Vollrath, M.: Observation (Version 3.0) [Computer software]. Technische Universität Braun-schweig (2019). https://www.tu-braunschweig.de/psychologie/verkehrspsychologie/software

Author Index

Printed in the United States
by Baker & Taylor Publisher Services